SEPARATION METHODS

New Comprehensive Biochemistry

Volume 8

General Editors

A. NEUBERGER

London

L.L.M. van DEENEN

Utrecht

ELSEVIER
AMSTERDAM · NEW YORK · OXFORD

Separation Methods

Editor

Z. DEYL

Prague

1984

ELSEVIER
AMSTERDAM · NEW YORK · OXFORD

ISBN for the series: 0-444-80303-3
ISBN for the volume: 0-444-80527-3

Published by:
Elsevier Science Publishers B.V.
PO Box 211
1000 AE Amsterdam
The Netherlands

Sole distributors for the U.S.A. and Canada:
Elsevier Science Publishing Co. Inc.
52 Vanderbilt Avenue
New York, NY 10017
USA

Library of Congress Cataloging in Publication Data
Main entry under title:

Separation methods.

(New comprehensive biochemistry; v. 8)
Includes index.
1. Separation (Technology) I. Deyl, Zdeněk.
II. Series.
QD415.N48 vol. 8 574.19'2s [574.19'285] 84-1502 [QD63.S4]
ISBN 0-444-80527-3

Printed in the Netherlands

Contents

Deyl (ed.) Separation Methods

CHAPTER 1

Principles and theory of chromatography

JOSEF NOVÁK

Institute of Analytical Chemistry, Czechoslovak Academy of Sciences, 611 42 Brno, Czechoslovakia

1.1 Basic terms

It is useful to begin the chapter on the theory of chromatographic separation methods with a definition of chromatography. However, several such definitions can be formulated according to various classification aspects. For the sake of accuracy a phenomenological definition, a molecular kinetic definition and various working definitions can be introduced. According to the first definition chromatography is understood as a *phenomenon* of differential migration of solute compounds in a system of two phases, of which one is stationary and the other mobile. According to the molecular kinetic definition, chromatography is taken as a continuous *process* of convective upsetting and diffusional reestablishment of equilibrium between the concentrations of the solute compound in the stationary and in the mobile phase of the chromatographic system. This process results in a differential migration of the solute compounds. According to the working definitions chromatography is a certain method (specifically a separation and analytical method and various methods of physicochemical measurements). From the point of view of the theory of chromatography we are particularly interested in the chromatographic process.

Whereas the realization of a chromatographic experiment is often surprisingly simple — a number of important chromatographic processes proceed spontaneously — the mechanism of the chromatographic process is relatively complex. A prerequisite of the proper understanding of the mechanism of chromatography is the concept of dynamic equilibrium between the concentrations of a solute in a system of two coexisting phases; more accurately, equilibrium between the concentrations of the solute should be understood as a result of the identity of its chemical potentials in the individual phases of the system. Even when assuming that such a system is stationary and in equilibrium, molecules of the solute permanently pass from one phase to the other, remaining for a certain time in one or other phase after each transition. As the process is random at this level, the individual time intervals of the occurrence of the solute molecules in a given phase are also random and, hence, very different. The mean time intervals of the occurrence of all solute molecules in each phase during a certain time are, however, constant under given conditions, and their ratio represents a basic factor of chromatographic retention. Thus, the ratio at which a given amount of the solute at equilibrium is distributed between the phases of the system is not determined by a static presence of the solute molecules in these phases but rather by the probability of their occurrence in the phases of the system. When, under these conditions, one phase moves with respect to the other, the solute molecules move together with the moving phase during their occurrence in that particular phase, but remain stagnant when in the stationary phase. Due to the statistical fluctuation some molecules of a given solute migrate a shorter or longer distance during a certain time interval than that corresponding to the mean time intervals of the occurrence of the molecules of this solute in the phases. This results, together with the longitudinal diffusion, in a spreading of the migrating zone of the solute. However, due to its statistical nature, this spreading increases only as the square root of the mean migration distance, so that, in the case of differential migration of zones of different solutes, the zones can be separated. This assumption of the mechanism of the chromatographic process will be formulated quantitatively in subsequent paragraphs of this chapter.

1.2 Classification of chromatographic systems and procedures

1.2.1 State of the aggregation of the coexisting phases

The traditional definition of the phases in a chromatographic system is often rather problematic. Whereas the term mobile phase is usually clear, specification of the chromatographic stationary phase is not always unambiguous. For instance, the whole content of the chromatographic column is sometimes considered as the stationary phase, but sometimes only those components of the packing that are functioning as sorbents of the solute compound are termed in this way. In the former case, the concept of chromatographic stationary phase apparently differs from the classical physical concept of the phase. Whereas in the physical conception the phase is a homogeneous part of the system, the chromatographic stationary phase may contain even more physical phases. In the latter case, the inert support of the sorbent is not considered to be the stationary phase, in spite of the fact that it represents a rather substantial physical phase of the system. However, when an active adsorbent plays the role of the sorbent support, it must then be considered as the chromatographic stationary phase. A problem then arises, viz. what part of the used adsorbent is really active with respect to the solute compound in the given system. Naturally, in a given chromatographic packing, chromatographic stationary phases cannot be unambiguously identified with physical phases. The above indeterminacies should be considered when classifying chromatographic systems according to the state of the aggregation of the phases; a summary of typical chromatographic systems according to this classification is presented in Table 1.1.

TABLE 1.1
Chromatographic systems

Stationary phase	Mobile phase	
	Liquid	Gas
Solid compound	LSC	GSC
Solid compound + liquid	LSLC	GSLC
Liquid	LLC	GLC

LSC, liquid-solid chromatography; GSC, gas-solid chromatography; LSLC, liquid-solid-liquid chromatography; GSLC, gas-solid-liquid chromatography; LLC, liquid-liquid chromatography; GLC, gas-liquid chromatography.

1.2.2 Physical arrangement of the system and the accomplishment of the chromatographic experiment

According to the physical arrangement chromatographic systems can be divided into planar and column ones. The planar arrangements are represented by systems of paper and thin layer chromatography. When further dividing the planar systems according to their physical arrangement we come to systems in the equilibration chamber and to the so-called sandwich systems. According to development procedures (flow of the mobile phase in the planar bed) the systems can be further classified as ascendent, horizontal, descendent and, occasionally, centrifugal; in orthogonal beds the development may proceed in one or more directions. When, during the development of the chromatogram, the composition of the mobile phase remains constant the development is termed isocratic, on the other hand, when the composition of the mobile phase varies, we speak of gradient development.

A more exact classification of column systems according to the physical arrangement leads to various types of packed and capillary columns. In column chromatography the use of several columns that can be suitably switched over, so that chromatographic fractions eluted from one column can be further chromatographed on other columns, is somewhat analogous to two-dimensional development in planar beds. In column chromatography the separation may proceed isocratically or with a programmed gradient of composition of the mobile phase, isothermally or with programmed changes of column temperature, and isobarically or with programmed changes of mobile phase pressure at the column inlet. The programming of the composition of the mobile phase is important practically only in liquid chromatography, whereas temperature and pressure programming is used primarily in gas chromatography.

In planar chromatographic systems the solute compounds are usually not eluted from the chromatographic bed but rather detected directly in it, whereas in modern column chromatography the solute compounds are gradually eluted with the mobile phase and detected in the effluent at the column outlet.

1.2.3 Development of the chromatogram

1.2.3.1 Frontal chromatography

A continuous supply of the analyzed material, or of its mixture with a non-sorbed mobile phase, into the column or into the planar bed results first in frontal chromatography and then in the saturation of the sorbent with all the components of the supplied material. After the development of the chromatogram, and during continuing supply of the mixture, the front of the least sorbed component is washed out first, followed by a mixture of the first component and the more strongly sorbed component etc., and, finally, after all the components of the mixture break through, a mixture identical in composition to that of the mixture supplied flows out of the column. By interrupting the supply of the analyzed mixture to the previously saturated column, and connecting the supply of the mobile phase alone, the opposite (desorption) frontal chromatogram arises. Initially, the mixture of all the components flows out of the column. After the least sorbed component has been eluted the mixture deprived of this component flows out of the column. After the further, more strongly sorbed component is eluted the mixture deprived of the first and second components flows out of the column. Finally, the most strongly sorbed component is washed out and only the supplied mobile phase leaves the column. Both versions of development of the frontal chromatogram are schematically and in an idealized form illustrated in Fig. 1.1.

1.2.3.2 Elution chromatography

Elution chromatography is simpler, and, with respect to the separation of an analyzed mixture, more effective. With this alternative a dose of the analyzed

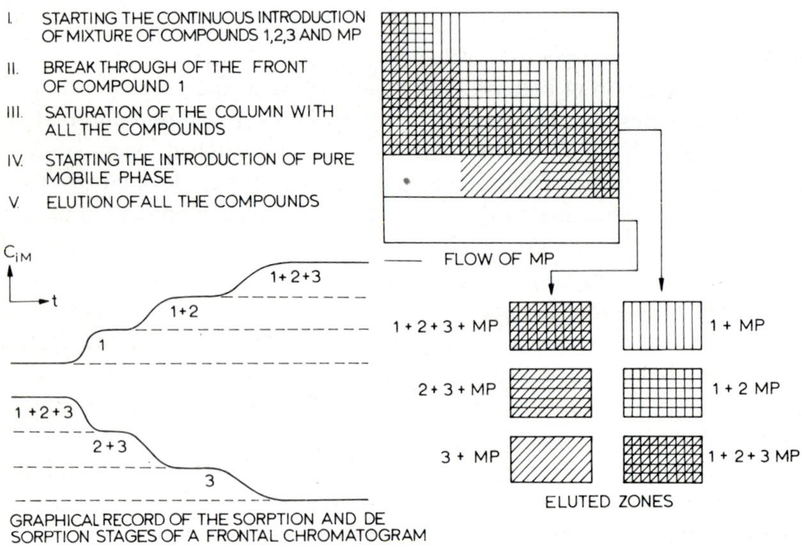

Fig. 1.1.

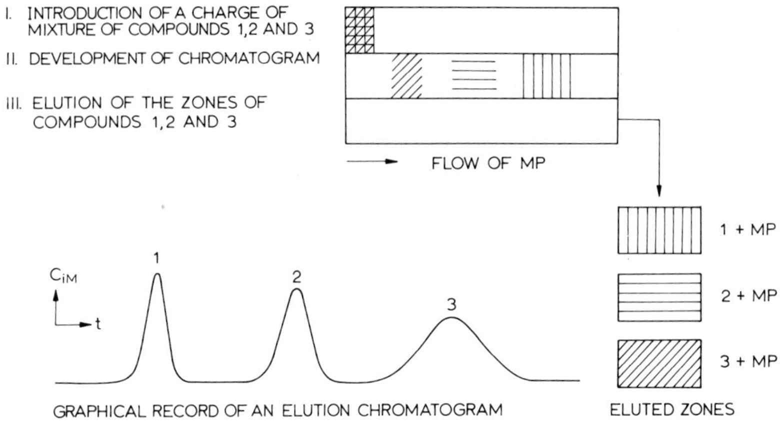

Fig. 1.2.

material is supplied to the column inlet or to the planar bed and is then washed with a non-sorbed mobile phase through the column. The development and differential migration of elution zones of individual components of the mixture thus take place. When the supply of the mobile phase continues the individual zones are gradually washed out of the column; the zone of the most weakly sorbed component is washed out first, followed by the zone of a more strongly sorbed component etc., and, finally, after the elution of the zone of the most strongly sorbed component, only the supplied mobile phase flows out of the column. A schematic illustration of the elution chromatography is presented in Fig. 1.2.

1.2.3.3 Displacement chromatography

When the stationary phase functions as an adsorbent and a compound that is adsorbed more strongly than any other component of the analyzed mixture serves as the mobile phase, the procedure otherwise similar to that used with elution chromatography is termed displacement chromatography. With this alternative the most weakly adsorbed component is displaced by the more strongly adsorbed component, this latter is then displaced by the more strongly adsorbed component, etc., resulting in a situation when the most strongly adsorbed component of the analyzed mixture is displaced by the supplied displacement agent. After the chromatogram has been developed, the zones of all the components migrate closely next to each other and, when the supply of the displacement agent continues, they leave the column in the order of increasing adsorption ability. In the case of elution chromatography (and in frontal chromatography when the mixture of the analyzed material is supplied together with the mobile phase) the eluted fractions are in fact mixtures of the solute compounds with the mobile phase, whereas in the case of displacement chromatography the individual zones are more or less the solute compounds alone. A scheme of displacement development is illustrated in Fig. 1.3.

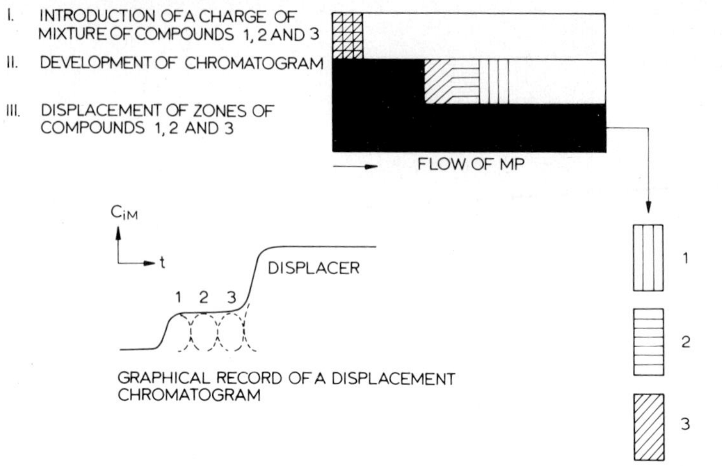

Fig. 1.3.

1.2.4 Mechanism of the distribution of the solute compound between the phases of the system

The mechanisms of sorption and/or the interaction of the solute with the mobile phase can be summarized as follows: *a*, physical dissolution in the phase; *b*, physical adsorption on the surface of the phase; *c*, chemical reaction in the bulk phase or on its surface (acido-basic equilibrium, formation of coordination complexes or chelates, association of ionic pairs, exchange of ions, precipitation); *d*, steric exclusion (molecular sieving effect, gel permeation); *e*, bioaffinity association.

The cases presented in paragraphs 1.2.1–1.2.4 can be mutually combined. The number of all possible combinations naturally exceeds the number of the real combinations, however, the number of real chromatographic systems and procedures is still very large. From the practical point of view, the alternatives of elution chromatography are most important. Therefore, with the exception of general problems, only elution chromatography will be discussed in this chapter.

1.3 Development of chromatography — a review

The oldest intentional chromatographic experiments were performed as frontal chromatography in a liquid-solid system and date from the beginning of the 19th century [1]. Elution chromatography (liquid-solid) was discovered at the beginning of the 20th century [2], but developed rapidly only after the discovery and theoretical explanation of liquid-liquid elution chromatography [3] in the forties and particularly after the discovery of elution gas chromatography [4–6] in the fifties. The pioneers in chromatography are noted in Table 1.2. A detailed description of the development of chromatography can be found in reviews by Ettre [7,8] and Zechmeister [9].

TABLE 1.2
Pioneers in chromatography

Stationary phase:	Solid sorbent		Liquid sorbent	
Mobile phase:	Liquid (LSC)	Gas (GSC)	Liquid (LLC)	Gas (GLC)
Elution development	M.S. Tswett (1906); R. Kuhn, A. Winter-stein and E. Lederer (1931)	E. Cremer (1951); J. Janák and M. Rusek (1953); H.W. Patton, J.S. Lewis and W.I. Kaye (1955)	A.J.P. Martin and R.L.M. Synge (1941)	A.T. James and A.J.P. Martin (1952); N.H. Ray (1954); B.W. Bradford, D. Harvey and D.E. Chalkley (1955)
Frontal development	D.T. Day (1897); A. Tiselius (1943); S. Claesson (1949)	C.S.G. Phillips (1953)	C.S.G. Phillips (1952)	C.S.G. Phillips (1954)
Displacement development	A. Tiselius (1943); S. Claesson (1949)	N.C. Turner (1943); C. Claesson (1946); N.M. Turkel'taub (1950); C.S.G. Phillips (1953)		

1.4 Theoretical models of chromatography

When describing the chromatographic process in terms of mathematics it is necessary to define a suitable (sufficiently realistic and yet mathematically tractable) model of chromatography. From the point of view of theoretical considerations the following models are of interest [10].

a. Model of 'ideal chromatography', assuming a piston flow of the mobile phase, infinitely rapid setting of equilibrium between the concentrations of the solute in the coexisting phases, and zero longitudinal diffusion of the solute.

b. Model of 'non-ideal chromatography', considering the actual velocity profile of the mobile phase flow, finite rate of of equilibration between the concentrations of the solute in the coexisting phases, and the actual longitudinal diffusion of the solute.

c. Model of 'linear chromatography', using a linear sorption isotherm for calculations.

d. Model of 'non-linear chromatography', using a non-linear sorption isotherm for calculations.

In this way four combined models of chromatography may be postulated: A, ideal linear; B, non-ideal linear; C, ideal non-linear; and D, non-ideal non-linear. Whereas the models B and D are real, the models A and C are apparently hypothetical. In spite of this even the latter two models are very useful from the theoretical point of view.

1.5 Description of models of linear chromatography with an incompressible mobile phase

1.5.1 Linear non-ideal chromatography

The mass balance of a solute in the infinitesimal volume of a chromatographic bed (column), delineated by two parallel sections of identical area A, drawn perpendicular to the direction of the mobile phase flow at distances z and $z + \mathrm{d}z$ from the beginning of the bed leads to the equation:

$$\frac{\partial c_{\mathrm{iM}}}{\partial t} + \frac{\phi_{\mathrm{S}}}{\phi_{\mathrm{M}}} \frac{\partial c_{\mathrm{iS}}}{\partial t} = -u \frac{\partial c_{\mathrm{iM}}}{\partial z} + D_{\mathrm{M}} \frac{\partial^2 c_{\mathrm{iM}}}{\partial z^2} + \frac{\phi_{\mathrm{S}}}{\phi_{\mathrm{M}}} D_s \frac{\partial^2 c_{\mathrm{iS}}}{\partial z^2} \tag{1}$$

where c_{iM} and c_{iS} are the mean (over the cross-section) concentrations (mass/volume) of the solute in the mobile and stationary phases, ϕ_{M} and ϕ_{S} are the fractions of the area A occupied by the mobile and the stationary phase, D_{M} and D_{S} are the diffusion coefficients of the solute in the mobile and stationary phases, u is the mean forward velocity of the mobile phase, averaged over the cross-section i.e., $u = F/\phi_{\mathrm{M}}$, where F is the volumetric rate of the mobile phase, t is time and z is the longitudinal distance from the beginning of the bed in the direction of the mobile phase flow. It follows from the right side of equation 1 that the given mass balance includes the convective transport of the solute in the mobile phase and the diffusional transport of the solute in the mobile and stationary phases of the system. For ϕ_M and ϕ_S it holds that:

$$\phi_S/\phi_M = A_S/A_M,$$

where A_S and A_M the are absolute parts of the area A, occupied by the stationary and mobile phases.

In the case of liquid-solid chromatography or gas-solid chromatography the value ϕ_M represents total porosity of the bed ε, so that $\phi_S = 1 - \varepsilon$ and $\phi_S/\phi_M = (1 - \varepsilon)/\varepsilon$.

Equation 1 has two unknown quantities, c_{iM} and c_{iS}, so that one additional independent equation is necessary for the solution. Such an equation can be derived on the basis of the concept of solute mass transfer between the phases of the system. The volume element of the bed $A\mathrm{d}z$ is also considered here. The interphase transfer of the solute is then given by the flow $J(\mathrm{M} \rightleftarrows \mathrm{S})$ through the total area of the phase interface in the volume element $A\mathrm{d}z$, and the actual direction and density of this flow are determined by the actual sense and degree of the deviation from equilibrium between the concentrations c_{iM} and c_{iS}. The difference between the actual solute concentration in phase 1 and such a concentration in this phase, which would be in equilibrium with the solute concentration in phase 2, is the driving force of the solute transfer, e.g., from phase 1 to phase 2. The solute flow through a unit area of the phase interface is given by the relation $\xi[c_{iM} - (c_{iS}/K)]$, where ξ is the mass transfer coefficient and K is the distribution constant defined as the equilibrium ratio of c_{iS} and c_{iM}, i.e.,

$$K = (c_{iS}/c_{iM})_{eq} \tag{2}$$

If κ is the area of the phase interface in unit bed volume, then for the flow $J(\mathrm{M} \rightleftarrows \mathrm{S})$ it holds that

$$J(\mathrm{M} \rightleftarrows \mathrm{S}) = \xi\kappa[c_{iM} - (c_{iS}/K)]\,A\mathrm{d}z \tag{3}$$

Changes in the concentration of the solute in the stationary phase occur due to transfer of the solute across the phase interface and longitudinal diffusion in the stationary phase. Thus, it may be written

$$A_S \frac{\partial c_{iS}}{\partial t}\,\mathrm{d}z = \xi\kappa\left(c_{iM} - \frac{c_{iS}}{K}\right)A\mathrm{d}z + A_S D_S \frac{\partial^2 c_{iS}}{\partial z^2}\,\mathrm{d}z \tag{4}$$

and after dividing by the volume $A\mathrm{d}z$ the equation

$$\frac{\partial c_{iS}}{\partial t} = \frac{\xi\kappa}{\phi_S}\left(c_{iM} - \frac{c_{iS}}{K}\right) + D_S \frac{\partial^2 c_{iS}}{\partial z^2} \tag{5}$$

is obtained. Equation 5 is the second equation required for the solution of the problem.

Let us now define the initial and boundary conditions for the case of column elution chromatography. At the beginning no solute is present in the column, i.e.

$$\text{at } t = 0 \text{ and } 0 < z < \infty,\ c_{iM} = c_{iS} = 0 \tag{6}$$

The solute is applied to the column in the form of a concentration pulse of the concentration $c_{iM,0}$ and duration δt, so that

$$\text{at } t > \delta t \text{ and } z = 0,\ c_{iM} = 0$$

$$\text{at } 0 < t < \delta t \text{ and } z = 0,\ c_{iM} = c_{iM,0}\,. \tag{7}$$

If the terms for the longitudinal diffusion of the solute in the stationary phase are neglected in equations 1 and 5, the following solution exists for the system of equations 1, 5, 6 and 7 [11,12]

$$c_{iM,L} = \frac{m_i}{\phi_M A u \sigma_t \sqrt{2\pi}} \exp\left[-\frac{(t_R - t)^2}{2\sigma_t^2}\right] \tag{8}$$

where $c_{iM,L}$ is the actual solute concentration in the mobile phase in the section $z = L$ (at the end of the column), t_R is the elution time of peak's maximum, σ_t is standard deviation of the time record of the elution peak and m_i is the total solute mass in the elution zone.
For t_R and σ_t in equation 8 it further holds

$$t_R = \frac{L}{u}\left(1 + \frac{\phi_S}{\phi_M} K\right) = \frac{L}{u}(1 + k) \tag{9}$$

$$\sigma_t^2 = \frac{2LD_M(1 + k)^2}{u^3} + \frac{2L\phi_M k^2}{\xi \kappa u} \tag{10}$$

where K is distribution constant defined by equation 2, k is the so-called capacity ratio defined as the equilibrium ratio of solute masses in the stationary and mobile phases, i.e., $k = (m_{iS}/m_{iM})_{eq}$, and L is the length of the column. Solution 8, together with equations 9 and 10, holds sufficiently accurately only in the case that $\delta t \ll t_R$ and $\sigma_t \ll t_R$. Relation 9 represents the basic equation of chromatographic retention.

1.5.2 Linear ideal chromatography

As already mentioned in paragraph 1.4, the concept of ideal linear chromatography is based on the model [13] which should have the following properties: (i) infinitely fast setting of equilibrium between the solute concentrations in the mobile and stationary phases; (ii) zero longitudinal diffusion of the solute in both phases; (iii) absolutely linear sorption isotherm; and (iv) piston flow of the mobile phase. In spite of the fact that this model is not real it is interesting as it provides for a fairly accurate description of chromatographic retention. Naturally, it does not yield any information about zone spreading, as the spreading factors have not been considered at all. The initial concentration profile of the solute would, under the conditions of linear ideal chromatography, proceed through the column without any change of its shape at such a rate at which the center of a broadening elution zone proceeds under the conditions of non-ideal chromatography (a more rigorous treatment [14] of the model of linear non-ideal chromatography shows that the retention time is not fully independent of spreading factors).

When the terms representing the longitudinal diffusion of the solute in the mobile phase in equation 1 is neglected and equation 5 is substituted by the following equation

$$\frac{\partial c_{iS}}{\partial t} = K\frac{\partial c_{iM}}{\partial t} \tag{11}$$

the relation

$$\frac{\partial c_{iM}}{\partial t}\left(1 + \frac{\phi_S}{\phi_M}K\right) = -u\frac{\partial c_{iM}}{\partial z} \tag{12}$$

is obtained, representing in principle the mathematical definition of linear ideal chromatography. The solution of equation 12 leads to the fundamental retention equation 9. According to the theorems about the properties of partial differentiations, and with respect to equation 12, it may be written

$$\left(\frac{\partial c_{iM}}{\partial z}\right)_t \Big/ \left(\frac{\partial c_{iM}}{\partial t}\right)_z = -\left(\frac{\partial t}{\partial z}\right)_{c_{iM}} = -\frac{1 + \dfrac{\phi_S}{\phi_M}K}{u} \tag{13}$$

and under the assumption that c_{iM} is invariant (which is one of the premises of ideal linear chromatography) it holds

$$\mathrm{d}t = [(1+k)/u]\,\mathrm{d}z,\ \int_0^{t_R}\mathrm{d}t = \frac{1+k}{u}\int_0^L \mathrm{d}z \text{ and}$$

$t_R = L(1 + \mathrm{k})/u$, where again $k = K\phi_S/\phi_M$

1.6 Simplified description of linear non-ideal chromatography

1.6.1 Retention equations

An exact solution of a completely general model of non-ideal linear chromatography has not yet been found. Therefore, approximate methods [15,16] which would make it possible to characterize this model on the basis of analysis of individual components of the mechanism of the chromatographic process were sought. Such an approach leads very simply to the basic equation of chromatographic retention and provides for the description of the individual spreading factors in terms of the physical features of the system. When limited only to the aspects of chromatographic retention this approach corresponds in general to LeRosen's concept of chromatography [17].

The migration rate of the center of the elution zone with respect to the rate of the mobile phase is determined by the mean probability of the occurrence of the solute molecules in the mobile phase, hence

$$\frac{t_{iM}}{t_{iM}+t_{iS}}=\frac{m_{iM}}{m_{iM}+m_{iS}}=\frac{u_i}{u}=R \tag{14}$$

where $t_{iM}/(t_{iM}+t_{iS})$ is the mean fraction of the total time spent by the solute molecules in the chromatographic bed (column), for which the solute molecules occur in the mobile phase, $m_{iM}/(m_{iM}+m_{iS})$ is the mean fraction of the total mass of the solute component within the chromatographic zone, which is present in the mobile-phase part of the zone, u_i is the mean forward velocity of the center of the elution zone and R is the so-called retardation factor (with certain reservations [18] identical with R_F used to express retention in systems of planar chromatography). As $t_{iM}+t_{iS}=t_R$, $t_{iS}/t_{iM}+m_{iS}/m_{iM}=k$ and $u_i=L/t_R$, the relation $t_R=L(1+k)/u$ is immediately obtained. For the ratio L/u it holds $L/u=t_M$, where t_M is the so-called dead retention time (retention time of a non-sorbed compound).

Equation 9 can thus be written in the form

$$t_R=t_M(1+k) \tag{15}$$

By multiplying this equation by the volumetric flow rate of the mobile phase the relation

$$V_R=V_M(1+k) \tag{16}$$

is obtained, where V_R is the retention volume of the solute compound and V_M is the dead retention volume, i.e., the retention volume of a non-sorbed compound. As $k=K\phi_S/\phi_M=KA_S/A_M$, and in a uniform bed (packing of the column) $A_S/A_M=V_S/V_M$, equation 16 may be rewritten as

$$V_R=V_M+KV_S \tag{17}$$

where V_S is the volume of the sorbent in the column, and V_M is generally identical with the geometrical void volume of the column. For the quantity R it apparently holds

$$R=\frac{u_i}{u}=\frac{L}{t_R u}=\frac{t_M}{t_R}=\frac{V_M}{V_R}=\frac{1}{1+k} \tag{18}$$

Retention characteristics represent chromatographic retention correctly only when they are expressed under the conditions at which retention takes place. As the phase volumes generally depend on pressure and temperature (particularly the mobile phase in gas chromatographic systems), data calculated from equations 9, 16, 17 and 18 are sufficiently representative only on the condition that values at the temperature and mean pressure in the column are substituted for the mobile phase flow rate and the volumes of both phases. This problem will be discussed in more detail in Section 7.

It follows from equation 17 that the distribution constant can be expressed by retention parameters using the relation

$$K = (V_{\mathrm{R}} - V_{\mathrm{M}})/V_{\mathrm{S}} \tag{18}$$

In gas chromatography it is often advantageous to work with the so-called specific retention volume [19] which is defined by the relation

$$V_{\mathrm{g}} = K\,273.15/T\rho_{\mathrm{S}} \tag{19}$$

where T is the absolute column temperature and ρ_{S} is the density of the sorbent.

1.6.2 Spreading of the chromatographic zone

In this paragraph it will be useful to consider the length standard deviation of the actual elution zone in the chromatographic bed instead of the time standard deviation σ_t (see equations 8 and 10). The length standard deviation is a function of the migration distance, i.e., the elution zone whose center has migrated a distance z has the length standard deviation σ_z. Further discussion will be limited to the case when $z = L$ and, hence, $\sigma_z = \sigma_L$, i.e., the situation at the end of the chromatographic column will be analyzed. When $\sigma_t \ll t_{\mathrm{R}}$, then between σ_L and σ_t the relation

$$\sigma_L = \sigma_t u_{\mathrm{i}} = \sigma_t u/(1 + k) \tag{20}$$

holds with sufficient accuracy. Spreading occurs due to several factors, each of them contributing, to a certain extent, to the final effect. Theory indicates that the squares of the standard deviations (variances) corresponding to the individual spreading factors are roughly additive [20]. However, there are cases in which some spreading factors are mutually dependent to such an extent that the respective variances combine in a different way. Seven spreading factors should be considered for a sufficiently detailed description of zone spreading in a general case of non-ideal linear chromatography in packed beds.

1. Non-uniformity of the mobile phase flow (A):

$$\sigma_L^2(A) = 2\lambda d_{\mathrm{p}} L \tag{21}$$

where λ is the so-called eddy diffusion coefficient and d_{p} is the diameter of bed particle.

2. Longitudinal solute diffusion in the mobile phase (B_M):

$$\sigma_L^2(B_M) = 2\gamma_M D_M L/u \tag{22}$$

where γ_M is the so-called obstructive factor for diffusion in the mobile phase ($\gamma_M \leqslant 1$).

3. Longitudinal solute diffusion in the stationary phase (B_S):

$$\sigma_L^2(B_S) = 2\gamma_S D_S L(1-R)/Ru \tag{23}$$

where γ_S is the obstructive factor for diffusion in the stationary phase.

4. Deviation from sorption equilibrium in the stationary phase in adsorption chromatography (C_{Sa}):

$$\sigma_L^2(C_{Sa}) = 2R(1-R)Lu/k_d \tag{24}$$

where k_d is the desorption rate constant (desorption is considered as a first order reaction).

5. Deviation from sorption equilibrium in the stationary phase in chromatography on a liquid sorbent applied on a macroporous support (C_{Sl}):

$$\sigma_L^2(C_{Sl}) = qR(1-R)d_f^2 Lu/D_S \tag{25}$$

where q is a geometrical factor and d_f is the effective thickness of the liquid sorbent film.

6. Deviation from sorption equilibrium in the flowing mobile phase (C_M):

$$\sigma_L^2(C_M) = \omega d_p^2 Lu/2d_M \tag{26}$$

where ω is a factor characterizing the geometrical structure of the packing.

7. Deviation from sorption equilibrium in the mobile phase inside the particles (C'_M).

In the pores of the particles the 'mobile' phase is stagnant, so that the contribution to zone spreading due to nonequilibrium in this portion of the mobile phase differs from that due to nonequilibrium in the flowing mobile phase. In the case that the particles are of spherical shape it holds [21] that

$$\sigma_L^2(C'_M) = \left[(1-\varphi_M R)^2/30\gamma'_M(1-\varphi_M)\right] d_p^2 Lu/D_M \tag{27}$$

where φ_M is the fraction of the mobile phase present in the inter-particle space (flowing mobile phase) and γ'_M is obstructive factor for diffusion in the stagnant 'mobile' phase in the pores inside the particle.

The mutual roles of the individual spreading factors and, hence, the combinations of the respective variances depend on the nature of the chromatographic system. The contributions of the non-uniformity of flow of the mobile phase and nonequilibrium in the flowing mobile phase are mutually compensated to a certain extent [22], and the resulting variance caused by these two factors, $\sigma_L^2(A, C_M)$, is given by the relation

$$\sigma_L^2(A, C_M) = \frac{1}{1/\sigma_L^2(A) + 1/\sigma_L^2(C_M)} \tag{28}$$

By increasing the velocity of the mobile phase $\sigma_L^2(A, C_M)$ reaches a constant value, i.e., approaches the $\sigma_L^2(A)$ value.

In chromatography on a liquid sorbent applied on a solid support the total variance $\Sigma\sigma_L^2$ can be described as

$$\sum \sigma_L^2 = \sigma_L^2(A, C_M) + \sigma_L^2(B_M) + \sigma_L^2(B_S) + \sigma_L^2(C_{S1}) + \sigma_L^2(C'_M) \tag{29}$$

When using a support with sufficiently large pores and/or a completely nonporous support, or in the case that the pores of a microporous support are completely filled with the applied liquid sorbent, the term $\sigma_L^2(C'_M)$ can be omitted. When the liquid sorbent forms a completely continuous film on the support (a situation which may occur in an ideal case when using a macroporous support or when using a capillary column), then q in the term $\sigma_L^2(C_{S1})$ has a value of 2/3, whereas in the case of a microporous support with the pores filled completely with the liquid sorbent, q in the term $\sigma_L^2(C_{S1})$ equals $1/30\gamma'_S$ and $d_f = d_p$, where γ'_S is the obstructive factor for diffusion of the solute in the liquid sorbent inside the pores.

For chromatography on packings without a liquid sorbent it may be written

$$\sum \sigma_L^2 = \sigma_L^2(A, C_M) + \sigma_L^2(B_M) + \sigma_L^2(B_S) + \sigma_L^2(C_{Sa}) + \sigma_L^2(C'_M) \tag{30}$$

In chromatography on solid adsorbents the term $\sigma_L^2(C'_M)$ always plays a significant role. Ion-exchange chromatography is a typical example of the application of the term $\sigma_L^2(C'_M)$. Equation 30 can also be applied to chromatography based on steric exclusion. The term $\sigma_L^2(C_{Sa})$ is either zero in this case or it may characterize a possible participation of adsorption.

Equations 29 and 30 hold both for gas and liquid chromatography. In the case of gas chromatography the term $\sigma_L^2(B_S)$ can always be neglected.

The relations for the individual variances and their combinations are unambiguous only when u and all the other parameters are constant along the migration path (L). However, this condition is fulfilled practically only in modern liquid column chromatography. In gas chromatography u and D_M change considerably along the column due to the high compressibility of the mobile phase, and in chromatography in planar systems the velocity of the mobile phase depends on the actual distance of the front of the chromatogram from the level of the elution liquid. In these cases the above relations are valid only with the limitation that they describe the situation in a certain site of the column or at a certain moment and the measured resulting variance represents only the average features of the system.

The variances caused by longitudinal diffusion are indirectly proportional to the velocity of the mobile phase, whereas the variances occurring due to deviations from equilibrium are directly proportional to this velocity. Thus, the graph relating the total variance ($\Sigma\sigma_L^2$) with u at a given L has the shape of a general hyperbola [12]; hence, at a certain (optimal) velocity of the mobile phase the value $\Sigma\sigma_L^2$ is minimal under the given conditions.

1.6.3 Concept of the theoretical plate

The model of the theoretical plate [3] is based on the concept that the chromatographic column consists of a series of segments in which equilibrium between the concentrations of the solute compound in the mobile and stationary phases is established under the given conditions. The natural continuous model is thus substituted by a hypothetical discontinuous model in which the height equivalent to a theoretical plate, H, is a parameter of spreading. In spite of the fact that the plate model is very unrealistic, the quantity H is a useful criterion of the separation efficiency of the chromatographic column. A mathematical treatment [23] of this model leads to a simple relation according to which the variance divided by length of migration path (column) is the height equivalent to a theoretical plate, i.e.

$$H = \sigma_z^2/z = \sigma_L^2/L \tag{31}$$

When the variance is expressed in units of time or volume ($\sigma_v = F\sigma_t$) then, under the above conditions ($\sigma_t \ll t_R$), it holds approximately that

$$H = \sigma_t^2 Ru/t_R = \sigma_v^2 Ru/V_R F \tag{32}$$

For the number of plates of the column, N, it holds that

$$N = L/H = (L/\sigma_L)^2 = (t_R/\sigma_t)^2 = (V_R/\sigma_v)^2 \tag{33}$$

It follows from equation 31 that the discussion of spreading factors in terms of length variance (see section 1.6.2) can easily be converted to the discussion in terms of H by dividing the corresponding equations by the quantity L.

1.7 Mobile phase flow

The flow of the mobile phase is determined by the structure of the chromatographic bed, rheological properties of the flowing liquid, and driving forces of the flow. A general description of the flow dynamics is represented by the Navier-Stokes [24] equation, together with the continuity equation. However, the solution of this combination for systems with such a complex geometry as that exhibited by chromatographic beds is not possible. Therefore, simpler systems based on an analogy between hydrodynamics and electrodynamics were sought. Darcy's law [25], defined by the relation

$$u = -\frac{B_o}{\varepsilon_o \mu}\frac{\mathrm{d}p}{\mathrm{d}z} \tag{34}$$

is the basis of this conception. In this relation B_o is the specific permeability constant, ε_o is the inter-particle porosity, μ is the viscosity of the liquid and $\mathrm{d}p/\mathrm{d}z$ is the pressure gradient in the direction of flow. For an empty capillary it holds that $B_o = r^2/8$, where r is the radius of the capillary. For packed beds it holds according to Kozeny-Carman's equation [26,27] that $B_o = \mathrm{d}_p^2 \varepsilon_o^3 / 180\ (1 - \varepsilon_o)^2$. In the case of incompressible liquid the quotient $-\mathrm{d}p/\mathrm{d}z$ may be substituted by the expression $(p_i - p_o)/L$, where p_i and p_o are the absolute pressures at the inlet and outlet of the column, and L is the column length. Thus, for chromatography with a liquid mobile phase it may be written

$$u = (B_o/\varepsilon_o \mu L)(p_i - p_o) \tag{35}$$

This relation holds for column systems, and, in a more general concept, also for planar systems; in the first case L is the length of the column, and in the second case L designates the distance of the front of the chromatogram from the level of development liquid. In gas chromatography the situation is more complex, due to the high compressibility of the mobile phase. It holds here that

$$u(p_o) = (B_o/\varepsilon_o \mu L)(p_i^2 - p_o^2)/2p_o \tag{36}$$

where $u(p_o)$ is the velocity expressed at pressure p_o, or

$$u(\bar{p}) = (B_o/\varepsilon_o \mu L)(p_i^2 - p_o^2)/2\bar{p} =$$

$$= u(p_o)\, p_o/\bar{p} = u(p_o)\, j \tag{37}$$

where $\bar{p}$ is the mean pressure in the column, $u(\bar{p})$ is the velocity expressed at pressure $\bar{p}$, and j is James-Martin's compressibility factor [5] defined by the relation

$$j = (3/2)\left[(p_i/p_o)^2 - 1\right] / \left[(p_i/p_o)^3 - 1\right] \tag{38}$$

With respect to equations 36 and 38 the basic retention equation (see equation 9) for gas chromatography can be defined more rigorously as

$$t_R = t_M + k\int_0^L \frac{dz}{u(z)} = t_M + \frac{kL}{u(p_o)j} = \frac{L(1+k)}{u(p_o)j} \tag{39}$$

Equation 39 has been derived under the assumption that k is independent of pressure, however, this condition need not always be fulfilled to a sufficient extent.

The pressure difference $p_i - p_o$, where p_o is usually the atmospheric pressure, is the driving force of the flow. Whereas in column systems p_i is determined by the source of the mobile phase and the corresponding regulatory device, in planar systems capillary forces function as driving forces. In non-horizontal arrangements they are, in addition, combined with the gravitational force and, in centrifugal arrangements, with the centrifugal force. A highly simplified treatment of the model of a planar system leads to the relation

$$u_f = (b/L_f) \pm G, \tag{40}$$

where L_f is the distance of the chromatogram front from the level of the developing liquid, b is a constant of the given system and G is the gravitational component (which has (+) for descending development, (−) for ascending development, and (0) for the horizontal position of the bed).

1.8 Sorption equilibrium and the distribution constant

1.8.1 Problem of sorption equilibrium in a migrating chromatographic zone

It is known from chemical thermodynamics that a system consisting of several components and phases is in equilibrium when all the chemical potentials of all the components in all the phases are identical. Such a situation may occur in the case of a closed isolated or thermostated system. However, the migrating chromatographic zone represents an open and non-stationary system which is usually thermostated. Nevertheless there is a region within the elution chromatographic zone that is very close to equilibrium during the migration of the zone. It is a narrow region in close proximity to the concentration maximum of the zone. As in the leading part (part ahead of the maximum) of the migrating zone, passage of the solute from the mobile to the stationary phase predominates (i.e., sorption of the solute occurs), whereas in the rear part of the zone the opposite occurs (i.e., desorption of the solute from the stationary to the mobile phase takes place), it may be assumed that it is just in the maximum of the zone where neither sorption nor desorption occur, hence, sorption equilibrium (i.e., the identity of the chemical potentials of the solute in both phases) exists there. It is thus apparent that for the formulation of the relations between chromatographic retention data and the thermodynamic properties of the chromatographic system, the retention data should be calculated so as to represent the course of the migration of the concentration maximum of the zone.

In the case of a symmetrical chromatographic zone (i.e., in the case of linear chromatography) the maximum of the zone is localized in its center, and its velocity is a constant fraction of the forward velocity of the mobile phase during each stage of the migration. Thus, in this case, it is relatively simple to experimentally define and determine retention data so as to make it possible to calculate data representing sorption equilibrium. It is, above all, the distribution constant, which can be calculated from equation 9 or some of its suitable modifications, that constitutes such a retention quantity. In the case of non-linear chromatography such a possibility does not exist, as the velocity of the maximum of an asymmetrical (due to non-linearity of the sorption isotherm) zone varies along the migration path with respect to the velocity of the mobile phase. There is no unambiguous relation between chromatographic retention and the distribution constant under these conditions; a different distribution constant corresponds to any position of the zone maximum along its migration path, so that only an effective mean value of the distribution constant, which is not defined accurately, is obtained by means of equation 9. Thus, further considerations about relations between chromatographic retention and the thermodynamic properties of the system will concern only examples of linear chromatography.

The longitudinal concentration profile of the zone in the column is usually not known, but the time course of the solute concentration in the effluent can be detected by a detector at the column outlet. In such a record the retention time of the center of gravity (first statistical moment) of the detected peak [28] corresponds to the retention time of the concentration maximum of the real zone. However, in most cases of linear chromatography these two retention times are practically identical, i.e. the retention times of the center of gravity and of the maximum of the peak detected are also identical.

1.8.2 Relations between the chromatographic distribution constant and the thermodynamic properties of the chromatographic system

The chemical potentials of the solute (i) in the stationary phase (sorbent) and in the mobile phase, μ_{iS} and μ_{iM}, are defined by the relations

$$\mu_{iS} = \mu_{iS}^{o} + \underline{R}T \ln a_{iS} \tag{41}$$

$$\mu_{iM} = \mu_{iM}^{o} + \underline{R}T \ln a_{iM} \tag{42}$$

where μ_{iS}^{o} and μ_{iM}^{o} are the standard chemical potentials, a_{iS} and a_{iM} are the activities of the solute in the sorbent and in the mobile phase, $\underline{R}$ is the universal gas constant, and T is the absolute temperature of the system. In the concentration maximum of the zone, i.e., in equilibrium, $\mu_{iS} = \mu_{iM}$, and it may be written

$$\mu_{iS}^{o} - \mu_{iM}^{o} = \Delta G_{sp}^{o} = -\underline{R}T \ln(a_{iS}/a_{iM})_{eq.} \tag{43}$$

where ΔG^{o}_{sp} is the standard differential molar Gibbs function of sorption and the subscript eq. indicates that the equilibrium ratio of the activities is involved. The expression $(a_{iS}/a_{iM})_{eq}$. represents the thermodynamic distribution constant, whose numerical value depends on the selection of the standard states for the solute in the sorbent and in the mobile phase. It should be pointed out here that the activity of a given component is defined by the ratio of its actual fugacity and the fugacity in the standard state. Thus, in the case of solute i in the sorbent and in the mobile phase, $a_{iS} = f_{iS}/f^{o}_{iS}$ and $a_{iM} = f_{iM}/f^{o}_{iM}$. From the general point of view standard states can be chosen quite arbitrarily, with the exception of the standard temperature, which is chosen as identical with the actual temperature of the studied system. However, the selection of standard states should be made with respect to the objective pursued; the selection of standard states can be considered as a strategy leading to a situation when the standard thermodynamic quantities suitably reflect those features of the studied system that are of interest. The selection of standard states includes (with a given method of the expression of solute concentrations in the phases of the system) the specification of the standard concentration and standard physical states of the solute in both phases and the convention(s) for the normalization of the activity coefficients of the solute in the condensed phase(s) of the system. Examples for liquid-liquid and gas-liquid chromatographic systems will be presented below.

Liquid-liquid system (LLC) The solute concentrations in both phases will be expressed in mole fractions, a hypothetical pure solute at infinite dilution in the solvent at the temperature and mean pressure of the system will be chosen as a standard concentration and standard physical state for the solute in both phases, and the activity coefficient of the solute in both phases will be normalized by the convention according to which $\gamma_i^* \rightarrow 1$ as $x_i \rightarrow 0$. The fugacities of the solute in the stationary and mobile phases are then $f_{iS} = \gamma^*_{iS} h_{iS} x_{iS}$ and $f_{iM} = \gamma^*_{iM} h_{iM} x_{iM}$, where γ_i^* is the activity coefficient characterizing the deviation from Henry's law, h_i is the Henry law constant, and x_i is the molar fraction of the solute in a given phase. The standard fugacities ($x_i^o = 1$ and $\gamma_i^* = 1$) will then be $f^o_{iS} = h_{iS}$ and $f^o_{iM} = h_{iM}$. By substituting from the above relations into equation 43 the relation

$$\Delta G^*_{sp}(\text{LLC}) = -\underline{R}T \ln(\gamma^*_{iS} x_{iS}/\gamma^*_{iM} x_{iM}) \tag{44}$$

is obtained. The quantity ΔG^*_{sp}(LLC) can also be expressed in terms of the activity coefficients of Raoult's law; these activity coefficients are designated γ^{θ}_{iS} and γ^{θ}_{iM}. Under common chromatographic conditions (high solute dilution) the activity coefficients γ^*_{iS} and γ^*_{iM} approach unity. The fugacities f_{iS} and f_{iM} can be expressed as $f_{iS} = \gamma^{\theta}_{iS} f^{\theta}_i x_{iS}$, and $f_{iM} = \gamma^{\theta}_{iM} f^{\theta}_i x_{iM}$ when using the convention $\gamma^{\theta}_i \rightarrow 1$ at $x_i \rightarrow 1$; f^{θ}_i is the fugacity of the pure liquid solute at the temperature and mean pressure in the column, and in equilibrium it holds $f_{iS} = f_{iM}$. Equation 44 can thus be rewritten as

$$\Delta G^*_{sp}(\text{LLC}) = -\underline{R}T \ln(\gamma^{\theta}_{iM}/\gamma^{\theta}_{iS}) \tag{45}$$

where the values of γ^{θ}_i in this case correspond to infinite solute dilution in the respective solvents. It follows from equation 45 that ΔG^*_{sp}(LLC) equals the difference between the partial molar excess Gibbs functions of infinitely diluted solute in the sorbent and in the mobile phase when using the above specified standard states and assuming unit γ^*_{iS} and γ^*_{iM}. Thus, it holds

$$\Delta G^*_{sp}(\mathrm{LLC}) = \Delta \overline{G}^{\mathrm{E}}_{iS} - \Delta \overline{G}^{\mathrm{E}}_{iM} \tag{46}$$

Gas-liquid system (GLC) The solute concentrations in both phases will be again expressed in mole fractions, the standard concentration and standard physical state of the solute in the stationary (liquid) phase will be defined in the same way as with the liquid-liquid system, and a hypothetical pure solute in a state of ideal gas at a unit pressure and at the temperature of the system will be chosen as a standard state for the solute in the mobile (gaseous) phase. Thus, f_{iS} and f_{iM} may be written as $f_{iS} = \gamma^*_{iS} h_{iS} x_{iS}$ and $f_{iM} = \nu_{iM} p x_{iM}$, where ν_{iM} is the fugacity coefficient (mean value) of the solute in the mixture with carrier gas, and p is the mean pressure in the column. The corresponding standard fugacities ($x^{\mathrm{o}}_{iS} = 1$ and $\gamma^*_{iS} = 1$; $x^{\mathrm{o}}_{iM} = 1$, $p^{\mathrm{o}} = 1$, and $\nu_{iM} = 1$) are $f^{\mathrm{o}}_{iS} = h_{iS}$ and $f^{\mathrm{o}}_{iM} = 1$, so that, according to equation 43,

$$\Delta G^*_{sp}(\mathrm{GLC}) = -\underline{R}T \ln(\gamma^*_{iS} x_{iS} / \nu_{iM} p x_{iM}) \tag{47}$$

Also here γ^*_{iS} approaches unity under common chromatographic conditions and, at the same time, it holds $\nu_{iM} p x_{iM} = \gamma^{\theta}_{iS} f^{\theta}_{i} x_{iS}$. Thus, equation 47 may be rewritten as

$$\Delta G^*_{sp}(\mathrm{GLC}) = -\underline{R}T \ln(1/\gamma^{\theta}_{iS} f^{\theta}_{i}) \tag{48}$$

where, in this case, γ^{θ}_{iS} corresponds to infinite dilution of the solute in the sorbent. Relation 48 shows that ΔG^*_{sp}(GLC) equals the sum of the standard molar Gibbs function of condensation of pure solute and the partial molar excess Gibbs function of infinitely diluted solute in the sorbent when using the above standard states and under the assumption that $\gamma^*_{iS} = 1$. Thus, it may be stated that

$$\Delta G^*_{sp}(\mathrm{GLC}) = \Delta G^{\theta}_{cd} + \overline{G}^{\mathrm{E}}_{iS} \tag{49}$$

ΔG^{θ}_{cd} refers to the transition of one mole of pure solute from the hypothetical state of an ideal gas at unit pressure to the liquid state at the overall pressure and temperature of the system.

For the distribution constant defined as the equilibrium ratio of the mass concentrations of the solute in the sorbent and in the mobile phase it holds that

$$K = \frac{m_{iS} V_{\mathrm{M}}}{m_{iM} V_{\mathrm{S}}} = \frac{n_{iS} V_{\mathrm{M}}}{n_{iM} V_{\mathrm{S}}} = \frac{x_{iS} v^{\mathrm{mol}}_{\mathrm{M}}}{x_{iM} v^{\mathrm{mol}}_{\mathrm{S}}} \tag{50}$$

where n_{iS} and n_{iM} are the substance amounts of the solute in the sorbent and in the mobile phase (in the chromatographic zone), and v_M^{mol} and v_S^{mol} are the molar volumes of the mobile phase and of the sorbent. When these molar volumes are expressed as $v_M^{mol} = M_M/\rho_M$ and $v_S^{mol} = M_S/\rho_S$, where M_M and M_S are the molar masses and ρ_M and ρ_S are the densities of the mobile phase and of the sorbent, then, with respect to relations 44, 45 and 50 it may be written

$$\Delta G_{sp}^{*}(\mathrm{LLC}) = -\underline{R}T \ln(K\gamma_{iS}^{*} M_S \rho_M / \gamma_{iM}^{*} M_M \rho_S) \tag{51}$$

$$K(\mathrm{LLC}) = \gamma_{iM}^{\theta} M_M \rho_S / \gamma_{iS}^{\theta} M_S \rho_M \tag{52}$$

For the case of gas-liquid chromatography the relation $v_S^{mol} = M_S/\rho_S$ is used again, but v_M^{mol} is expressed as $v_M^{mol} = Z_M \underline{R}T/p$, where Z_M is the compressibility factor (mean value) of carrier gas. According to relations 47, 48 and 50 the following equations expressing $\Delta G_{sp}^{*}(\mathrm{GLC})$ and $K(\mathrm{GLC})$ are obtained:

$$\Delta G_{sp}^{*}(\mathrm{GLC}) = -\underline{R}T \ln(K\gamma_{iS}^{*} M_S / \nu_{iM} Z_M \underline{R}T \rho_S) \tag{53}$$

$$K(\mathrm{GLC}) = \nu_{iM} Z_M \underline{R}T \rho_S / \gamma_{iS}^{\theta} f_i^{\theta} M_S \tag{54}$$

It should be pointed out here that there is unity having a dimension of pressure ($f_{iM}^{o} = 1$) in the numerator of the fraction behind the logarithm in equations 47, 48 and 53.

1.8.3 Dependence of the standard differential molar Gibbs function of sorption and the chromatographic distribution constant on temperature and pressure

By applying the generally valid definitions of the temperature and pressure dependence of chemical potential to equation 43 relations

$$\left(\frac{\partial \Delta G_{sp}^{o}}{\partial p}\right)_{T,\,(\mathrm{comp})} = \Delta V_{sp}^{o} \tag{55}$$

$$\left[\frac{\partial}{\partial T}\left(\frac{\Delta G_{sp}^{o}}{T}\right)\right]_{p,\,(\mathrm{comp})} = -\frac{\Delta H_{sp}^{o}}{T^2} \tag{56}$$

$$\left(\frac{\partial \Delta G_{sp}^{o}}{\partial T}\right)_{p,\,(\mathrm{comp})} = -\Delta S_{sp}^{o} \tag{57}$$

are defined, where ΔV_{sp}^{o}, ΔH_{sp}^{o} and ΔS_{sp}^{o} are the differential standard molar volumes, enthalpies and entropies of the solute in the system. The standard states of these derived quantities are determined by the selection of the standard states for ΔG_{sp}^{o} (according to the selection of the standard states ΔG_{sp}^{o} either is, or is not, a function of composition).

The temperature and pressure dependences of the distribution constant can be easily derived from the temperature and pressure dependences of the right hand sides of equations 45 and 48. According to the well-known thermodynamic definitions it may be written for LLC systems

$$\left[\frac{\partial \ln K(\mathrm{LLC})}{\partial p}\right]_{T,\,\mathrm{comp.}} = -\frac{\bar{V}_{\mathrm{iS}} - \bar{V}_{\mathrm{iM}}}{\underline{R}T} + \beta_{\mathrm{S}} - \beta_{\mathrm{M}} \tag{58}$$

$$\left[\frac{\partial \ln K(\mathrm{LLC})}{\partial T}\right]_{p,\,\mathrm{comp.}} = -\frac{\bar{H}_{\mathrm{iS}} - \bar{H}_{\mathrm{iM}}}{\underline{R}T^2} + \alpha_{\mathrm{M}} - \alpha_{\mathrm{S}} \tag{59}$$

where $\bar{H}_{\mathrm{i}}$ and $\bar{V}_{\mathrm{i}}$ are the partial molar enthalpies and partial molar solute volumes in the phases of the system, α_{M} and α_{S} are the coefficients of thermal expansivity of the mobile and stationary liquids and β_{M} and β_{S} are the coefficients of the compressibility of the mobile and stationary liquids at the temperature and total (mean) pressure in the system. For GLC systems it holds

$$\left[\frac{\partial \ln K(\mathrm{GLC})}{\partial p}\right]_{T,\,\mathrm{comp.}} = \left[\frac{\partial \ln(\nu_{\mathrm{iM}} Z_{\mathrm{M}})}{\partial p}\right]_{T,\,\mathrm{comp.}} - \frac{\bar{V}_{\mathrm{iS}}}{\underline{R}T} + \beta_{\mathrm{S}} \tag{60}$$

$$\left[\frac{\partial \ln K(\mathrm{GLC})}{\partial T}\right]_{p,\,\mathrm{comp.}} = \left[\frac{\partial \ln(\nu_{\mathrm{iM}} Z_{\mathrm{M}})}{\partial T}\right]_{p,\,\mathrm{comp.}} + \frac{\bar{H}_{\mathrm{iS}} - H_{\mathrm{i}}^{\mathrm{g}}}{\underline{R}T^2} + \frac{1}{T} - \alpha_{\mathrm{S}} \tag{61}$$

where $H_{\mathrm{i}}^{\mathrm{g}}$ is the molar enthalpy of pure solute vapors at the temperature of the system and at a very low pressure. Under common gas chromatography conditions the coefficients ν_{iM} and Z_{M} are practically of unit value, so that the first terms in the right hand sides of equations 60 and 61 can be neglected.

It follows from the discussion in this paragraph that only *standard* differential thermodynamic functions can be calculated from any chromatographic distribution constant defined in whatever way. Also, it is necessary to always specify the choice of the standard states for the solute in both phases of the system. Without specifying the standard states the data on the thermodynamic functions calculated from chromatographic retention data lack any sense. When choosing certain standard states it may happen that the standard differential Gibbs function is identical with another form of the differential Gibbs function, or includes such a form; situations described by equations 46 and 49 may serve as examples. The same also holds true for standard differential volumes, entropies and enthalpies (compare Section 1.8.3). However, every particular situation requires a special treatment.

When the definitions 55–57 are applied to $\Delta G_{\mathrm{sp}}^{*}(\mathrm{LLC})$ and $\Delta G_{\mathrm{sp}}^{*}(\mathrm{GLC})$ defined by equations 51 and 53, by using equations 58–61 and on the condition that γ_{i}^{*}, ν_{iM} and Z_{M} are of unit value it is possible to write

$$\left[\frac{\partial \Delta G_{\mathrm{sp}}^{*}(\mathrm{LLC})}{\partial p}\right]_{T,\,\mathrm{comp.}} = \Delta V_{\mathrm{sp}}^{*}(\mathrm{LLC}) = \bar{V}_{\mathrm{iS}} - \bar{V}_{\mathrm{iM}} \tag{62}$$

$$\left[\frac{\partial \Delta G^*_{sp}(\mathrm{GLC})}{\partial p}\right]_{T,\,\mathrm{comp.}} = \Delta V^*_{sp}(\mathrm{GLC}) = \bar{V}_{iS} \tag{63}$$

$$\left\{\frac{\partial}{\partial T}\left[\frac{\Delta G^*_{sp}(\mathrm{LLC})}{T}\right]\right\}_{p,\,\mathrm{comp.}} = -\frac{\Delta H^*_{sp}(\mathrm{LLC})}{T^2} = -\frac{\bar{H}_{iS} - \bar{H}_{iM}}{T^2} \tag{64}$$

$$\left\{\frac{\partial}{\partial T}\left[\frac{\Delta G^*_{sp}(\mathrm{GLC})}{T}\right]\right\}_{p,\,\mathrm{comp.}} = -\frac{\Delta H^*_{sp}(\mathrm{GLC})}{T^2} = -\frac{\bar{H}_{iS} - H_i^g}{T^2} \tag{65}$$

The standard differential sorption volumes ΔV^*_{sp} and the standard differential sorption enthalpy ΔH^*_{sp} are hence practically identical with the actual differential sorption volumes and enthalpies.

As $\Delta \bar{V}_i^E = \bar{V}_i - \bar{V}_i^{id}$, $\Delta \bar{H}_i^E = \bar{H}_i - \bar{H}_i^{id}$, $\bar{V}_i^{id} = V_i^L$, $\bar{H}_i^{id} = H_i^L$, and $H_i^L - H_i^g = \Delta H^\theta_{cd}$, where $\Delta \bar{V}_i^E$ and $\Delta \bar{H}_i^E$ are the partial molar excess volume and the partial molar excess enthalpy of the solute in a given solvent, $\bar{V}_i^{id}$ and $\bar{H}_i^{id}$ are the partial molar volume and partial molar enthalpy of the solute in ideal solutions, V_i^L and H_i^L are the molar volume and molar enthalpy of the pure liquid solute, respectively, and ΔH^θ_{cd} is the standard molar condensation enthalpy of the pure solute, all under the conditions of the system, it holds that

$$\Delta V^*_{sp}(\mathrm{LLC}) = \bar{V}_{iS} - \bar{V}_{iM} = \Delta \bar{V}^E_{iS} - \Delta \bar{V}^E_{iM} \tag{66}$$

$$\Delta V^*_{sp}(\mathrm{GLC}) = \bar{V}_{iS} = \Delta \bar{V}^E_{iS} + V_i^L \tag{67}$$

$$\Delta H^*_{sp}(\mathrm{LLC}) = \bar{H}_{iS} - \bar{H}_{iM} = \Delta \bar{H}^E_{iS} - \Delta \bar{H}^E_{iM} \tag{68}$$

$$\Delta H^*_{sp}(\mathrm{GLC}) = \bar{H}_{iS} - H_i^g = \Delta \bar{H}^E_{iS} + \Delta H^\theta_{cd} \tag{69}$$

For the standard differential entropy of sorption ΔS^*_{sp} it holds

$$\left(\frac{\partial \Delta G^*_{sp}}{\partial T}\right)_{p,\,\mathrm{comp.}} = -\Delta S^*_{sp} = \frac{\Delta G^*_{sp} - \Delta H^*_{sp}}{T} \tag{70}$$

It should be pointed out that $\bar{G}_i^{id} \neq G_i^L$ and $\bar{S}_i^{id} \neq S_i^L$, where $\bar{G}_i^{id}$ and $\bar{S}_i^{id}$ are the partial molar Gibbs function and partial molar entropy of the solute in an ideal solution and G_i^L and S_i^L are the molar Gibbs function and molar entropy of the pure liquid solute, respectively. However, when using the above mentioned choice of standard states it also holds that $\Delta G^*_{sp}(\mathrm{LLC}) = \Delta \bar{G}^E_{is} - \Delta \bar{G}^E_{iM}$ and $\Delta G^*_{sp}(\mathrm{GLC}) = \Delta \bar{G}^E_{iS} + \Delta G^\theta_{cd}$, so that even here

$$\Delta S^*_{sp}(\mathrm{LLC}) = \frac{\Delta \bar{H}^E_{iS} - \Delta \bar{G}^E_{iS} - (\Delta \bar{H}^E_{iM} - \Delta \bar{G}^E_{iM})}{T} = \Delta \bar{S}^E_{iS} - \Delta \bar{S}^E_{iM} \tag{71}$$

$$\Delta S^*_{sp}(\mathrm{GLC}) = \frac{\Delta \bar{H}^E_{iS} - \Delta \bar{G}^E_{iS} + \Delta H^\theta_{cd} - \Delta \bar{G}^\theta_{cd}}{T} = \Delta \bar{S}^E_{iS} + \Delta S^\theta_{cd} \tag{72}$$

where ΔH_{cd}^{θ} is the standard molar condensation enthalpy of the pure solute (compare comments to the quantity ΔG_{cd}^{θ} below equation 49).

1.9 Chromatographic resolution

Chromatographic resolution is defined as the distance between the concentration maxima of two elution zones, expressed in the units of the mean standard deviation of these zones. When considering a chromatographic record plotted in coordinates with the detector response as a function of the solute concentration in the column effluent on the ordinate, and the time elapsed from the start of the chromatographic run on the abscissa, the resolution (RS) of the peaks of compounds 1 and 2 having the retention times $t_{R2} > t_{R1}$ and the standard deviations σ_{t1} and σ_{t2} can be described by the relation

$$RS = (t_{R2} - t_{R1})/\bar{\sigma}_t = 2(t_{R2} - t_{R1})/(\sigma_{t2} + \sigma_{t1}) \tag{73}$$

If the peaks are of roughly the same height and symmetrical, an almost complete separation of them can be attained at $RS = 4$. However, with peaks having considerably different heights larger RS values are required for the same separation effect to be attained.

On the basis of the relations discussed in sections 1.6.1 and 1.6.2 it can be easily derived that

$$t_{R2} - t_{R1} = \frac{L}{u}\left(\frac{1}{R_2} - \frac{1}{R_1}\right) = \frac{L}{u}(k_2 - k_1) \tag{74}$$

$$\bar{\sigma}_t = \frac{(HL)^{1/2}}{2u}\left(\frac{1}{R_2} + \frac{1}{R_1}\right) = \frac{(HL)^{1/2}}{2u}(k_2 + k_1 + 2) \tag{75}$$

Equations 74 and 75 show that, whereas the distance between the concentration maxima of two migrating zones increases linearly with their migration distance, their standard deviations increase only as the square root of the length of the migration distance. This fact represents the basic principle of chromatographic separation. By combining equations 73, 74 and 75, the equation

$$RS = 2\left(\frac{L}{H}\right)^{1/2}\frac{R_1 - R_2}{R_1 + R_2} = 2\left(\frac{L}{H}\right)^{1/2}\frac{k_2 - k_1}{k_2 + k_1 + 2} \tag{76}$$

is obtained after rearrangement.

Equation 76 makes it possible to calculate the number of theoretical plates necessary for a required resolution of the peaks of components 1 and 2. As the

TABLE 1.3
Summary of classical theories of chromatography

J.N. Wilson (1940)	Mathematical treatment of the model of ideal linear chromatography
A.J.P. Martin and R.L.M. Synge (1941)	plate theory of chromatography
D. De Vault (1943); J. Weiss (1943)	Improvement of Wilson's treatment of the model of ideal linear chromatography
H.C. Thomas (1948)	Mathematical treatment of the model of non-ideal linear chromatography (neglecting the longitudinal solute diffusion)
L. Lapidus and N.R. Amundson (1952)	Detailed mathematical treatment of the model of non-ideal linear chromatography
E. Glueckauf (1954)	Diffusion (continuous) model of chromatography
J.C. Giddings and H. Eyring (1955)	Statistical model of chromatography
J.J. Van Deemter, F.J. Zuiderweg and A. Klinkenberg (1956)	Simplification and rationalization of the Lapidus and Amundsen treatment of the model of non-ideal linear chromatography
M.J.E. Golay (1958)	Theory of capillary gas chromatography
J.C. Giddings (1959)	Generalized nonequilibrium (non-ideal) theory of chromatography

number of theoretical plates $N = L/H$, it may be written according to equation 76

$$N_{RS} = \left(\frac{RS}{2} \frac{R_1 + R_2}{R_1 - R_2} \right)^2 \tag{77}$$

or

$$N_{RS} = \left(\frac{RS}{2} \frac{k_2 + k_1 + 2}{k_2 - k_1} \right)^2 \tag{78}$$

where the required resolution is substituted for RS in the right hand sides of the equations. Equation 77 is suitable in planar chromatography (R can be substituted by R_F) and equation 78 is useful in column chromatography (k can be substituted by $(t_R - t_M)/t_M$.

The expression $(k_2 - k_1)/(k_2 + k_1 + 2)$ in equation 76 can be expressed in terms of the distribution constants or relative retentions:

$$\frac{k_2 - k_1}{k_2 + k_1 + 2} = \frac{K_2 - K_1}{K_2 + K_1 + 2(\phi_M/\phi_S)} = \frac{\alpha_{21} - 1}{\alpha_{21} + 1 + (2/k_1)} \tag{79}$$

where $\alpha_{21} = k_2/k_1$. The expression $(k_2 + k_1 + 2)/(k_2 - k_1)$ in equation 78 can naturally be expressed in a similar way. The effect of the capacity properties of the column on its separation ability may well be seen from the middle member of relation 79.

1.10 Development of theories of chromatography

The classical theories of chromatography developed roughly from 1940 to 1960. During these two decades views about the possibilities and limitations concerning the exact description of the chromatographic process crystallized. Further theoretical works were devoted primarily to verification, extension and utilization of the existing theoretical knowledge. A representative review of the theories of chromatography is given in Table 1.3.

References

1 Day, D.T. (1897) Proc. Am. Phil. Soc. 36, 112.
2 Tswett, M.S. (1906) Ber. Dtsch. Bot. Ges. 24, 384.
3 Martin, A.J.P. and Synge, R.L.M. (1941) Biochem. J. (London) 35, 1358.
4 Cremer, E. and Prior, F. (1951) Z. Elektrochem. 55, 66.
5 James, A.T. and Martin, A.J.P. (1952) Biochem. J. (London), 50, 679.
6 Janák, J. and Rusek, M. (1953) Chem. Listy 47, 1190.
7 Ettre, L.S. (1971) Anal. Chem. 43, 20A.
8 Ettre, L.S. (1975) J. Chromatogr. 112, 1.
9 Zechmeister, L. (1967) in Chromatography (Heftman, E. ed.) 2nd Edn, Reinhold, New York, p. 3.
10 Keulemans, A.I.M. (1957) Gas Chromatography (Verver, C.G. ed.) Reinhold, New York, p. 99.
11 Lapidus, L. and Amundson, N.R. (1952) J. Phys. Chem. 56, 984.
12 Van Deemter, J.J., Zuiderweg, F.J. and Klinkenberg, A. (1956) Chem. Eng. Sci. 5, 27.
13 Wilson, J.N. (1940) Am. Chem. Soc. 62, 1583.
14 Wičar, S., Novák, J. and Rakshieva, N.R. (1971) Anal. Chem. 43, 1945.
15 Giddings, J.C. (1958) J. Chem. Educ. 35, 588.
16 Giddings, J.C. (1959) J. Chem. Phys. 31, 1462.
17 Le Rosen, A.L. (1945) J. Am. Chem. Soc. 67, 1683.
18 Giddings, J.C., Stewart, G.H. and Ruoff, A.L. (1960) J. Chromatogr. 3, 239.
19 Desty, D.H., Glueckauf, E., James, A.T., Keulemans, A.I.M., Martin, A.J.P. and Phillips, C.S.G. (1957) Nomenclature Recommendations: Vapour Phase Chromatography (Desty, D.H. ed.) Butterworths, London, 1957, p. XI.
20 Chandrasekhar, S. (1943) Rev. Mod. Phys. 15, 1.
21 Giddings, J.C. (1961) Anal. Chem. 33, 962.

22 Giddings, J.C. and Robinson, R.A. (1962) Anal. Chem. 34, 885.
23 Glueckauf, E. (1955) Trans. Faraday Soc. 51, 34.
24 Navier, M. (1827) Mem. Acad. Sci. 6, 389.
25 Darcy, H. Les Fontaines Publiques de la Ville de Dijon, Paris, 1856.
26 Kozeny, J. and Wiener, S.B. (1927) Akad. Wiss. 136, 271.
27 Carman, P.C. (1937) Trans. Inst. Chem. Engin. (London) 15, 150.
28 Conder, J.R. and Young C.L. (1979) Physicochemical Measurement by Gas Chromatography, Wiley, Chichester, New York, Brisbane, Toronto, p. 68.

Deyl (ed.) Separation Methods

CHAPTER 2

Principles and theory of electromigration processes

JIŘÍ VACÍK

Department of Physical Chemistry, Charles University, Albertov 2030, 12840 Prague, Czechoslovakia

2.1 Principles of electromigration separation methods [1]

The separation and characterization of individual components of complex mixtures is of considerable importance in the advancement of many areas of science and the separation procedures are therefore widely applied. For the separation of mixtures of compounds in solutions of electrolytes electromigration processes can be applied. In the past numerous electromigration methods have been worked out that differ in the way the fundamental separation principle — the differing migration velocity of different charged particles in the electrical field — is made use of. The classification of electromigration methods varies according to the criteria used. Some of the possible classifications are presented in Table 2.1. The comparison of the course of separation as the result of the experimental arrangement is schematically shown in Fig. 2.1.

In the zone electrophoresis method the sample is placed in a definite area of the separation column filled with the electrolyte. After the electrical field has been imposed onto the system, individual particles migrate, according to their effective mobilities with different speeds, towards the respective electrodes (and, concomitantly, both positively and negatively charged particles are separated). The constituents of the mixture are separated into distinct zones, that, however, are not sharp; their width increases with the increasing separation time and, consequently, the maximum compound concentration within the zones decreases.

Due to the necessity of zone stabilization the separation is carried out in suitable stabilizing media (paper, starch, cellulose, etc.). Interactions with the stabilizing media like adsorption, sieving effects or ion exchange can influence considerably the resulting separation. Zone electrophoresis can be compared to elution zone chromatography.

The moving boundary method is, at its starting phase similar to the zone electrophoresis method. The sample to be separated is, however, present in such a large part of the separation column, that complete separation into individual zones is

TABLE 2.1
Some classifications of electromigration methods [1]

Classification criterion	Electromigration methods
Separated amount	Analytical Micropreparative Preparative
Method materialization	One-dimensional Multi-dimensional (two-dimensional) Continuous
Method of zone stabilization	Free flow (in solution) On sorbents Capillary Gel
Separation principle	Migration in a single separation medium Migration in different separation media Migration and interaction with the sorbent Migration, interaction with the sorbent, diffusion and detection reaction
Potential applied	Low voltage High voltage
Experimental arrangement (starting and limiting conditions of separation)	Zone Moving boundary Isotachophoresis Focusing methods Combined methods
The size (and charge polarity) of separated particles	Electrophoresis * (separation of colloids) Iontophoresis (separation of ions in true solutions) Cataphoresis (separation of cations) Anaphoresis (separation of anions)

* This classification was used in the older literature; more recently the term electrophoresis was generalized, so that it now means the movement of any charged particle of the solution in the electrical field and, what is more, it is frequently used today to denote all the different electromigration methods.

not possible. After the application of the electric field the compounds migrate stepwise from the front of the original complex zone into the electrolyte, the first being the fastest component, followed by a mixed zone of two, then three fast components, etc. On the contrary the slowest components of the mixture are delayed behind the rare boundary of the original mixed zone, the last being the slowest component of the mixture, preceded by the mixture of the two slowest compounds, etc. The shape of the boundary is considerably influenced by the potential gradient profile along the separation column. The moving boundary method is analogous to frontal chromatography.

The characteristic feature of isotachophoresis is the fact that in a single experi-

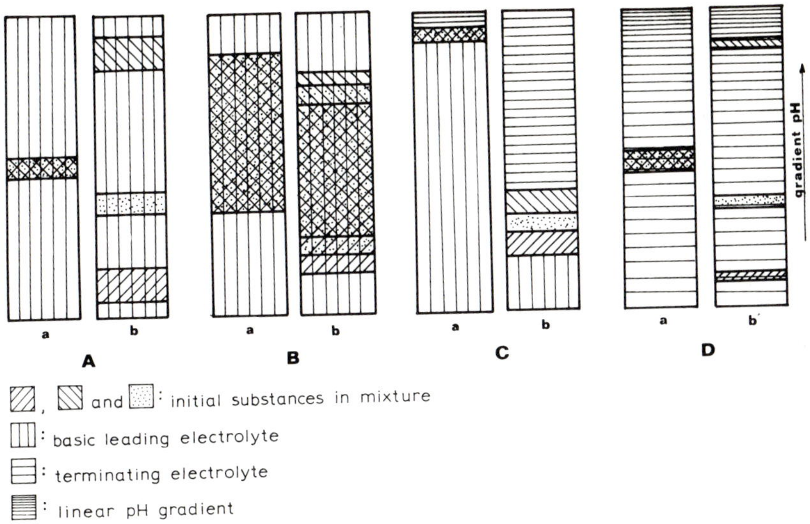

Fig. 2.1. Schematic representation of the four electrophoretic methods: (A) zone electrophoresis; (B) moving boundary electrophoresis; (C) isotachophoresis; and (D) isoelectric focusing. (a) The beginning of the experiment; (b) separation of a mixture of the substances.

ment it is possible to separate only different particles of identical polarity. The following description refers to the separation of anions; the situation in the separation of cations is analogous and can be easily derived. The sample to be separated is located on the separation column between two different electrolytes — the leading and the terminating electrolyte. The anions of the leading electrolyte (at the beginning of the experiment they occupy practically the whole volume of the separation column) exhibit a higher electrophoretic mobility than any one of the separated anions. On the contrary the mobility of the anions of the terminating electrolyte is lower than the mobility of any anion of the separated sample. After the electric field has been imposed onto the system, the individual components of the sample start to separate and finally reach a 'steady state'. Characteristically in the 'steady state' the individual components of the mixture are separated according to their effective mobilities into individual sharply separated zones which, however, follow intimately one after the other. All zones move with identical speed — from which characteristic the name of the method is derived. The concentrations of separated compounds in individual zones are adjusted (in coincidence with the regulatory function) according to the concentration of the leading electrolyte. The compounds, which in the original sample are present in high concentration, are diluted during separation while compounds present in small amounts are concentrated during the isotachophoretic process. According to the different relative proportion of individual compounds in the separated mixture the zones of individual compounds of appropriate length are formed; the zone length (at constant compound concentration throughout the whole zone) can be used for quantitation. Isotachophoresis is analogous to displacement chromatography.

The term sgn z (it holds that: sgn $z = 1$ for $z > 0$; sgn $z = -1$ for $z < 0$; and sgn $z = 0$ for $z = 0$) reflects the fact that particles with the positive charge are moving in the sense of the vector $\vec{E}$, while particles with the negative charge are moving the opposite direction. The proportionality constant in this equation (U_{iz}) is called (actual) mobility, and represents the velocity that would be exhibited by the particle of the i-th component with the relative charge z in the electric field of unit intensity.

2.2.2 Mobility

Actual mobility $U_{i,z}$ of any ionic form (component) of the i-th compound is, generally saying a complicated function of the distribution of all ions in the solution [3,4].

For the characterization of a particular component the limiting (absolute) mobility $(U_{i,z})^0$ can be used that is defined as mobility of this component at a given temperature in a solution in which the concentration of all components approaches zero. It holds

$$U_{i,z} = \gamma^*(U_{i,z})^0$$

where γ^* represents a correction factor whose value can be theoretically derived.

Besides the actual mobilities $U_{i,z}$ and the limiting mobilities $(U_{i,z})^0$ of the i-th component with the relative charge z the magnitude $(U_i)_{\text{eff}}$ is also in use. This magnitude characterizes the mobility of the i-th compound as a whole.

If the i-th compound is a weak electrolyte, it takes part in the appropriate protolytic equilibria with the solvent, e.g., water. The total concentration c_i of this compound is given by the sum $c_{i,z}$ of all ionic forms ($z \neq 0$) and the concentration $c_{i,0}$ of the electroneutral molecules ($z = 0$) of this i-th compound. It holds that $c_i = \sum_z c_{i,z}$. The proportion of the i, z-th component in the i-th compound, expressed, e.g., by the molar fraction $x_{i,z} = \frac{c_{i,z}}{c_i}$ cannot be arbitrarily changed since this proportion is determined by the appropriate equilibrium [2,5]. Each particle of the i-th compound in its participation in protolytic equilibria passes during the separation though all ionic forms (including the form with zero relative charge) — it remains in each of these forms for a period of time that is proportional to its molar fraction. In spite of the different actual mobility it is not possible to separate the individual ionic forms. The compound moves, as a whole, with a speed that is proportional to the effective mobility; the individual components contribute, to this effective mobility, a part that is proportional to the product of their proportion in the given compound and actual mobility. It holds that

$$(U_i)_{\text{eff}} = \sum_z x_{i,z} \cdot U_{i,z} \text{ [6]} \qquad (3)$$

If the i-th compound also participates in complex-forming or other equilibria, then this fact has to be respected in the equation for $(U_i)_{\text{eff}}$.

From this viewpoint of the electrophoretic separation it is interesting to compare the behavior of separated compounds in the particular case in which they take part only in protolytic equilibria, and also in the case where they participate in other equilibria. Since the solvent is present in the whole column, the compound enters into the protolytic equilibria throughout the entire column, and thus the compound moves as a single zone. If, however, the compound takes part in complex-forming equilibria, and if both the particles and ligands are present in a certain part of the column (e.g., in the place where the sample is loaded), then breakage of the complex may occur, and separate zones of the components of the complex are formed.

The following relations exist between the mobility and other magnitudes.

(a) Between $U_{i,z}$ and the diffusion coefficient $D_{i,z}$

$$U_{i,z} = \frac{|z| \cdot D_{i,z} \cdot F}{\mathrm{RT}} \tag{4}$$

(b) Due to the inverse proportionality between $D_{i,z}$ and particle diameter of the moving globular particle $r_{i,z}$, it holds that

$$U_{i,z} = \frac{|z| \cdot e}{6\pi \eta r_{i,z}} \tag{5}$$

where e is the electron charge and η the viscosity coefficient.

(c) For the mobility of globular colloid particles it holds that:

$$U_{\mathrm{col}} = \frac{\varepsilon \xi}{6\pi \eta} \tag{6}$$

where ε is the permittivity of the media and ξ is the electrokinetic potential.

If the colloid particle is not globular, the numerical factor in the denominator changes from 4 to 8 depending on the shape of the particle.

(d) between $U_{i,z}$ and the ionic conductivity $\lambda_{i,z}$ it holds that

$$\lambda_{i,z} = |z| \cdot F \cdot U_{i,z} \tag{7}$$

where F is Faraday's constant.

Due to the temperature dependence of the viscosity coefficient the mobility is also strongly temperature dependent.

Besides the ionic conductivities $\lambda_{i,z}$, the molar electrolyte conductivity Λ and specific conductivity κ of the solution are also introduced. The following relations can be written:

$$\Lambda = \sum_i \sum_z \lambda_{i,z} = \sum_i \sum_z |z| \cdot F \cdot U_{i,z} \tag{8}$$

$$\kappa = \sum_i \sum_z c_{i,z} \cdot \lambda_{i,z} = \sum_i \sum_z c_{i,z} \cdot |z| \cdot F \cdot U_{i,z} \tag{9}$$

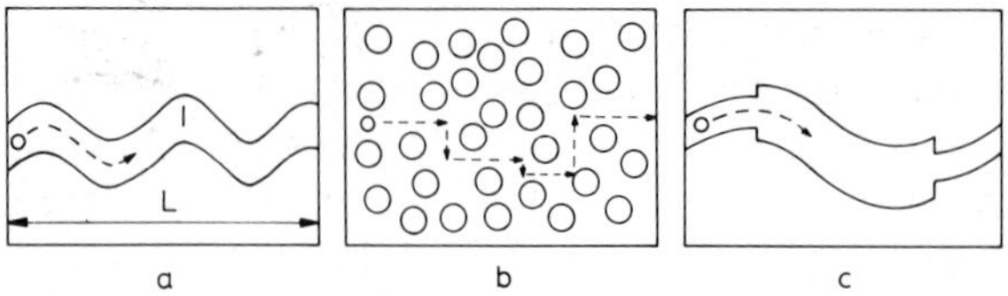

Fig. 2.2. Schematic representation of velocity gradient for (a) hydrodynamic flow, (b) electro-osmotic flow in an open column, and (c) electro-osmotic flow in a closed column.

The stabilization of zones arising during the electrophoretic separation can be materialized in different ways, such as by increasing the electrolyte viscosity, by forming a density gradient in the electrolyte, by using stabilizing media, etc. As stabilizing media, compact porous materials with an intrinsic capillary microstructure, packed columns, or capillary columns can be used. The influence of the sorbent upon the experimentally determined mobility of a compound can be characterized by a correction factor γ, that represents a proportionality constant between the effective mobility $(U_i)_{\mathrm{eff}}$ of the i-th compound measured in the stabilizer-free buffer, and the mobility of the same compound $(U_i)_{\mathrm{macro}}$ measured during the separation in porous media.

$$(U_i)_{\mathrm{eff}} = \gamma (U_i)_{\mathrm{macro}} \tag{10}$$

Assuming that it deals only with lengthening of the path that has to be traversed by the particle in the porous media, the correction factor can be considered identical with the square of the tortuosity factor γ_t [7–9] defined as $\gamma_t = (\frac{l}{L})$ where l is the actual path to be traversed by the migrating particle in porous media of total length L. This situation, as well as other ideas about the structure of porous media [9] originating in the barrier theory [10], and leading to other definitions of the correction factor, are schematically presented in Fig. 2.2.

If, besides the lengthened path of the migrating particles, their interaction with the sorbent (adsorption) has also to be considered, then this influence must also be included in the correction factor [2,11] by means of the R_E factor, which is defined in analogy to the R_F factor in chromatography.

Mobility is frequently expressed in relative terms [8,9] with respect to the mobility of a standard compound S. For the i-th compound it then holds that

$$(U_{i,s})_{\mathrm{rel}} = \frac{(U_i)_{\mathrm{macro}}}{(U_{\mathrm{stand}})_{\mathrm{macro}}} \tag{11}$$

and assuming the same interaction of the i and s compounds with the sorbent we can write

$$(U_{i,s})_{\mathrm{rel}} = \frac{(U_i)_{\mathrm{eff}}}{(U_{\mathrm{stand}})_{\mathrm{eff}}} \tag{12}$$

Depending on the standard selected it can be that $(U_{i,s})_{\mathrm{rel}} \lesseqgtr 1$

2.2.3 *Diffusion velocity*, $\vec{v}_{\mathrm{dif}}$

$\vec{v}_{\mathrm{dif}}$ always comes into consideration when the chemical potential $\mu_{i,z}$ of any component is not constant throughout the whole (by other words if there are concentration gradients in the column arising from the separation of the sample into distinct zones). This velocity is given by Fick's law, which can be used in the following form:

$$(\vec{v}_{i,z})_{\mathrm{dif}} = \frac{D_{i,z} \cdot \mathrm{grad}\, \mu_{i,z}}{RT} \qquad \mu_{i,z} = \mu^{\mathrm{o}}_{i,z} + RT \ln c_{i,z} \tag{13}$$

where $\mu^{\mathrm{o}}_{i,z}$ represents the standard chemical potential of the i, z-th component.

2.2.4 *Velocity of convection*, $\vec{v}_{\mathrm{conv}}$

$\vec{v}_{\mathrm{conv}}$ depends on the reason of this flow. During electrophoretic separations hydrodynamic flow resulting from the pressure difference between both ends of the separation column, flow caused by capillary forces in the stabilizing porous media, or electro-osmotic flow mainly come into operation. Flows caused by the thermal gradient are usually listed separately as thermal flows.

2.2.5 *Hydrodynamic flow*

Hydrodynamic flow depends on the profile of the separation column. If the separation column has a cylindrical shape then, due to the internal friction, there arises a transversal velocity gradient as demonstrated schematically in Fig. 2.3.

Another velocity distribution occurs when the hydrodynamic flow is caused by capillary forces in a stabilizing flat bed sorbent, when evaporation of the electrolyte from the surface can occur. If this evaporation is not too intensive, it is equal over the whole surface of the sorbent. Concomitantly, with the buffer evaporation from the surface, the buffer solution is supplied from both electrode vessels due to surface forces. Thus, a longitudinal flow originates the velocity of which is a function of the distance. This flow can be made use of in focusing methods [1,9,12,13].

2.2.6 *Electro-osmotic flow*

Electro-osmotic flow arises when an electric field is imposed into the system. Due to the electric field the spatial charge of the diffusion part of the electric double layer [3] moves towards the oppositely charged electrode. This causes a unidimensional flow of ions in the intimate neighbourhood of the column's walls, which also causes the solution in the column to move. The distribution of the velocities of the electro-osmotic flow depends on the experimental arrangement. A schematic representation of this is presented in Fig. 2.3.

The velocity of the osmotic flow through a single capillary with a length of L and

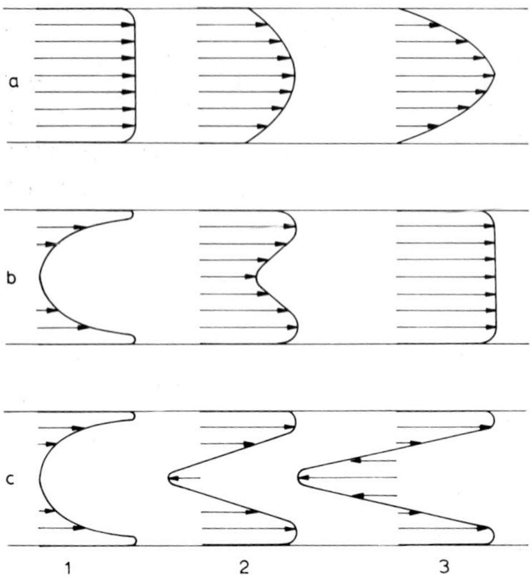

Fig. 2.3. Schematic representation of (a) tortuosity, (b) retardation and (c) constrictive effects [10] during migration of an ion in the stabilized medium.

a diameter of S can be expressed by the following relation:

$$\vec{v}_{os} = \frac{\varepsilon\xi\vec{E}}{\eta} = -\frac{\varepsilon\xi\varphi}{\eta L} = -\frac{\varepsilon\xi I}{\eta\kappa S} \tag{14}$$

Also, the volume velocity of the osmotic flow, $\vec{w}_{os}$, defined as the volume of the solution transported due to electroosmotic in unit time, is used

$$\vec{w}_{os} = \vec{v}_{os} \cdot S = -\frac{\varepsilon\xi\varphi S}{\eta L} = -\frac{\varepsilon\xi I}{\eta\kappa} \tag{15}$$

All these relations are also valid for the osmotic flow through porous media without correction.

2.2.7 *The velocity of the thermal flow,* $\vec{v}_{therm}$

A characteristic feature of all electrophoretic separation procedures is the fact, that when the electric current is passing through a solution, Joule's heat is generated. If R represents the electrolyte resistance in a column of length L and diameter S, through which, within a time period t, passes the electric current I, then for Q it holds that

$$Q = RI^2t = \frac{LI^2t}{\kappa S} \tag{16}$$

The heat is generated within the whole volume of the solution, but is removed only through the walls of the separation column, thus resulting in a transversal temperature gradient. The temperature gradient is the reason of the thermal flow. In the unordered thermal flow the direction towards lower temperature predominates, and therefore particles are moved to places with lower temperature. For the thermal flow velocity $\vec{v}_{\text{therm}}$ it holds that

$$\vec{v}_{\text{therm}} = -D^{x} \operatorname{grad} T$$

where D^{x} is the thermal diffusion coefficient.

2.2.8 The distribution of the potential gradient

Another characteristic feature of the electrophoretic separation is the fact that the conductivity of the solution (and therefore also the electric field intensity $\vec{E}$) is in any place and in any time point given by all ions present in the solution. It is well known [14–16] that the distribution of the potential gradient (and, during the separation in stabilizing media, also the shape of the separation isotherm) can considerably influence the zone shape of individual compounds. Two extremes can occur.

(a) The specific conductivity of the electrolyte κ_{el} is such that it is not influenced by the separated sample. Then, the potential gradient is constant along the whole length of the column, L. (The potential gradient can in any position be expressed as $\frac{\varphi}{L}$ which simplifies the evaluation of the experiment considerably.) In this case the shape of the zones is not influenced by the potential gradient. This situation is assumed in all types of zone electrophoresis.

(b) The specific conductivity of the solution in the place where the sample is located (κ_c) is considerably different from the specific conductivity κ_{el} of the electrolyte. Then, two variants can occur. If $\kappa_c > \kappa_{\text{el}}$, then the potential gradient within the zone is lower than in the surrounding media. The rare boundary is then focused (if an ion is delayed behind the rare boundary it occurs in a place of a higher potential gradient and moves back into the zone), and the frontal boundary becomes diffuse (if an ion overtakes the frontal boundary, it moves to a place with a higher potential gradient and is moved even further away). If $\kappa_c < \kappa_{\text{el}}$, then the situation is reversed (the frontal boundary is focused and the rare boundary is diffuse).

References

1 Vacík, J. (1979) in Electrophoresis, a Survey of Techniques and Applications, Deyl, Z., Everaerts, F.M., Prusik, Z. and Svendsen, P.J. (eds.) Elsevier, Amsterdam, p. 23.

2 Vacík, J. (1979) in Electrophoresis, a Survey of Techniques and Applications, Deyl, Z., Everaerts, F.M., Prusik, Z. and Svendsen, P.J. (eds.) Elsevier, Amsterdam, p. 1.

3 Moore, W.J. (1972) Physical Chemistry, Prentice-Hall, Englewood Cliffs, N.J., or other textbook of physical chemistry.
4 Onsager, K. and Fuoss, R.M. (1932) J. Phys. Chem. 36, 2689.
5 Everaerts, F.M., Beckers, J.L. and Verheggen, T.P.E.M. (1976) Isotachophoresis, Elsevier, Amsterdam.
6 Tiselius, A. (1930) Nova Acta Regiae Soc. Sci. Ups. Ser. 4, 4, 7.
7 Kunkel, H.G. and Tiselius, A. (1951) J. Gen. Physiol. 35, 89.
8 Vacík, J. (1979) in Electrophoresis, a Survey of Techniques and Applications, Deyl, Z., Everaerts, F.M., Prusik, Z. and Svendsen, P.J. (eds.) Elsevier, Amsterdam 1979, p. 39.
9 Ostrowski, W. (1979) in Electrophoresis, a Survey of Techniques and Applications, Deyl, Z., Everaerts, F.M., Prusik, Z. and Svendsen, P.J. (eds.) Elsevier, Amsterdam, p. 69.
10 Mc Donald, H.J., Lappe, R.J., Marbach, E., Spitzer, R.H. and Urbin, M.C. (1955) Ionography, Electrophoresis in Stabilized Media, Chicago, IL.
11 Fidler, Z., Vacík, J., Dvořák, J. and Grubner, O. (1962) J. Chromatogr. 7, 228.
12 Macheboeuf, M., Rebeyrotte, P., Dubert, J.M. and Brunerie, M. (1953) Bull. Soc. Chim. Biol. 35, 334.
13 Waldman-Meyer, H. (1972) Biochim. Biophys. Acta 261, 148.
14 Vacík, J. (1971) Collect. Czech. Chem. Commun. 36, 1713.
15 Vacík, J. and Fidler, V. (1971) Collect. Czech. Chem. Commun. 36, 2125.
16 Vacík, J. and Fidler, Z. (1971) Collect. Czech. Chem. Commun. 36, 2342.

Deyl (ed.) Separation Methods

CHAPTER 3

Gas chromatography

MILOS NOVOTNY and DONALD WIESLER

Department of Chemistry, Indiana University, Bloomington, IN 47405, USA

3.1 Introduction

While the early investigations on gas chromatography (GC) had already taken place during the 1940's in the form of gas-solid systems, only the introduction of gas-liquid chromatography in 1952 by James and Martin [1] marked the beginning of a very fruitful period in the history of organic analytical chemistry. A few exceptions notwithstanding, it is primarily the gas-liquid (partition) mode of chromatography that has major importance in the analysis of biological compounds; adsorption coefficients of such compounds are clearly too large to permit an easy and quantitative elution from most adsorption columns at reasonable column temperatures.

Whereas the James-Martin historical paper on gas-liquid partition chromatography had involved analysis of fatty acids, most applications of the following decade were primarily outside the biochemical and biomedical areas. As the theoretical studies of that period significantly advanced understanding of the column processes and new detection techniques became gradually available, the designs of reliable commercial gas chromatographs dramatically improved in the following period. In particular, the availability of highly sensitive ionization detectors in the early 1960's set the stage for determination of trace quantities of high-boiling organic substances.

The most fundamental period for the gas-phase investigations of biological compounds, however, were the 1960's. With the development of stable stationary phases and inert solid supports, technological improvements in column design were achieved with an ever increasing pace. While introducing thin-film packed columns to GC, VandenHeuvel et al. [2] had already demonstrated in 1960 that several steroid compounds could be chromatographed without structural alterations at temperatures close to 300°C. However, publication of numerous additional studies was needed in the subsequent years to remove the 'psychological barrier' that most biological investigators have had with respect to uses of gas-phase analytical techniques.

Due to their biological function, most endogenous metabolites and other biochemically important substances possess a variety of polar groups in their molecules.

The presence of amino, carboxyl, hydroxyl, carbonyl, etc., groups in organic molecules increases their boiling points very substantially, and thus, the probability of undesirable interactions (namely, irreversible adsorption and catalytical decomposition) in the GC analytical system. Some very polar biomolecules cannot be volatilized, in a conventional way, without a loss of their chemical individuality. However, as shown repeatedly during the last two decades, suitable derivatives can frequently be prepared that are sufficiently volatile and stable for GC investigations. Even as seemingly non-volatile substances as carbohydrates, amino acids or alkaloids can now be chromatographed successfully as derivatives.

Derivatization of biological molecules itself has become a very important direction in the field of bioanalytical chemistry. In fact, today's implications have gone well beyond the original intent. With an increasing search for new reactions at the microscale, or even improvement in the yields of well-established reactions, it is hardly surprising that some derivatization techniques originally developed for GC find an increasing use in mass spectrometry, high-performance liquid chromatography, laser spectroscopy, etc.

A search for suitable derivatization methods in GC of biological compounds was initiated in the late 1950's, and it still continues today. Much has been accomplished throughout that period. Some derivatizations are now trivial and routine, while other reaction schemes proved satisfactory only in a limited number of laboratories. 'Universal reagents' are still hard to find; closest to that definition would be some silylation agents. Silylation was for the first time utilized by Langer et al. [3] for carbohydrates, and by Luukkainen et al. [4] for steroid compounds. However, the problems of different types of functional groups in one molecule frequently necessitate multiple derivative formation; amino acids, prostaglandins and steroids are typical examples of this. On the other hand, numerous quantitation studies with various reagents that were conducted over the years, have considerably reduced many initial difficulties. More potent and predictable reaction approaches are becoming gradually available. Numerous examples of this will be demonstrated throughout this chapter.

A very significant step forward in the field of bioanalytical chemistry was the development of combined gas chromatography/mass spectrometry (GC/MS). This combination is a very powerful extension of both analytical methods: while gas chromatography provides an effective resolution of the individual mixture components, identification of these separated compounds can frequently be accomplished through their characteristic 'molecular fingerprints', mass spectra. The point of importance is that these two methods have a considerable overlap in their sensitivities. Developing capabilities of modern mass spectrometers are being matched by the gradual improvements in chromatographic columns (e.g., the recent utilization of capillary columns) and sampling techniques.

While the idea of combining gas chromatographs with mass spectrometers had already occurred in several laboratories in the late 1950's, solution of the technological problems associated with interfacing the two methods was needed to stimulate general interest in this powerful analytical approach. Specifically, the development of 'molecular separators' (devices that cause removal of the carrier gas prior to mass spectrometry) was essential to progress in this area [5,6]. Much important work on GC/MS was accomplished by chemists interested in biological problems, and identification of numerous metabolically important substances during the late 1960's could not have been possible without the GC/MS combination. The subsequent development of mass fragmentography [7] was the next very significant step, as it added another dimension to an effective utilization of a mass spectrometer. Through this development, high-sensitivity quantitative measurements of selected compounds in relatively complex matrices became feasible.

Numerous additional advances in GC/MS occurred throughout the last decade. Mass spectrometers have now become almost routine, reliable instruments. Improvements in design of both sector and quadrupole instruments is today reflected in greater spectral resolution and sensitivity parameters. Versatility of the GC/MS combined instruments has been dramatically improved by better interfacing techniques and an increased use of capillary columns. The chemical ionization methods have become important for work with relatively unstable molecules; there is a significant rationale for their increasing use in biochemical research.

A decade ago, reliable acquisition and processing of mass-spectral data in GC/MS was a bottleneck of the overall procedure. An increasing use of the modern computer technology for instrument control and data treatment has been of paramount importance. Today's instruments are quite capable of recording reliably the mass spectra of individual mixture components, as well as comparing this information automatically with a library of many thousands of previously recorded spectra. 'Selective chromatograms' can further be generated through the method of mass chromatography [8].

Further advances in GC/MS are still to be expected. In addition, our capabilities of the structural elucidation of biological compounds will undoubtedly be strengthened through a wider utilization of other on-line ancillary techniques, such as GC combined with Fourier-transform IR spectroscopy.

While the capillary column was invented by Golay [9] as early as 1956, its major impact on chromatography has only been felt throughout the last decade. The license restrictions combined with inadequate column technology clearly prevented an earlier, wider use of capillary GC. With availability of glass capillary columns and the stable films of organic phases on the surface of glass [10,11], the potential of capillary GC in biochemical analysis was gradually recognized [12–15]. In 1971, A.J.P. Martin stated [16] "...I feel sure that in time capillary columns will replace packed columns. They are intrinsically capable of higher resolutions and speeds."

As most mixtures of biological origin are exceedingly complex, the role of capillary columns in biochemistry is currently beyond dispute. Thus, it is hardly surprising to see so many applications of this powerful technique rapidly emerging.

Capabilities of capillary GC extend well beyond the analysis of complex mixtures. The use of capillary columns is reflected in a more effective trace analysis, unique detection capabilities, and a more reliable GC/MS. While much important biochemical work has already been done with glass capillary columns, development of flexible and highly inert fused silica columns [17] has further been increasing the popularity of capillary GC.

The ionization detectors used in GC possess unparalleled measurement sensitivity, and are frequently the major reason for using GC in biochemical and biomedical investigations. While the flame ionization detector has been extremely useful as a sensitive, universal detector, there are numerous analytical problems where detection selectivity is required. A mass spectrometer employed in a mass-fragmentographic mode provides almost the ultimate in both selectivity and high sensitivity. However, the use of other selective devices, such as the electron capture, thermionic or flame-photometric detectors, has been steadily on the increase. Whereas some of these detectors were primarily developed to overcome the problems of sample complexity, an increasing use of capillary GC does not appear to diminish their importance. On the contrary, the columns of greater resolving power and higher degree of inertness further expand the utility of such detectors.

The most frequent use of selective detectors in biochemical analysis is associated with the formation of derivatives which are particularly suitable for such detection. Earlier investigations with the electron capture detector [18,19] after perfluorination of certain polar compounds are perhaps the best-known examples. The general approach of introducing detectable moieties into otherwise uncharacteristic (in terms of detection) molecules is now found with increasing frequency in the literature.

Reliable determinations of the compounds of interest in biological samples were frequently developed in the past due to advances in column technology and GC instrumentation. In addition, sample preparation methods are at least equally important. As our understanding of the sample complexity gradually improves, better extraction, purification, deconjugation, fractionation, etc., are developed. Various forms of liquid chromatography are usually employed for the sample purification and fractionation prior to the GC analysis; improvements in these steps are crucially dependent on advances in chromatographic packing technology. As such advances take place, more rapid, simple and reliable GC techniques emerge for biological compounds.

At this time, empirical approaches to sample isolation in biochemical GC are most common. However, numerous attempts have already been made to develop more universal sample fractionation and purification schemes. Ultimately, these will be automated, resulting in greater sampling frequency and precision.

During its existence for three decades, gas-liquid chromatography has found numerous applications to biochemical problems. The advances in GC technique discussed above have caused much progress in biochemistry of relatively small molecules that are among the important components of body fluids and tissues. It is now commonly appreciated that the discovery of new steroidal compounds, drug metabolites, unusual natural products, prostaglandins, etc., would have been considerably more difficult without GC and GC/MS. This important role of new techniques is still evident today, as exemplified by a recent discovery of lignans [20,21] in physiological fluids.

In some instances, knowledge of structure and function of important biopolymers has also been enhanced through a GC analysis of their characteristic fragments. Until recently, resolution of optical isomers [22,23] has been an unchallenged domain of capillary GC. These directions will undoubtedly be continued in the future.

During the 1960's, biochemical GC frequently had a major objective to measure, with adequate sensitivity, a limited number of compounds in a given sample. This emphasis has now substantially changed for two reasons: (a) as the resolution of GC columns has dramatically improved, most researchers find the determination of entire 'profiles' of metabolites more attractive to pursue; and (b) whenever high sensitivity is needed in routine determinations, various immunoassay methods seem preferable now.

A brief inspection of the current literature on biochemical analysis reveals that the applications of GC in this field are still on the rise. However, it also indicates that the utilization of high-performance liquid chromatography (HPLC) is becoming even greater. The relative merits of these two powerful analytical methods for biological compounds are of some concern. Whenever applicable, we shall investigate this question in the following text, while dealing with the chromatographic analysis of the individual compound classes.

This chapter will discuss the role of contemporary GC in biochemistry. Due to the very rapid proliferation of this technique into various branches of science that could be considered 'biological', it is virtually impossible to review comprehensively all applications. Thus, considerable emphasis will be placed on the most important developments of the last decade, stressing the principles and fundamental directions. The major classes of metabolically important compounds will be treated individually, while the most useful applications will be referred to. In general, man-made chemicals (such as pesticides, drugs and their metabolites) have been excluded from this discussion.

3.2 Modern instrumentation of gas chromatography

3.2.1 General considerations

Technological achievements of the last decade have had a profound effect on design features of contemporary gas chromatographs. Besides advances in the most important part of these systems (the gas-chromatographic column itself) such technological progress has affected each vital part of these instruments: sampling and detection systems, control of temperature and carrier-gas flow, and signal acquisition and processing. In particular, progress in microelectronics and computer technology now provides numerous new possibilities for reliable instrument control and signal handling. Microprocessor-controlled gas chromatographs have become the common items in the instrument industry. Associated signal-processing and data-handling systems have become considerably 'smarter' over the last several years, and it is fair to say that their general capabilities are frequently under-utilized by an average user.

As the demands for method automation and sophisticated data conversion gradually increase, the general capabilities of these new devices will be met.

For a great number of years, the GC column was a major limitation to the utilization of the instrumental capabilities of this method. For example, migration of the unstable column products (so-called 'bleeding') into a GC detector during the analysis seriously impairs any quantitative investigations, and limits highly sensitive detection and signal amplification alike. Thus, reliable GC analyses become feasible only when stable columns are employed. Similarly, highly adsorptive columns with significant 'thresholds' for sensitive compounds seriously limit the capabilities of high-sensitivity detectors for trace analysis, and in addition they frequently contribute to the lack of retention reproducibility. Thus, it is not surprising that the modern technology of capillary GC columns has resulted in some very significant improvements concerning the above points.

The chromatographic theories developed in the earlier years of GC were essential to predictions of column performance under various circumstances that may occur during the sample analysis. Verification of these predictions was only possible while using the carefully designed instruments. Such instruments are permitted to contribute only a minimum amount of the overall chromatographic band dispersion; resolution of high-efficiency columns must not be negated by a poor design of an injection port, detector cell, heating units, etc.

Obviously, designs of capillary gas chromatographs must be more carefully executed than those of packed column instruments. The chief reasons for this are the very low flow-rates used and the overall small volumes of capillary columns. Under such circumstances, the units connecting the column to either the inlet or detector parts must virtually be absent of any dead volumes. Inlet systems with 'clean geometry' are also required to introduce the sample as the narrowest possible band into the first column section. A constant dilemma of the manufacturers of modern instruments has been whether to design 'universal instruments' or those usable just for certain column types. It seems now that the production of 'dedicated' capillary instruments is becoming common. Alternatively, instruments can be provided with multiple inlet and detector capabilities. Numerous laboratories also successfully modified the earlier versions of instruments into capillary gas chromatographs.

Numerous experts in the field had already predicted proliferation of special-purpose GC instruments some time ago. While unique conditions of certain analyses would seem to support such a trend, GC instruments dedicated to a particular type of determination are still very rare.

As the workload in routine laboratories increases, automation of GC determinations becomes mandatory. Automated sampling devices are now commercially available that permit GC analyses on a repetitive basis: at an appropriate time, the sample is injected into the system, temperature program is initiated, automatically changed, terminated, etc., while the system again recycles itself into the initial analysis conditions for the following sample. Most of the autosamplers are based on a pneumatically operated syringe, and provide for an intermediate syringe wash to prevent a carry-over from one sample to another. Besides the time-saving factor, it is widely known that these automatic injection devices usually improve sampling reproducibility as compared to the manual injection procedure.

The above considerations are not necessarily specific for biochemical GC. There are additional considerations of the inertness of a GC system when dealing with relatively unstable biological compounds. Since long ago, most investigators have preferred working with all-glass systems and deactivated solid supports. Likewise, contemporary capillary GC uses exclusively glass and fused-silica capillaries rather than metal columns. Reliable glass-to-metal seals are now available as various polyimide or graphite ferrules that can withstand temperatures over 350°C. Availability of various thermally stable materials has also decreased sample contamination and reduced the occurrence of spurious peaks and baseline instability in high-temperature GC.

3.2.2 Operating conditions

In any chromatographic separation problem, a proper choice of operating conditions is essential. These include primarily the choice of carrier gas and its flow-rate, temperatures of the injection port and the detector, as well as the column itself. For difficult separations, a proper temperature programming rate must also be adjusted.

A carrier gas of low viscosity is preferred for high-efficiency and fast analyses. Although price considerations would favor gases like nitrogen and argon, helium or hydrogen provide considerably better results, as demonstrated in Fig. 3.1, in which the plate-height versus average gas velocity curves have been plotted for the same solute (an ester of C_{18} fatty acid) and temperature, but four different carrier gases [24]. These dramatic differences are caused by the variations in the solute's diffusion coefficient in different carrier gases; both the optimum gas velocities and the slopes

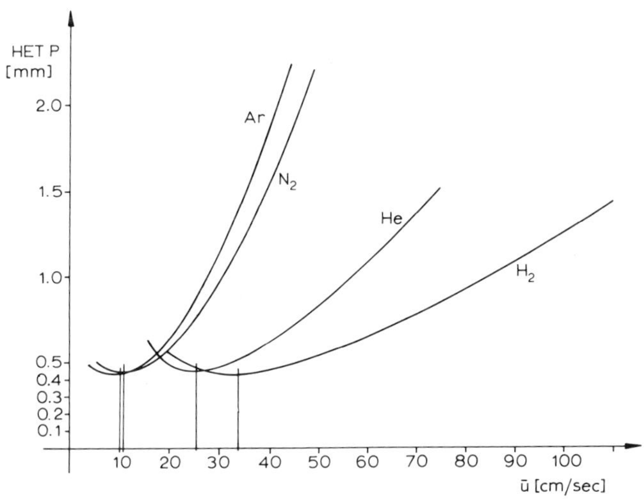

Fig. 3.1. Plate-height versus linear velocity plots for the same solute, with four different carrier gases. Reproduced from reference [24]. HETP, height equivalent to a theoretical plate.

of the corresponding van Deemter curves are affected. The situation is most favorable for hydrogen. While hydrogen has been frequently avoided in the past because of the explosion hazards, many investigators now use it with capillary columns where the flow rates are very low and the hazard considerably less significant. Some of the most impressive capillary chromatograms demonstrated in the literature were obtained with hydrogen as a carrier gas.

The choice of carrier gas is sometimes dependent on the detector in use, as is the case, for example, with the thermal conductivity cell, the electron capture detector, or in GC/MS combination. For an optimum performance of certain detectors (dependent on either the principle or a particular design), an extra gas must be added at some point between the column end and the point of detection. Such 'make-up gas' arrangements are particularly common in capillary GC.

Purity of the carrier gas is very important in modern GC equipment designated for trace analysis. Consequently, it is essential that the gas purifiers, such as the traps containing various adsorbents, be inserted in the gas line before the injection port. The same requirement usually applies for purification of the combustion gases for the flame ionization detector. The role of these adsorbent traps is to remove even the trace quantities of water, oxygen and organic impurities present in commercial gas cylinders, and thus minimize both the system contamination and chemical alteration of an injected sample.

The flow rates used in GC analyses are determined by the column type; while conventional 2 mm, i.d., packed columns may have carrier-gas flow rates between 50 and 100 ml/min, a typical flow for capillary columns is 1 ml/min or less. Wide-bore (0.5–0.7 mm, i.d.) capillaries or the support-coated open tubular columns need intermediate flow rates, corresponding to their inner diameters. During temperature-programmed runs, the flow rates decrease proportionally to increased gas viscosities if the columns are operated at a constant pressure. To overcome this inconvenience, otherwise incurring departures from the optimum gas velocity and decreased speed of analysis, flow controllers are usually installed into the gas chromatographs. While the flow-controlling devices are easy to use for typical packed column flow settings, very low flow rates associated with capillary GC work are harder to control. Thus, the pressure control rather than the flow control is commonly used in capillary GC; owing to the high capillary column permeabilities, the flow changes due to viscosity increases at higher temperatures are not so dramatic.

Proper flow adjustments for any detector-related gases are essential to quantitative GC, as both fluctuations or long-term drifts will effect performance of both concentration and mass-flow sensitive detectors. For example, it was shown [25] that even the fluctuations in the atmospheric pressure could cause some deviations in peak areas with the flame ionization detector. Yet another systematic study of detection variables [26] reinforces the importance of instrumental control with the flame detectors. Wherever highly quantitative results are expected, frequent calibrations with appropriate standards are urgent.

Temperature programming is essential for maximizing the resolution of complex mixtures and other hard-to-separate components. Reproducibility of temperature programs is determined by both the quality of oven design and temperature-controlling devices. In contemporary instruments, mechanically-actuated temperature programmers have been replaced by electronic devices; this has resulted in greater accuracy and precision of temperature control and reproducibility of the measured retention parameters. The theory of temperature-programmed GC with regard to retention has been described extensively [27,28].

Flow programming has been suggested [29,30] as a viable alternative to temperature programming. Its chief advantage is in eluting later mixture components at relatively low temperatures, while the thermal decomposition of a stationary phase and the subsequent 'bleeding' can be avoided. Both linear and non-linear flow programs can be accomplished through an appropriate manipulation of the column inlet pressure. While some applications of flow programming, or a combined flow/temperature programming, are demonstrated in the literature, such devices are relatively uncommon. The main disadvantage of flow programming is an increase of the column's resistance to mass transfer at higher carrier gas velocities, and the consequent loss of separation efficiency.

3.2.3 Multiple-column systems

Complexity is a common feature of biological mixtures. Numerous components of biological samples may be spread over a wide boiling-point range. Yet, the compounds of interest are typically present in a dilute form, in either a large amount of a solvent or a derivatization agent. With a limited sample work-up, the solvents, reagents, impurities, etc., are all introduced into the GC-system, volatilized, and sent through a chromatographic column into the detector. In such instances, the less volatile components of a biological extract deposit in the injection port or the first section of the GC column.

The above circumstances may frequently lead to undesirable consequences that are well known to numerous scientists in biology and medicine. These primarily include decreased column lifetimes, formation of artifact peaks, sample decomposition, and impaired detection capabilities. As pointed out by Horning et al. [31], such difficulties lead to two divergent opinions about GC methods: (a) that GC is primarily a way to introduce a very limited number of purified components into the detector for a high-sensitivity measurement; and (b) that the sample should ideally be used in a relatively non-fractionated form, while the column should separate as many components as possible.

Both views have their own individual problems. Extensive sample purifications may result in uncontrolled sample losses together with unreasonably tedious and time-consuming procedures for routine analysis. In the second case, chromatography of relatively crude mixtures often leads to a decreased reliability of multicomponent analyses on a repetitive basis. If there is a generally acceptable 'middle course', GC will become considerably more popular in biochemical investigations than it has been thus far. In order to systematically approach the problems, the contemporary goals and uses of biochemical GC should briefly be re-examined.

Certain attractive features of a highly sensitive detection of single biological compounds by the electron capture detector or other selective detectors have lately been somewhat decreased by the availability of competitive methods (e.g., radioimmunoassay, HPLC/electrochemical detection or mass fragmentography) which require considerably less sample clean-up. In other instances, the multicomponent analytical approaches have been increasingly emphasized, where the solution of the above-mentioned problems appear eminently worthwhile. A combination of chemical and technological improvements should ultimately lead to vastly improved GC analytical capabilities. The 'chemical approaches' consist of designing simple and highly reproducible fractionation schemes for various classes of biological compounds. The 'technological approaches' advocate direct sampling of biological materials or their crude extracts into an analytical system where the interfering substances will be removed on-line. Various multiple-column systems, discussed in the following pages, are largely a result of the latter philosophy of sample treatment.

Multiple-column systems were previously explored in the petroleum industry and some process-control situations. In the former case, typical petrochemical samples share some similarities with biochemical samples in terms of complexity; while the GC column typically receives a total sample, only certain portions of it may be of interest. Thus, selected parts of a column effluent can be pneumatically switched over to a second column for an optimum analysis, while the residual 'uninteresting' substances (heavy ends) are being rapidly removed through backflushing. In the case of process GC analysis, such backflushing is essential to the speed of analysis required from these industrial analyzers; indeed, a similar situation is often found in a clinical laboratory.

Additional uses of multiple-column systems are nearly unlimited in their analytical scope. They permit effective measurements of trace components in the immediate vicinity of major components of mixtures through a 'heart-cutting' approach [32]. Similarly, excessive amounts of reagents or solvents (including water) can effectively be removed from the compounds of interest while using a double-column arrangement. Through the use of columns with different physical or chemical characteristics, selected sample components can be effectively resolved from each other. A term 'multidimensional chromatography' is frequently used to describe such techniques.

Various multiple-column systems have been widely described in the chromatography literature. In terms of the principles described, little can be added today to the column systems described earlier [33]. Nevertheless, numerous technical problems and the general pneumatic complexity of multicolumn systems have limited both commercialization and practical uses. While the need for such systems in biomedical GC has been stressed [31], the general instability and relatively high boiling points of biological molecules put extraordinarily high demands on the current GC technology. A few examples have been demonstrated in the literature that illustrate both the current capabilities and potential of multicolumn approaches. These applications may involve various combinations of packed and capillary columns.

An interesting approach to the simplification of chromatographic compound profiles is the use of a subtraction column. Such a column, containing a highly selective material, precedes the analytical separation column in a serial arrangement. When a complex sample (containing substances with various functionalities) passes through the first column, the compounds with selective reactivity will be adsorbed or significantly retarded on such a column, while the remaining components will pass unhindered. Selective retention of *n*-hydrocarbons on a molecular sieve or olefins on silver nitrate-coated supports are now considered classical examples of this behavior. More recently, Picker and Sievers [34] used a short column loaded with a europium (III) coordination polymer sorbent to trap selectively various oxygenated compounds from a complex mixture of urinary volatile metabolites.

Double-column GC systems, utilizing a short packed column and an analytical capillary column with an intermediate trapping system, have been used in a variety of situations: (a) sampling and enrichment of the trace quantities of analytes from a dilute solution; (b) removal of solvents and derivatization reagents for improving performance of capillary columns and certain GC detectors; (c) direct injection of aqueous samples with subsequent removal of the water peak; and (d) backflushing applications. The general pneumatic arrangement of a typical system, as described by Schomburg et al. [35] is shown in Fig. 3.2. The trapping device situated between the packed column and the capillary (analytical) column served to minimize band spreading in the capillary GC stage.

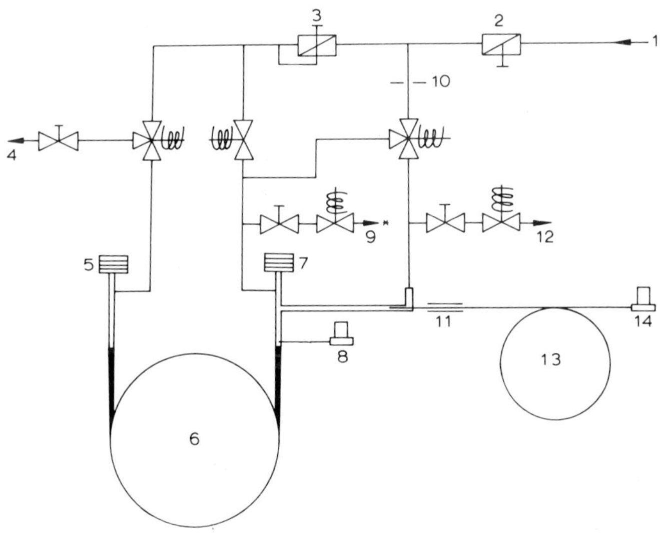

Fig. 3.2. System for double-column chromatography with intermediate trapping and re-injection, suitable also for direct injection of aqueous solutions. 1, carrier gas; 2, pressure regulator; 3, flow controller; 4, vent for back-flushing; 5, injection port for heart-cut and back-flushing; 6, precolumn (packed); 7, injection port for aqueous solutions; 8, control flame ionization detector for pre-separation; 9, vent for cutting; 10, leak for make-up gas; 11, trap; 12, outlet of splitter; 13, glass capillary column; 14, flame ionization detector for main separation. Reproduced from [35].

An example [35] of reagent removal is shown in Fig. 3.3; a cut was made from packed-column chromatogram A in order to remove an excess (more than 95%) of silylating agent and some later eluting peaks. Chromatogram B shows separation of the trapped components on a capillary column, while chromatogram C displays the whole mixture under the high-resolution condition. An analytical system similar to that shown in Fig. 3.2 was used in this application; the sample under investigation contained mainly silylated isomeric sugars. While the component resolution shown in chromatogram C would have been adequate for analysis, removal of the silylation agent was needed to protect the capillary column. Additional analytically interesting applications of multiple-column systems exist. If two analytical columns are connected in parallel, a sample injected and equally split into two portions can be analyzed on the columns with two stationary phases of different selectivities. This is currently feasible even with the capillary columns. Observations of retention increments under such conditions could be of value in qualitative studies. Serially coupled columns used at different temperatures [36] may be of a similar advantage.

A successful utilization of various multicolumn systems will be strongly dependent on various technological and engineering advances in equipment design. Increasing utilization of highly inert capillary columns necessitates reliable and inert interfacing parts to preserve the current performance of such columns. While inert valves operating reliably at sufficiently high temperatures have been difficult to manufacture, the pneumatic column-switching techniques pioneered by Deans [32] may not be straightforward to use under routine circumstances. For example, very few quantitative aspects of multiple-column systems have been investigated thus far.

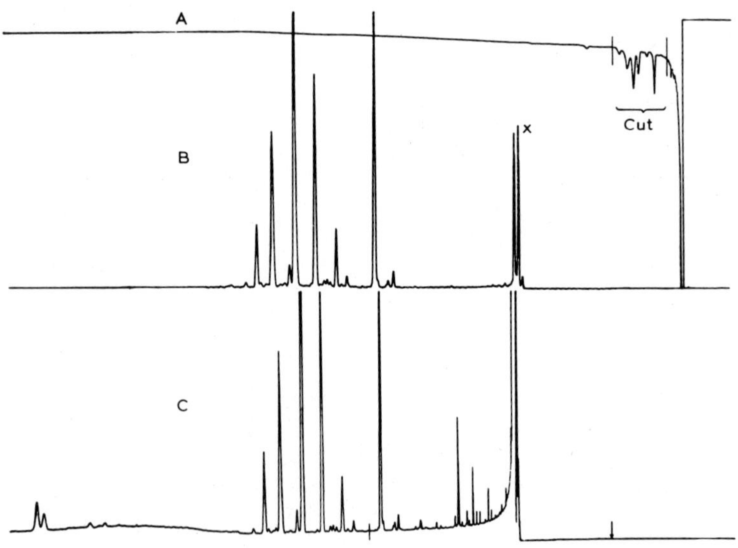

Fig. 3.3. Removal of excess reagent through a double-column arrangement. See text for explanation. Reproduced from [35].

The current techniques in multidimensional chromatography have recently been reviewed [37,38].

Multicolumn GC techniques can undoubtedly be of advantage in numerous analytical situations. Applications of these techniques in biochemical analysis have been thus far limited. Although some commercial units of this kind appeared recently, most investigators work with conventional modified instruments.

3.2.4 Sampling systems

Sample introduction is undoubtedly the most critical point in the GC analysis. A large band dispersion occurring at the point of sampling cannot be compensated for by a chromatographic column. While the volumes of injection ports are not so critical in work with packed columns, design of an inlet system for capillary GC techniques is considerably more crucial. Design considerations must include both the overall volume and flow geometry of such inlet systems. Inadequacies in these directions may negate the column performance and quality of the overall GC analysis.

Inertness of the inlet systems is yet another very important consideration in the gas-phase separations of relatively labile biological compounds. Long ago [39], decomposition of steroid compounds was observed in the metal inlet system; consequently, all modern instruments feature all-glass sampling systems. In addition, the mode of sample vaporization, contact with the surfaces of an injector and the rubber septum may all negatively influence analytical results. Extent of compound decomposition and irreversible adsorption immediately subsequent to sample introduction can be particularly severe when analyzing subnanogram quantities. With increasing demands for trace analytical work, the technology of sampling has significantly advanced over the last decade. The very high performance of the present day's capillary columns has particularly contributed to the need for well-designed (and often sophisticated) inlet systems, so that such columns can be taken advantage of in analytical work. For these reasons, the following discussion will primarily be centered around the capillary GC inlet system.

Most biochemically important GC determinations are typical trace analysis problems: small amounts of determined compounds are encountered in relatively large volumes of solvents used for their extraction from physiological fluids or tissue material. In addition, various derivatization agents often serve as media for sample prior to their injection into a gas chromatograph. In order to comply with the sensitivity requirements of various GC detectors, samples must frequently be concentrated into smaller volumes. An alternative — injection of large samples — has very negative effects on the chromatographic process and detection. Large amounts of some solvents or derivatization agents drench the first part of a chromatographic column, uncovering adsorptive sites and decreasing the column lifetime. The column instability may also manifest itself in migration of a stripped stationary phase into the detector, and the resulting drift or contamination. Passage of large solvent peaks through certain detectors (e.g., various selective flame detectors) may cause complications associated with a loss of signal and/or increase of electrical noise. Such problems may be unique to different detector types. For example, it has been a common experience that deposits of silicon oxide on the flame detector electrodes, resulting from decomposition of silylation agents and/or silicone liquid phase bleeding into the flame, decrease the detector signal. In a different example, sensitivity of the photo-ionization detector may be decreased due to a film deposition on the optical window originating from the column bleed; chemically-bonded stationary phases are highly recommended for this type of detection. In yet another case, such a film could result in a reduced formation of electrons within the electron capture detector, or for that matter, any other ionization detector based on the presence of a radioactive source.

Many of the above biochemical GC detection problems can be solved by: (a) sample concentration through solvent evaporation prior to injection; (b) an alternative use of detectors that are relatively insensitive to the used solvents; (c) a solid sampling procedure; or (d) using a concentration precolumn. Naturally, some of these approaches can be combined to achieve optimum results.

Solvent removal through evaporation is feasible only for compounds of limited volatility. Most compounds of biological interest as well as their more volatile derivatives qualify for this, provided that volatile solvents are used (hexane, dichloromethane, ether, carbon disulfide, etc.). While concentrations down to 50–100 μl are feasible with miniature concentric vials, the volumes of samples injected into a gas chromatograph should be no more than a few microliters. In cases of the trace determinations, this is not a satisfactory practice, as it would be more sensible to use a greater sample aliquot. As discussed below, several remedies to this general problem have been suggested.

While concentrating the compounds of analytical interest, contaminants are unfortunately concentrated as well. Thus, it is essential to use very pure solvents in order to avoid high blanks in trace analysis. Extensive procedures for solvent purification have frequently been described in pertinent articles. Obviously, some solvents are easier to purify than others. Various steps in a sample work-up and the overall number of manipulations will necessarily increase the possibility of contamination. Various phthalates and other plasticizers are among the ubiquitous contaminants that are well-known to many analysts. Even contact of solution with human skin may result in interfering peaks [40].

Extraneous peaks may also originate from various parts of a gas chromatograph constructed from polymeric materials, such as regulating valve membranes, seals, septa, etc. Other sources of possible contamination include trace impurities in the carrier gas and insufficiently removed lubricants on the inner wall of connecting tubes. Purcell et al. [41] studied various sources of extraneous peaks and suggested methods of removal. Gas-sealing technology has improved considerably over the years: high-temperature silicone septa and stable ferrules made from various materials (e.g., polyimides or graphite) are now readily available.

Syringe injection into a hot inlet system remains the most common way of sample introduction in GC analysis. It is, however, the most frequent source of error in quantitative analysis. While the contemporary microsyringes measure quite accurately microliter or submicroliter volumes of liquids, a quantitative transfer into the gas chromatograph, i.e., injection from the cold state into the thermal zone is a quite disturbing matter. As shown with different injection techniques, discrimination in the hot tip of a syringe needle can occur due to a fractional distillation of samples [42]. The analyst's skill, type of syringe, inlet temperature, etc., can all influence quantitation. These problems are hardly even compensated for by the judicious use of appropriate internal standards. Some remedies to this general problem such as the use of 'dry sampling' techniques or on-column cool injection, are described below. Furthermore, pneumatic alternatives to the syringe injections appear to be worth developing in future studies.

Solid or 'dry sampling' procedures have frequently been sought in biochemical GC analysis for the numerous reasons discussed above. In such procedures, the measured sample solution is brought into contact with the surface of glass wool, metal gauze, capsule or a glass tube [e.g., 43–47], while a more volatile solvent is evaporated at appropriate temperature, and the high-boiling sample components are finally desorbed thermally from the matrix into the column. Various versions of this principle permit automation.

The two most common automated solid sampling procedures will be described here, although additional versions may also be feasible. With the devices described by Menini and Norymberski [46] and Tinti [47], the samples are placed on a rotating tray with a certain number of compartments that are sequentially allowed to fit the inlet opening of a gas chromatograph. The samples are introduced onto the column space by either gravity, or through magnetic manipulation. Several versions of this principle have become commercially available over the years.

A capsule autosampling device was originally described by Otte and Jentzsch [48]. The sample to be analyzed is deposited inside a metal capsule with conventional syringe. After gently evaporating the solvent, the capsules are cold-welded shut and sequentially loaded into the auto-sampler magazine. At an appropriate time, a capsule is introduced into the hot injector zone and pierced by a sharp thorn. The vaporized sample is subsequently purged from the capsule onto a chromatographic column for a determined period of time. Finally, the used capsule is rejected, while the inlet is re-sealed and prepared for the next analysis.

The solid sampling device described above offers advantages over the conventional syringe approach because of the easy solvent removal and increased precision. Due to the relatively large volumes associated with these sampling techniques, work with capillary columns demands peak compression following the thermal desorption process. As shown in the literature [49], this step is easily accomplished through cryogenic trapping of the vaporized sample at the capillary inlet.

Possibilities of sample losses on the surface of matrices used in solid sampling are of some concern. While such phenomena were not observed in work with urinary volatiles [49] and steroids [50], Lines et al. [51] found some response non-linearity with pesticide samples that appeared to originate from the excessive sample retention on the capsule wall. Silylation of the capsule material visibly improved quantitation. Chemical composition of the capsule material could be crucial in some applications. More recently, an all-glass version of the solid autosampler was reported by Miller and Bertsch [52].

A solid sampling device that has lately been popular with biomedical scientists is the so-called 'falling needle injector'. This device was originally described by van den Berg and Cox in 1972 [53]; it is primarily intended for sampling into capillary columns. Fig. 3.4 shows the falling needle injector in two different positions: sample loading, and its introduction into the column. A magnetic manipulator controls the position of a 'needle' (a thin glass rod attached to an iron plunger) within the

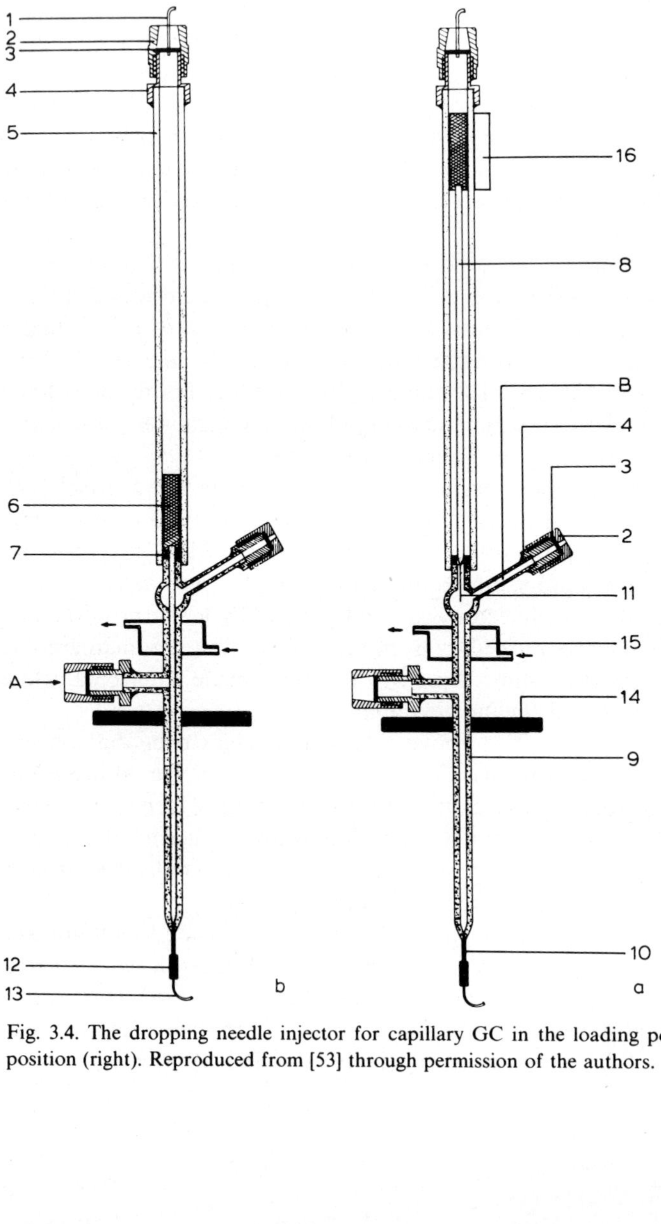

Fig. 3.4. The dropping needle injector for capillary GC in the loading position (left) and the injecting position (right). Reproduced from [53] through permission of the authors.

all-glass system. In the sample loading mode, the needle is raised to the upper part, while the sample is slowly (and, if necessary, repeatedly) deposited on the tip. The pneumatics of the system is designed in such a way that the carrier gas flowing slowly around the needle tip evaporates the low boiling solvent, while the needle is in the upper position. A subsequent sudden lowering of the needle into the heated zone of the injector results in sample vaporization, while the needle is effectively flushed with the carrier gas. Verzele et al. [54] have recently assessed the relative standard deviation associated with the repeated injections of 'difficult' nitrogen-containing compounds to be less than 1% for this type of sampling device.

The solvent removal problem can also be solved through the use of a packed precolumn. In this mode of operation, a dilute sample is introduced into a small column containing enough packing to separate the volatile solvent (or a derivatization agent) from the sample of interest; at sufficiently low temperatures, the high-boiling sample components are retained on this packing, whereas the solvent is removed through a valve system in a way similar to the 'heart-cutting' technique described in the previous section. A subsequent heating of the precolumn (with the valve closed) will desorb sample into the analytical column.

Although the precolumn devices were described quite early [55–58], their more frequent utilization is relatively recent. With a proper design, they can also be employed in capillary GC. Here, the precolumn systems often assume an additional role through acting as 'sample scrubbers'. It is a common experience in work with biological materials that non-volatile materials deposit in the sample ports and columns themselves during the injection process. Frequent injections of such materials may contaminate the system to such a degree that the following analyses become unreliable for a variety of reasons (contamination peaks, baseline drift, adsorption of small samples at the 'dirty inlet', etc.). The extent of these undesirable effects is strongly dependent on a sample type, extent of purification, a type of column, etc. While using the common on-column injection with glass packed columns, the analysts need to occasionally replace a length of contaminated packing and/or the silylated glass wool with fresh materials. Similarly, the first section of a capillary column is periodically cut off after a number of analyses in order to regain the original performance. However, if a precolumn is used in front of an analytical column, these problems can be minimized or eliminated altogether; precolumns can be treated as disposable items.

Due to the rapidly growing importance of capillary columns in bioanalytical applications, special attention will now be devoted to sampling techniques associated with capillary GC. Small samples are typical for this type of chromatography and, consequently, a direct introduction of such samples is an apparent technological problem. In most biochemically interesting applications (typically, trace analysis problems), there is no general discrepancy between the demands of such analysis and the performance and sensitivity of capillary separation techniques. However, the manipulation of samples presents difficulties, as reliable methods for measurement, disposal, and introduction of nanoliter volumes are not readily available. Ironically, in many capillary GC applications, the solvent serves only as a 'sample vehicle'; we measure microliter volumes of the analyzed solutions and inject them into a capillary gas chromatograph, while subsequently trying to minimize or eliminate the solvent's negative influence on the column and detection processes.

Historically, indirect sampling approaches have been used since the very earliest stages of capillary GC. In order to provide the detector with easily measurable solute quantities without overloading the capillary column, the injected sample is dynamically split. Here, a vast majority of this sample is vented outside the system, while only a small fraction (typically, less than 1%) of the total sample is allowed to enter the column's inlet. A particular design of a splitting injector and its geometry can critically influence quantitation. As summarized by Schomburg et al. [59], the requirements for sample homogenization within the carrier gas immediately after the sample introduction are very strict; some splitter designs lead to very poor reproducibility.

Only a few biochemical applications are currently reported in the literature that use splitting injectors. The main reason for this is undoubtedly the fact that splitting is highly undesirable in the situations where the total amount of available sample is severely limited (i.e., in most trace analysis problems); relatively concentrated solutions must be used with this sampling technique. In addition, the splitters are believed to be non-quantitative with respect to sample components of different boiling points (the so-called 'splitter discrimination'). Finally, the requirements of an instantaneous vaporization with this injector type necessitates very high inlet temperatures for typical derivatives of biological compounds; legitimate concerns about sample stability may thus arise. In spite of these drawbacks, some workers still prefer splitting injectors, because they easily produce narrow injection bands. For example, a device with low splitting ratios, designed by German and Horning [60], was successfully used in the analysis of steroids and other biologically important compounds.

Since most GC biochemical applications relate to trace components, techniques for direct sampling onto capillary columns have frequently been under investigation. Initially, Grob and Grob [61] reported a simple direct (splitless) injection technique in which the less volatile trace components of a sample are condensed at the capillary column inlet owing to a relatively low column temperature. While minimum band spreading occurs with the less volatile components, the large peak of a more volatile solvent is permitted to pass through the column with little retention. The entire sample can be utilized following such a procedure.

Whereas a general description of the splitless method appears trivial, there are several technical aspects of this sampling mode which can be quite critical. As detailed in subsequent publications [62,63] on the subject, the column temperature, boiling point of the used solvent, sample dilution, etc., can all be extremely critical in achieving success with this injection method. The underlying mechanism for this is the 'solvent effect' [62], under which a partial solvent condensation occurs just outside the hot injector, while the trace components are being effectively purged from the hot injector and concentrated at the tail part of the large solvent zone. This temporarily enhanced sorption capacity (an increased film thickness) at the column inlet improves symmetry and quantitation of the following peaks.

A number of biochemical and environmental applications of the splitless injection technique can now be found throughout the literature. An example of the solvent effect is demonstrated in Fig. 3.5 on a chromatographic profile of silylated substances from the ultrafiltration serum of a uremic patient [64]; a successful implementation of the solvent effect is usually documented by the square appearance of the solvent peak, as seen in this figure. An additional advantage of the splitless injection is that high inlet temperatures are not required, as desorption from the injector can be a relatively slow process.

A major disadvantage of the splitless injection is that some solvents may have a profound negative effect on the column lifetime. Among them are the common derivatization agents that are used so extensively in biochemical applications. Consequently, various versions of the above-mentioned solid sampling techniques will undoubtedly find a more increased utilization. A key detail in the adaptation of such techniques to capillary GC is 'thermal focusing' of injected samples at the very first part of the capillary column. Obviously, a cool zone must be employed with such small columns to trap ('reinject') the sample components that are purged from the hot injector at relatively slow rates; otherwise, considerable band broadening and a loss of column efficiency would occur. Sample focusing at room temperature is usually satisfactory for most heavier components.

Various precolumn concentration procedures used in conjunction with thermal focusing are desirable for biochemical applications. Novotny and Farlow [65] developed a simple technique, where an off-line injection of a relatively large dilute sample onto a small precolumn results in effective concentration. In this procedure, a small volume of deactivated solid support is packed into the glass liner of an injection system. After the volatile solvent is removed, the liner is quickly introduced into the injection port, and the desorbed sample is trapped for several minutes in the cool column. Chromatograms obtained through the following temperature program-

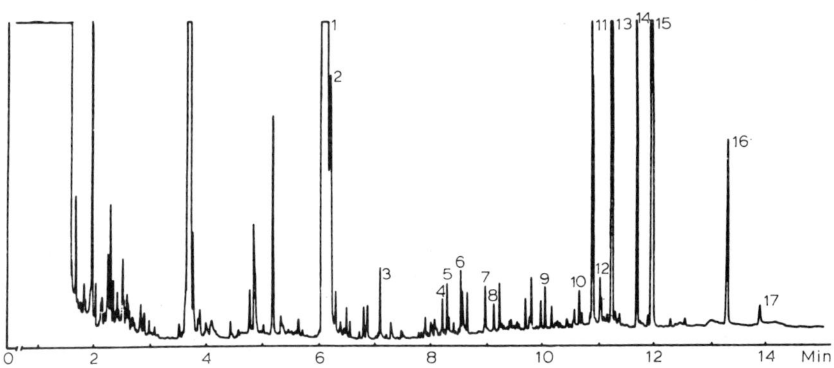

Fig. 3.5. A capillary GC profile of the silylated components from a uremic serum, using the splitless injection technique. Reproduced from [64] with permission of the Institute for Chromatography, Bad Dürkheim, F.R.G.

ming demonstrate almost a complete removal of the solvent zone, while quantitation of the later components is virtually unaffected. Such sample concentration/injection methods are particularly suitable in determination of trace components from relatively dilute media (blood samples, cerebrospinal fluid, amniotic fluid, etc.).

Additional solvent removal techniques have been described in conjunction with capillary GC of biological compounds. Besides the above-mentioned falling needle injectors, deLeeuw et al. [66] reported coating silylated hydroxy acids from a dilute solution onto the surface of a ferromagnetic conductor. After its placement inside the GC injector, this material can be rapidly heated (in a matter of milliseconds) in a high frequency field, desorbing the sample into a GC column. These authors claimed better than 1% injection reproducibility and no visible sample degradation while using this sampling method. However, need for such a rapid volatilization is questionable in view of the results obtained by Vogt et al. [67] through a conventional precolumn heating. In the latter system, solvent removal from up to 250 μl was accomplished in an on-line injector arrangement.

It is felt that the precolumn sampling techniques deserve much attention in future studies, as they can serve a double function in biochemical investigations: (a) removal of solvents or derivatization agents; and (b) protection of the analytical column from non-volatile impurities. Chemical nature of the precolumn packing can also be varied to suit a particular sample type. Further investigations aiming at the optimization and automation of the precolumn sampling techniques appear desirable.

An important development of the last several years is the design of on-column injectors for capillary GC [42,68,69]. Naturally, a direct sample deposition on the top of a packed column has been practiced for many years to minimize chances of thermal decomposition at the time of sample vaporization. To achieve a similar objective with capillary columns, some technological problems had to be overcome. Specifically, a very thin syringe needle is needed for a direct deposition of the liquid sample inside capillary columns with typical inner diameters in the range of 0.2–0.4 mm. Such thin, long needles cannot reliably be guided through a rubber septum into the column. Thus, septumless injectors have been designed in which the needle is guided through a valve into a length of the column; a version of this injector can be seen in Fig. 3.6. While depositing the liquid sample, the cooling zones (as shown in this figure) are supposed to insulate thermally the sampling process from the oven temperature, as well as to ascertain a quantitative sample transfer. The factors affecting sample deposition have been examined in some detail [42]. This sampling procedure is obviously very technique oriented and cannot be readily automated for routine work.

Submicroliter samples are needed for the capillary on-column injection technique. To counter a general scepticism following the development of this injection technique, excellent quantitative results have been reported [42]. The major advantages of this injection procedure seem to be in minimizing thermal decomposition of labile compounds as well as the lack of sample discrimination toward the later eluting components. This latter problem, observed frequently with the vaporizing injector

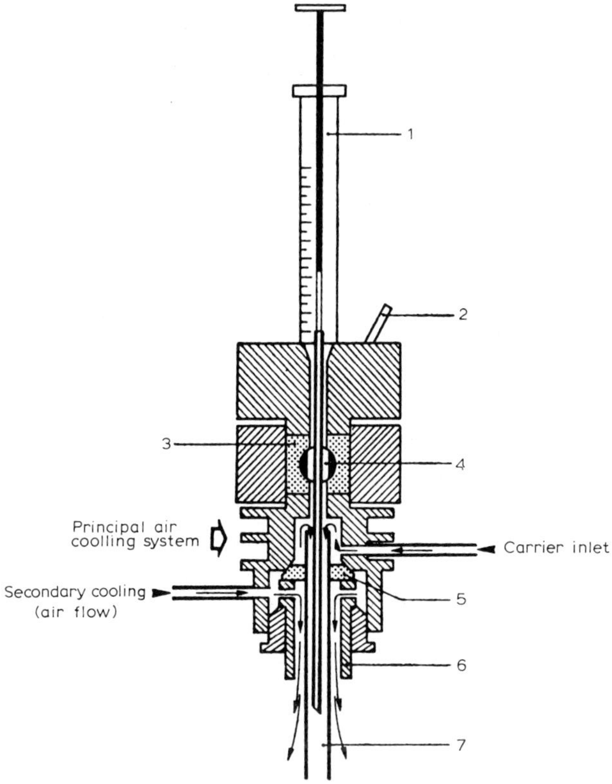

Fig. 3.6. A capillary on-column injector, including the secondary cooling system. 1, microsyringe; 2, valve lever; 3, valve seal; 4, stainless-steel rotating valve; 5, column seal; 6, cooling jacket; 7, capillary column. Reproduced from [42].

types, has been greatly reduced with the on-column injection. It readily appears that such sample discrimination has its origin in both the non-quantitative transfer from a hot syringe needle, and the sample contact with adsorptive surfaces of the vaporizing injectors.

The major disadvantages of the on-column injection approach are associated with drenching the column inlet directly with the introduced liquids, and a repeated deposition of non-volatile impurities. As a solution to the former problem, the development of immobilized stationary phases appears appropriate. However, removal of non-volatile sample impurities requires more judicious sample clean-up, a potential source of compound losses that may well counterbalance the advantages of the sampling method. While technical improvements are still needed, the on-column injection in capillary GC has undeniable advantages for the analyses of biological materials.

Sample introduction techniques, in general, remain the least developed part of the GC systems, in spite of many efforts to improve them over the last two decades.

Naturally, the overall successful quantitative analysis is a multifaceted problem of sample choice, quantitative recovery during extractions and purification steps, choice of internal standards, signal recording technology, etc. Most of these topics are beyond the scope of this chapter, as they are not deemed to be particularly characteristic of biochemical GC. However, sample treatment prior to its introduction into a gas chromatograph will be the subject of a later discussion.

3.3 Chromatographic columns

3.3.1 Phase systems

Most GC separations of biological compounds necessitate relatively high temperatures. Thus, the general capabilities of chromatographic columns that operate at temperatures up to about 350°C have been continuously sought. An increasing use of glass and fused silica capillary columns as well as the availability of relatively inert solid supports have now considerably increased the range of analyzed compounds. On the other hand, these developments in high-temperature GC have been somewhat less convincing during recent years, as more investigators gradually switch to the other viable alternative, analytical HPLC.

Since typical biological mixtures are exceedingly complex, adequate chromatographic resolution is imperative for both identification and quantitation purposes. Improved resolution is feasible through either increasing the column efficiency (number of theoretical plates), or phase selectivity. Alternatively, a combination of both can be practiced. The number of theoretical plates, N_{req}, required for adequate resolution of two adjacent peaks (98% separation of the peak areas) is related to the column selectivity (relative retention, α) and to the capacity ratio, k, according to the well-known equation derived by Purnell [70]:

$$N_{req} = 16\left(\frac{\alpha}{\alpha - 1}\right)^2\left(\frac{k+1}{k}\right)^2$$

It has been obvious that for values of k typically used in GC analysis, and for values of α lower than 1.05, the use of high-efficiency (capillary) columns becomes mandatory. Values of α can be related to the difference in the Gibbs' free energy between two compounds according to [71]:

$$\Delta(\Delta G) = \Delta G_1 - \Delta G_2 = RT \ln \alpha$$

Such differences can indeed be small for very similar compounds, such as certain isomers. However, exploitations of some very selective interactions of solute molecules with the stationary phase have now been documented; the separation of optical isomers on optically active stationary phases, as will be discussed in more detail in one of the following sections, is definitely the best example of such a situation. Less dramatic, but perhaps equally convincing are various other separations needed in biochemical work now; resolution of *cis/trans* isomers, or double-bond isomers, among others.

While the column selectivity alone can frequently be instrumental in resolving pairs of compounds with very small differences in molecular structure, it is of only limited value in chromatography of complex mixtures. This argument was appropriately expressed by Giddings [72] many years ago, in that "changes in selectivity may do little more than scramble the already crowded chromatogram, with new overlaps replacing the old; the only certain means of improving resolution is through an increase in the number of plates". On the other hand, the efficiency alone is sometimes limited in scope and, ideally, the best column should have *both* a high plate number and a unique retention property. Examples of this are relatively few, as the column selectivity (polarity, in a broader sense) is frequently incompatible with the high temperatures required for biochemical GC analyses. A notable compromise has been reached in steric resolution of the amino acid derivatized enantiomers: while the earlier used dipeptide optically-active stationary phases [73–75] basically had the necessary selectivity, the thermal limitations prevented analysis of all biologically interesting amino acids. The availability of optically-active silicone polymers [76,77] has changed the situations.

While choosing a stationary phase for a particular separation problem, compromises must frequently be reached between thermal stability, the lower operating range, detection conditions, selective effects, etc. In terms of material availability, the chromatographer's choice is typically limited to industrial products that are intended for other than chromatographic use. However, in an increasing number of instances, various substrates have specifically been synthesized for gas chromatography. Alternatively, the existing industrial polymers have been modified or purified. In an exceptional case, a C_{87} symmetrical hydrocarbon was prepared and suggested as an 'absolute' nonpolar standard phase [78].

While judging retentive properties for various biological compounds, a classification suggested by Horning et al. [79] in 1963 for steroids may still have a wider utility. Thus, the 'nonselective' stationary phases (mostly non-polar silicone polymers) separate compounds according to their boiling points, but generally fail in distinguishing finer structural details of chromatographed molecules. However, such phases frequently possess the capability to resolve molecules of the same molecular weight, but different geometry (group positional isomers, *cis/trans* isomers, axial and equatorial steroids, etc.). The nonselective phases of very high thermal stabilities are an excellent choice for screening the unknown samples. They provide effectively the first orientation on a composition of a given mixture.

The 'selective' stationary phases were further divided in the scheme proposed by Horning et al. [79] for the steroid separations into the types selective to alcohols, ketones, multiple bonds, etc. Here, various selective interactions between the column and the separated solutes are utilized. Most importantly, various polyglycols, polyesters, polyphenyl ethers, polyamides and polyimides belong to these categories. In addition, substitution of an alkyl group in a polysiloxane polymer by a more polar moiety leads to more favorable retentive properties for polar compounds. Such substitutions typically involve β-cyanoethyl, fluoroalkyl and phenyl groups. Certain polar silicone polymers (both gums and silicone fluids) combine effectively thermal stability with a degree of selectivity.

It should be pointed out that 'column selectivity' should frequently be understood in a very broad sense. Reliable predictions are frequently difficult. Different temperature dependence of partition coefficients for various solutes may further modify such predictions. 'Mixed' retention mechanisms are also observed due to interactions between the solid supports and chromatographed molecules.

A few clear-cut cases of selectivity notwithstanding, the selection process on different column types has been, in many laboratories, largely empirical, distinguishing 'the knowledgeable chromatographers' from the rest. However, the importance of having numerous stationary phases available will tend to dissipate with wider use of highly efficient capillary columns. Thus, with a notable exception of certain highly selective columns, there may not be a general need for more than a few stationary phases.

Notable attempts have been made toward a systematic classification of stationary phases in GC. The column classification system conceived by Rohrschneider [80], and further developed by McReynolds [81], does provide a valuable guide in the column selection process. Most commercial phases have now been characterized. More quantitative and elaborate approaches toward the characterization of liquid phases in GC involve solubility parameters and other thermodynamic considerations [82,83].

Thermal stability of a polymeric stationary phase is an important consideration, in that it frequently determines the column lifetimes in high temperature GC as well as the retention reproducibility. The degree of polymer crosslinking usually determines whether a given stationary phase will have a gum-like or a fluid-like consistency. Perhaps the silicone polymers are, within the field of biochemical GC, the best known examples of this behavior. An extensive crosslinking of polyglycols has also led to an improved thermal stability of the stationary phases for polar molecules [84]. A frequent penalty for improved thermal stability is an increased melting point of the phase, but this is not so critical in most biochemical applications. An example is shown in Fig. 3.7 with a mixture of triglycerides separated on a polyphenyl ether sulfone stationary phase [85], which appears to be most effective in the range of 200–400°C.

Incorporation of carborane moieties into silicone chains also leads to improved thermal properties of such 'semiorganic' polymers. The stationary phases based on carborane polymers have been available commercially since the early 1970's, when the first analytical results were also reported [86] for some biological compounds. Although such phases were believed to be stable up to about 400°C, only moderately successful separations have been reported. While the carborane 'cages' can be differently incorporated into siloxane polymers to influence their consistency, functional groups (phenyl, cyanoethyl, etc.) may also be varied to influence selectivity.

Some useful stationary phases are listed in Table 3.1 together with their thermal

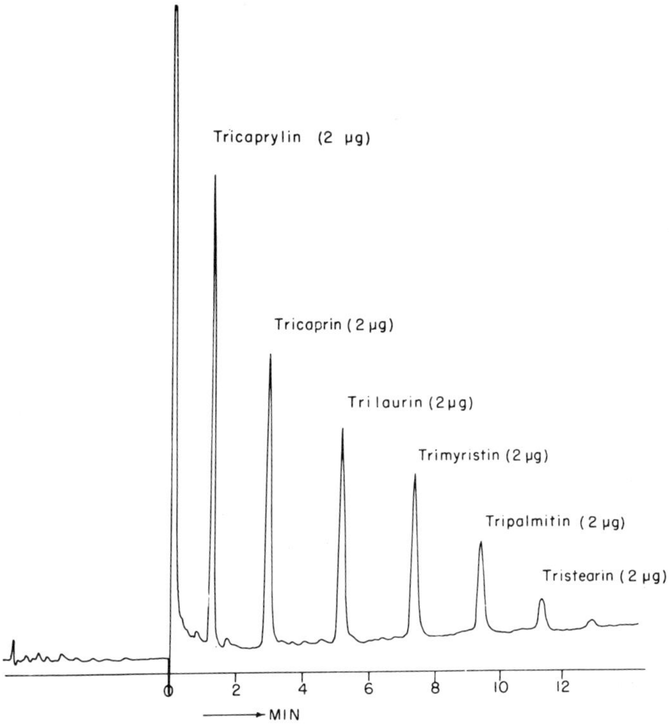

Fig. 3.7. A high-temperature separation (230–3759°C) of triglycerides. Reproduced from [85].

properties. The list is by no means a comprehensive one, as many more GC liquid phases can be found in commercial catalogues. The estimations of upper temperature limits are approximate (and somewhat more conservative than those found in the commercial literature), as the actual thermal stabilities are greatly dependent on a particular column technology, solid support modification, carrier gas purity, etc. Also, there is frequently a very substantial difference between packed and capillary columns.

While Table 3.1 lists mostly the common stationary phases, the substrates of very high selectivity may have some very specific uses in biochemical analysis. Optically active stationary phases will be mentioned in connection with steric resolution problems in a later section of this chapter. Yet another group of promising selective phases are the liquid crystalline substances. They have been of chromatographic interest for some time. Nematic liquid crystalline phases that exhibit an ordered molecular arrangement within a certain temperature range are particularly useful in the separation of isomeric compounds. The primary cause of resolution appears to be the length-to-breadth ratio of different molecules; for example, the rod-like molecules are strongly retained due to a greater probability of charge-transfer interactions with polycyclic molecules [87]. Some of these interactions are suffi-

TABLE 3.1

Commonly used liquid stationary phases

Stationary phase		Estimated temperature limit (°C)	Remarks
Chemical type	Commercial name		
Methylsilicone fluid	OV-101 or SP-2100	320	General-purpose phase
Methylsilicone gum	SE-30 or OV-1	350	For high-boiling compounds, various derivatized bio-chemicals
Methylphenylsilicone gum	SE-52 (5% phenyl substitution)	320	Slight polarity
	SE-54 (1% vinyl and 5% phenyl substitution)	320	
Methylphenylsilicone fluid	OV-3 (10% phenyl) OV-17 (50% phenyl) SP-2250	300	Moderately polar phase for a number of applications
Carborane/silicone polymer	Dexsil 300 (methyl substitution)	400	Triglycerides and other large non-polar molecules
	Dexsil 400 (methyl/phenyl substitution)	400	
Polyphenyl ether	Polysev	250	
Polymetaphenoxylene	Poly-MPE	350	
Polyphenyl sulfone	Poly-S-179	400	Polar phase for high-boiling compounds
Polyethylene glycols	Carbowax	220	For polar, volatile compounds
Cyanoalkyl silicone	OV-275	220	Useful for derivatized sugars and fatty acid esters
	SP-2300	230	
	SP-2310	230	
Modified polyethylene glycol	FFAP (Free fatty acid phase)	220	For polar compounds, fatty acids
Polyamide	Poly-A	210	Polar phase
Polyimide	Poly-I	250	Polar phase

ciently strong to override volatility and reverse the expected order of elution. An example of separation on a liquid crystalline phase is shown in Fig. 3.8 for model steroid molecules. The applications of such phases in biological separations are currently rare.

Polarities and selectivities of chromatographic columns can be greatly adjusted through the utilization of mixed stationary phases. Binary mixtures are usually used, with the two phases intimately blending with each other. In this way, certain

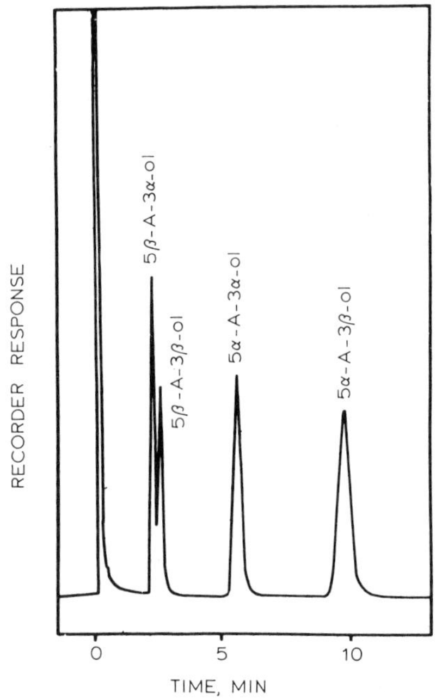

Fig. 3.8. Chromatographic resolution of 5α/β-androstan-3α/β-ols on a liquid crystalline stationary phase. Reproduced from [87] with permission of the American Chemical Society.

advantages of selectivity can be expressed in a given column, while minimizing the problem of limited thermal stability. Although the mixed-phase approach was proposed long ago, and even applied to useful biochemical separations [88–90], additional theoretical and practical aspects of this method were the subject of more recent, extensive studies by Laub and Purnell [91,92]. Solute retention is predictable, since it was shown [91] that for a stationary phase composed of a binary (A + S) mixture, the infinite dilution partition coefficient, K_R, is given as:

$$K_R = \phi_A K^0_{R(A)} + \phi_s K^0_{R(s)}$$

where $K^0_{R(A)}$ and $K^0_{R(s)}$ represent the corresponding partition coefficients in pure liquids, and ϕ is a volume fraction. Optimization strategies can be applied in the mixed-phase approach, in order to provide adequate selectivities and minimize analysis time. Given an appropriate column technology, the mixed-phase approach is applicable to both packed and capillary columns.

Besides the stationary-phase selectivity, the phase ratio (volume of the mobile phase/volume of the stationary phase) is an important consideration in practical chromatography. Thus, within the available 'arsenal' of GC columns, the efficiency increases in the following direction: packed column < micropacked column < support-coated open tubular column (or wide-bore capillary column) < conventional capillary column. Simultaneously, the column capacity (proportional indirectly to the phase ratio) decreases. Column selection for a given analytical problem must take this into consideration.

The analytical quality of the solid support in GC is a prerequisite of successful separations at low sample concentrations. The positive role of this material is to support the necessary amount of a stationary phase. Excessive surface activity is undesirable, as it frequently leads to irreversible adsorption phenomena and losses of the chromatographed solutes. To minimize these negative phenomena is an integral part of modern column technology.

A decade ago, an extensive discussion of the solid support treatment, support coating techniques, column packing methods, etc., might have been a very important section of a chapter similar to this one. However, a degree of uniformity brought by the manufacturing procedures into this area makes such a discussion non-essential now. Today, the great majority of users purchase well-deactivated solid supports, finished packings, or even the pre-packed columns with guaranteed performance.

Surface silylation of solid supports, glass columns, inserts, or even glass-wool spacers and glassware for the sake of surface deactivation remains highly recommended in biochemical GC. An alternative approach to surface deactivation is the method of Aue et al. [93], in which thermal treatment of polymer-coated supports results in a partial linkage of the macromolecule to the surface. This approach has been successfully employed with both packed and capillary columns.

While the developments in capillary GC were slow in coming in the late 1960's, many researchers then considered the support-coated open tubular and micropacked columns to be viable alternatives to the conventional capillaries. Although some interesting results were reported about 10 years ago [94,95] on the performance of such columns, they were largely overshadowed by the rapid advances in technology of wall-coated columns. The limited column permeability of micropacked columns and an excessive surface activity of support-coated open tubular columns are the major drawbacks of these column types. However, they may still offer a suitable compromise between sample capacity and column efficiency in certain special instances.

3.3.2 Capillary columns

Golay's invention [96] of the capillary column is considered today as one of the most important milestones in the entire history of separation methods. While capillary GC has now become a very common analytical technique, a very long induction time of some 20 years preceded its wide utilization.

Capillary columns yield efficiencies of one to two orders of magnitude higher than the conventional packed columns. Consequently, they are primarily used for resolution of exceedingly complex mixtures. Efficiencies of between 10 000 (an upper limit for packed columns) and 250 000 theoretical plates are feasible with today's column technology; such columns are capable of resolving up to several hundred components in a single run.

Today's utilization of capillary columns extends far beyond their original purpose, i.e., high separation efficiency. Rapid, yet moderately efficient separations are achieved with short capillary columns. In addition, there are some clear advantages in using capillary GC for trace analysis; these pertain to both the column inertness and detection aspects.

The capillary columns used in earlier investigations were almost exclusively made out of stainless steel. Glass, an otherwise superior material for the column manufacturing, was long neglected in spite of the early, successful use in hydrocarbon separations. It was not until the early 1970's that the advantages of glass inertness in biochemical separations were more clearly recognized [12]. Mastering technology of glass capillary columns, an exceedingly difficult task for a beginner, was initially a very crucial step for those laboratories which had realized the potential for dramatically improved separations and decided to embark on new developments. The number of enthusiastic followers has grown substantially since then. A successful transition from packed columns to capillary GC has now taken place in numerous biochemical laboratories. Both systematic and empirical studies of the chemistry of glass surfaces have made further advances in column technology feasible. Commercialization of glass capillary columns has brought certain uniformity into capillary GC and has further resulted in substantially improved instrument designs.

An important development occurred in 1979 with the first description of the flexible fused silica columns by Dandeneau and Zerenner [17]. This has helped considerably to remove the 'psychological barrier' toward capillary GC for those being uncomfortable with the relatively fragile glass columns.

Poor wettability of glass columns was a major obstacle to the reliable column technology in the past. With the advent of surface corrosion techniques [11,97,98], it has become feasible to coat uniform stationary films of even polar liquids on the inner column surface. Similarly, coated dispersions of solids [99] can aid in a uniform distribution of the stationary liquids for efficient separations.

Chemical modification of surface silanol groups on the glass surface affects both the spreading of liquid films [97] and the residual adsorptive properties of glass capillary columns. Thus, the surface wettability problems can largely be solved through different individual surface treatment techniques, or their combination. Most of the recent developments in this area have been reviewed [100].

Even though the glass surface is generally believed to be relatively inert, removal of the 'residual activity' of such a surface is highly important in GC of subnanogram samples. As a result of this residual activity, certain labile compounds tend to produce asymmetrical peaks, or they may 'disappear' in the column altogether. Various catalytic sites on the surface may further degrade sensitive solutes. Naturally, such effects can result in a loss of chromatographic resolution and poor quantitation.

The column deactivation techniques are highly important in that the trace analysis of certain compounds should not be limited by the column to a greater degree than by the sensitivity of our detection devices. The lack of deactivation may become particularly visible while separating polar solutes (primary amines, acids, alcohols, etc.) on non-polar columns. Various developments of recent years have significantly reduced these column deactivation problems. The main deficiencies appear to be related to the concentrations of various metals on the surface and an excessive reactivity of various surface hydrated structures, such as the silanol groups. Several approaches have been suggested to overcome these difficulties [100].

Although there are similarities between the surface properties of glass and fused silica columns, somewhat different approaches to their column technology are needed. Thus, the surface corrosion procedures ('geometrical modification') that are commonly used in glass column technology are not applicable to the thin-walled fused silica columns. The density of polar (reactive) groups on the surface of silica is also considerably less than on conventional glass; this provides a somewhat limited scope for extensive chemical modification to induce the surface wettability. However, the notable absence of metal oxides (the usual glass ingredients) on the fused silica surface makes the task of column deactivation considerably easier.

Preparation of the so-called 'immobilized stationary phases' has been a recent, important trend in capillary GC. Following the work by Madani et al. [101], Blomberg and Wännman [102] prepolymerized various siloxanes prior to their coating inside the column and attachment to the glass surface. The results of such a procedure are the capillary columns with non-extractable stationary phases; this has important implications in connection with certain newer sampling techniques (e.g., splitless and on-column sample injections), since excessive solvent peaks do not affect the film integrity negatively at the column inlet. In addition, extraneous materials depositing in the column after a number of injections can effectively be washed away with a variety of solvents to regenerate the columns.

The stationary-phase immobilization can also be achieved with mechanically-deposited siloxanes and subsequent crosslinking [103] of the polymer chains through a radical-aided reaction, as the following examples [104] demonstrate:

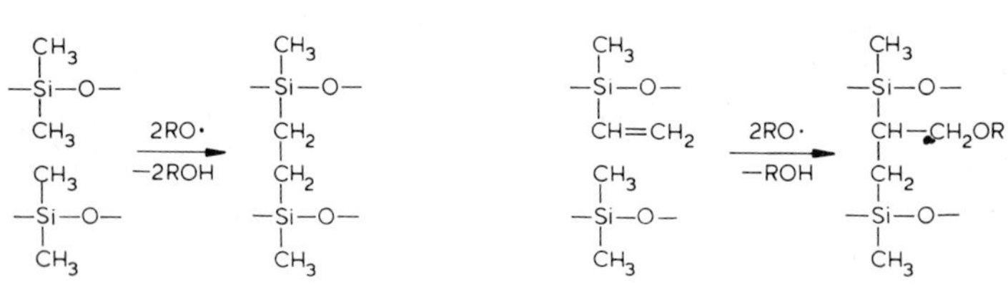

Again, the crosslinked polymers can form insoluble films on the column wall. A proper combination of the deactivation and immobilization procedures is needed to ascertain the columns of highest quality. A recent example from the literature [105] shows the state-of-the-art in this area (Fig. 3.9); a series of multi-functional compounds (including both acid and basic substances) have been chromatographed here with excellent results.

For the sake of completeness, it should be added that an important part of the capillary column technology is the method for depositing a uniform layer of the stationary phase on the wall. However, a coating technique is successful only if

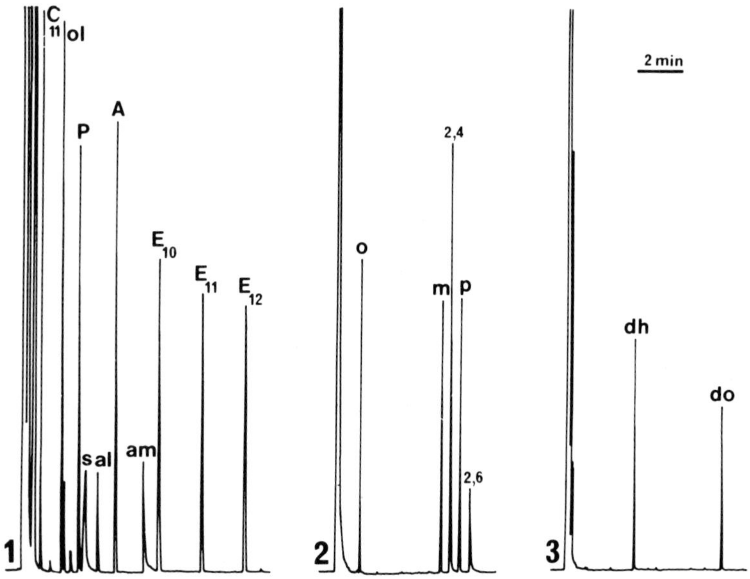

Fig. 3.9. Gas chromatograms (FID) of test mixtures on AR-glass capillary columns. (1) Grob test mixture on OV-215. Initial temperature 70°C, programmed at 5°C/min. Peaks: C_{11} = undecane; ol = octanol; P = 2,6-dimethylphenol; s = 2-ethylhexanoic acid; al = nonanal; A = 2,6-dimethylaniline; am = dicyclohexylamine; E_{10}, E_{11} and E_{12} = C_{10}, C_{11} and C_{12}-acid methyl esters. (2) Nitrophenol test on SE-52. Initial temperature 100°C, programmed at 7°C/min. Peak assignment; o, m, p = *ortho*, *meta*, *para*-nitrophenol; 2,4, 2,6 = 2,4-dinitrophenol and 2,6-dinitrophenol. (3) Diamine test on SE-30. Initial temperature 60°C, programmed at 7°C/min. Peaks: dh = 1,6-diaminohexane; do = 1,8-diaminooctane. Reproduced from [105].

preceded by an appropriate surface treatment; if this is not the case, the deposited film eventually breaks up into droplets.

The effect of capillary inner diameter on column efficiency is quite predictable: the column efficiency increases as the diameter decreases. However, this increased performance is at the expense of sample capacity. Capillary columns that are most commonly used today have inner diameters between 0.2 and 0.3 mm. While the sample capacity corresponding to such column dimensions is adequate for the combined GC/MS, wide-bore capillary columns are required for most remaining peak identification techniques. The wide-bore (0.5–0.7 mm, i.d.) columns may tolerate up to microgram amounts. The column technologies for the wide-bore and conventional capillary columns frequently differ, as an extensive geometrical modification of the column inner surface is needed for the former column type.

It is clear that the major importance of capillary GC in biochemical analysis will primarily remain in the analyses of complex mixtures. An example of such a complex mixture is shown in Fig. 3.10. The sample represents a variety of volatile secondary metabolites obtained from human urine [106]. It is quite clear that even the most efficient packed column could do little with such a sample. It should also

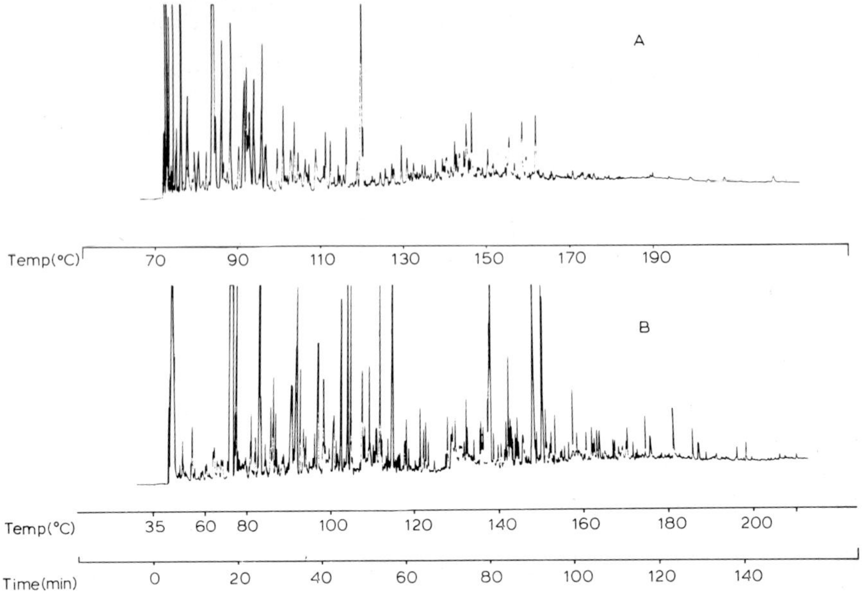

Fig. 3.10. Chromatograms of urinary volatiles of a normal man as recorded on glass capillary columns of different polarity: (A) 38 m column coated with Emulphor; and (B) 80 m column coated with SF-96 silicone fluid. Reproduced from [106] by permission of the American Association for Clinical Chemistry.

be noted that this complexity is not exceptional for various biological samples, as it will be repeatedly shown throughout this chapter.

An effective utilization of very efficient capillary columns needs appropriate instrumentation. In addition, periodical testing of the system and column performance is essential to maintain quantitative data. Various compounds that are difficult to chromatograph are utilized as very sensitive 'molecular problems' in such tests. If acid-base properties of the column surface are of importance, a simple mixture of dimethylaniline and dimethylphenol [107] may be a convenient sample for testing. However, additional 'molecular probes' may be more appropriate for certain cases. It should be emphasized that the peak symmetry and quantitative elution should be maintained at the level of desired analyses.

3.4 Detection methods

3.4.1 General considerations

Most biological GC determinations need high sensitivity of detection because of the small quantities involved. The use of ionization detectors has long been essential in this direction. As a general-purpose (universal) detector, the flame ionization detector has been unchallenged in its versatility; it is sensitive, yet relatively independent

can be pointed at in a complex chromatogra
overlooked as peaks of a negligible size or
detector.

Most GC detectors acquire a special analy
columns. The general sensitivity aspects were
examples will be demonstrated below. Suit
column work is determined by: (a) small volu
compatible with the conditions of separatio
response. Strictly considered, only a few dete
the exception of flame detection devices, dea
the detector itself are the most serious pr
reduced by adding extra carrier gas at the co
some sacrifice in the detection sensitivity of t

3.4.2 Selective detectors

The flame-based GC selective detectors deri
emission (flame photometric detectors), or
subsequent to the combustion in a flame (
Recent advances in the detector principles
biochemical uses, will now briefly be review

The response of the flame photometric
subsequent to combustion of certain organi
initial work on this principle by Brody and
with selective detection of sulfur and phosp
other elements is also feasible with different

The use of the flame photometric detecto
to the emission of S_2 spectral species at 39
sulfur-containing volatiles in physiological
patients [111]. A word of caution conce
compounds may result in a diminished c
species [112]. Hence, the need for maxin
responsive to nanogram amounts of sulfur-
increases with the square of sulfur conten
detector in the detection of phosphorus con
similar capability of the thermionic detecto

The earlier utility of the thermionic dete
by a lack of adequate technology and certa
selectivity on the secondary ionization pro
to the presence of an alkali metal. Althoug
detection mechanisms, no straightforward
various elements exist at present. Phospho
some less common elements have been r
operating conditions and detector designs.
expand the current utility of the thermion

on a number of experimental variables that may occur during a GC analysis. The detector can easily be used at high temperatures and in temperature-programmed runs.

The flame ionization detectors (as well as the other flame detectors) can be used equally well with packed and capillary columns. Different considerations may apply to other detector types. The well-known classification of chromatographic detectors into the concentration-sensitive and the mass-flow-sensitive types is highly relevant in this respect. A response enhancement [108] to the mass-flow-sensitive detector types is given as

$$E = \frac{t_{R(p)}}{t_{R(c)}} \sqrt{\frac{N_c}{N_p}}$$

where $t_{R(p)}$ and $t_{R(c)}$ are the retention times, and N_p and N_c the numbers of theoretical plates for a packed column and a capillary column, respectively. Assuming approximately equal retention times and 1–2 orders greater N for a capillary column than a packed column, some differences in sensitivity are observed; it agrees with the fact that the peak areas with thin, sharp capillary peaks will be more easily detectable than the packed-column peaks with a larger bandwidth. However, the situation is substantially different with the concentration-dependent detectors:

$$E = \frac{V_{R(p)}}{V_{R(c)}} \sqrt{\frac{N_c}{N_p}}$$

The retention volume, V_R, is a product of the retention time and the volumetric flow-rate. As the flow-rates in capillary GC are typically only 1 ml/min, a considerable response enhancement is realized (one to two orders of magnitude).

The above considerations, when translated into the practice of GC detection and quantitation mean that detectors such as the electron capture detector or the photoionization detector will greatly benefit from the columns of reduced flow-rates (provided that the detection cells can be manufactured correspondingly smaller). Further advantages of capillary columns include considerably reduced bleeding rates during the high-temperature operation as well as the already discussed column inertness. These practical gains may frequently be decisive in practical applications.

The most sensitive GC detectors are sample-destructive. Thus, if there is a need for further sample investigation, effluent splitting becomes essential. In a typical situation, a small portion of the effluent is led into such a detector, while the remainder is trapped for additional physical or chemical studies.

From many GC detection principles suggested over the years, only a few have been judged sufficiently useful and reliable for commercialization. Table 3.2 lists the most commonly used GC detectors; both universal and sample-selective detectors are included together with approximate sensitivity figures.

TABLE 3.2

Properties of some GC detectors

Detector	Selectivity mode
Flame-ionization	Universal
Thermionic	Nitrogen-selective
	Phosphorus-selecti
Flame-photometric	Sulfur-selective
	Phosphorus-selecti
Electrolytic conductivity	Halogen-selective
	Nitrogen-selective
Ultraviolet	Aromatic Groups
Photoionization	Partially enhanced to certain molecule compared to flame (not truly selective
Electron-capture	Affinity to low-en electrons

While the universal detectors, such as
degree, the photoionization detector, are
biological mixtures (sample screening).
applications. This is primarily due to th
possibilities to overcome sample comple
interfering molecular species. Their utili
procedure. While most biological molecu
a selective detection, incorporation of
into the molecules will change that. Thi
in ultratrace analysis in small samples
the detection limits can often be lower
The scope of applications in this area
methods still remain to be explored.

The mass spectrometers used as high
the ultimate in selectivity and sensitivi
(multiple ion detection) or mass chron
ago [7,8] provide unique possibilities
organics. In particular, the use of i
standards is extremely advantageous i
and trace endogenous biochemicals; a
under determination separately from
chromatographic resolution.

In general, selective detectors can p
the complex chromatograms obtaine
frequently enhanced sensitivity is an in

The modes of thermionic detection with particular importance in biochemical GC concern nitrogen- and phosphorus-containing compounds. Nitrogen compounds are abundant in various biological samples; high sensitivity (10^{-13} g/sec, typically 50-times more sensitive [114] than the conventional flame ionization detectors) and a significant selectivity factor (10^3–10^4) make this detector highly attractive for various investigations. Derivatizations can further introduce detectable moieties into organic molecules of interest, such as with the example of methoximation of steroids [115] and other carbonyl compounds. Similarly, formation of dimethylthiophosphinic esters from biological steroids, homovanillic acid, and other polar compounds [116,117] permits a highly sensitive and selective detection with the thermionic detector adjusted for the phosphorus-sensitive mode. As this detection mode is approximately 10-times more sensitive than the nitrogen version, femtogram amounts of certain compounds can be detected.

Today's thermionic detectors are quite reliable and quantitative devices. A key to long-term stability and quantitative reproducibility seems to have been the utilization of an externally-heated alkali source [118]. The explanation of nitrogen detection based on the interaction of cyano radicals with the excited rubidium atoms [118] appears quite plausible, while the nature of response to phosphorus compounds still seems obscure.

Since both the flame-photometric and thermionic detectors are the flame-type detectors, similar rules apply for their coupling with capillary columns as for the flame ionization detector; no special modifications are required.

The electron capture detector is perhaps the most sensitive detection device currently available in organic analysis. Its importance in biomedical applications hardly needs to be emphasized. Its general utility has been strengthened during the last decade through overcoming the earlier drawbacks of the radioactive source instability and response non-linearity. In addition, a proper understanding of the detection mechanisms [119,120] has been a most welcome advance. A monograph dealing exclusively with various theoretical and practical aspects of the electron capture detection was recently published [121].

While only a few biological molecules are strong gas-phase electron absorbers, numerous types of derivatives (mostly halogenated compounds) have been developed and used over the years in biomedical investigations. Different classes of biologically important compounds can be measured with high sensitivity following this general approach. The individual cases of successful analytical use will later be treated in more detail in connection with derivatization procedures and the application sections. Temperature dependence of the electron capture response could also be used in qualitative studies.

There are several attractive features of coupling the electron capture detector with capillary columns. Many interfering compounds frequently occur in complex biological mixtures, while many of the derivatization agents which are used are general enough to react with the compounds of interest and contaminants alike. Unless laborious methods of sample 'clean-up' are employed, the final measurements may be unreliable. Extensive sample purification may be needed in assuring specificity of a packed-column analysis, but uncontrolled losses of trace compounds could occur. Obviously, interfering compounds could more readily be resolved from a sample of interest with capillary columns.

The extreme sensitivity of the electron capture detector is considerably more utilized with highly inert glass or fused silica columns, where minimum sample loss occurs and the separated electron-absorbing molecules can easily be detected and quantified. As capillary columns generate considerably less bleeding than the packed columns, the electron capture detector can be easily used under temperature programming conditions.

Coupling capillary columns to the electron capture detector is not without problems. Unlike the flame detectors, relatively large detector cells are common due to certain constructional features of the ionization detectors housing a radioactive source. The use of a scavenger gas or a detector miniaturization are the currently used remedies to this problem. The miniaturization approach is obviously more attractive due to the concentration-sensitive nature of the detector, as already discussed above.

With further advances in the detector design, subpicogram determinations of many biologically important compounds should readily become feasible. An example [122] of derivatized biological molecules with a very strong affinity for thermal-energy electrons is demonstrated in Fig. 3.11; the thyroid hormones (T_3 and T_4) are easily detected here as diheptafluorobutyryl derivatives. The procedure allowed tagging the molecules at two different sites, while the natural presence of iodine atoms in these hormones further enhanced their electron-capturing properties.

It is now commonly believed that the potential of the electron capture detector has not yet been adequately explored. Recent developments in response sensitization [123,124] seem to support that notion. Further improvements in this area are likely to come with advancing knowledge of the ion-molecule reactions.

The photoionization detector, one of the oldest GC ionization detectors, has recently received renewed attention because of the novel sealed-source technology developed by Driscoll et al. [125,126]; this new detector version has seemingly reduced the earlier difficulties with controlling too many parameters during quantitative measurements. The changes in conductivity, following the photoionization effect, are measured in a confined space (the adjacent ‘detector cell’), while the fate of the generated molecular ions is of no particular concern.

While the photoionization detectors are not really selective in a true sense, an enhanced response to certain compound types is frequently observed, as shown in detection of certain drugs [126]. The detector response is considerably enhanced for aromatics, carbonyl compounds and solutes containing heteroatoms, as compared to aliphatic compounds of approximately the same molecular weight. Thus, with an appropriate solvent choice, the photoionization detector can be a more suitable device than the flame ionization detector. Simultaneously, the background signal due to non-specific substances is also significantly lowered. Some control of response can also be exercised through selection of the light sources with different energy [125].

Quite importantly, various photoionization detectors should have appreciably

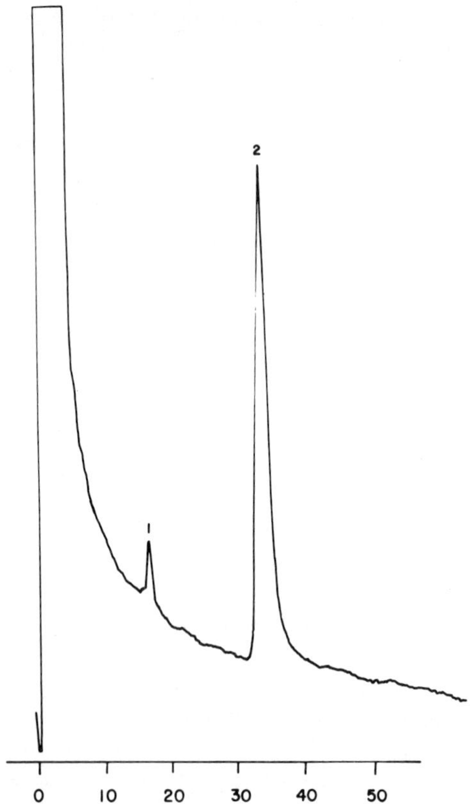

Fig. 3.11. Gas chromatography of *N*,*O*-diheptafluorobutyryl methyl ester derivatives of the dialyzed thyroid hormones, as detected by the electron capture detector. Peak 1, approximately 1 pg T_3; peak 2, approximately 8 pg T_4. Reproduced from [121].

higher sensitivity than the flame ionization detector. Since the photoionization detector is a concentration-sensitive device, a decrease of flow-rate can increase the signal substantially for the same compound mass. Thus, coupling a miniaturized photoionization detector to a capillary column has undoubtedly much promise for even greater sensitivities. Some preliminary work in this direction has already been reported [127].

Some additional selective detectors have been described, but their use in biochemical GC has been minimal thus far. Among them, most notably, belong various optical spectroscopic detectors as well as various element-specific detectors based on the solute combustion and measurement of electrolytic conductivity [128]. While little has happened during the last decade with further development of the latter detector types, various gas-phase optical devices remain among the most interesting detectors for future studies. Element-specific plasma devices [129], UV absorption

[130] and fluorescence [131] detectors are certainly worth further investigations and applications.

3.5 Solute identification techniques

3.5.1 Retention studies

Certain relationships between the molecular parameters and chromatographic retention are very useful in qualitative studies. Prior to the development of GC/MS and other ancillary techniques, retention measurements were critically important in any identification efforts. This is well-documented by many laborious data collections in the earlier literature that are of little use in today's efforts. Whereas it is established that no serious structural elucidation can nowadays be based solely on solute retention studies, their utility *in combination* with various ancillary techniques should not be underestimated.

The additivity of functional group contributions to the overall solute retention, shown for the first time in chromatography by A.J.P. Martin [132], has been particularly developed for fatty acids and steroid structures [133–136]. Certain useful information can be derived from these systematic studies even today. While predictions of the chromatographic retention, as based on various empirical and systematic observations, could be quite useful, only the match of *both* retention and spectral properties between a suspected compound and the authentic sample is an acceptable proof.

The Kováts retention index [137], using *n*-alkanes as a series, is most commonly used for both internal and external data comparisons. With certain precautions in mind, this retention system is usable for temperature-programmed runs. Within the field of biological investigations, fatty acid esters [133] and steroidal hydrocarbons [134] were also used to standardize the retention data within a compound class. However, the use of the so-called 'methylene units' [135] is basically identical with the Kováts system.

With an increasing use of capillary GC, retention data will undoubtedly gain more popularity. This is due to the fact that the modern capillary columns offer retention measurement precision which is greatly superior to the previous situation. In addition, the great separating power of such columns permits resolution of various isomeric compounds. While these isomers elute at different retention times, even the best mass spectrometric equipment has difficulty in distinguishing isomerism. Thus, various positional isomers, *cis/trans* pairs, diastereoisomers, etc., give rise to differences in chromatographic mobility.

Whereas the Kováts retention system was primarily developed for the purpose of relating *any* organic molecule to the set of *n*-alkane standards, its utilization may not be always preferable over the use of other standard series. In particular, during the temperature-programmed runs, appreciable deviation may occur due to different surface-related phenomena and solute partition trends of *n*-alkanes and other compounds. As shown in more recent retention studies on polyaromatic molecules [138] and certain nitrogen-containing compounds [139], the use of an 'internal standard series' is far more advisable; following this procedure decreased the deviations of repetitive runs typically below ± 0.25 index units.

Further utility of relations between structure and retention will be dependent on: (a) acquisition of a significant number of reference compounds for comparative purposes; and (b) advancing retention prediction capabilities. A recent successful use of modern computational techniques [140] to predict retention of aromatic compounds from certain molecular parameters appears indicative of this trend.

Comparing solute retention on the stationary phases of different polarity further extends identification capabilities. For this reason, injection and detection systems were reported [141] in which the injected sample is split and its fractions are simultaneously recorded on two different columns under the same thermal conditions. Since most biological GC investigations involve some sample derivatization, the 'peak shift' techniques [142] can also be used more frequently for identification purposes.

3.5.2 Ancillary techniques

The meaning of 'ancillary techniques' can be interpreted in a broader sense to include a variety of sample manipulations in a pre- or post-column arrangement; they all ultimately serve to enhance the qualitative information content on the individual solutes, or even provide their structural elucidation directly. Although a variety of techniques, both 'chemical' and instrumental in nature, have been suggested and used at different times, only some approaches find a wider utilization at present. While certain high-capacity GC columns permit the solutes to be trapped at the column exit for further investigations, the on-line acquisition of spectral properties directly from the column effluent is more convenient and popular. Furthermore, with capillary columns that yield numerous fractions separated by seconds or less, effluent trapping would be very tedious at best. Unfortunately, the fact that efficient GC columns necessitate sample size considerably less than micrograms tends to rule out many powerful structural tools of organic chemistry as ancillary techniques (proton or ^{13}C-NMR, conventional IR techniques, X-ray crystallography, etc.).

At least several micrograms are needed for further characterization of trapped chromatographic peaks. The micropreparative GC separations have been more popular with smaller than larger molecules; this is, naturally, related to the general difficulties of high-temperature GC, contamination problems arising from the stationary-phase decomposition, as well as poor recoveries due to the aerosol formation. To comply with the sample size requirements of various identification techniques, it is either possible to perform repeated injections and trappings, or, at some sacrifice of efficiency, to overload the packed analytical columns. Micropreparative separations are also feasible with the wide-bore capillary columns, but such procedures appear technically demanding.

Once a fraction is trapped at the GC column end, it also becomes feasible to employ a variety of chemical techniques. A variety of approaches that are more or less specific for certain compounds or compound classes have been utilized [143]. A further investigation of trapped GC fractions through thin-layer chromatography was earlier demonstrated [144,145], but this technique is seldom utilized in practice. Here, a variety of chemical reactions could be employed for specific structural information. Once more, it is believed that an increasing emphasis on improved resolution of biological mixtures through capillary GC has now rendered some of the above approaches obsolete because of the sample size problems.

Currently, the two most powerful ancillary techniques are undoubtedly mass spectrometry (MS) and the Fourier-transform infra-red (FTIR) spectroscopy. While the former is now nearly a state-of-the-art technique, the latter is being very rapidly developed. Importantly, the two techniques are very complementary to each other in yielding a specific type of structural information. In addition, both MS and FTIR spectroscopy can now be effectively coupled with high-resolution capillary columns.

Today, the role of a combined gas chromatograph/mass spectrometer in biochemical analysis is widely evident, as some of the key developments of the last 10–15 years could not have been done without it. Importantly, such an instrument is not merely a mechanical combination of the two methods, but a unique tool in itself that maximizes and combines certain, most essential components of the two methods.

Once the technical problems of directly interfacing a gas chromatograph and a mass spectrometer were solved during the 1960's, the combined instrument was immediately applied to structural elucidation tasks within a variety of natural compounds. Until the early 1970's, most work was primarily carried out with packed columns. In much of the recent work, capillary columns are preferred, as the task of structural elucidation is considerably easier while dealing with the spectra of pure compounds rather than their mixtures. However, while identifying suspected compounds (with the known fragmentation patterns) in a complex mixture, the methods of mass fragmentography or mass chromatography can also be effective with packed columns [146,147].

Structural elucidation of organic compounds by MS is dependent on the acquisition of a reproducible and easily 'readable' mass spectrum following the ionization process. The ionization efficiency and the best possible design of MS ion optics are vitally important to high-sensitivity measurement. Acquiring mass spectra routinely from nanogram quantities of the separated components is quite typical of the current state of methodology. However, much higher sensitivities may become feasible in the near future due to newer ionization techniques, such as, for example, laser photoionization [148].

The extent of ion fragmentation is important in relating a given mass spectrum to possible organic structures. Under the widely used electron impact ionization, the organic molecules fragment readily, but yield only sometimes recognizable fragment ions. Absence of the parent (M^+) ion is quite typical for biological molecules under these circumstances. This problem may frequently lead to an initial error in spectral

interpretation. A usu
a review, see Ref. 14
success. In spite of t
today a common tec
High mass-spectra
obtained with conven
cal in terms of atom
obtained in a GC/M
the typical values obt
the exact molecular
ments, sensitivities a
GC/MS instruments.
Computers have t
have largely corrected
a better instrument c
the main reasons fo
automatic identificati
decade aim undoubt
become more reliable
normalization has al
available from the ins
While a spectrum
dealing with a new st
principles'. Once a sp
the growing libraries
from biological and
computer search for
spectrum from a part
have been accumulat
compound types may
As IR absorption
groups in organic mo
structural techniques.
past. This situation h
technology. Thus, wi
[150], the combination
for structural work. T
major mixture compo
Whereas it is clear
high-resolution MS or
for structural studies
many structural puzzle
tion of trace biologica
some of these determi

individuality' by Williams [157]. Importance of the 'balanced' body fluid and tissue constituents to one's health and physical condition has also been emphasized by Pauling [158] in his concept of 'orthomolecular medicine'; it has been intuitively felt by many investigators that the capabilities of metabolic profiling would be underutilized if used only as a means of disease recognition. Importantly, with an increasing knowledge of the complex metabolic network, some possibilities should occur for the early diagnosis of a disease, or an individual's tendency toward metabolic defects.

Both conceptual misunderstandings and technological difficulties still occur in the field of metabolic profiling. However, with the rapidly improving analytical capabilities, much progress has already occurred. The inborn errors of metabolism have been increasingly recognized with such techniques, and additional disease conditions are currently under study in numerous laboratories; an excellent account of these activities, until 1977, is provided by Jellum [15], with particular emphasis on the metabolism of organic acids.

The advances in sample fractionation methods, sample derivatization approaches, and the instrumentation of GC and GC/MS, in particular, are fundamental to metabolic profile research. Biological variation that is inherent to the samples of physiological fluids or tissues should not be obscured by an excessive imprecision of measurement techniques. Thus, reliable sampling and sample treatment procedures (including as much automation as is feasible) should precede the use of sophisticated GC and GC/MS techniques.

A typical complexity of samples encountered in the metabolic profile investigations requires very efficient chromatographic columns. While the principal aim of such investigations is to separate, identify and measure quantitatively as many constituents as possible, capillary GC appears the best to facilitate this goal. The information content of chromatographic profiles can further be enhanced through a parallel use of selective detectors. As shown in Fig. 3.12, a complex urinary profile of volatile constituents detected by the flame ionization detector is complemented by a recording from the nitrogen-sensitive detector [114]. Similarly, a parallel use of the electron capture detector for high-sensitivity measurements has been advocated [159].

Another approach to overcoming the mixture complexity problem in metabolic profiling is using mass chromatography as advocated by Sweeley et al. [146,147]. Provided that the metabolites of interest are selected for a study, the GC/MS/computer system can generate selective profiles while plotting the chosen fragment ions as a function of time (a mass chromatogram). Thus, quantitative comparisons are feasible for certain compounds of interest, while the instrument ignores 'uninteresting' metabolites. The general capabilities of such a procedure and the GC/MS system have been demonstrated with human urinary acid profiles [146]. Other uses of GC/MS systems in metabolic screening efforts have been reported [147,160–163]. Computer systems adjacent to GC/MS units will undoubtedly continue to be recognized as powerful means of search for anomalous metabolites and new metabolic disorders.

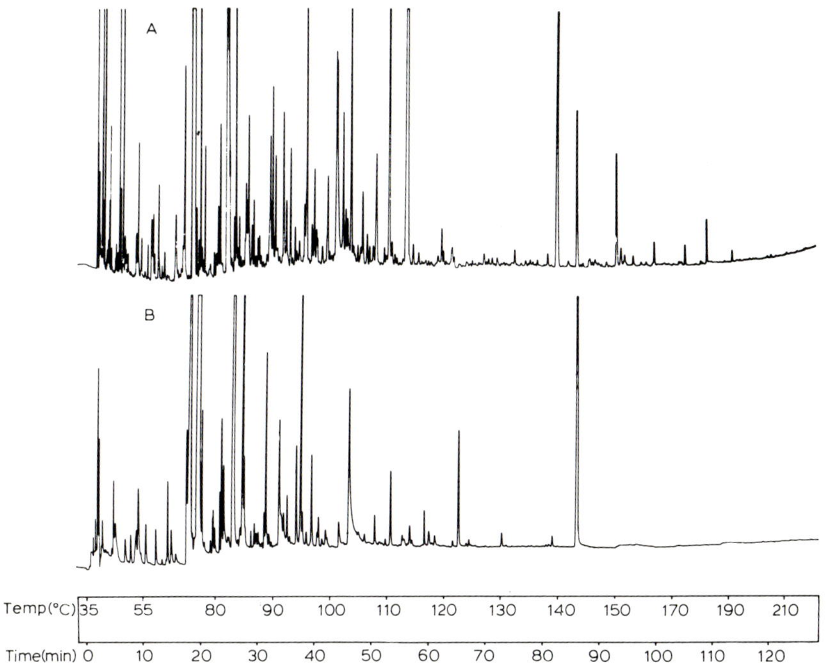

Fig. 3.12. Capillary chromatograms of volatiles from 24 h urine of a normal male detected by (A) flame ionization detector, and (B) the nitrogen sensitive thermionic detector. Reproduced from [113].

An example of a search for unusual metabolites related to a human disorder is well illustrated with the case of hereditary progressive deafness in a large Norwegian family, studied by Jellum et al. [164]. While no other obvious clinical symptoms were associated with the deafness cases, capillary GC/MS demonstrated the presence of two unusual metabolites, 3-hydroxyisovaleric acid and 3-methylcrotonylglycine. Both compounds appear to be intermediates in the metabolism of leucine, and as Fig. 3.13 demonstrates, a metabolic loading experiment with leucine revealed an enzymatic deficiency in patients with hereditary progressive deafness.

While the metabolic conditions that involve enzyme deficiencies may be relatively easy to establish through the modern techniques, more subtle alterations in metabolism are likely to require even better techniques. Most such studies have been thus far limited by a lack of quantitative capabilities as well as a correct interpretation of complex metabolic interactions. The above-mentioned 'biochemical individuality' complicates the finding of the 'normal values' of metabolic patterns in humans.

The two important avenues for furthering metabolic profiling research appear to be: (a) selection of suitable animal models of human disease conditions; and (b) an increased use of computer-aided studies to extract the important metabolic information from the 'biochemical noise' in large human profile sets.

In the former approach, the genetic variation, diet, environmental conditions, etc., can be carefully controlled, while the effects of disease induction and progression pioneered by Gil-Av and his associates [22], initially involved separation of the usual volatile amino acid derivatives (e.g., various esters of *N*-acylamino acids) on dipeptide selective phases [73–75]. The nature of steric interaction between the dipeptide phases and separated amino acids was discussed [74,179].

The major problem of the dipeptide stationary phases is their thermal instability and the lack of column technology leading to highly efficient separations. With a notable exception of the report by König and Nicholson [180] from 1975, most of the reported 'peptide columns' had neither the efficiency consistent with modern capillary GC, nor could they cover the necessary thermal range for eluting all naturally occurring amino acids. Although the boiling points of amino acid derivatives could be lowered through employing *N*-pentafluoropropionyl rather than the traditional *N*-trifluoroacetyl compounds, this has not been generally sufficient to overcome the column instability problems.

The major breakthrough in the GC enantiomer separation has been the work of Bayer and associates [23,76], who synthesized a silicone-based chiral phase, stable up to 240°C. As shown in Fig. 3.14, a racemic mixture of 19 protein amino acids can be separated [23] on a glass capillary column coated with Chirasil-Val, a chiral polysiloxane phase. The phase was synthesized through coupling L-valine-tert-butylamide to a copolymer of dimethylsiloxane and carboxyalkylmethylsiloxane.

Subsequent communications by other research groups involve modifications of certain commercially available stationary phases. It has been shown [77] that a cyano

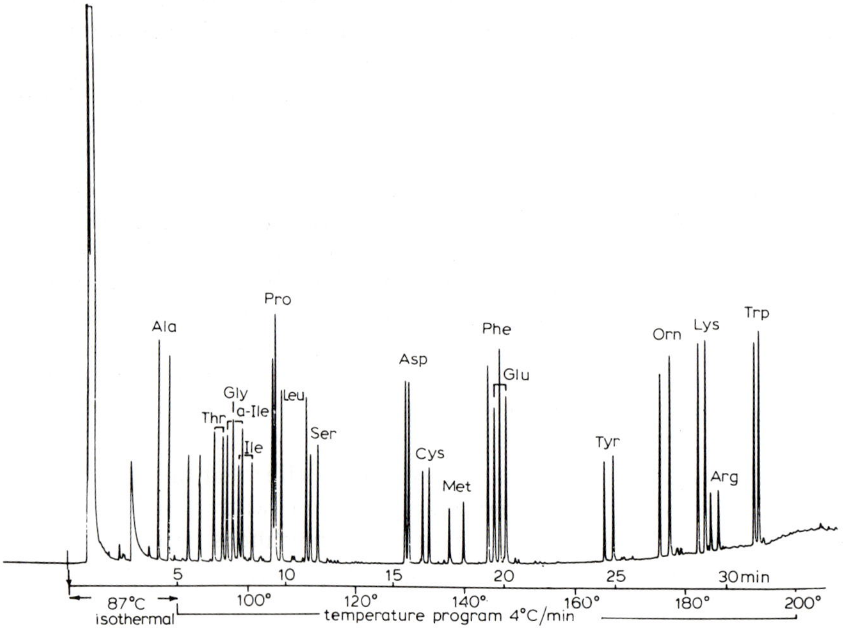

Fig. 3.14. Capillary GC separation of a racemic mixture of 19 protein amino acids on an optically active stationary phase. Reproduced from [23].

silicone could be chemically altered to yield a highly selective substrate for the separation of the D,L-enantiomers of the common amino acids and amino alcohols. Additional modifications were recently reported by König et al. [181–183] to cover a wider range of applications.

While the amino acids are popular for use in the stereoselectivity studies, there is a need to resolve other racemic mixtures. To this date, some progress has been indicated with amino alcohols [183], amines [77,183], hydroxy acids [181,182] and carbohydrates [183]. It appears that the 'tailor-made' substrates will be essential to cover a wider range of applications. Additional applications are likely to emerge in time, while the area of chiral separations is likely to remain one of the more challenging and interesting directions in chromatography.

An interesting and potentially quite useful sidelight of the chiral separations is 'enantiomer labeling' [23]. This involves the use of an unnatural enantiomer as an internal standard to ascertain accuracy of the determination of other amino acids (or, generally speaking, additional optically-active compounds). Such a standard is added at the initial stage of a sample work-up, so that any errors due to non-quantitative derivatization, decomposition, column problems, etc., are compensated for through a corresponding response to this unusual internal standard. In general, capillary GC with a chiral phase is needed in such separations. With only small amounts of materials available in some biological investigations, additional sensitivity and selectivity may be obtained with a nitrogen-sensitive, or the electron capture, detector.

3.8 Derivatization methods

3.8.1 General aspects

The scope of biochemical GC would be quite limited without sample derivatization. Yet, a simple chemical conversion of the compounds of interest into suitable derivatives is frequently all that is required for successful chromatography. Benzoic acid, typical of many naturally occurring substances which, because of their polarity, are not well suited for gas-chromatographic analysis, is readily converted by treatment with methanol in the presence of an acidic catalyst (such as boron trifluoride) into the thermally stable methyl benzoate, which is more volatile than the parent acid and becomes easier to chromatograph. As another example, GC properties are favorably affected when the typical alcohol cholesterol is converted to the silyl ether by simple treatment with chlorotrimethylsilane in pyridine.

Examination of the current literature reveals, however, that more elaborate derivatization reagents and methods than these are frequently used in the GC analysis of complex mixtures, and that still more elaborate ones continue to be developed, suggesting that simplicity of operation and volatility, thermal stability, and easier resolution of derivatives are only a few of the qualities desired for many a gas-chromatographic analysis.

It is the objective of the following discussion to trace the development of currently used reagents for the derivatization of the more commonly encountered functional groups and combinations thereof, with emphasis on the special areas of applicability for each method as well as its limitations.

In general, desired qualities for a derivatization method, other than those mentioned above may include the following: speed; quantitativeness; functional group selectivity; formation of products stable toward hydrolysis, oxidation, etc.; formation of products with enhanced response in certain detectors; availability of structural information from retention data; formation of products whose mass spectra yield ample structural information; formation of a single product per constituent whenever possible.

3.8.2 Derivatization of alcohols and phenols

3.8.2.1 Silylation agents

The ease with which alcohols and phenols can be converted to silyl ethers as compared with alkyl ethers has made silylation, generally, the derivatization method of choice for hydroxy compounds. Furthermore, intermolecular attractive forces involving trialkylsilyl (R_3Si-) groups are relatively small; the conversion of a typical alcohol to its trimethylsilyl (TMS) ether results in an increase of 72 mass units per molecule with only slight decrease in volatility.

From the equation:

$$ROH + R'_3SiX \rightarrow ROSiR'_3 + HX$$

it is clear that a base (commonly pyridine or triethylamine) should be included in the silylation mixture if the R'_3SiX chosen yields an acid HX strong enough to damage substrate molecules . Even with pyridine present, the strong silylating agent chlorotrimethylsilane (TMCS) was reported [184], in a study of silylation of carbohydrates, to give less quantitative results than mixtures of TMCS and hexamethyldisilazane (HMDS, $(Me_3Si)_2NH$) in 1 : 1 to 1 : 4 ratios. In this study, HMDS alone was found ineffective as a silylating agent. Mixtures of HMDS and TMCS, commercially available with or without pyridine as TRI-SIL [185], are reputed to be the most widely used silylating agents today.

The search for a compound R_3SiX strongly electrophilic (but yielding a nonacidic HX) led to the silyl iminol ethers, the prototype of which is *N,O*-bis-trimethylsilylacetamide (BSA) [186]. Reactions with proton donors such as alcohols are thermodynamically driven by the formation of the carbonyl group of the corresponding amide:

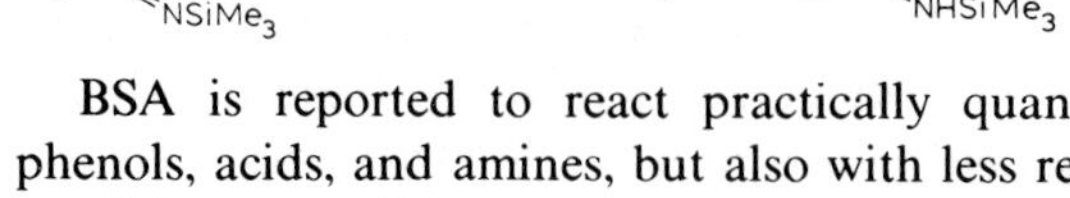

BSA is reported to react practically quantitatively, not only with alcohols, phenols, acids, and amines, but also with less reactive amides and imides. Reaction conditions are mild.

Though less effective, the product *N*-trimethylsilylacetamide (MSA) is also a silylating agent; hence a typical mixture, having been treated with BSA, will contain MSA and acetamide as byproducts. That is generally not a serious problem, though, since both materials are relatively volatile and are eluted early. Yet, herein lies the advantage of *N,O*-bis-trimethylsilyltrifluoroacetamide (BSTFA) [187]: substitution of a trifluoromethyl group for a methyl group leads to byproducts (trifluoroacetamide and its *N*-trimethylsilyl derivative) which, because of greater volatility, present less danger of overlap with the peaks of concern.

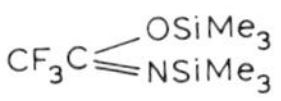

BSTFA

For still greater volatility of both reagent and byproduct (*N*-methyl-trifluoroacetamide), *N*-methyl-*N*-trimethylsilyltrifluoroacetamide (MSTFA) was developed [188]. Although not a silyl iminol ether like the two preceding compounds, it has been found to be an effective silylator of steroids [189] and has been used in the profiling of urinary acids [190]. In both cases, the result was conversion, not only of hydroxyl groups to silyl ether groups, but also of keto groups to silyl enol ether groups.

Where strength of silylating agent is the most important consideration, trimethylsilylimidazole (TSIM) has been the reagent of choice. Another reagent which takes advantage of a nucleophilic group's ready displacement without formation of a

TSIM

strong acid, TSIM is capable of derivatizing most O-H bonds (but not N-H bonds), including such inaccessible ones as those on C-17 of pregnane derivatives [189]. In contrast to MSTFA, it does not affect enolizable ketones [191].

silyl enol ethers of:

2,4-pentanedione methyl acetoacetate

Recently, the trimethylsilyl enol ethers of 2,4-pentanedione and methyl acetoacetate, themselves prepared by treatment of the corresponding beta-dicarbonyl compound with TMCS and imidazole (presumably TSIM is the silylating agent), were introduced [192] as rapid silylating agents for alcohols. Reaction is reported to be complete within a few minutes at room temperature without catalytic assistance. Advantages of some of the previously mentioned reagents seem to be combined in these: volatility of byproducts, and reactivity toward relatively unreactive hydroxyl groups such as the tertiary hydroxyl of linalool.

Obviously, a wealth of reagents exists for the conversion of hydroxyl to silyloxyl groups. All the above reagents, however, lead to the same product, a trimethylsilyl ether, which possesses qualities that may be regarded as not totally satisfactory, in contrast to the obvious asset of volatility.

For one, the mass spectra of trimethylsilyl derivatives may not be suitably informative. Frequently, the intensities of molecular-ion and useful fragment peaks are low compared with those of the ubiquitous Me_3Si^+ ($m/e = 73$), Me_2SiOH^+ ($m/e = 75$), and $Me_3SiOSiMe_2^+$ ($m/e = 147$) peaks, obscuring clues as to the identity of low concentration components. Polyhydroxy compounds, such as certain steroids, tend to undergo fragmentation by successive losses of trimethylsilanol (Me_3SiOH) molecules, exhibiting series of peaks 90 mass units apart, with attendant uncertainty as to whether the molecular-ion peak is the highest-mass member of the series or something invisible beyond that.

In this regard, the low retention increment imparted to a molecule by a trimethylsilyl group may be viewed as a liability, for it is unsafe to use retention time as a measure of the number of trimethylsilyl groups. This objection could, of course, be circumvented through the use of higher alkylsilylating agents. However, caution must be exercised not to incorporate too much bulk into the silylating agent, to avoid excessive retention and, for steric reasons, non-quantitative derivatization.

Alkyldimethylsilanol RMe_2SiOH (R = ethyl or propyl) is not eliminated from the corresponding silyl ethers as readily as is trimethylsilanol, permitting more mass-spectral information to be gained from the high-mass end of the spectrum [193]. Hence, with the use of ethyl- or propyldimethylsilylimidazole or -trifluoroacetamide, for example, some of the objections to the trimethylsilylating reagents are overcome with only modest sacrifices in volatility and reactivity.

To the extent that susceptibility of the silyl ether to hydrolysis may be considered a drawback, use of the silylating agent *tert*-butyldimethylsilyl chloride/imidazole/dimethylformamide to yield the sterically crowded *tert*-butyldimethylsilyl (TBDMS) ethers offers a satisfactory alternative [194]. The base peak for this class of ethers is reported to be nearly always M-57, loss of the *t*-butyl group being the most important fragmentation. Hence, uncertainty regarding the nature of the molecular ion is alleviated in this manner.

Sterically crowded reagents may create the undesirable effect of a substrate molecule yielding both completely and partially silylated derivatives, but advantage may be taken of this by silylation of a sample with a sterically crowded reagent and, again, with a conventional trimethylsilylating agent; contrasting the sets of mass spectra obtained in the two cases may yield clues regarding the positions of the hydroxyl groups, and in some cases the stereochemistry. With this end in mind, a study has been made [195] of the reactivities of a number of reagents:

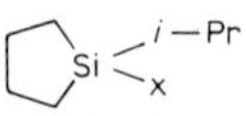

,
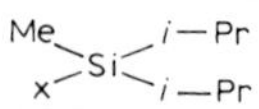

, TBDMSX, $(i\text{-}Pr)_3SiX$, and
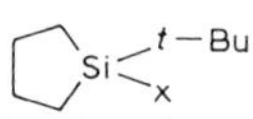

(in order of increasing steric requirement) toward hydroxyl groups in various positions in steroid molecules.

Another strategy for locating hydroxyl groups in steroids involves a 'sandwich technique' [196]: injection of the product of preliminary treatment of the sample with TSIM or BSTFA along with ethyl- or propyldimethylsilylimidazole results in the displacement of trimethylsilyl groups by the larger alkylsilyl groups only from phenolic positions. This is useful, in particular, for the detection of estrogens in mixtures.

If steric crowding is not required, but the appearance of a base peak near the molecular ion is considered important, allyldimethylsilyl chloride/imidazole ($CH_2 = CH\text{-}CH_2SiMe_2X$) is available [197]. The base peak for the resulting silyl ethers is commonly M-41 (loss of the allyl group).

It is claimed [198], however, that the advantage of being able to spot the M-57 peak of a *t*-butylsilyl derivative may be outweighed by the disadvantage of losing detailed structural information about the substrate, the mass spectrum displaying the fragmentation pattern of the M-57 peak rather than of the molecular ion itself. Therefore, chlorodimethoxymethylsilane [$(MeO)_2MeSiCl$] has been proposed as a silylating agent capable of generating the sorts of fragmentation patterns exhibited by the trimethylsilyl ethers along with the stability of the TBDMS ethers toward hydrolysis.

Sensitivity with certain selective detectors may be an important reason for derivatization. At very low concentrations, where electron-capture detection (ECD) merits serious consideration, it frequently becomes useful to choose a derivatizing reagent containing a number of halogen atoms. In preliminary studies aimed at the quantitative determination of such trace constituents as insect ecdysones [191], several agents RMe_2SiCl ($R = CF_3CH_2CH_2$-, $C_3F_7CH_2CH_2$-, and C_6F_5-) were synthesized and allowed to react with cholesterol. Of these, the pentafluorophenyldimethylsilyl (shortened to flophemesyl) derivatives showed the best ECD response along with being stable to nucleophilic attack at silicon. The investigators asserted that no simple relationship has been found between molecular structure and ECD sensitivity; although chlorinated compounds are more ECD-sensitive than their fluorinated counterparts, the latter offer the advantage of greater volatility.

Several desirable features — ECD-sensitivity and stability toward hydrolysis — were combined in one agent with the introduction of *tert*-butylpentafluorophenylmethylchlorosilane [199].

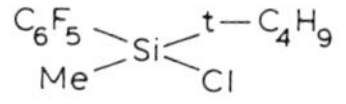

tert—butylpentafluorophenylmethylchlorosilane

3.8.2.2 Other derivatization agents

Despite the appearance of there being a silylating agent for any purpose, acylation with acetic anhydride and similar compounds is, of course, an older method and is still widely used in derivatization of alcohols and phenols. Of more current importance as derivatives are the perfluoroalkanecarboxylates; whenever high volatility is sought, trifluoroacetic is preferred over acetic anhydride, while pentafluoropropionic and heptafluorobutyric anhydrides $(RCO)_2O$ ($R = C_2F_5$, C_3F_7) are useful for preparing electron-capturing derivatives.

Trifluoroacetyl derivatives are, in fact, reported [200] to exhibit shorter retention times at lower temperatures than the corresponding trimethylsilyl derivatives, a

trifluoroacetylimidazole

property that should make trifluoroacetylation of especially non-volatile hydroxy compounds like carbohydrates particularly advantageous. However, the use of either trifluoroacetic anhydride or trifluoroacetylimidazole (expected to be a very strong acylating agent on the basis of what has been said about trimethylsilylimidazole), led to multiple and irreproducible peaks when applied to a variety of sugars. This difficulty has been surmounted through the use of *N*-methyl-bis-trifluoroacetamide (MBTFA), $(CF_3CO)_2NCH_3$ [201] (contrast silylating agent MSTFA), which allows mixtures of polysaccharides of up to four units to be analyzed using a polar GC column, with a sharp single peak observed per constituent.

A special category of acylation is the reaction of hydroxy compounds with derivatives of phosphorus acids, the objective being the formation of phosphate, phosphonate or phosphinate esters so that advantage can be taken of the special sensitivity of thermionic detectors toward phosphorus compounds. With volatility of the products being an important consideration, attention [202] was directed to the acid chlorides and dimethylamides of dimethylphosphinic and dimethylthiophosphinic acids. Of these, dimethylthiophosphinic chloride (Me_2PSCl) gave the consistently best yields of the thiophosphinate esters, and was designated the reagent of choice.

Special methods exist for the simultaneous derivatization of two hydroxyl groups, particularly when located on adjacent carbon atoms. Ketal formation — reaction of a diol with an aldehyde or ketone to form a 1,3-dioxolane derivative — is the most obvious approach; for example, 20,21-dihydroxycorticosteroids have been converted to acetonides [203].

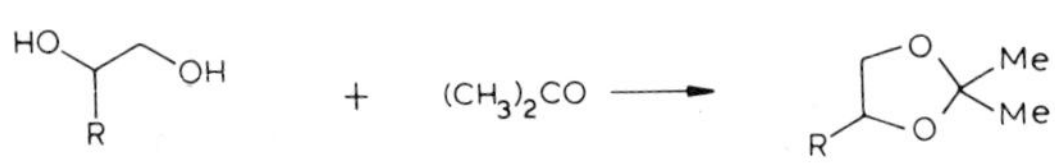

A somewhat more elaborate derivatization through ketal formation permits simultaneous derivatization of three functional groups of 17,21-dihydroxy-20-oxo-steroids such as cortisone with formaldehyde [204].

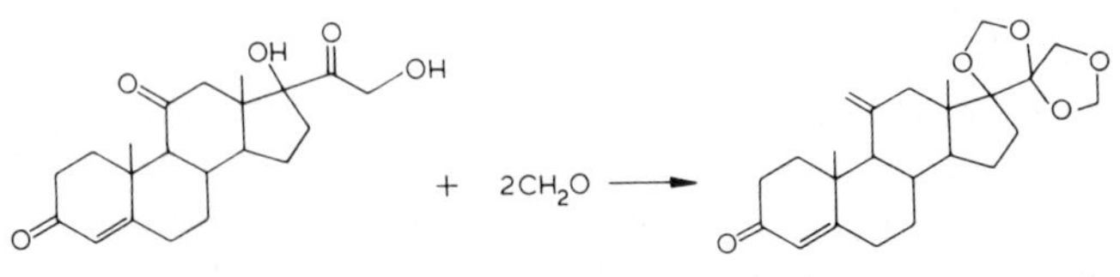

Ketalization, however, requires acidic catalysis, with its potential for damage to some substrates. Boronic acids [$RB(OH)_2$] have accordingly been advanced [205] as diol derivatization agents capable of functioning under neutral conditions. The products are cyclic boronates, analogous to the ketals:

Some boronates are susceptible to hydrolysis, but on the whole, gas-chromatographic properties are good and mass-spectral molecular ions are generally easily seen.

In addition, 1,3-diols can be converted to boron analogs of 1,3-dioxanes. Among a number of other compound types which have converted to boronates are β- and γ-amino alcohols and α- and β-hydroxy carboxylic acids.

It is interesting to consider the derivatization of aldosterone with methyl- or butylboronic acid: four functional groups per molecule are tied up by a single molecule of boronic acid [206]:

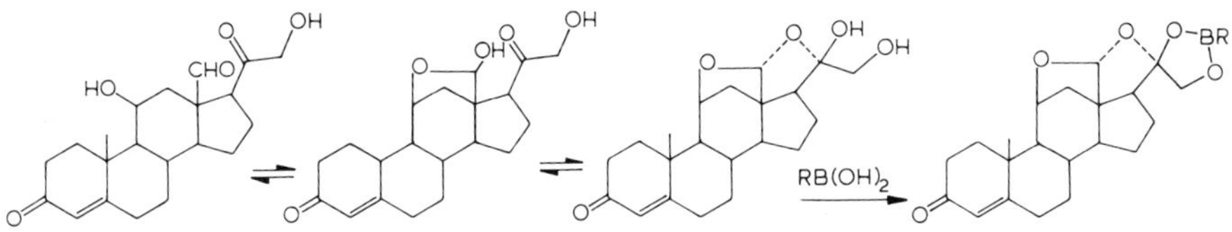

3.8.3 Derivatization of carboxylic acids

The current literature is replete with accounts of the determination of one fatty acid or another in numerous natural samples. It seems that the most common way to render fatty acids amenable to GC analysis is to use the time-honored method of acid-catalyzed direct esterification. The traditional acid catalysts H_2SO_4 and HCl are still used, but one also notes considerable reference to boron trifluoride and trichloride; for BF_3, in particular, the shorter time (relative to the mineral acids) needed to effect complete esterification has been noted [207].

Methyl esters can be obtained directly from triglycerides by transesterification with methanol; traditional catalysts in transesterification to form methyl esters have been acids (e.g., sulfuric) and sodium methoxide. However, if the sodium methoxide contains any sodium hydroxide, some of the triglyceride will be hydrolyzed to the carboxylate salt, which cannot be esterified. On this account, *meta*-trifluoromethylphenyltrimethylammonium hydroxide (m-CF_3-$C_6H_4NMe_3^+$ OH^-) was offered as a transesterification catalyst [208]; methanolysis is reported to occur rapidly at room temperature.

The selection of this reagent was based partly on the observation [209] that, because various quaternary ammonium carboxylates undergo thermal decomposition to methyl esters, such esters could be formed in situ in good yield by the simple injection of a methanolic solution of the quaternary ammonium carboxylate above 250°C. Thus, the use of the quaternary ammonium hydroxide in the transesterification scheme described above permits the simultaneous conversion of both glycerides and fatty acids to methyl esters.

In a modification of the above method, methyl propionate is added as a scavenger for excess hydroxide ion which might cause degradation of any polyunsaturation in the fatty acid chain [210].

The reaction $RCOO^- + R'I \rightarrow RCOOR' + I^-$ has also been employed [211]. Quantitative results in less than 10 min under mild, non-acidic conditions are claimed for the conversion of any acidic compound to a primary alkyl ester when the tetramethylammonium salt is treated with the alkyl iodide in a highly polar solvent system such as anhydrous methanolic *N,N*-dimethylacetamide.

The phase-transfer principle has likewise been applied to alkylation of carboxylate salts [212]. Acids sensitive to hydrolysis, like acetylsalicylic acid, were found to react with alkyl iodides after neutralization with tetrabutylammonium bicarbonate in methylene chloride.

For quantitative, instantaneous methylation of a wide spectrum of acidic substances, it is hard to replace diazomethane, handled as an ether solution to mitigate greatly the inherent risks of explosiveness and toxicity. From the equation:

$$RCOOH + CH_2N_2 \rightarrow RCOOCH_3 + N_2$$

it is apparent that the cessation of nitrogen bubbling signals the completeness of the reaction. Diazomethane is readily prepared by the addition of NaOH or KOH to any of a number of compounds $CH_3N(NO)Z$, where 'Z' is an electron-withdrawing group (for example *N*-nitroso-*N*-methylurea, $CH_3N(NO)CONH_2$).

Trimethylsilyldiazomethane, Me_3SiCHN_2, is regarded as a stable and safe substitute for diazomethane [213]. It is believed that the expected product, $RCOOCH_2SiMe_3$, is formed first, but quickly converted to the methyl ester (with displacement of $Me_3SiCH_2OCH_3$) by methanol present in the reaction mixture.

Yet another family of alkylating agents for fatty acids is composed of the alkyl acetals of dimethylformamide, $Me_2NC(OR)_2$ [214]. These may be the most convenient reagents for introducing alkyl groups higher than methyl. Reaction proceeds quickly and quantitatively under neutral or slightly basic conditions.

An important reason for choosing an alkylating group other than methyl is to add sufficient bulk to a small acid molecule to keep it from being eluted too early. A popular alkylating agent for this purpose is phenyldiazomethane, $C_6H_5CHN_2$, made analogously to diazomethane by treatment of *N*-benzyl-*N*-nitroso-*para*-toluenesulfonamide with KOH. Benzyl esters ($C_6H_5CH_2OCOR$) are formed in this way; those with volatilities up to about that of benzyl lactate [215] are satisfactorily eluted from typical columns.

Mixtures benzylated with phenyldiazomethane, however, display a number of GC peaks belonging to substances other than benzyl esters. It has been claimed [216] that a more satisfactory route to benzyl esters involves benzylation of the benzyltrimethylammonium salts of the acids with benzyl chloride.

To yield quantitative results, volatile acids in dilute aqueous solutions have been converted to phenacyl esters (217) by the action of *para*-bromophenacyl bromide on the metal carboxylates, using a crown ether to render the carboxylate anions more

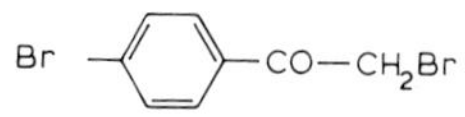

para—bromophenacyl bromide

nucleophilic. *N*-Chloromethylsuccinimide [218], in the presence of a crown ether and in a polar solvent, exhibits similar alkylating ability toward metal carboxylates and permits the extension of the volatility range of the acids to include stearic, palmitic, etc.

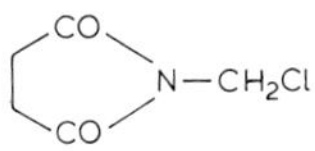

N– chloromethyl – succinimide

For electron-capture detection, hexafluoroisopropyl esters, $RCOOCH(CF_3)_2$, can be prepared by conventional esterification [219]. These esters are claimed to have the additional advantage of earlier elution than methyl esters. Another group giving good electron-capture detector response is the pentafluorobenzyl group, $C_6F_5CH_2$-; this group can be attached to a carboxyl group by phase-transfer alkylation of a metal carboxylate with pentafluorobenzyl bromide [220]. Excess alkylating agent, which disturbs the electron-capture detector, is permitted to react with the hydroxyl group of an aminophenol, the product being removed by acid extraction.

The often difficult problem of derivatizing carbon-carbon double bonds has been attacked with respect to unsaturated fatty acids or their methyl esters in a number of ways. Acquiring knowledge of the position of a double bond is the primary reason for any modification thereof; were it not for the ease with which double bonds migrate upon electron impact, their positions should be ascertainable through study of the mass spectra of the methyl esters. One solution to this problem has been the conversion of the double bond to a diol group which is derivatized by one of the methods described earlier for alcohols [221]. Of these methods, trimethylsilylation has been found superior to the special diol-derivatization methods, as the most diagnostically useful mass spectral fragmentation of the carbon chain takes place adjacent to a trimethylsilyloxyl group:

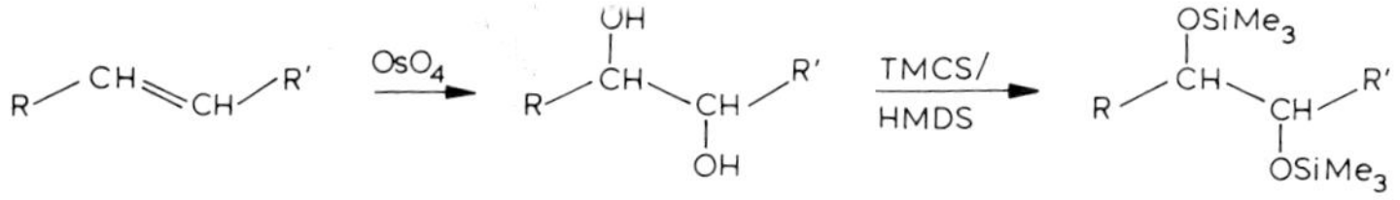

The double bond has also been converted to the chlorohydrin (R-CHCl-CHOH-R′) group prior to trimethylsilylation, or, for electron-capture detection, pentafluorophenyldimethylsilylation [222]. Of course, positional isomer formation here is inevitable.

Oxidative cleavage of the double bonds is practical only for mono-unsaturated acids [221]. Ozonolysis with reductive workup has been used [223], the resulting aldehydes being derivatized as the 1,3-dioxanes:

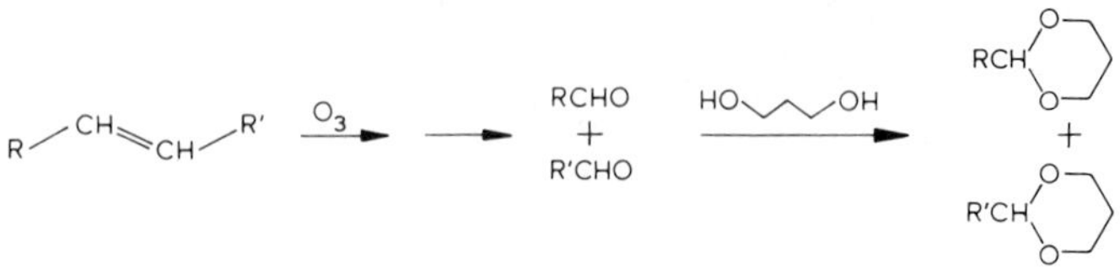

The oxymercuration-demercuration sequence has been applied to the analysis of 1,4-dienoic acids [224]:

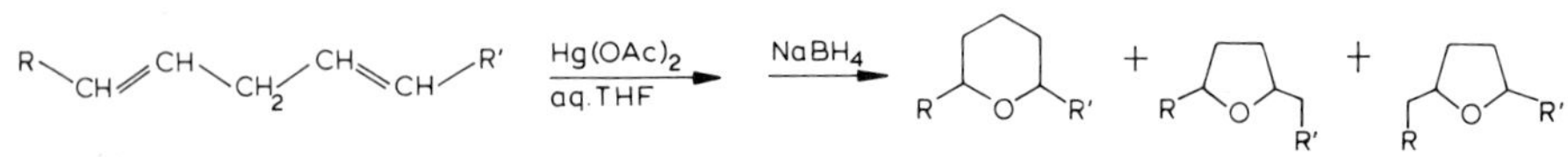

Despite the formation, again, of multiple derivatives, facile α cleavage of the side chains from the tetrahydropyran or tetrahydrofuran rings aids in mass-spectral interpretation.

Hydroxy acids are frequently the subject of GC analysis, examples being lactic, citric, malic acids, etc., bile acids and phenolic acids, The hydroxyl group in such acids is commonly silylated or acylated by methods previously described. Derivatization may be carried out in two stages, i.e., with prior methylation of the carboxyl group; alternatively, since the common silylating agents esterify carboxyl groups about as readily as they etherify hydroxyl groups, a simple silylation of a mixture of hydroxylated and unhydroxylated acids normally suffices.

Hydroxyl and carboxyl groups alike of bile acids have been converted to methyl ether and ester groups in one step by, first, formation of the alkoxide and carboxylate groups using as a base the sodium derivative of dimethylsulfoxide ($NaCH_2$-$SOCH_3$) in dimethylsulfoxide and subsequent methylation with methyl iodide [225].

For α- and β-hydroxy acids, treatment with methyl- or butylboronic acid [205] gives cyclic boronates:

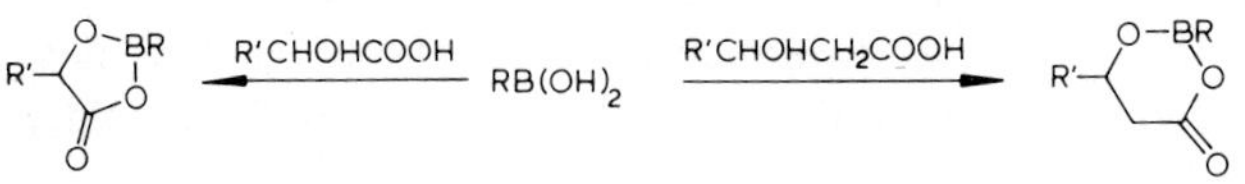

Thus, a method complementary to silylation of hydroxy acids is available.

Diazomethane would be a reasonable candidate for simultaneous methylator of carboxylic and phenolic functions (both acidic), were it not for its reputation for erratic behavior toward phenolic groups. However, satisfactory results were reported [226] in the conversion of acids such as homovanillic to the corresponding *ethyl* ether-esters with diazoethane, CH_3CHN_2. This reagent is specific for carboxylic and

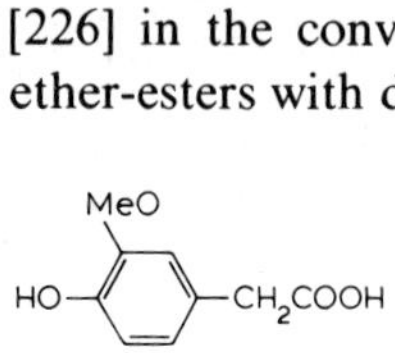

homovanillic acid

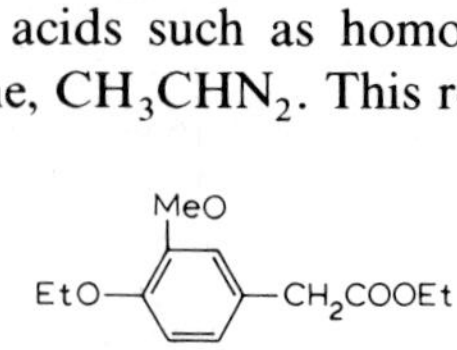

ethyl ether–ester of homovanillic acid

phenolic groups, and has the additional advantage over diazomethane that, since aryl methyl ethers, unlike aryl ethyl ethers, are common in nature, doubt as to whether a methoxy group is part of the natural product or has been introduced in the derivatization is greatly diminished.

3.8.4 Derivatization of aldehydes and ketones

For most compounds in which the carbonyl group of an aldehyde or ketone is the only functional group, polarity is not such as to make derivatization normally necessary. Yet, it is a rare mixture of urinary acids, steroids, prostaglandins and similar substances, that is completely free of constituents bearing a ketone function. The initial question concerning these compounds, then, is: "what effect do the common hydroxyl and carboxyl group derivatizing reagents have upon the carbonyl group?"

Aldehydes and ketones containing α hydrogen atoms exist in tautomeric equilibrium with the corresponding enols

which are silylatable, but under the usual silylating conditions only to a varying degree. Hence, this reaction is commonly regarded as more nuisance than help, which means that in analysis of complex mixtures it has been customary to convert ketone functions into non-enolizable derivative functions prior to silylation. However, it has been expressed on at least one occasion (see, for example, the urinary profiling studies of Pfeifer and Spiteller [190]) that derivatization of ketones by silyl enol ether formation is the preferable approach, as derivatives elute earlier than the subsequently discussed methoximes and can be formed in one step. The objection regarding incompleteness of silylation can, it is claimed, be overcome by extended heating.

Derivatives of ketones made by substitution of nitrogen for oxygen ($R_2C=O + NH_2Y \rightarrow R_2C=NY$, where Y = N, O, Ar, etc.) have long been known. Since volatility of the derivative is important, the methoximes (Y = OMe) have emerged as the generally most useful carbonyl derivatives [227,228]. So-called MO-TMS derivatives, then, are commonly made of biological acids, steroids, prostaglandins, etc., by treatment with methoxyamine (CH_3ONH_2) hydrochloride and pyridine, followed by silylation with trimethylsilylimidazole or a similar compound.

If hydroxylamine rather than methoxyamine is used, the result is an oxime, a less satisfactory compound, because it contains a polar hydroxyl group ($R_2CO + NH_2OH \rightarrow R_2C{=}NOH$); silylation of the hydroxyl groups of the substrate, however, also causes silylation of the oxime hydroxyl group, creating a satisfactory derivative.

Oximes may also be converted by an exchange reaction into methoximes and higher alkyloximes. It was shown [229], in a study involving a number of steroids, that the rate of displacement of the hydroxyl group of an oxime by a methoxyl group is an indicator of the position of the keto group, with 3-oxosteroid oximes undergoing this exchange reaction especially rapidly.

For electron-capture detection, pentafluorobenzyloxyamine has been described [230]. 2-Chloroethoxyamine [231] has been recommended as an oximating agent of high reactivity, yielding chlorooximes; the chloro-compound mass-spectral pattern is alleged to be helpful in identification.

Advantage is taken of the reactivity of α-dicarbonyl compounds toward aromatic 1,2-diamines in a novel derivatization method for α-keto acids [232], as illustrated for pyruvic acid:

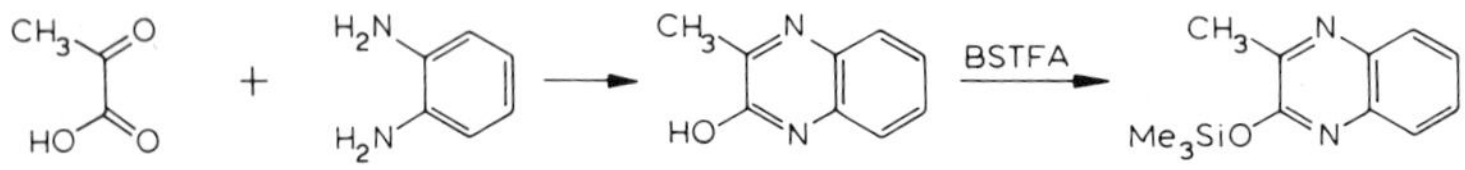

The condensation product, before silylation, is a quinoxalinol. One advantage of this method over methoximation is that only one product is formed; in methoximation *syn*- and *anti*-isomers of unsymmetrical oximes are observed:

syn- and *anti*-isomers of unsymmetrical oximes

3.8.5 Derivatization of amines and amino acids

A first approximation states that whatever reacts with alcohols by cleavage of the O-H bond also reacts with amines by cleavage of the N-H bond. But amines, being more basic than alcohols, are acylated more readily. On the other hand, alcohols are more readily silylated than are amines as a rule, while trimethylsilylamines are much less stable than are trimethylsilyl ethers of comparable structure. Hence, it seems reasonable, when selecting a derivatization reagent for a mixture of amines, to consider acylating agents first: trifluoroacetic, pentafluoropropionic or heptafluorobutyric anhydride, or if avoidance of acidic byproducts is essential, a perfluoroacylimidazole [233].

Primary and secondary amines have been converted to dithiocarbamates, of which only those derived from primary amines undergo elimination of alkanethiol in the injection port to yield isothiocyanates [234]:

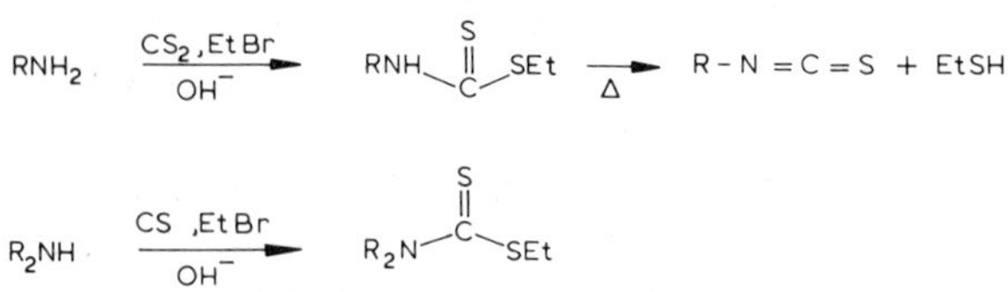

A means of distinguishing between primary and secondary amines is thus available.

A modern adaptation of the classic Hinsberg method, whereby primary and secondary amines are converted to sulfonamides using benzenesulfonyl chloride (but only the secondary sulfonamides remain non-neutralized by sodium hydroxide), was used in a derivatization scheme for volatile secondary amines [235]. An attractive feature of this method is the formation of sulfur-containing derivatives which respond to the flame photometric detector.

Among the amines most commonly studied, because of their biological importance, are those which also contain hydroxyl groups, such as the catecholamines (epinephrine, dopamine and related substances). One might employ an all-purpose silylating agent in a one-step procedure, but it appears preferable to add trimethylsilylimidazole first, taking advantage of its reported unreactivity toward amino groups [233] while being a very effective silylator of hydroxyl groups, and then acylating the resulting silyloxy amine with one of the above-mentioned fluorinated acylating agents, or the especially reactive *N*-methyl-bis-trifluoroacetamide [201].

Catecholamines are particularly polar and difficult to extract from aqueous media. An acylating agent convenient to use directly in aqueous solution is methyl chloroformate [236], Cl-CO-OMe. Amino and phenolic hydroxyl groups are converted to carbamate (MeO-CO-NHR) and carbonate (MeO-CO-OAr) groups simultaneously, with aliphatic hydroxyl groups subsequently silylated.

When the hydroxyl and amino groups are located on adjacent carbon atoms, a stable heterocycle can be formed by the addition of a chloromethylsilylating agent [237,238]:

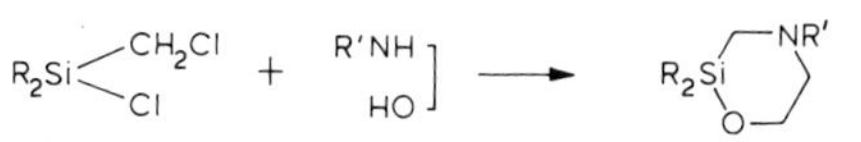

In the above example, the amino group is derivatized by alkylation of the nitrogen atom, far less commonly encountered than acylation. Another example of derivatization by alkylation is a reductive alkylation method, using formaldehyde and sodium borohydride, developed for airborne amine pollutants [239]:

A novel specific arylation of primary amines, occurring rapidly under mild conditions, is based upon the enhancement of aromatic nucleophilic displacement when nitro groups are situated in *ortho* positions [240]:

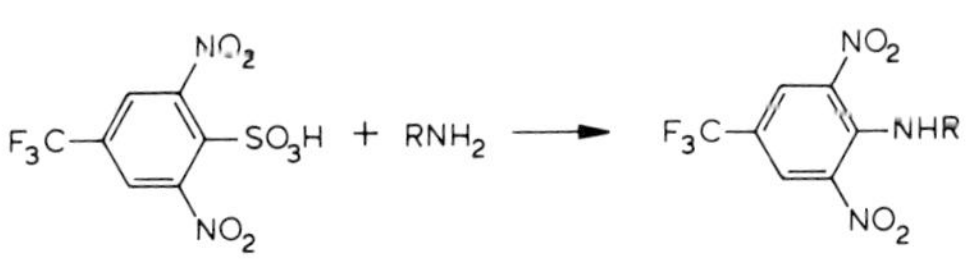

The great amount of attention given to analysis of amino acids and peptides, particularly those related to proteins, has generated a number of methods especially appropriate for this class of compounds. A survey of amino acid GC analyses reported in recent years reveals that most workers tend to use a combination of methods already discussed: carboxyl groups are commonly transformed into alkyl esters (propyl, butyl, etc., more so than methyl), while amino groups are generally acylated using one of the perfluorinated acylating agents or an alkyl chloroformate.

It has been mentioned [241] that these few derivatization steps are sufficient to render a peptide of up to six units (consisting of amino acid units of medium polarity) analyzable by GC. It is important to note in this context that acylation of a primary amine RNH_2 yields a product RNHCOR′ which still has a polar nitrogen-hydrogen bond. Permethylation of an acylated peptide (i.e., conversion of all N-H bonds to N-Me bonds) using, for example, $NaCH_2SOCH_3$ followed by methyl iodide, extends the range of acceptable volatility to perhaps ten to twelve amino acid units. Still more volatility can be imparted by subsequent reduction of the amino and ester groups to amino and alcohol (RCO-NMe ⁓⁓ COOR′ → RCH_2-NMe ⁓⁓ CH_2OH which is silylated) groups using lithium aluminum hydride, $LiAlH_4$. In fact, lithium aluminum deuteride was used in the reduction to facilitate mass spectral interpretation [241].

Trideuterioborane, BD_3 (as the tetrahydrofuran complex) has been recommended [242] as superior to $LiAlD_4$ in the reduction step; among other attributes, the ability of BD_3 to reduce carboxyl groups without the necessity of prior esterification was mentioned.

In another study [243] an alternative to *N*-acylation was presented: initial permethylation of the amino group using formaldehyde and the imino group-specific reducing agent, sodium cyanoborohydride ($NaBH_3CN$) (cf. Ref. 239). Subsequent acetylation proved necessary, nonetheless, to derivatize any hydroxyl and sulfhydryl groups present in the amino acid.

As is true of other bifunctional compounds mentioned earlier, amino acids have been derivatized using special ring-forming reactions. Such a reaction is thiohydantoin formation [244,245]:

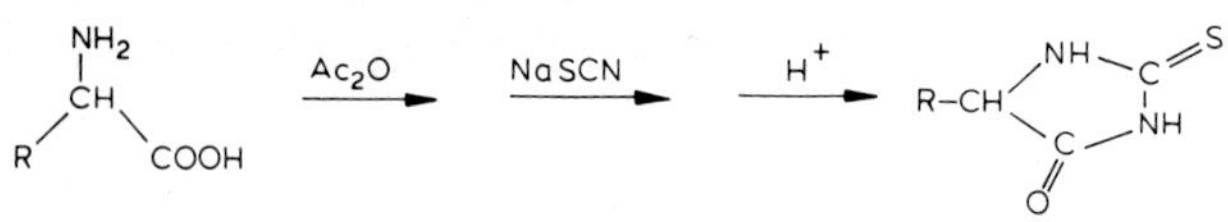

Trimethylsilylation of a number of thiohydantoins gave products (silylated at both nitrogen atoms) with generally good GC characteristics; a complication was the formation of tris-silylated byproducts, created by additional silylation of the enol form of the thiohydantoin.

Amino acids have likewise been converted into oxazolidinones in a rapid reaction with 1,3-dichlorotetrafluoroacetone [246,247], reminiscent of acetonide or boronate formation:

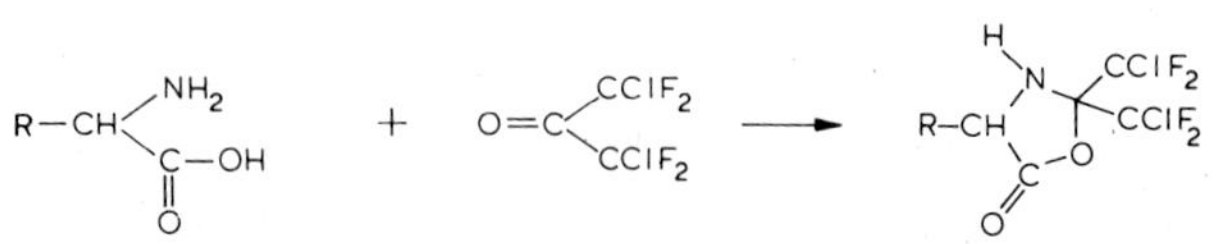

The mild conditions permit such sensitive amino acids as cystine, histidine and tryptophan to be analyzed properly after further derivatization by acylation.

3.8.6 *Derivatization for the separation of enantiomers*

The requirement that a reagent be chiral in order to exert differential action upon a pair of enantiomers is normally met in one of two ways: (1) treatment of the mixture with a conventional derivatizing agent and separating the products on a suitable chiral stationary phase; (2) treatment of the mixture with a chiral derivatizing agent and separating the products on a conventional stationary phase.

The first option, originally directed toward the separation of amino acids and peptides via their *N*-acyl esters, has been expanded to encompass carbohydrates, hydroxy acids, and additional substances [248–250]; there has likewise been a trend toward converting hydroxyl and carboxyl functions to nitrogen derivatives (e.g., urethanes, *tert*-butylamides, and the like), the objective being better resolution. A typical chiral stationary phase might consist of the *tert*-butylamide of a chiral amino acid, acylated with a long-chain group like lauroyl ($C_{11}H_{23}CO$-).

For preparation of diastereomeric esters from a racemic acid mixture, there is the classic naturally occurring alcohol, (−)-menthol, or, if greater volatility is sought, optically active 2-butanol. (+)-3-Methyl-2-butanol reportedly leads to good separation of diastereomeric esters from amino acids, hydroxy acids and branched acids [251].

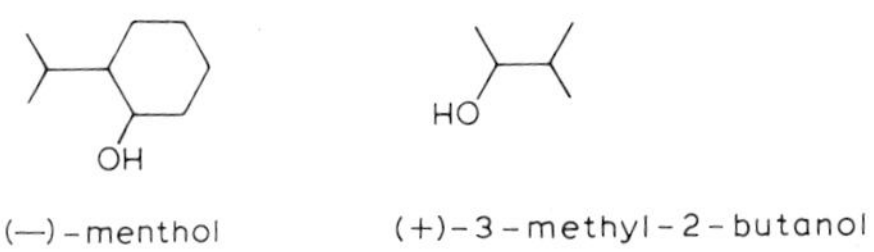

Where resolution of a particular amino acid via the *N*-fluoroacyl (+)-2-alkyl ester is not feasible, an alternative strategy involves the formation of a diastereomeric acyl dipeptide ester mixture through condensation with a second, chiral amino acid unit [252]:

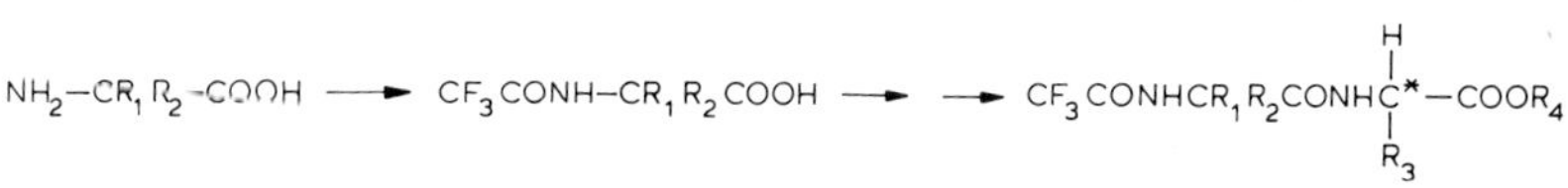

Employment of a chiral acylating agent is another possibility; resolution of 3,4-dihydroxyphenylalanine (DOPA) and its α-methyl derivative was accomplished [253] using (*S*)-α-methoxy-α-trifluoromethylphenylacetyl chloride as acylating agent.

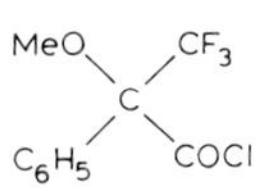

(*S*)—α—methoxy—α—trifluoromethylphenylacetyl chloride

3.9 *Sample preparation*

A typical biological sample may contain hundreds or even thousands of different compounds while only a small fraction of these is of analytical interest at a given time. It is thus essential to remove the interfering compounds prior to a GC analysis. A selective preconcentration of the substances of interest can ideally be accomplished at the same time.

In a broader sense, the removal of interfering molecular species is not unique to GC analyses; it has been widely practiced with many other biochemical and clinical determinations. For example, biomacromolecules can be denatured and precipitated through a variety of methods, or alternatively removed by gel-permeation chromatography prior to the sample analysis by GC. Additional purification methods may employ the acidobasic properties of either the interfering molecules, or a sample itself (pH manipulations, ion-exchange chromatography, ion-pairing extraction, etc.). Polar and nonpolar molecules can often be separated from each other in an easy manner as based on their different solubilities.

Solvent extractions of the biological compounds of interest are by far most common. The appropriate solvents provide acceptable media for sample storage or further concentration. While aqueous injections into a gas chromatograph have been described on previous occasions, they are limited in scope and should generally be discouraged. The choice of solvent can be crucial to a determination, as there are some uncompromising requirements that must be met in securing good results.

First, the solvents should be relatively free of trace impurities while dissolving the compounds of interest effectively. The solvent trace impurities, while not detectable in the bulk, can produce spurious peaks after the concentration of the volume of extracted sample. Whenever an extensive solvent clean-up is either impossible or impractical, appropriate sample blanks should frequently be run. With today's capabilities of high-resolution GC and spectral identification, occasional solvent impurities are tolerated as convenient 'markers'. A good sample solubility is required to minimize possible losses due to sample adsorption on the glassware or, simply, its unnoticed precipitation.

A single solvent-solvent extraction step seldom satisfies the purification requirements for a GC determination; multiple extractions are often needed for a more complete sample purification. However, such multiple extractions and transfers of a sample from one medium to another may result in further uncontrolled sample losses. Selective fractionations are used with advantage wherever some unique features of the analyzed molecules exist; among the most typical approaches, researchers have extensively used ion-pairing extractions of various ionic samples and the digitonin or Girard separations of steroidal compounds.

In order to achieve a more efficient purification, the compounds of interest are sometimes converted to their derivatives [254–256], which are easier to separate from the ballast compounds. General guidelines for purification of biological samples would be difficult to propose. The extent of sample fractionation is strongly dependent on the type of biological material (urine, blood, cerebrospinal fluid, tissue, etc.), the nature of solutes, and their concentrations. The sensitivity and precision requirements will also vary. Any attempts to devise 'universal procedures' for analyzing biological materials may entail difficulties similar to those experienced in the environmental chemistry field [257,258].

Numerous small molecules present in biological fluids and tissues are frequently encountered as a part of larger molecular complexes, or with a different degree of biological conjugation. For example, numerous metabolites that the human body excretes into the urine are conjugated as glucuronides or sulfates. The attachment of various molecules to phosphate groups throughout the body metabolism is also quite common. In order to analyze these metabolites, a deconjugation step is usually employed. Thus, it is quite common to cleave glucuronides, sulfates, etc., either enzymatically or 'chemically'. The chemical approaches consist of a treatment with an acid or alkaline medium. Similar to this is a commonly used saponification of various large lipid molecules to yield fatty acids for a GC determination. The enzymatic approaches are generally believed to be more gentle toward sensitive structures, although the occurrence of endogenous inhibitors of the enzymes used could seriously complicate the overall analytical task. Alternatively, certain biological conjugates can now be subjected to high-temperature GC without cleavage [259,260].

Formation of artifacts in GC analysis is perhaps encountered more frequently than many investigators acknowledge. While some artifacts could be formed even prior to the GC terminal analysis, the most critical point seems to be the rapid vaporization following the sample injection. A wider utilization of the on-column sampling techniques is desirable.

'Classical adsorbents' such as alumina or silica gel have been used extensively in the past to separate the individual compounds of interest from the ballast material or, alternatively, groups of compounds from each other. Short columns or thin layers of adsorbents are used to 'filter' the compounds of interest from the rest of the material. Similarly, ion-exchange resins can be employed to rid the sample of unwanted basic or acidic constituents. More recently, numerous efforts have been made to replace these conventional chromatographic materials with new alternative packings that cause less irreversible adsorption and sample losses. Among them, modified polydextrane gels [261,262], Amberlite XAD resins [262–265] and the new reversed-phase HPLC 'sample filters' [266,267] have been most notably used. A few illustrative examples will be mentioned below.

Different types of Amberlite XAD resins with unique physical and chemical properties have been available. Their use for adsorption of polar organic molecules directly out of biological samples has been demonstrated with pharmaceuticals [263], plant nucleotides [264], plant growth hormones [265], and various steroids [262,268]. In the last case, ample evidence is now available that the use of these resins causes substantially better recoveries of more polar steroid metabolites [262,268] than the previously employed solvent extractions. Setchell et al. [262] used the organic resins and modified dextranes for a complete fractionation of urinary steroid conjugates. Their general procedure (Fig. 3.15) involves the initial sample adsorption and several additional chromatographic steps combined with conjugate hydrolysis prior to the sample derivatization and GC analysis. While the overall procedure appears complex, it permits an effective analysis of five different groups of metabolically important compounds from one sample.

Hydroxypropyl derivatives of cellulose and Sephadex allow further reactions [261] to prepare a wide range of lipophilic ion-exchangers. Because such materials possess relatively high capacity and selectivity toward certain sample types, their further extensive use as group separation media may become popular. Separation of steroid conjugates [262,269], oximated 3-ketosteroids [256], and the metabolites of contraceptive steroid drugs [270] were all achieved with excellent sample recoveries. Such nonadsorptive media are essential in picogram-level determinations of steroids in small samples of blood.

Extensive purification procedures prior to the GC analysis may result in sample losses and a considerable analytical error. To provide meaningful results, sample recovery has to be established throughout the entire sample work-up procedure. The use of labeled compounds is generally considered adequate for checking the losses of standard compounds. Understandably, there is an increasing tendency toward the development of simplified sample preparation procedures. With a wider utilization of high-resolution chromatography, the number of steps in a sample preparation may eventually decrease. HPLC is also likely to find an increasing applicability as a very efficient and fast sample purification and fractionation method in the future. An example of this is shown in Fig. 3.16 [271]; although the fractionated sample (the

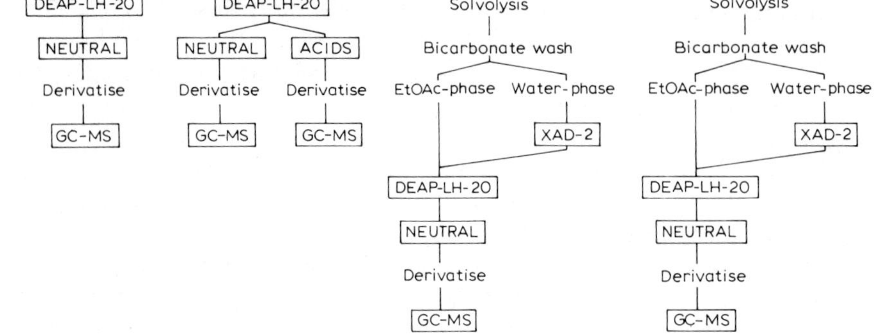

Fig. 3.15. General scheme for the analysis of urinary steroid conjugates. Reproduced from [262] with permission of Pergamon Press.

basic fraction of marijuana smoke condensate) represents a different sample matrix, it is easy to imagine a similar role of HPLC with a variety of complex biological mixtures. Of particular note is the use of high-performance exclusion chromatography for separating low-molecular-weight samples from the complex matrices of larger molecules, as demonstrated by Majors and Johnson [272].

While working with ever smaller biological materials, the biochemists increasingly face the problem of manipulating very small samples (e.g., a few microliters of a fluid, a tissue biopsy, a microsomal suspension, etc.). In order to determine trace quantities of organic compounds in such materials, it is necessary to develop appropriate manipulation techniques and skills at a microscale. It is essential in such determinations that a very substantial fraction of the total sample be utilized in a GC sample injection. Such efforts are not yet sufficiently documented in the literature. However, a report by Dünges and Kiesel [273] has shown that microliter amounts of physiological fluids can be adequately handled while utilizing up to 40% of the total sample content for the actual analytical measurement.

Finally, a note of caution is needed to secure proper handling of biological specimens for GC analysis. Many of the recommendations as well as the rules for GC analysis are similar to the requirements for other clinical determinations (collection rules, sample storage, transportation, etc.), but special needs for GC may sometimes arise. For example, while certain foreign compounds (preservatives, dietary artifacts, therapeutic drugs, etc.) may not matter in conventional determination, they may be a problem in GC analysis. A publication by Jellum [15] discusses this matter in some detail.

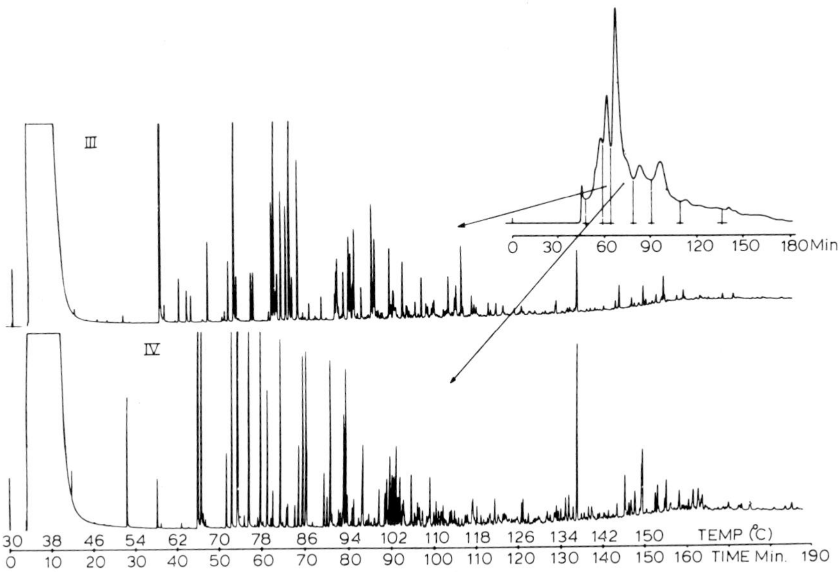

Fig. 3.16. A typical analysis of HPLC fractions by capillary GC. Reproduced from [271] with permission of the American Chemical Society.

3.10 Selected applications (see Note following References, p. 147)

3.10.1 Steroids

3.10.1.1 General

Implications of steroids in a variety of biological processes provide a continuous impetus for improved analytical techniques. In particular, the hormonal regulation of physiological processes that remains one of the most investigated areas of modern biomedical research requires measurement techniques of considerable sophistication. The sensitivity of such measurements is being continuously challenged by the requirements to determine ever smaller quantities of steroid hormones and their metabolites in various body fluids and tissues.

Since the steroids are among the most 'versatile' molecules that nature has designed, they are also known to affect a variety of metabolic processes. Formation of numerous steroidal isomers is possible through such metabolic interactions. Thus, a number of active hormones synthesized by the organs can be metabolized during their action into various oxygenated products which are subsequently excreted by the body. Various biological conjugation processes (e.g., formation of glucuronides and sulfates) further add to the seemingly high complexity of steroid metabolic pathways. Owing to this multitude of metabolic processes, there are a number of steroids to be determined analytically. Hence, the need for efficient separation methods, such as capillary GC.

Various analytical methods of extremely high sensitivity are now available to measure minute concentrations of steroid hormones in blood. GC is frequently preferred over other measurement principles because of its high sensitivity and reliability. Since some of these GC procedures are technically involved, they are used more often in biomedical research laboratories than in a routine clinical environment.

Many additional steroid compounds are encountered in relatively complex mixtures. An increasing use of GC for the separation of biological sterols has been noticed during the last decade. The materials of interest may include bacteria, algae, various plants, marine animals, mammalian tissues, etc. Various dietary aspects of sterols and their metabolites, including bile acids, have recently been studied to a large extent.

The following sections will summarize the most important aspects of steroid applied investigations involving GC. A special emphasis will be placed on methodologically interesting cases.

3.10.1.2 Steroid hormones in blood and tissue

Advances in GC separation and ionization detection techniques during the 1960's generated a great deal of interest and justifiable excitement for endocrinologists, due to the new possibilities of following hormones and their metabolites in body circulation. In particular, the development of electron-absorbing derivatives

[18,19,274] led to optimistic estimates that during a decade or so most clinical and endocrinological laboratories would widely utilize GC with electron-capture detection. However, this trend was substantially affected by the later availability of competitive protein-binding analysis and radioimmunoassay techniques. Consequently, the development of gas-phase analytical methods for hormone measurements has continued at a slower pace. However, it should be emphasized that there are still compelling reasons for further progress in this area.

In order to understand the dynamics of hormone production, secretion, and its effects on the target tissues, high-sensitivity measurements in blood and tissues are required. Because the levels of circulating hormones are generally low and the volumes of blood or tissue available for the analysis tend to be small, subnanogram-level determinations are necessary. The sensitivity of many ionization detectors, including the flame ionization detector and the electron-capture detector, do have this capability.

However, the *measurement specificity* is yet another important consideration in assuring the desired results; conventional GC detectors can hardly satisfy this criterion if the measurements are performed in crude biological extracts. In this view, it is hardly surprising to read the statement by Adlercreutz and Luukkainen [275] from 1968 that "...at present the most pressing need in the GLC study of estrogens is not to find better derivatives for GLC or more sensitive instruments, but to develop more specific and convenient purification methods in order to permit higher specificity in estimates of the compounds." Indeed, the monograph of Eik-Nes and Horning [274] contains many references and descriptions of methods for reliable measurements of estrogens, androgens, progestins and corticosteroids in plasma and tissue samples. Given the technological limitations of GC in the 1960's, most precision values quoted throughout the book (usually 10–15% coefficient of variation) are surprisingly good. The greatest problem was indeed sample purification and particularly the time of sample clean-up needed to arrive at satisfactory results. For this reason alone, it is hardly surprising that these approaches found little utilization in the routine clinical laboratories.

Indeed, many such laboratories have performed urinary analyses of hormone metabolites rather than plasma measurements. While the urinary metabolite determinations may provide valuable information of its own, it is often argued that they are not entirely representative of the hormonal secretion due to a variety of reasons (incompleteness of sample collection, altered renal function, multiple metabolic sources of a urinary metabolite, etc.).

After the successful development of the radioimmunoassay [276,277], enzyme immunoassay [278] and competitive protein-binding methods [279,280] for most hormones of interest, it was generally felt that the necessary clinical criteria of precision, accuracy, specificity and sensitivity can be met without the tedious GC methods. However, this notion has been challenged more recently. While such techniques are undoubtedly more useful with considerably larger molecules, cross-reactivity [283] toward molecules as similar, structurally, as testosterone and 5α-dihydrotestosterone is regularly observed. Thus, unless the tedious purifications are once more considered prior to the actual assays, significant errors may result [283]. The necessary specificity at picogram levels appears difficult to assure through immunoassays. In addition, the preparation of antisera or binding proteins can be a relatively complicated task. These facts have now caused a number of leading endocrinological laboratories to reconsider GC as a potentially more valuable alternative.

Several major advances that made GC a more attractive tool in blood hormone analysis took place during the last 10–15 years: (a) development of more efficient, fast, and selective ways to purify plasma extracts, as based on the availability of various lipophilic gels and HPLC; (b) availability of highly efficient capillary columns to reduce the cases of co-elution of hormones with other mixture components (also, increasing detection sensitivity in most instances); and (c) a wider utilization of the mass spectrometer as a highly specific and sensitive GC detector. Several examples will now be shown to reinforce these points.

Baba et al. [283] developed a mass-fragmentographic technique for plasma testosterone measurements with sensitivity down to a few picograms. A comparison with the radioimmunoassay method showed the GC method to be considerably more precise. The good accuracy of mass-fragmentographic methods is greatly facilitated by the possibility of using hormones labeled with stable isotopes as internal standards. A recent study of Tetsuo et al. [284], measuring endogenous estradiol in rat plasma and uterine cytosol, is quite representative of the state-of-the-art in low-level steroid measurements. Their method involved isolation with lipophilic gels, capillary GC and single-ion monitoring with a medium-resolution mass spectrometer. The detection limit was estimated to be 4–5 pg/ml estradiol; Fig. 3.17 shows the peak corresponding to an equivalent of 0.12 ml of female rat plasma.

Improved chromatographic techniques can vastly simplify hormone measurements even with more conventional detectors, such as the electron-capture, or even the flame-ionization detector. Wehner and Handke [285] used a very simple sample purification procedure with high recovery of plasma progesterone and its good quantitation as a 3,20-di-*O*-pentafluorobenzyloxime derivative at picogram concentrations. Measurements of 18-hydroxycorticosterone in human peripheral blood as a heptafluorobutyrate were performed by Wilson et al. [286]. Levels of testosterone, estrone and estradiol in male rat plasma were described by Maskarinec et al. [287,288] in relation to treatment with Δ^9-tetrahydrocannabinol and cannabis extracts; heptafluorobutyryl derivatives were used for the electron capture detection.

Whereas the primary objective of a great many GC plasma steroid determinations is to measure one or just a few compounds, modern GC permits determination of the entire profiles of substances. Novotny et al. [115] utilized glass capillary columns combined with sample preconcentration to develop multiple profiles of different steroid conjugates from 1–5 ml plasma samples. Capillary GC and mass-fragmentographic detection were used by Axelson and Sjövall [256] to monitor selectively the profiles of 3-ketosteroids in plasma from women in different stages of pregnancy.

Multicomponent analytical capabilities of modern GC are quite useful in monitoring quantitative metabolic changes in tissues, cell cultures, etc. Uptake, binding

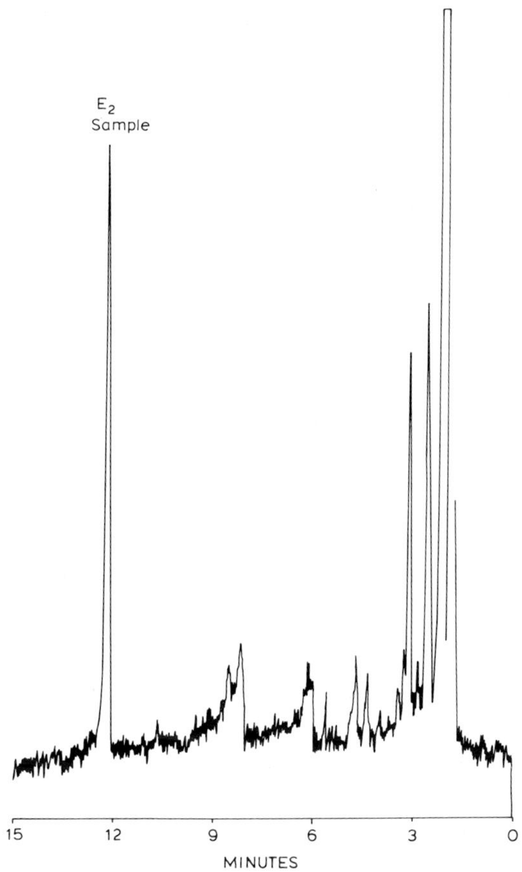

Fig. 3.17. GC/MS analysis of estradiol in a sample of female rat plasma. Reproduced from [284].

and cellular metabolism of steroid hormones can be followed in the isolated organs from laboratory animals. Typically, precursors of the measured steroids are introduced into the perfused organs, cell homogenates, etc., and allowed to interact with the corresponding enzymes for a certain period of time. The steroid metabolites are then extracted from the tissue, derivatized and determined by GC. Adequate sensitivity is needed in work with small laboratory animals. An example of such investigations was demonstrated by Maume et al. [289]; several corticosterone metabolites were shown to be synthesized by the rat adrenal cell culture after stimulation with ACTH. The current sensitivity and reliability of capillary GC will undoubtedly permit various future investigations in small tissue biopsies or blood samples.

3.10.1.3 Urinary steroids

Determination of urinary steroids has been of continued interest because: (a) certain unique metabolic information becomes available that is not easy to obtain otherwise; (b) accumulation of hormone metabolites in the urine frequently results in a relatively easy determination; and (c) in work with small laboratory animals, the steroid measurements in the urine as opposed to plasma, are considerably easier; they permit working with individual animals without sacrificing them and, thus, complicate to a lesser degree various designs of biological experiments. Importantly, the acquisition of 24 h urine samples tends to minimize fluctuations due to the 'bursts' of hormones at different times of the day. However, in all of the above considerations it remains true that the concentrations of the urinary metabolites have to be ultimately related to the hormone levels in circulation.

The gas-phase analytical techniques have been used for the analysis of urinary steroids for a long time. The determinations of urinary estrogens, progesterone metabolites, 17-ketosteroids and, to a lesser degree, corticosteroid metabolites, with packed-column GC are extensively documented in the earlier monographs on the subject [274,290]. Various sample treatments, approaches to conjugate hydrolysis, and volatile derivatives have been described. Among those steroids, aldosterone stands out as a uniquely difficult substance to derivatize and determine.

Whereas the earlier work used GC for the estimation of only selected urinary steroid constituents, most current work favors the multicomponent (metabolic profiling) approach. The preparation of 'mixed derivatives', such as, for example, the methoxime-trimethylsilyl derivatives of Gardiner and Horning [291], in principle facilitates conversion of all metabolites containing hydroxy and carbonyl groups. Through the introduction of capillary columns to steroid analysis [12,292,293], it became possible to separate the complex mixtures of such derivatives.

The remarkable utility of capillary GC in urinary steroid analysis has been widely documented throughout the last decade. The urinary profiles may reflect much of the total steroid metabolism within the body, as the chromatograms typically display peaks ranging from 17-ketosteroids, through their oxygenated products, various pregnane derivatives and corticosteroid metabolites, to some heavily oxygenated substances (e.g., pregnane hexols or heptols, if present in a given sample). Provided that suitable identification techniques and standard compounds become available in the future, much new metabolic information can be revealed from the currently available profiling methodology. Examples of this are seen with the recently identified cortoic acids [294] and C_{26}-pentols in the urine [268].

The quantitative comparisons of steroid urinary profiles may reveal much useful information that is currently sought in modern biomedical research. Thus, while capillary GC/MS techniques have been used to identify the individual urinary metabolites, peak-height comparisons were shown to facilitate characterization of the steroids typical of human newborns [295], studies of various endocrinological disorders [296–299], breast cancer [300] and diabetes mellitus [268]. As an example, Figure 3.18 shows a comparison of typical profile differences between normal and diabetic human males [268]; briefly, peaks 2 and 3 (androsterone and etiocholanolone) are depressed in the diabetic, while peaks 48–56 (cortisol metabolites), and peak 66 (a C_{26}-pentol), are characteristically elevated.

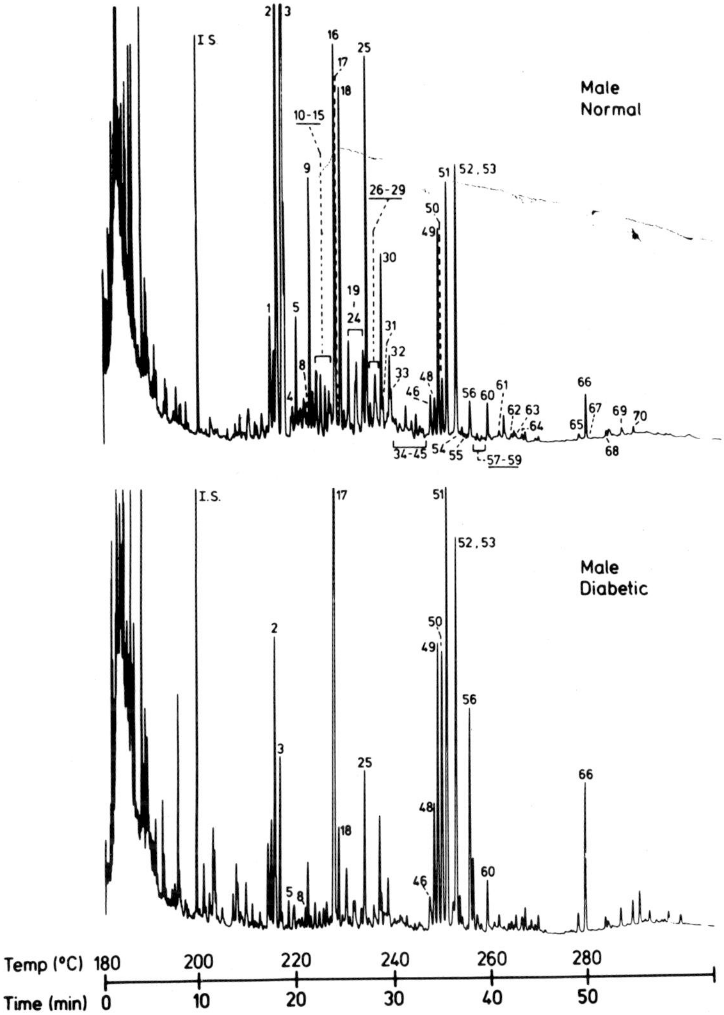

Fig. 3.18. Representative urinary steroid profiles of a normal male versus a diabetic male. Reproduced from [268].

As urinary steroid metabolites have different degrees of biological conjugation, analysis of the separate conjugate profiles may be informative of certain physiological and pathological situations. Development of methods for corresponding conjugate analysis thus appears desirable; while direct conjugate analyses were attempted [259,260], group separations followed by deconjugation and capillary GC or GC/MS are currently preferred [262].

Although the present use of urinary steroid profiling techniques appears to be largely confined to biomedical research, their gradual acceptance in clinical diagnosis and preventive medicine (to establish the risk of biochemical endocrine disorders) is very likely in time. Biological implications of the current work are indeed numerous.

3.10.1.4 Sterols

Sterols are under continuous investigation because of their role as precursors of hormones and their function in the central nervous system and membranes. Even though other methods are available for cholesterol determination, GC measurements are still fairly common when high sensitivity is required. Many other sterols are abundant in nature; determination of sterol profiles by GC is often of chemotaxonomic value in investigating bacteria, plants, marine organisms, etc. In addition, the products of sterol oxidation have received some attention for their cytotoxicity, mutagenicity and carcinogenic potential [301,302].

There are no particular problems with derivatization of sterols. The hydroxy group at C_3 can be readily converted to various derivatives; trimethylsilyl and acyl derivatives are by far the most common. While the derivatization renders sterol molecules more stable, it has been demonstrated [303] that the inert capillary columns will also elute underivatized sterols. The analytical challenge in sterol separations is, however, caused by the frequent need to resolve structurally similar compounds. This has been clearly demonstrated already in the earlier studies of Knights [304], who utilized an elaborate separation scheme for a more complete resolution of plant sterols.

While the basic steroidal skeleton and presence of the polar group at C_3 are preserved for a variety of natural sterols, the minor structural changes in the side-chain contribute little to the chromatographic mobility of these compounds. For example, the presence of a double bond in such a large molecule will cause a relatively minor retention effect. An effective resolution needs either stationary phase of high selectivity, or a high column efficiency; for example, the difficult pair cholesterol/cholestanol can only be resolved with a capillary column [305]. A double-bond position and minor changes in the side-chain stereochemistry present further separation challenges. Thus, it is hardly surprising that a number of investigators have been using capillary GC with an increasing frequency for the separations of biological sterols. Further efforts to use more polar thermostable phases have also been evident [305–308]. As shown by Brooks et al. [309,310], certain sterol resolution problems can also be overcome through the use of cholesterol oxidase; in samples of marine invertebrates and sponges, the enzymatic conversion of Δ^5- and 5α-3β-hydroxysteroids leads to products that are easier to resolve.

The above-referenced studies on biological sterols published during the last decade are representative of the general capabilities of capillary GC for this class of compounds. The use of combined GC/MS leads generally to structural elucidation, but careful comparisons of retention characteristics are in order due to the extensive occurrence of isomeric compounds. In this view, extensive retention studies of various sterols [311,312] are still justified. An example of the general capabilities of capillary GC is shown in Fig. 3.19, comparing the sterol profiles of tobacco and marijuana plant materials [306]; most of the resolved components were identified through GC/MS and retention time of authentic compounds.

Chromatography and quantitative evaluation of sterol esters in both plant and animal materials has been of much interest for some time. GC separations of sterol esters necessitate high column temperatures (300–350°C). Kuksis [313] pioneered these separations. A more comprehensive review of this area will be given in a discussion of lipids in one of the following sections.

3.10.1.5 Bile acids

Determinations of bile acids as the criteria of hepatobiliary system functioning have been on the increase. Gas-liquid chromatography played a key role in the earlier analytical separations in this field and various metabolic studies carried out subsequently. Extensive studies of Sjövall and co-workers during that period have been reviewed in the chapters on this subject [314–316]. Bile acids are typically polyfunctional compounds that need the development of reliable derivatization approaches prior to GC. As these derivatization methods gradually improve, an increasing number of investigators will participate in various studies of bile acid metabolism. Since the thermal stability of certain bile acid derivatives appears limited [219], there is a general feeling that modern HPLC methods will eventually replace GC. However, HPLC approaches to the analysis of bile acids are still in a developing stage.

The initial complications of the analysis of the bile acids are due to their natural occurrence as various biological conjugates (e.g., as glycine, taurine and various

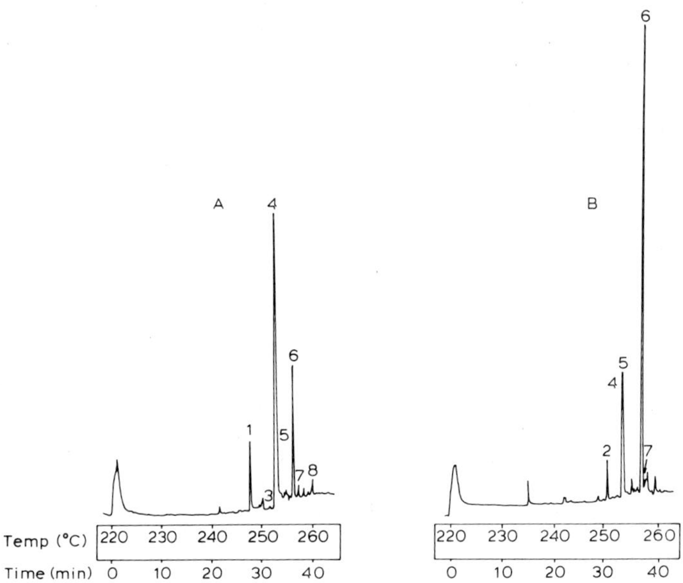

Fig. 3.19. Comparative runs of tobacco (A) and marijuana (B) sterol fractions by capillary GC. Reproduced from [306] with permission of Holden-Day, Inc.

polysulfated forms); these must be reliably cleaved prior to GC. Extensive investigations [267,317–320] were reported on the subject of quantitative extraction, conjugate fractionation and cleavage of isolated conjugates. An effective isolation is frequently a tedious task, as the biological materials for analysis are frequently difficult to extract. These include serum, bile, duodenal contents or fecal matter. The removal of interfering substances usually involves selective extractions, ion-pairing and acid-base manipulations, as well as various forms of chromatography. A sufficient degree of specificity may further be provided by the use of computerized GC/MS [320].

Formation of suitable derivatives from bile acids has also been a subject of numerous studies. Since the earlier procedures have been reviewed [314–316], a continuous interest in the subject of derivatization appears to indicate that there is still a need for improvement. The initial derivatization of the acidic function to form methyl esters seems to be almost universally employed, while the remaining functional groups (hydroxy and keto) can be converted to a number of derivatives. A number of earlier and more recent studies have endorsed silylation, although various types of acylation are also common [219]. Permethylation has also been advocated recently [225], but no biological separations were demonstrated using this approach.

The mixtures of bile acids isolated from biological materials can be exceedingly complex. A recent interest in capillary GC of these compounds [321,322,225,267] is thus justified. Interestingly, even a partial derivatization has been advocated [323] to increase resolution of various bile acids which are not adequately resolved when all polar groups are fully covered. A need for reliable identification and characterization techniques is reflected in the systematic investigations of chromatographic retention and mass-spectral studies of various bile acid derivatives [219,322,324,325].

Obviously, much remains to be learned about the occurrence of bile acids in physiological fluids, their relation to steroid metabolism [326], carcinogenesis [327], general liver metabolism [328], etc. A good example of what a new, powerful methodology can achieve is shown in a paper by Almé et al. [320], who used GC/MS to resolve a complex urinary mixture of highly oxygenated bile acids and identified about 30 components. Such profiling techniques are likely to have increasing utilization in the investigation of various cholestatic states, liver diseases and the defects of steroid and bile acid metabolism.

3.10.2 Lipoid substances

3.10.2.1 General

The term ‘lipid’ is frequently used to denote a wide variety of natural products; ‘fatty’, ‘oily’ or ‘waxy’ substances of animal and vegetable origin that are easily soluble in organic solvents readily satisfy this loose definition. Thus, such diverse compounds as fatty acids and their derivatives, triglycerides, sterols, phosphatides and sphingolipids, carotenoids, bile acids, vitamins A, D, E, and K, long-chain alcohols, terpenes, etc., may be included. Studies of these substances by GC are indeed numerous; a comprehensive survey of this field is beyond the scope of this contribution. The role of chromatography in lipid investigations has been reviewed [329–331].

The lipids are among the most important components of human diet and occur widely in nature. However, it is the biochemical role of lipids as the basic components of various cellular membranes and the lipid-protein complexes (lipoproteins) that bring them into the focus of highly important scientific activities. Major clinical interest has concentrated on blood lipid chemistry as related to atherosclerosis, lipid storage diseases, diabetes, and other metabolic conditions.

Investigations and determinations of various types of lipids usually employ a number of analytical techniques of which GC is just one approach. The large lipid mixtures are first fractionated into different classes (cholesteryl esters, triglycerides, phospholipids, etc.), while various forms of liquid chromatography are typically used to separate further the individual molecular species from each other. A controlled chemical degradation may subsequently be applied to generate molecular fragments, such as fatty acids, that are amenable to GC.

While contemporary GC is frequently powerful enough to resolve various saturated and unsaturated molecules, the selectivity inherent to 'argentation chromatography' in the liquid phase is often needed to resolve these complex mixtures prior to a GC investigation. Most investigations of lipid materials clearly benefit from a multi-technique approach. An example is shown in Fig. 3.20, where reversed-phase HPLC was first employed [332,333] to achieve separation of a triglyceride mixture on the basis of solubility (A) and, secondly, through a specific interaction of the previously trapped and rechromatographed peak with the silver ions added to the mobile phase (B); finally, a fraction of this material, following the second chromatographic step, was saponified, methylated and determined by GC (C).

3.10.2.2 Intact lipids

Fairly large (but relatively non-polar) lipids such as triglycerides and cholesterol esters can be directly chromatographed in the gas phase. Such separations need

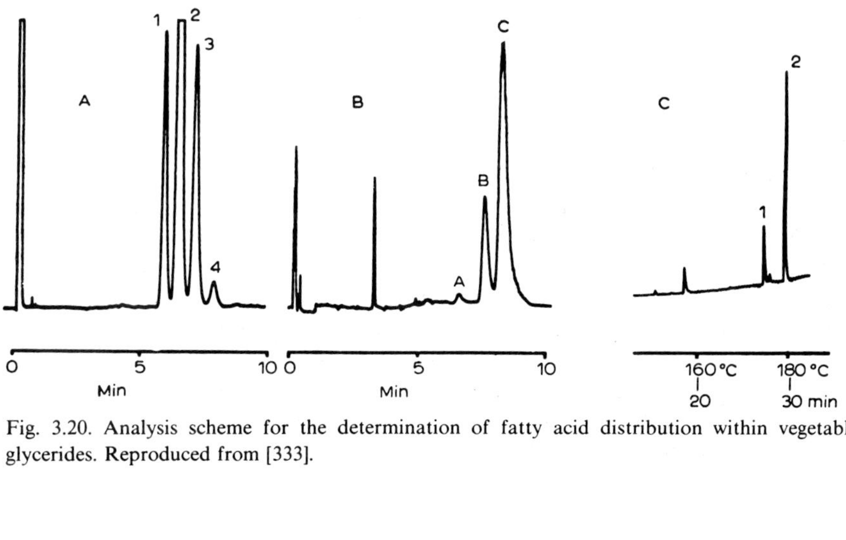

Fig. 3.20. Analysis scheme for the determination of fatty acid distribution within vegetable oil triglycerides. Reproduced from [333].

column temperatures in excess of 350°C and highly stable stationary phases. While the early work by Kuksis [313] was done with short packed columns, more recent investigations with thin-film glass capillary columns [86, 334] show some improvements in resolution of triglycerides. However, capillary columns alone do not seem to provide here the kind of 'breakthrough methodology' as commonly observed with other classes of endogenous metabolites. This is due to the fact that a degree of unsaturation and other minor differences in large triglyceride molecules contribute little to the overall retention at such high temperatures. In addition, degradation of triglyceride molecules during GC has been observed [335], and the best sampling techniques and column technology must be used. Potentially, HPLC appears to be a method of choice [333,336] for many future studies in this area.

Most clinical determinations of triglycerides are based today on non-chromatographic principles; total triglycerides are routinely evaluated. Nevertheless, high-temperature GC profiles were shown to be useful in biomedical investigations involving various cases of hyperlipidemia [337,338]. It is possible that the ratios of different triglycerides and cholesteryl esters could provide some metabolic information.

Additional large lipid molecules have been directly analyzed by GC. Typical studies of this type include the determination of Coenzyme A esters [339], intact triacylglycerols of human serum lipoproteins [340], tocopherols as trimethylsilyl derivatives [341], and ceramides as both trimethylsilyl [342,343] and boronate [344] derivatives.

GC can also be used to separate partial hydrolytic products of certain larger molecules, such as phospholipids. An enzymatic hydrolysis readily yields the products (e.g., monoglycerides, ceramides and diglycerides) that are amenable to GC after a proper derivative formation [345,346]. A dephosphorylation of phospholipids by an enzyme prior to GC was also used by Kuksis et al. [337,347] to assess the effect of certain drugs on hyperlipidemic conditions.

Alternatively, pyrolysis in combination with GC for a direct analysis of glycerophospholipids was described by Kuksis et al. [348] and Horning and coworkers [349].

3.10.2.3 Fatty acids

The major application described in the James-Martin landmark paper on gas-liquid chromatography is the separation and quantitation of fatty acids (FA). Indeed, it has remained as one of the major applications to this date. It has been estimated [350] that some 25% of all papers published in the field of GC involve, in one way or another, FA or their derivatives. A vast range of samples have been analyzed for FA: animal and plant oils, foodstuffs, bacterial products, glandular secretions of animals, and various physiological fluids and tissues, just to mention a few. As more and more information is being sought concerning the composition of various lipids, the applications are expected to increase even further in the future.

Determination of lower FA (typically, C_1–C_5) are of particular importance in microbiology, characterization of fermentation products and related industrial problems. The packed-column technology has been fairly well established in this area, although special precautions are needed concerning treatment of the column packing and a type of stationary phase [350]. When porous polymers are used as the packing materials, a direct analysis of aqueous media is feasible [351]. The preparation of esters with an aromatic moiety is sometimes preferred to improve separation from interfering materials [352] or to increase specificity with a gas-phase UV detector [130]. These approaches have already been discussed in the previous section on derivatization.

A majority of GC applications in the lipid analysis field involve FA analyses in the C_{12}–C_{24} range, as these are the major components of various oils and fats. Interest in the higher homologs has been limited for both methodological and biochemical reasons. While few methodological differences exist among different applications, the number of investigations is currently overwhelming. It is most common to analyze such FA as their more volatile esters, although Zoccolillo et al. [353] demonstrated separation of C_8–C_{18} free acids with an acidic stationary phase coated on glass capillary columns. Alternatively, a phosphoric acid additive to a conventional phase [354] can accomplish similar goals. On the other hand, the preparation of volatile esters is extremely easy, yielding very stable derivatives and extending considerably the volatility range.

Resolution problems in the GC analysis of FA esters prompted many investigators in the past to develop capillary techniques to search for more selective stationary phases, and ultimately, to combine both approaches whenever required. In fact, FA esters were among the first substances (beyond hydrocarbons) that were successfully chromatographed on stainless steel capillary columns [355]. The most difficult separations involve different geometrical isomers, and the presence and positions of unsaturated carbon-carbon bonds. Such separations are non-trivial and justify the effort of numerous laboratories to solve these problems.

Clearly, the polyester stationary phases advocated earlier (79) in the field of lipid separations possess only a limited thermal stability. Cyanopropyl silicones [356,357] now provide selective and more stable stationary phases for the separation of FA esters. However, the column selectivity alone falls short of some resolution requirement, while capillary GC seems to be the most profitable route to pursue. The state-of-the-art separation of FA methyl esters is reflected in Fig. 3.21 [358], where many of the minor structural differences in the chromatographed FA molecules yield readily distinguishable peaks. Acquisition of suitable standards as well as extensive correlations of FA molecular parameters with retention characteristics [359] are now clearly needed.

Characterization of unsaturated FA, as well as those FA that possess some substitution in the chain, is clearly a difficult task. As it was indicated in the previous section on derivatization, a continual suggestion of new derivatives from different laboratories over the years point to some deficiencies in this area [221]. In addition, better methods are needed to recover quantitatively polyunsaturated acids from various lipids [210]. Lipid unsaturation is of much current interest in relation to medicine and dietary research.

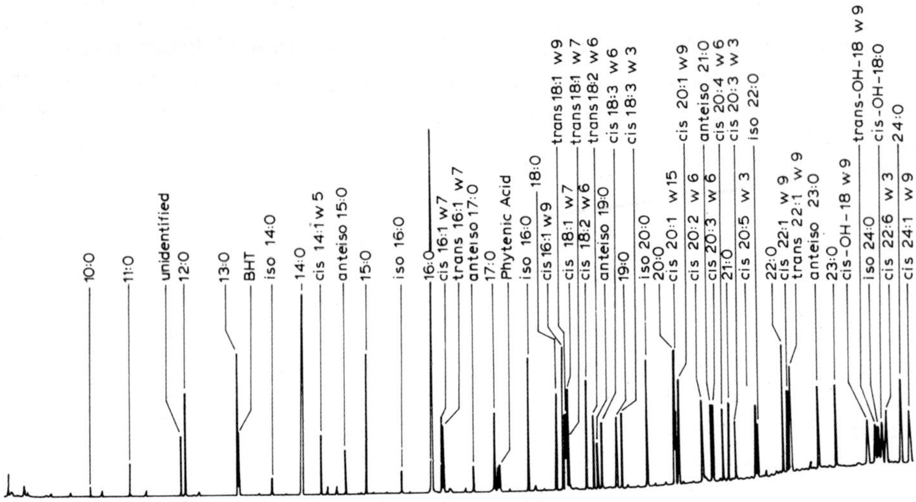

Fig. 3.21. High-efficiency separation of fatty acid methylesters (standard compounds) on a glass capillary column. Reproduced from [358] with permission of Lipid Research, Inc.

Numerous applications are now encountered where FA chromatographic profiles of a human physiological fluid or tissue are correlated to certain pathological conditions. A few representative examples will now be mentioned that include both free (non-esterified) FA and the saponified lipids. The identification of a methyl-branched FA (phytanic acid) in plasma of the patients with Refsum's disease [360] is now a widely known example of the power of GC in studying various metabolic defects. The profiles of FA from brain tissue lipids were investigated for various neurological disorders [361,362] and in experimental animals [363]. Tichy et al. [364] determined FA in different lipids isolated from the cerebrospinal fluid; while the FA profiles in cerebrospinal fluid differ from those in blood serum, no obvious correlations between the FA composition and human neurological complications were established at this time.

Lin and Horning [341,365] carried out comparative investigations of serum long-chain acids (in both free and 'bound' state) of patients during the post-stroke period versus normal individuals; a marked decrease in the linoleic/oleic acid ratio was observed for the stroke patients. As capillary GC becomes a widely accepted routine method, its use for rapid screening of various disorders of the lipid metabolism will become more obvious to a number of clinical laboratories. Some progress in this direction has already been indicated [366–368]. As reported by Haan et al. [368], screening of up to 50 samples a day for FA from serum and adipose tissue biopsies is entirely feasible.

Determination of the lipid profiles from the skin surface and sebum [369] may also hold some promise for various biomedical purposes. If increased sensitivity is required, special derivatives can be prepared [220] for the electron-capture detector or other sensitive devices.

The profiles of cellular FA also appear to be valuable in taxonomic studies on microorganisms [370,371]. In the method of Moss and Dees [371], the whole cells are saponified, with a subsequent derivatization and GC to determine more or less characteristic profiles of the straight-chain, branched, cyclopropane and hydroxy acids. Numerous applications of GC in clinical microbiology have been reviewed in a book by Mitruka [372].

3.10.3 Acidic metabolites

Concentrations of various carboxylic acids in human body fluids reflect some of the major metabolic processes of the body. These metabolites apparently originate from lipid and amino acid metabolism; the major metabolic defects are frequently associated with unbalanced concentrations of these acidic substances. One of the most widely occurring conditions of this kind is ketoacidosis in diabetic disease; high concentrations of the so-called 'ketone bodies' (3-hydroxybutyric acids, acetoacetic acid and others) are the traditional hallmarks of ketoacidosis. Many additional acidurias were discovered (particularly during the last 15 years) in major part due to the availability of GC and GC/MS techniques. Acidurias are among the serious medical conditions that are usually a result of genetic aberration (enzyme deficiencies), but environmental factors or nutritional deficiency could occasionally be involved. These conditions are characterized by either (a) drastically enhanced excretion of normal metabolic intermediates, or (b) excretion of unusual metabolites that are produced from the accumulated intermediates via alternate biochemical pathways. Many acidemic conditions have now been documented in the literature, and the role of GC in such medical discoveries has been adequately stressed in the recent reviews of Jellum [15] and Tanaka and Hine [373].

While the profiling methods in general have proved to be of great clinical value in diagnosing acidurias and related conditions, methodological difficulties still persist. For example, the wide range of organic acids that are present in blood or urine (aliphatic or hydroxy acids, keto acids, aromatic metabolites, di- and tricarboxylic acids, etc.) complicates their comprehensive isolation and acquisition of 'a complete profile'. Thus, many procedures that have been developed for the isolation of organic acids involve a compromise solution.

Two primary isolation procedures for organic acids in physiological fluids involve either (a) solvent extraction (usually with diethyl ether and/or ethyl acetate), or (b) anion-exchange chromatography. Alternatively, the latter can be substituted by the adsorption process on silica [374].

Solvent extraction methods usually involve partition of a salt-saturated, acidic sample aliquot with an appropriate solvent. Alternatively, the sample may be subjected to a derivatization step and a clean-up extraction prior to the acidic extraction [15]. A major drawback of the extraction procedures is an incomplete recovery of more polar acid metabolites. However, the method is simple and, consequently, more popular. The anion-exchange approach, developed initially by Horning [375], has been subjected to frequent modifications in other laboratories

[376–379] in order to increase isolation yields and decrease interferences. The methods described by Thompson and Markey [376] and Gates et al. [377] are quite representative of these necessary modifications. The inorganic acid interferences and the losses of more volatile acids during lyophilization, as well as difficulties in obtaining a completely dry residue [378] for derivatization appear to be the most serious drawbacks of the ion-exchange approach. The procedure is also too time-consuming for routine investigations.

Derivatization of acidic extracts can also be a source of analytical problems. Apparently, the presence of different functional groups necessitates multiple derivative formation in certain cases. Incomplete coverage of all functional groups prior to GC can lead to uncontrolled decomposition of certain molecules. As evident from the discussion in the previous section on the derivatization methods, the acidic polyfunctional compounds still need refinement of the associated techniques. Apparently, many laboratories have devoted a considerable effort to solving these problems, as evidenced by many reports on derivatization. Those acids which possess ketonic groups in their molecules are often first selectively derivatized by forming the oxime, methoxime or ethoxime derivative [380]. Subsequent treatment with a methylation and/or silylation agent renders various acids sufficiently volatile for GC profiling. Other derivatives, such as 2-quinoxalinols and *O*-trimethylsilylquinoxalinols [381], have also been successfully employed in biomedical investigations. The influence of a derivatization technique on mass-spectral properties should also be strongly considered.

Capillary GC is essential to deal with the very complex mixtures of urinary acidic metabolites. However, even the best columns of this type do not provide a complete resolution in some instances. As shown by Grupe and Spiteller [382], thin-layer chromatography and HPLC can be used to fractionate various urinary acids. As our separation techniques gradually improve, numerous new metabolites can be found for which no biochemical and physiological roles are known at present. Thus, investigations of unknown human biochemical pathways appear feasible through this new high-resolution methodology. However, such studies are considerably complicated by the variation in individual metabolic patterns. A significant rationale exists for studies in model animals where both genetic and dietary factors can be controlled, as shown by the recent investigations of the germ-free conditions and diabetes in rodents [168,383,384].

Among the elegant demonstrations of the power of GC techniques in the studies of human acidemias is the prenatal diagnosis of the defective metabolism of methylmalonic acid by Nakamura et al. [385], and the study of the progressive genetic deafness carried out by Jellum and co-workers [164]. However, less obvious metabolic alterations need good quantitative techniques. Correction of special metabolic conditions through medication or dietary modification can also be followed quantitatively. A good illustration of this is Fig. 3.22 [386] where the metabolic patterns of a maple syrup disease patient have been determined under different circumstances. Similarly, the complex profiles of urinary acidic metabolites in various diabetic conditions were quantitatively followed [387–390].

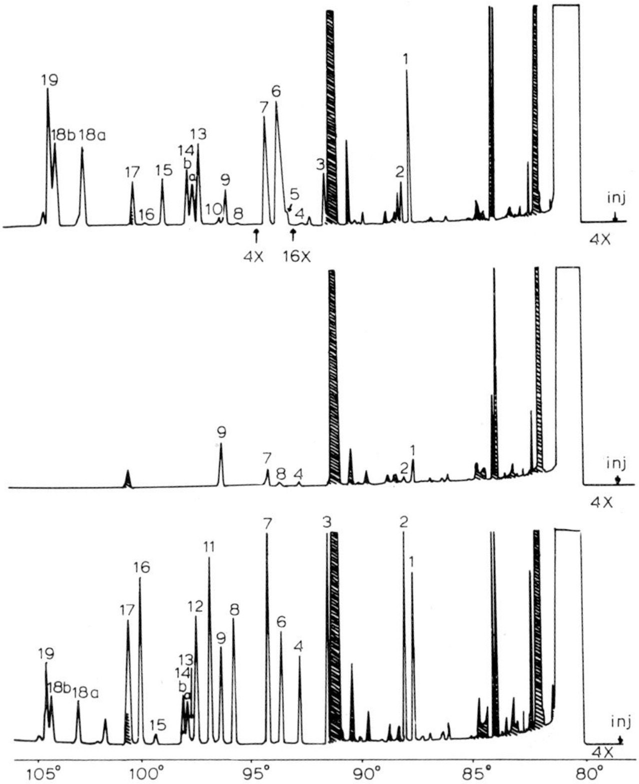

Fig. 3.22. GC separation of keto and hydroxy acids from the urine of a patient with maple syrup urine disease. Top chromatogram, the patient before dietary treatment; middle chromatogram, the same patient after two days on a diet; bottom chromatogram, a mixture of reference compounds. Peaks: 1, lactic acid; 2, 2-hydroxyisobutyric acid; 3, 2-hydroxybutyric acid; 4, pyruvic acid; 5, 3-hydroxyisobutyric acid; 6, 3-hydroxybutyric acid; 7, 2-hydroxyisovaleric acid; 8, 2-ketobutyric acid; 9, malonic acid (internal standard); 10, 2-methyl-3-hydroxybutyric acid; 11, 2-hydroxy-*n*-valeric acid; 12, methylmalonic acid; 13, 3-hydroxyisovaleric acid; 14a and b, 2-ketoisovaleric acid; 15, acetoacetic acid; 16, 2-hydroxyisocaproic acid; 17, 2-hydroxy-3-methylvaleric acid; 18a, L-2-keto-3-methylvaleric acid; 18b, D-2-keto-3-methylvaleric acid; 19, 2-ketoisocaproic acid. Reproduced from [386].

Interesting discoveries have recently surfaced due to the modern capabilities of resolving the optical isomers of certain acidic products of metabolism. As pointed out by Kamerling et al. [391] many reactions of the Krebs cycle and other vital pathways are often highly stereospecific. Many observed acidurias, thus, involve chiral molecules, which, given a proper derivatization and the efficiency of capillary GC, can be resolved. The representative cases studied by Kamerling and co-workers include D-glyceric acidemia [392], a permanent D-lactic aciduria [393], and the assigned L-configuration of 2-hydroxybutyrate [394]. Stereospecific interactions during the metabolism of methionine were also noted by Kaji et al. [395]. It is hoped that more investigations will come out in the near future on these interesting ways to unravel the seemingly complex metabolic interactions.

3.10.4 Carbohydrates

Although carbohydrates are among the most polar and non-volatile substances of biochemical interest, the use of GC in their analysis has been quite successful. They are almost a perfect example of the utility of sample derivatization. Based on the early common observations of organic chemists that methylated sugars can be distilled, the first reports on derivatization and GC of carbohydrates appeared quite early in the relevant literature. McInnes et al. [396] demonstrated GC of methylated sugars in 1958, while Sweeley and co-workers [184,397] introduced silylation for the same purpose. By the mid-1960's, it was not surprising to see a GC separation of fairly large disaccharide molecules [398] such as sucrose or cellobiose.

Carbohydrates occur widely in nature and there are numerous analytical problems where GC is highly applicable to their separation and quantitation. Thus, the reports on its use include the analyses of bacteria, yeast, fungi, various plants, and animal tissues and physiological fluids. The overall analytical problems here are somewhat reminiscent of those encountered in the investigations of proteins and complex lipids; likewise, carbohydrates have to be obtained first by cleavage of the larger molecular forms, then derivatized and determined by GC. Structural studies of various biopolymers frequently necessitate specific strategies, while the substances related to carbohydrates, such as sugar phosphates, nucleosides, glycosides, etc., that are present as degradation products in a sample, are frequently amenable to GC. The combination GC/MS has often been used for structural elucidation of sugar-containing natural products [399]. Among the most important applications, a recent exploration of the role of the carbohydrates in immunological interactions is likely to sustain interest in high-sensitivity GC quantitation methods.

Determination of carbohydrates in biological materials by GC was reviewed by Clamp et al. [400] who present extensive tables of the previously published applications as well as the retention characteristics of various carbohydrate derivatives. The survey of applications strongly indicates that the use of trimethylsilyl derivatives is by far the most popular. The other very common procedures are acetylation and methylation. Because of the procedural simplicity, these main derivatives continue to be most commonly used.

The major difficulty in the GC separation of various sugar forms is the formation of multiple peaks from one mixture component. Within the monosaccharides alone, multiple peaks arise from anomeric and ring isomerization forms in pyranose and furanose rings. This tends to contribute to confusion in dealing with complex sugar mixtures. Alternative derivatization procedures have been sought to overcome this difficulty. Among them, reduction to alcohols has been proposed [401], but this procedure may not distinguish certain aldoses and ketoses if present in the same sample [402]. Single derivatives are also formed in the procedure of Szafranek et al. [403] based on the classical degradation of aldoses, followed by acetylation; in a typical sample containing polyols and aldoses, the former are analyzed as peracetyl derivatives and the latter as peracetyl aldonitriles [404]. However, the procedure is not satisfactory for ketoses [402].

In the procedure of Laine and Sweeley [405], methoximation and trimethylsilylation are combined to avoid variation in the ratios of α- and β-anomers of the pyranose and furanose forms. This approach was used to handle successfully 33 sugar derivatives, including aldoses, ketoses, polyols, acidic forms and *N*-acetylated amino sugars, on a fused silica capillary column [406]. Obviously, the problems of multiple peak formation can be overcome through the use of highly efficient capillary columns. The application of capillary columns in carbohydrate analysis, used probably for the first time by Tesarik [407] in 1972, has now become widespread.

GC in biomedical investigations is hardly expected to replace the well-established methods for the determination of glucose and other major carbohydrates. Its strength is seen rather in connection with the profiling efforts. The urinary sugars are typically investigated in a variety of diabetic conditions, renal failure, changes of intestinal permeability, and certain errors of metabolisms. For such investigations, simple and rapid approaches are generally sought. Normal excretion ranges of 17 polyols and aldoses were determined in the urine of humans of different age by Pfaffenberger et al. [408]. The same methodology was also applied to determine carbohydrates in the autopsy samples of human lenses as related to cataractous conditions [404]. Recent urinary carbohydrate analyses also include investigation of certain conjugates in cancer patients [409] and pathological processes in bone [410]. Glycolipids, glycoproteins, and mucopolysaccharides are now frequently studied for a variety of metabolic conditions. Gehrke et al. [411] have recently claimed that fucose, mannose and galactose levels in human serum glycoproteins are reliable predictors in ovarian and small-cell lung cancer.

Concentrations of major carbohydrates in other physiological fluids are usually sufficiently high to permit a reliable profile analysis with the flame-ionization detector. Thus, the GC carbohydrate analyses have been described for plasma [412,413] as well as the seminal fluid from both normal and sterile men [164]. Several attempts have been made to relate the polyol concentrations in the human cerebrospinal fluid to certain pathological conditions [414–416]. If higher sensitivities are needed in the carbohydrate determinations, it is of advantage to consider perfluoroacyl derivatives and the electron-capture detector [402].

3.10.5 Biological amines

Amines are among the organic substances which are most difficult to gas-chromatograph. Primary amines are particularly known for their notorious irreversible adsorption effects and subsequent peak tailing. Useful quantitative chromatographic results can only be obtained under the conditions of superior column technology or, alternatively, if the amines are suitably derivatized. Most amines of biological interest possess additional troublesome sites in their molecules that must be blocked prior to GC; for example, catecholamines have the chemical properties of phenols, alcohols and amines. Consequently, multiple derivatizations are often sought to assure sufficient volatility and/or stability of these compounds during GC. Further-

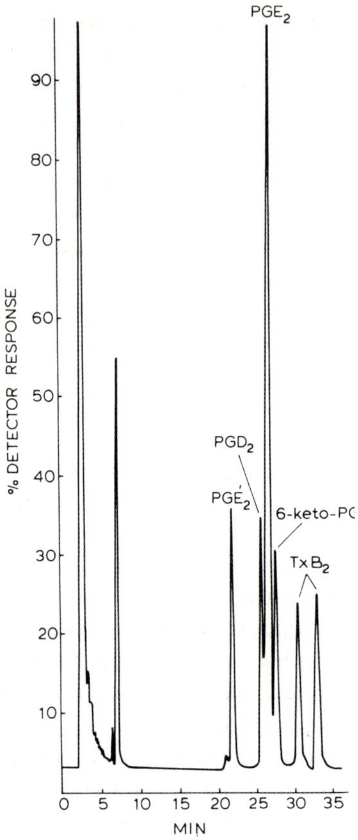

Fig. 3.24. Separation of principal stable metabolites of PGH_2 as methyl ester pentafluorobenzyloxime capillary column. Reproduced from [464].

time. However, more modern HPLC techniques based on the reversed-phase chromatography are now increasingly endorsed by the protein chemists.

Since the first report on GC of amino acids [474], the method has long had the principal advantage of sensitivity. The other, often suggested advantages of GC, i.e., a greater speed of analysis as well as the lower cost and greater versatility of the equipment, are highly debatable. The main disadvantages of GC for the amino acid analysis are the volatility and stability problems that must be overcome through the preparation of suitable derivatives. Derivatization of multifunctional compounds, such as amino acids, remains an important problem. While most effort has gone toward developing a quantitative approach for derivatization of the major protein amino acids, the situation is far from ideal even today. The scope of GC applications thus appears limited, and with the growing potential of HPLC techniques in mind, probably restricted in the future to special cases. While many publications on the topic have appeared, and still continue to appear in the literature, the present review will emphasize only unique directions in the GC analysis of amino acids and related substances.

Due to their zwitterionic character, the amino acids are difficult to convert quantitatively and uniformly into suitable GC derivatives. Numerous methods have now been reported toward meeting this difficult goal. From the 19 amino acids contained commonly in protein hydrolyzates, the trifunctional compounds are particularly difficult to handle in a quantitative fashion. The problems here result from a different reactivity of various functional groups as well as only a limited solubility of certain amino acids in the reaction media.

Arginine and histidine are particularly notable for their derivatization problems [475–477]. Sometimes, even when proper derivatives can eventually be made, they can be easily hydrolyzed or catalytically degraded in an insufficiently inert GC system.

Numerous volatile derivatives that have been reported in the literature over the years shall not be comprehensively reviewed in this chapter. The general approaches toward derivatization of various functional groups from the organic-chemical point of view have already been discussed. As much of the previous work in this area has already been reviewed by Husek and Macek [478], MacKenzie [479] and Jaeger et al. [480], only the key points and new directions will be emphasized here.

It is most desirable that a given amino acid should form a single-derivative peak after treatment with a single derivatization agent. Unfortunately, that is not the case for many important determinations. Thus, for example, permethylation [481] and the formation of *N*-dimethylaminomethylene alkyl esters [482] appeared limited to only some amino acids. Persilylation of all amino acid functional groups with potent silyl donors [483] comes perhaps closest to definition of 'a universal reaction', but even here some problems are encountered: (a) derivatization can be time-consuming; (b) multiple derivatives are occasionally formed even under precautions; (c) Si-N bonds are moisture-sensitive; and (d) truly quantitative derivatization is difficult to achieve for all protein amino acids.

Multiple derivatizations are now generally accepted as being necessary for a reliable quantitative analysis of amino acids. Among those, the most popular appears to approach esterification of carboxyl groups, followed by acylation of amino, hydroxyl, and mercapto groups, as well as the remaining nitrogen functionalities. Thus, the various esterification schemes reported typically involve C_3–C_5 alcohols, while acylation can be accomplished to form acetates, trifluoroacetates, pentafluoropropionates or heptafluorobutyrates [478–480]. Again, histidine and arginine remain among the compounds that are most resistant to a complete derivatization [484,485].

A useful alternative to esterification procedure appears to be the formation of oxazolidinone derivatives using 1,3-dichlorotetrafluoroacetone, as reported recently by Husek and co-workers [246,247]; the resulting heterocyclic nitrogen is further acylated with heptafluorobutyric anhydride. This procedure, which is applicable to 20 common amino acids, was found to be both rapid and sensitive [247].

Although much previous work utilized packed columns, now glass or fused-silica

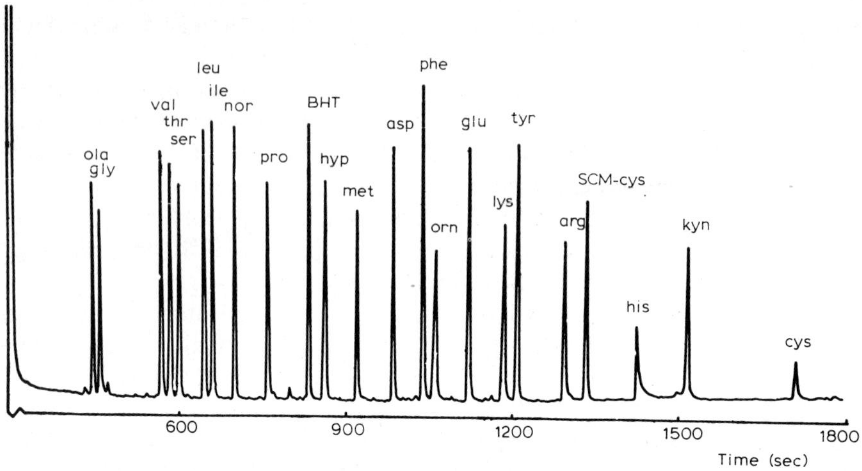

Fig. 3.25. Standard chromatogram of the amino acid calibration mixture (*N*-heptafluorobutyryl isobutyl esters) on a glass capillary column. Reproduced from [486].

capillary columns offer fast, simple separations, in addition to the well-known inertness of such columns. A typical chromatogram of a standard mixture is shown in Fig. 3.25 [486], while another chromatogram of an amino acid hydrolyzate from β-lactoglobulin is demonstrated in Fig. 3.26 [484]. Precision values of $\pm 5\%$ or better

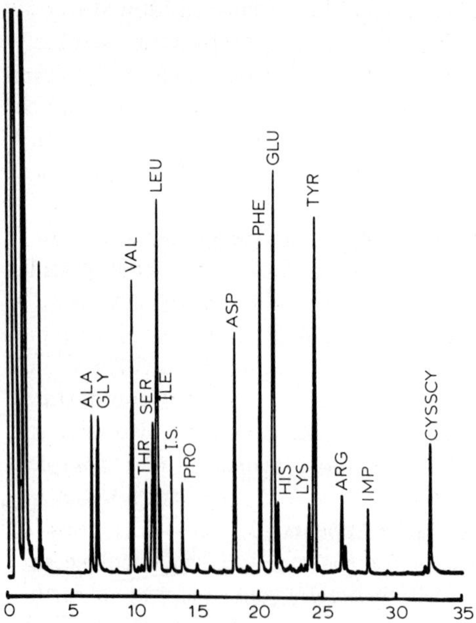

Fig. 3.26. Capillary GC of the amino acids from hydrolyzed insulin (120 ng); the impurity (IMP) is probably hexosamine. Reproduced from [484] with permission of Academic Press.

seem to be typical [487,488] for capillary GC or amino acid derivatives. A considerable number of publications now exist where the protein hydrolyzates have been reliably analyzed by GC.

Additional, non-protein amino acids need to be occasionally analyzed by GC for a variety of reasons. A representative reference to such applications is a comprehensive study of fifty biologically interesting amino acids as *N*-heptafluorobutyryl isobutyl esters by Siezen and Mague [489].

The sequencing methods and determination of C-terminal and N-terminal amino acids are now widely used in biochemical research. The identification and quantitation of the characteristic degradation products can be accomplished by the gas-phase analytical methods. Thus, GC of both dinitrophenyl and various hydantoin amino acid derivatives has now been widely documented. Separation of thiohydantoins [244,245], phenylthiohydantoins [490,491] and methylthiohydantoins [492] generally requires additional silylation for the sake of volatility. Furthermore, acyl derivatives of similar substances have also been reported [493,494]. The most obvious advantage of GC for determination of the Edman degradation products is sensitivity which is particularly important in the sequence analysis of only minute amounts of proteins and peptide hormones.

Determinations of free amino acids in biological fluids and the tissue constitute yet another important area of GC applications. Depending on the nature of a given biological material, the sample preparation methods can be considerably more involved here than in the treatment of protein hydrolytical products.

In order to remove other interfering molecular species, typically present in biological fluids or tissue extracts, free amino acids are often subjected to ion-exchange or adsorption chromatography clean-up steps. An additional degree of specificity is ascertained through the use of high-efficiency capillary columns. An extensive review of all the different reasons for the determination of free amino acids is beyond the scope of this chapter, so that some applications mentioned below just provide a representative cross-section of the field.

Numerous genetic defects of the amino acid metabolism in humans are now well-documented [495]. Thus, under different circumstances, the gas-phase analytical method may have some significance for monitoring single amino acids or their profiles in physiological fluids. According to Davis et al. [496], GC can be effectively used to monitor mono- and diiodotyrosines for the early prediction of thyroid disease. Cooper et al. [497] used GC to observe certain elevations of acylamino acids in the cerebrospinal fluid of patients with hepatic encephalopathy. These compounds were directly related to neurological impairment. Ramsdell and Tanaka [498] followed conjugation of glycine with various metabolically important substances, while Mussini et al. [499] measured the levels of methylated amino acids in muscle. Other representative GC applications [485,487,488,500–502] of the recent years attest to the utility of GC techniques in blood and urine analysis for a variety of problems.

High detection sensitivity is frequently needed in biochemical research pertaining to amino acid metabolism. While the gas-phase detection methods are generally known for sensitivity, there are large differences in the capabilities of various ionization detectors. These sensitivities have been roughly compared for the chromatography of heptafluorobutyryl isobutyl esters by Bengtsson and Oldam [503] to be about 1–5 ng for the flame ionization detector, while the detection limits appear to be 100 times and 1000 times more favorable for the electron-capture detector and mass fragmentography, respectively.

An increase of sensitivity with some GC detectors can largely be attributed to their selective response. As shown by Adams et al. [487], a substantial response enhancement is realized while using the nitrogen-selective thermionic detector. The availability of various fluorine-containing derivatives is highly beneficial in work with the electron capture detector as well as the high-sensitivity mass-spectral measurements. For example, Petersen and Vouros [504] have estimated the detection limit for heptafluorobutyryl methyl esters of the thyroid hormones to be around 500 fg.

The uses of GC to study enzymatic reactions in the amino acid metabolism have also been common. In such studies, measuring either a decreasing substrate concentration or an increase of the reaction product, GC can frequently offer greater sensitivity than other analytical methods. Alternatively, stereospecificity of some enzymatic reactions can be distinguished [505] if the resolution of optical isomers through GC is employed. Recent examples of the enzyme activity determinations are those concerning tryptophan pyrolase [506] and glutamic acid decarboxylase [507] in brain tissue.

An increasing number of laboratories are now involved with the chromatographic resolution of the optical isomers. Although the general aspects of steric resolution and the necessary derivatization approaches were already discussed in the previous sections of this chapter, it should be stressed here that the capabilities of such methods have been primarily demonstrated with amino acids and small peptides. In particular, the method of enantiomeric labeling (adding the unnatural enantiomers into the analyzed mixtures [480] as internal standards) may prove to be quite important in various biochemical studies.

The merits of gas-phase analytical techniques in the analysis of various protein degradation products are now widely recognized. Chromatographic and mass-spectrometric approaches to protein sequencing originate from the well-known studies of Stenhagen [508] and Vetter [509]. It has now been repeatedly shown that numerous peptides, when appropriately derivatized, are sufficiently volatile; for example, Thomas et al. [510] demonstrated as early as 1968 that some molecules as large as dodecapeptides can be chromatographed after permethylation. As certain difficulties were originally observed with more polar peptides, only recent advances in derivatization techniques enabled successful general investigations. For both the reasons of volatility and characteristic mass-spectral fragmentation, multiple derivatization techniques are frequently necessary, as shown, for example, in the studies by Nau and Biemann [241,511].

While newer enzymatic procedures facilitate protein cleavages into characteristic peptide fragments, the overall purpose of the gas-phase investigations of such peptides remains the determination of the amino acid sequences and their homolo-

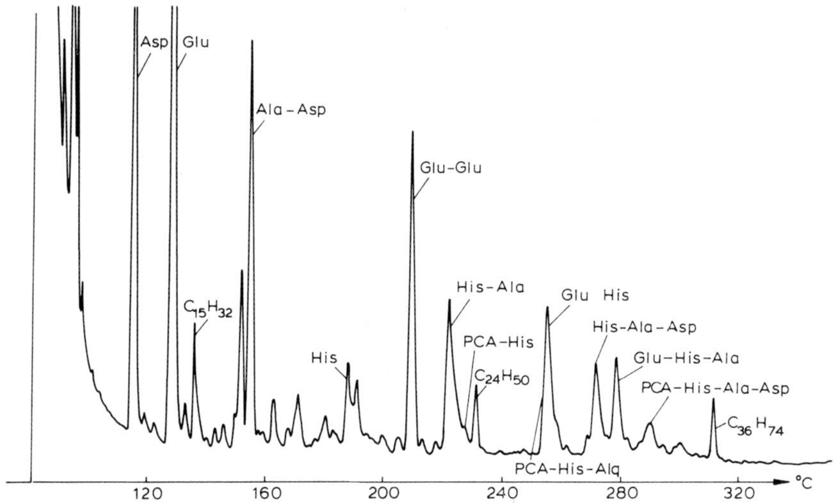

Fig. 3.27. Gas chromatogram of a mixture of *O*-trimethylsilylated trifluoro-dideuteroethyl polyamino alcohol derivatives from a limited acid hydrolysis of 1.0 μmol of the amino terminal peptides from the carboxypeptidase inhibitor of potatoes. Reproduced from [514] with permission of Academic Press.

gies. While the modern mass-spectroscopic methods, including novel ionization techniques such as SIMS (secondary-ion mass spectrometry) and FAB (fast atom bombardment), can now effectively analyze some very large peptides, predictions from retention data remain important for a complete structural elucidation [511,512]. In addition, steric differences within the analyzed peptides may often give rise to a distinct chromatographic behavior.

Discussion of various strategies in the peptide structural work through the gas-phase methods is beyond the scope of this chapter. Sensitivity and accuracy of identification are the most important assets of such techniques. An example of the peptide GC is shown in Fig. 3.27, where a profile of *O*-trimethylsilylated trifluoro-dideuteroethyl polyamino alcohols is eluted at relatively high column temperature [511]; the sample was obtained from the carboxypeptidase inhibitor of potatoes.

According to Nau and Biemann [511], samples between 2 nmol and 12 μmol of the original peptides could be safely analyzed. The high-sensitivity aspect of GC methods may also have some importance for a variety of clinically important peptides, although the current approaches using HPLC and spectrofluorimetric detection are likely to become more popular.

References

1 James, A.T. and Martin, A.J.P. (1952) Biochem. J. 50, 679.
2 VandenHeuvel, W.J.A., Sweeley, C.C. and Horning, E.C. (1960) J. Am. Chem. Soc. 82, 3481.

3 Langer, S.H.P., Pantages, P. and Wender, I. (1958) Chem. Ind. (London), 1664.
4 Luukkainen, T., VandenHeuvel, W.J.A., Haahti, E.O.A. and Horning, E.C. (1961) Biochim. Biophys. Acta 52, 599.
5 Watson, J.T. and Biemann, K. (1965) Anal. Chem. 37, 844.
6 Ryhage, R. (1964) Anal. Chem. 36, 759.
7 Hammar, C.G., Holmstedt, B. and Ryhage, R. (1968) Anal. Biochem. 25, 532.
8 Hites, R.A. and Biemann, K. (1968) Anal. Chem. 40, 1217.
9 Golay, M.J.E. (1957) Anal. Chem. 29, 928.
10 Grob, K. (1968) Helv. Chim. Acta 51, 718.
11 Novotny, M. and Tesarik, K. (1968) Chromatographia 1, 332.
12 Novotny, M. and Zlatkis, A. (1971) Chromatogr. Rev. 14, 1.
13 Novotny, M. (1978) Anal. Chem. 50, 16A.
14 Jennings, W.G. (ed.) (1981) Applications of Glass Capillary Gas Chromatography, Marcel Dekker, Inc., New York.
15 Jellum, E. (1977) J. Chromatogr. 143, 427.
16 Martin, A.J.P. (1971) Clin. Clim. Acta 34, 131.
17 Dandeneau, R. and Zerenner, E.H. (1979) J. High Resolut. Chromatogr. 2, 351.
18 Clark, S.J. and Wotiz, H.H. (1963) Steroids 2, 535.
19 Landowne, R.A. and Lipsky, S.R. (1963) Anal. Chem. 35, 532.
20 Stitch, S.R., Toumba, J.K., Groen, M.B., Funke, C.W., Leemhuis, J., Vink, J. and Woods, G.F. (1980) Nature 287, 738.
21 Setchell, K.D.R., Lawson, A.M., Mitchell, F.L., Adlercreutz, H., Kirk, D.N. and Axelson, M. (1980) Nature 287, 740.
22 Gil-Av, E., Feibush, B. and Charles-Sigler, R. (1966) Tetrahedron Lett. p. 1009.
23 Frank, H., Nicholson, G.J. and Bayer, E. (1978) Angew. Chem. Intern. Ed. 17, 363.
24 Schomburg, G., Dielmann, R., Borwitzky, H. and Husman, H. (1978) in 'Chromatography 1978' (Schomburg, G. and Rohrschneider, L., eds.), p. 157, Elsevier, Amsterdam.
25 Bocek, J., Novak, J. and Janak, J. (1970) J. Chromatogr. Sci. 8, 226.
26 Grant, D.W. and Clarke, A. (1971) Anal. Chem. 43, 1951.
27 Habgood, H.W. and Harris, W.E. (1964) Anal. Chem. 36, 663.
28 Grant, D.W. and Hollis, M.G. (1978) J. Chromatogr. 158, 3.
29 Zlatkis, A., Fenimore, D.C., Ettre, L.S. and Purcell, J.E. (1965) J. Gas Chromatogr. 3, 75.
30 Nygren, S. (1977) J. Chromatogr. 142, 109.
31 Horning, E.C., Pfaffenberger, C.D. and Moffat, A.C. (1972) Anal. Chem. 44, 2.
32 Deans, D.R. (1968) Chromatographia 1, 18.
33 Horváth, C. (1967) in The Practice of Gas Chromatography (Ettre, L.S. and Zlatkis, A., eds.), p. 129, Interscience Publishers, New York.
34 Picker, J.E. and Sievers, R.E. (1981) J. Chromatogr. 217, 275.
35 Schomburg, G., Husmann, H. and Weeke, F. (1974) J. Chromatogr. 99, 63.
36 Pretorius, V., Smuts, T.W. and Moncrieff, J. (1978) J. High Resolut. Chromatogr. 1, 200.
37 Bertsch, W. (1978) J. High Resolut. Chromatogr. 1, 85.
38 Bertsch, W. (1978) J. High Resolut. Chromatogr. 1, 187.
39 Lipsky, S.R. and Landowne, R.A. (1961) Anal. Chem. 33, 818.
40 Lurie, A.O. and Villee, C.A. (1966) J. Chromatogr. 21, 113.
41 Purcell, J.E., Downs, H.D. and Ettre, L.S. (1975) Chromatographia 8, 605.
42 Galli, M. and Trestianu, S. (1981) J. Chromatogr. 203, 193.
43 Chamberlain, J., Knights, B.A. and Thomas, G.H. (1963) J. Endocrinol. 26, 367.
44 Collins, W.P., Sommerville, I.F. (1964) Nature 203, 836.
45 Lurie, A.O. and Villee, C.A. (1966) J. Gas Chromatogr. 4, 160.
46 Menini, E. and Norymberski, J.K. (1965) Biochem. J. 95, 1.
47 Tinti, P. (1966) J. Gas Chromatogr. 4, 140.
48 Otte, E. and Jentzsch, D. (1971) in Gas Chromatography 1970 (Stock, R. and Perry, S.G., eds.) p. 218, The Institute of Petroleum, London.

49 McConnell, M.L. and Novotny, M. (1975) J. Chromatogr. 112, 559.
50 Cramers, C.A. and Vermeer, E.A. (1975) Chromatographia 8, 479.
51 Lines, D.S., Brain, K.R. and Ross, M.S.F. (1976) J. Chromatogr. 117, 59.
52 Miller, R. and Bertsch, W. (1981) in Capillary Chromatography (Kaiser, R.E., ed), p. 267, Fourth International Symposium, Hindelang 1981, Hüthig, Heidelberg.
53 van den Berg, M.J. and Cox, T.P.H. (1972) Chromatographia 5, 301.
54 Verzele, M., Redant, G., Qureshi, S. and Sandra, P. (1980) J. Chromatogr. 199, 105.
55 Abel, K. (1964) J. Chromatogr. 13, 14.
56 Evrard, E. (1967) J. Chromatogr. 27, 40.
57 Evrard, E. and Couvreur, J. (1967) J. Chromatogr. 27, 47.
58 Jahnsen, V.J. (1967) Anal. Chem. 39, 141.
59 Schomburg, G., Behlau, H., Dielmann, R., Weeke, F. and Husmann, H. (1977) J. Chromatogr. 142, 87.
60 German, A.L. and Horning, E.C. (1972) Anal. Lett. 5, 619.
61 Grob, K. and Grob, G. (1969) J. Chromatogr. Sci. 7, 584.
62 Grob, K. and Grob, K., Jr. (1974) J. Chromatogr. 94, 53.
63 Yang, F.J., Brown, A.C., III. and Cram, S.P. (1978) J. Chromatogr. 158, 91.
64 Schutjes, C.P.M., Vermeer, E.A., Rijks, J.A. and Cramers, C.A. (1981) in Capillary Chromatography (Kaiser, R.E., ed.) p. 687, Fourth International Symposium, Hindelang 1981, Hüthig, Heidelberg.
65 Novotny, M. and Farlow, R. (1975) J. Chromatogr. 103, 1.
66 de Leeuw, J.W., Maters, W.L., vanden Ment, D. and Boon, J.J. (1977) Anal. Chem. 49, 1881.
67 Vogt, W., Jacob, K., Ohnesorge, A.-B. and Obwexer, H.W. (1979) J. Chromatogr. 186, 197.
68 Grob, K. and Grob, K., Jr. (1978) J. Chromatogr. 151, 311.
69 Galli, M., Trestianu, S. and Grob, K., Jr. (1979) J. High Resolut. Chromatogr. 2, 366.
70 Purnell, J.H. (1959) Nature (London) 184, 2009.
71 Karger, B.L. (1967) Anal. Chem. 39, 24A.
72 Giddings, J.C. (1965) in Gas Chromatography 1964 (Goldup, A., ed.) p. 2, The Institute of Petroleum, London.
73 König, W.A., Parr, W., Lichtenstein, H.A., Bayer, E. and Oró, J. (1970) J. Chromatogr. Sci. 8, 183.
74 Parr, W., Yang, C., Bayer, E. and Gil-Av, E. (1970) J. Chromatogr. Sci. 8, 591.
75 Parr, W., Pleterski, J., Yang, C. and Bayer, E. (1971) J. Chromatogr. Sci. 9, 141.
76 Frank, H., Nicholson, G.J. and Bayer, E. (1978) J. Chromatogr. 146, 197.
77 Saeed, T., Sandra, P. and Verzele, M. (1979) J. Chromatogr. 186, 611.
78 Riedo, F., Fritz, D., Tarján, G. and Kováts, E. sz. (1976) J. Chromatogr. 126, 63.
79 Horning, E.C., VandenHeuvel, W.J.A. and Creech, B.G. (1963) in Methods of Biochemical Analysis (Glick, D., ed.) Vol. XI, Interscience Publishers, New York.
80 Rohrschneider, L. (1966) J. Chromatogr. 22, 6.
81 McReynolds, W.O. (1970) J. Chromatogr. Sci. 8, 685.
82 Novák, J., Ružičková, J., Wičar, S. and Janák, J. (1973) Anal. Chem. 45, 1365.
83 Risby, T.H., Jurs, P.C. and Reinbold, B.L. (1974) J. Chromatogr. 99, 173.
84 Verzele, M. and Sandra, P. (1978) J. Chromatogr. 158, 111.
85 Schwartz, R.D., Mathews, R.G., Ramachandran, S., Henly, H.S. and Doyle, J.E. (1975) J. Chromatogr. 112, 111.
86 Novotny, M., Segura, R. and Zlatkis, A. (1972) Anal. Chem. 44, 9.
87 Zielinski, Jr., W.L., Johnston, K. and Muschik, G.M. (1976) Anal. Chem. 48, 907.
88 Haahti, E.O.A., VandenHeuvel, W.J.A. and Horning, E.C. (1961) Anal. Biochem. 2, 344.
89 Sjövall, J. (1962) Acta Chem. Scand. 16, 1761.
90 Okishio, T. and Nair, P.P. (1966) Anal. Biochem. 15, 360.
91 Laub, R.J. and Purnell, J.H. (1975) J. Chromatogr. 112, 71.
92 Chien, C.-F., Kopecni, M.M. and Laub, R.J. (1980) Anal. Chem. 52, 1402.
93 Aue, W.A., Hastings, C.R. and Gerhardt, K.O. (1974) J. Chromatogr. 99, 45.
94 Cramers, C.A., Rijks, J.A. and Bocek, P. (1971) Clin. Chim. Acta 34, 159.

95 Cramers, C.A., Rijks, J. and Bocek, P. (1972) J. Chromatogr. 65, 29.
96 Golay, M.J.E. (1960) U.S. Patent 2,920, 478.
97 Novotny, M. and Bartle, K.D. (1974) Chromatographia 7, 122.
98 Schieke, J.D., Comins, N.R. and Pretorius, V. (1975) J. Chromatogr. 112, 97.
99 German, A.L., Pfaffenberger, C.D., Thenot, J.-P., Horning, M.G. and Horning, E.C. (1973) Anal. Chem. 45, 930.
100 Lee, M.L. and Wright, B.W. (1980) J. Chromatogr. 184, 235.
101 Madani, C., Chambaz, E.M., Rigaud, M., Durand, J. and Chebroux, P. (1976) J. Chromatogr. 126, 161.
102 Blomberg, L. and Wännman, T. (1979) J. Chromatogr. 186, 159.
103 Grob, K., Grob, G. and Grob, Jr., K. (1981) J. Chromatogr. 211, 243.
104 Lipsky, S.R. and McMurray, W.J. (1982) J. Chromatogr. 239, 61.
105 Blomberg, L., Buijten, J., Markides, K. and Wännman, T. (1982) J. Chromatogr. 239, 51.
106 Novotny, M., McConnell, M.L., Lee, M.L. and Farlow, R. (1974) Clin. Chem. 20, 1105.
107 Grob, K. and Grob, G. (1971) Chromatographia 4, 422.
108 Guiochon, G. (1978) Anal. Chem. 50, 1812.
109 Brody, S.S. and Chaney, J.E. (1966) J. Gas Chromatogr. 4, 42.
110 Zlatkis, A., Lichtenstein, H.A., Tishbee, A., Bertsch, W., Shunbo, F. and Liebich, H.M. (1973) J. Chromatogr. Sci. 11, 299.
111 Chen, S., Mahadevan, V. and Zieve, L. (1970) J. Lab. Clin. Med. 75, 622.
112 Perry, S.G. and Carter, F.W.G. (1971) in Gas Chromatography 1970, (Stock, N., ed.) p. 38, Institute of Petroleum, London.
113 Karmen, A. and Guiffrida, L. (1964) Nature (London) 201, 1204.
114 Hartigan, M.J., Purcell, J.E., Novotny, M., McConnell, M.L. and Lee, M.L. (1974) J. Chromatogr. 99, 339.
115 Novotny, M., Maskarinec, M.P., Steverink, A.T.G. and Farlow, R. (1976) Anal. Chem. 48, 468.
116 Jacob, K., Falkner, C. and Vogt, W. (1978) J. Chromatogr. 167, 67.
117 Vogt, W., Jacob, K., Ohnesorge, A.-B. and Schwertfeger, G. (1980) J. Chromatogr. 199, 191.
118 Kolb, B. and Bischoff, J. (1974) J. Chromatogr. Sci. 12, 625.
119 Lovelock, J.E. (1974) J. Chromatogr. 99, 3.
120 Lovelock, J.E. and Watson, A.J. (1978) J. Chromatogr. 158, 123.
121 Zlatkis, A. and Poole, C.F. (1981) (eds.), Electron Capture: Theory and Practice in Chromatography, Elsevier, Amsterdam.
122 Petersen, B.A., Hanson, R.N., Giese, R.W. and Karger, B.L. (1976) J. Chromatogr. 126, 503.
123 Grimsrud, E.P. and Miller, D.A. (1978) Anal. Chem. 50, 1141.
124 Wizner, M.A., Singhawangcha, S., Barkley, R.M. and Sievers, R.E. (1982) J. Chromatogr. 239, 145.
125 Driscoll, J.N., Ford, J., Jaramillo, L.F., Becker, J.H., Hewitt, G., Marshall, J.K. and Onishuk, F. (1978) Am. Lab. 10, 137.
126 Driscoll, J.N., Ford, J., Jaramillo, J.F. and Gruber, E.T. (1978) J. Chromatogr. 158, 171.
127 Jaramillo, J. and Driscoll, J.N. (1979) J. High Resolut. Chromatogr. 2, 536.
128 Hall, R.C. (1978) CRC Crit. Rev. Anal. Chem. 7, 323.
129 McLean, W.R., Stanton, D.L. and Penketh, G.E. (1973) Analyst (London) 98, 432.
130 Novotny, M., Schwende, F.J., Hartigan, M.J. and Purcell, J.E. (1980) Anal. Chem. 52, 736.
131 Burchfield, H.P., Wheeler, R.J. and Bernos, J.B. (1971) Anal. Chem. 43, 1976.
132 Martin, A.J.P. (1949) Biochem. Soc. Symp. 3, 4.
133 Woodford, F.P. and vanGent, C.M. (1960) J. Lipid Res. 1, 188.
134 VandenHeuvel, W.J.A. and Horning, E.C. (1962) Biochim. Biophys. Acta 64, 416.
135 VandenHeuvel, W.J.A., Gardiner, W.L. and Horning, E.C. (1965) J. Chromatogr. 19, 263.
136 VandenHeuvel, W.J.A., Gardiner, W.L. and Horning, E.C. (1967) J. Chromatogr. 26, 387.
137 Kováts, E. sz. (1958) Helv. Chim. Acta 41, 1915.
138 Lee, M.L., Vassilaros, D.L., White, C.M. and Novotny, M. (1979) Anal. Chem. 51, 768.
139 Novotny, M., Kump, R., Merli, F. and Todd, L.J. (1980) Anal. Chem. 52, 401.

140 Whalen-Pedersen, E.K. and Jurs, P.C. (1981) Anal. Chem. 53, 2184.
141 Kugler, E., Halang, W., Schlenkermann, R., Webel, H. and Langlais, R. (1977) Chromatographia 10, 438.
142 Knights, B.A. (1964) J. Gas Chromatogr. 2, 160.
143 Merritt, Jr., C. (1969) in Ancillary Techniques of Gas Chromatography, (Ettre, L.S. and McFadden, W.H., eds.), p. 325, Wiley Interscience, New York.
144 Janák, J. (1964) J. Chromatogr. 15, 16.
145 Kaiser, R. (1964) Z. Anal. Chem. 205, 284.
146 Sweeley, C.C., Young, N.D., Holland, J.F. and Gates, S.C. (1974) J. Chromatogr. 99, 507.
147 Vrbanac, J.J., Braselton, Jr., W.E., Holland, J.F. and Sweeley, C.C. (1982) J. Chromatogr. 239, 265.
148 Rhodes, G., Opsal, R.B., Meek, J.T. and Reilly, J.P. (1983) Anal. Chem. 55, 280.
149 Munson, B. (1977) Anal. Chem. 49, 772A.
150 Garlock, S.E., Adams, G.E. and Smith, S.L. (1982) Am. Lab., December issue, p. 48.
151 Yang, F.J. (1981) J. Chromatogr. Sci. 19, 523.
152 Hrivnac, M., Frischknecht, W. and Cechova, M. (1976) Anal. Chem. 48, 937.
153 Horning, E.C. and Horning, M.G. (1970) Methods Med. Res. 12, 369.
154 Gates, S.C. and Sweeley, C.C. (1978) Clin. Chem. 24, 1663.
155 Anderson, N. and Anderson, N.G. (1977) Proc. Natl. Acad. Sci. U.S.A. 74, 5421.
156 Jellum, E. (1982) J. Chromatogr. 239, 29.
157 Williams, R.J. (1956) Biochemical Individuality, John Wiley and Sons, New York.
158 Pauling, L. (1968) Science 160, 265.
159 Zlatkis, A., Poole, C.F., Brazell, R., Lee, K.Y., Hsu, F. and Singhawangcha, S. (1981) Analyst (London) 106, 352.
160 Jellum, E., Stokke, O. and Eldjarn, L. (1971) J. Scand. Clin. Lab. Invest. 27, 273.
161 Jellum, E., Stokke, O. and Eldjarn, L. (1972) Clin. Chem. 18, 800.
162 Baty, J.A. and Wade, A.P. (1974) Anal. Biochem. 57, 27.
163 Jellum, E., Helland, P., Eldjarn, L., Markwardt, V. and Markhöfer, J. (1975) J. Chromatogr. 112, 573.
164 Jellum, E., Storseth, P., Alexander, J., Helland, P., Stokke, O. and Teig, E. (1976) J. Chromatogr. 126, 487.
165 Rhodes, G., Holland, M.L., Wiesler, D., Novotny, M., Moore, S.A., Peterson, R.G. and Felten, D.L. (1982) Experientia 38, 75.
166 Rhodes, G., Holland, M.L., Wiesler, D., Novotny, M., Moore, S.A., Peterson, R.G. and Felten, D.L. (1982) J. Chromatogr. 228, 33.
167 Alasandro, M., Wiesler, D. and Novotny, M., submitted for publication.
168 Holland, M.L., Rhodes, G.R., Dalle Ave, M., Wiesler, D. and Novotny, M. (1983) Life Sci. 32, 787.
169 McConnell, M.L., Rhodes, G., Watson, U. and Novotny, M. (1979) J. Chromatogr. 162, 495.
170 Rhodes, G., Miller, M., McConnell, M.L. and Novotny, M. (1981) Clin. Chem. 27, 580.
171 Zlatkis, A., Lee, K.Y., Poole, C.F. and Holzer, G. (1979) J. Chromatogr. 163, 125.
172 Hare, P.E., Tishbee, A. and Gil-Av, E. (1980) J. Am. Chem. Soc. 102, 5115.
173 LePage, N., Lindner, W., Davies, G., Seitz, D.E. and Karger, B.L. (1979) Anal. Chem. 51, 433.
174 Pirkle, W.H., Shreiner, J.L. and Hamper, B.C. (1981) J. Am. Chem. Soc. 103, 3964.
175 Kuo, J.C. and Yeung, E.S. (1982) J. Chromatogr. 229, 293.
176 Gil-Av, E., Charles, R. and Fischer, G. (1965) J. Gas Chromatogr. 17, 408.
177 Halpern, B. and Westley, J.W. (1965) Biochem. Biophys. Res. Commun. 19, 361.
178 Pollock, G.E., Oyama, V.I. and Johnson, R.D. (1965) J. Gas Chromatogr. 3, 174 (1965).
179 Parr, W. and Howard, P.Y. (1973) Anal. Chem. 45, 711.
180 König, W.A. and Nicholson, G.J. (1975) Anal. Chem. 47, 951.
181 König, W.A., Sievers, S. and Schulze, U. (1980) Angew. Chem. Int. Ed. Engl. 19, 910.
182 König, W.A. and Sievers, S. (1980) J. Chromatogr. 200, 189.
183 König, W.A., Sievers, S. and Benecke, I. (1981) in Capillary Chromatography (Kaiser, R.E., ed.) p. 703, Fourth International Symposium Hindelang 1981, Hüthig, Heidelberg.
184 Sweeley, C.C., Bentley, R., Makita, M. and Wells, W.W. (1963) J. Am. Chem. Soc. 85, 2497.

185 Catalog, Pierce Chemical Company, Rockford, Illinois.
186 Klebe, J.F., Finkbeiner, H. and White, D.M. (1966) J. Am. Chem. Soc. 88, 3390.
187 Stalling, D.L., Gehrke, C.W. and Zumwalt, R.W. (1968) Biochem. Biophys. Res. Commun. 31, 616.
188 Donike, M. (1969) J. Chromatogr. 42, 103.
189 Gleispach, H. (1974) J. Chromatogr. 91, 407.
190 Pfeifer, P. and Spiteller, G. (1981) J. Chromatogr. 223, 21.
191 Morgan, E.D. and Poole, C.F. (1974) J. Chromatogr. 89, 225.
192 Veysoglu, T. and Mitscher, L.A. (1981) Tetrahedron Lett., 1303.
193 Miyazaki, H., Ishibashi, M., Itoh, M. and Nambara, T. (1977) Biomed. Mass. Spectr. 4, 23.
194 Kelly, R.W. and Taylor, P.L. (1976) Anal. Chem. 48, 465.
195 Quilliam, M.A. and Westmore, J.B. (1978) Anal. Chem. 50, 59.
196 Miyazaki, H., Ishibashi, M., Itoh, M., Yamashita, K. and Nambara, T. (1977) J. Chromatogr. 133, 311.
197 Phillipou, G. (1976) J. Chromatogr. 129, 384.
198 Harvey, D.J. (1980) J. Chromatogr. 196, 156.
199 Poole, C.F., Singhawangcha, S., Chen Hu, L.E., Sye, W.-F., Brazell, R. and Zlatkis, A. (1980) J. Chromatogr. 187, 331.
200 Sullivan, J.E. and Schewe, L.R. (1977) J. Chromatogr. Sci. 15, 196.
201 Donike, M. (1973) J. Chromatogr. 78, 273.
202 Jacob, K. and Vogt, W. (1978) J. Chromatogr. 150, 339.
203 Bailey, E. (1967) Steroids 10, 527.
204 Kirschner, M.A. and Fales, H.M. (1962) Anal. Chem. 34, 1548.
205 Anthony, G.M., Brooks, C.J.W., MacLean, I. and Sangster, I. (1969) J. Chromatogr. Sci. 7, 623.
206 Gaskell, S.J. and Brooks, C.J.W. (1978) J. Chromatogr. 158, 331.
207 Metcalfe, L.D. and Schmitz, A.A. (1961) Anal. Chem. 33, 363.
208 McCreary, D.K., Kossa, W.C., Ramachandran, S. and Kurtz, R.R. (1978) J. Chromatogr. Sci. 16, 329.
209 Robb, E.W. and Westbrook, J.J. (1963) Anal. Chem. 35, 1644.
210 Williams, M.G. and MacGee, J. (1982) J. Chromatogr. 234, 468.
211 Greeley, R.H. (1974) J. Chromatogr. 88, 229.
212 Arbin, A., Brink, H. and Vessman, J. (1979) J. Chromatogr. 170, 25.
213 Hashimoto, N., Aoyama, T. and Shioiri, T. (1981) Chem. Pharm. Bull. (Tokyo) 29, 1475.
214 Thenot, J.-P., Horning, E.C., Stafford, M. and Horning, M.G. (1972) Anal. Lett. 5, 217.
215 Doms, E.K. (1977) J. Chromatogr. 140, 29.
216 Korte, W.D. (1981) J. Chromatogr. 214, 131.
217 Barcelona, M.J., Liljestrand, H.M. and Morgan, J.J. (1980) Anal. Chem. 52, 321.
218 Lindner, W. and Truger, W. (1979) J. Chromatogr. 168, 171.
219 Edenharder, R. and Slemr, J. (1981) J. Chromatogr. 222, 1.
220 Gyllenhaal, O., Brötell, H. and Hartvig, P. (1976) J. Chromatogr. 129, 295.
221 Dommes, V., Wirtz-Peitz, F. and Kunau, W.-H. (1976) J. Chromatogr. Sci. 14, 360.
222 Startin, J.R., Gilbert, J. and McWeeny, D.J. (1978) J. Chromatogr. 152, 495.
223 Lakshminarayana, G., Rao, K.V.S.A., Sita Devi, K. and Kaimal, T.N.B. (1981) J. Am. Oil Chem. Soc. 58, 839.
224 Lie Ken Jie, M.S.F. and Lam, C.H. (1976) J. Chromatogr. 129, 181.
225 Barnes, S., Pritchard, D.G., Settine, R.L. and Geckle, M. (1980) J. Chromatogr. 183, 269.
226 Chauhan, M.S. and Dakshinamurti, K. (1982) J. Chromatogr. 227, 323.
227 Fales, H.M. and Luukkainen, T. (1965) Anal. Chem. 37, 955.
228 Thenot, J.-P. and Horning, E.C. (1972) Anal. Lett. 5, 21.
229 Axelson, M. (1978) Anal. Biochem. 86, 133.
230 Nambara, T., Kigasawa, K., Iwata, T. and Ibuki, M. (1975) J. Chromatogr. 114, 81.
231 Nambara, T., Iwata, T. and Kigasawa, K. (1976) J. Chromatogr. 118, 127.
232 Langenbeck, U., Möhring, H.-U. and Dieckmann, K.P. (1975) J. Chromatogr. 115, 65.

233 Horning, M.G., Moss, A.M., Boucher, E.A. and Horning, E.C. (1968) Anal. Lett. 1, 311.
234 Blotny, G., Kusmierz, J., Maliński, E. and Szafranek, J. (1980) J. Chromatogr. 193, 61.
235 Hamano, T., Mitsuhashi, Y. and Matsuki, Y. (1980) J. Chromatogr. 190, 462.
236 Gyllenhaal, O., Johansson, L. and Vessman, J. (1980) J. Chromatogr. 190, 347.
237 Hammar, C.-G. (1978) Biomed. Mass Spectrom. 5, 25.
238 Poole, C.F., Sye, W.-F., Singhawangcha, S. and Zlatkis, A. (1980) J. Chromatogr. 199, 123.
239 Chiavari, G. and Giumanini, A.G. (1981) J. Chromatogr. 206, 555.
240 Doshi, P.S. and Edwards, D.J. (1979) J. Chromatogr. 176, 359.
241 Nau, H. (1976) J. Chromatogr. 121, 376.
242 Frank, H., Das Neues, J.H.C. and Bayer, E. (1978) J. Chromatogr. 152, 357.
243 Severin, Th. and Popp-Ginsbach, H. (1978) J. Chromatogr. 147, 473.
244 Rangarajan, M., Ardrey, R.E. and Darbre, A. (1973) J. Chromatogr. 87, 499.
245 Dwulet, F.E. and Gurd, F.R.N. (1977) Anal. Biochem. 82, 285.
246 Hušek, P., Felt, V. and Matucha, M. (1979) J. Chromatogr. 180, 53.
247 Hušek, P. (1982) J. Chromatogr. 234, 381.
248 Ôi, N., Horiba, M. and Kitahara, H. (1980) J. Chromatogr. 202, 299.
249 König, W.A., Benecke, I. and Sievers, S. (1982) J. Chromatogr. 238, 427.
250 Chang, S.-C., Charles, R. and Gil-Av, E. (1982) J. Chromatogr. 238, 29.
251 König, W. and Benecke, I. (1980) J. Chromatogr. 195, 292.
252 Flores, J.J., Bonner, W.A. and Van Dort, M.A. (1977) J. Chromatogr. 132, 152.
253 Gal, J. and Ames, M.M. (1977) Anal. Biochem. 83, 266.
254 Cox, R.I. and Bedford, A.R. (1964) Steroids 3, 663.
255 Brooks, C.J.W., Chambaz, E. and Horning, E.C. (1967) Anal. Biochem. 19, 234.
256 Axelson, M. and Sjövall, J. (1976) J. Chromatogr. 126, 705.
257 Ehmann, A., McKinney, W.J., Reinsfelder, R.E., Saliman, P.M. and Silveir, E.J. (1979) Anal. Chem. 51, 985A.
258 Golton, W.C. (1979) Anal. Chem. 51, 987A.
259 Jaakonmaki, P.I., Yarger, K.A. and Horning, E.C. (1967) Biochim. Biophys. Acta 137, 216.
260 Shackleton, C.H.L. (1981) Clin. Chem. 27, 509.
261 Almé, B. and Nyström, E. (1971) J. Chromatogr. 59, 45.
262 Setchell, K.D.R., Almé, B., Axelson, M. and Sjövall, J. (1976) J. Steroid Biochem. 7, 615.
263 Ibrahim, G., Andryanskas, S. and Bastos, M.L. (1975) J. Chromatogr. 108, 107.
264 Nieman, R.H., Pap, D.L. and Clark, R.A. (1978) J. Chromatogr. 157, 303.
265 Andersson, B. and Andersson, K. (1982) J. Chromatogr. 242, 353.
266 Shackleton, C.H.L. and Whitney, J.O. (1980) Clin. Chim. Acta 107, 231.
267 Setchell, K.D.R. and Worthington, J. (1980) Clin. Chim. Acta 125, 135.
268 Alasandro, M., Wiesler, D., Rhodes, G. and Novotny, M. (1982) Clin. Chim. Acta, 126, 243.
269 Axelson, M. and Sjövall, J. (1979) J. Chromatogr. 186, 725.
270 Sahlberg, B.-L., Axelson, M., Collins, D.J. and Sjövall, J. (1981) J. Chromatogr. 217, 453.
271 Merli, F., Wiesler, D., Maskarinec, M.P., Novotny, M., Vassilaros, D.L. and Lee, M.L. (1981) Anal. Chem. 53, 1929.
272 Majors, R.E. and Johnson, E.L. (1978) J. Chromatogr. 167, 17.
273 Dünges, W. and Kiesel, K. (1979) in Advances in Chromatography 1979 (Zlatkis, A., Ettre, L.S. and Kováts, E. sz., eds.) p. 697, Chromatography Symposium, Houston.
274 Eik-Nes, K.B. and Horning, E.C. (1968) Gas Phase Chromatography of Steroids, Springer Verlag, New York.
275 Adlercreutz, H. and Luukkainen, T. in Ref. 274, p. 143.
276 Skelley, D.S., Brown, L.P. and Besch, P.K. (1973) Clin. Chem. 19, 146.
277 Cameron, E.H.D. and Scarisbrick, J.J. (1973) Clin. Chem. 19, 1403.
278 Wisdom, G.B. (1976) Clin. Chem. 22, 1243.
279 Murphy, B.E.P. (1967) J. Clin. Endocrinol. Metab. 27, 973.
280 Demetriou, J.A. and Austin, F.G. (1970) Clin. Chem. 16, 111.

281 Björkhem, I., Blomstrand, R. and Lantto, O. (1975) Clin. Chim. Acta 65, 343.
282 Winkler, V.W., Strong, J.M. and Finley, R.A. (1977) Steroids 29, 739.
283 Baba, S., Shinohara, Y. and Kasuya, Y. (1979) J. Chromatogr. 162, 529.
284 Tetsuo, M., Eriksson, H. and Sjövall, J. (1982) J. Chromatogr. 239, 287.
285 Wehner, R. and Handke, A. (1979) J. Chromatogr. 177, 237.
286 Wilson, A., Mason, P.A. and Fraser, R. (1976) J. Steroid Biochem. 7, 611.
287 Maskarinec, M.P., Novotny, M., Shipley, G., Brown, D.J. and Forney, R.B. (1978) Toxicol. Appl. Pharmacol. 45, 617.
288 Maskarinec, M.P., Shipley, G., Novotny, M., Brown, D.J. and Forney, R.B. (1979) Experientia 34, 88.
289 Maume, B.F., Millot, C., Mesnier, D., Patouraux, D., Doumas, J. and Tomori, E. (1979) J. Chromatogr. 186, 581.
290 Lipsett, M.B. (ed.) (1965) Gas Chromatography of Steroids in Biological Fluids, Plenum Press, New York.
291 Gardiner, W.L. and Horning, E.C. (1966) Biochim. Biophys. Acta 115, 524.
292 Novotny, M. and Zlatkis, A. (1970) J. Chromatogr. Sci. 8, 347.
293 Völlmin, J.A. (1970) Chromatographia 3, 223.
294 Monder, C. and Bradlow, H.L. (1980) Recent Progr. Horm. Res. 36, 345.
295 Anderson, R.A., Defaye, C., Madani, C., Chambaz, E.M. and Brooks, C.J.W. (1974) J. Chromatogr. 99, 485.
296 Pfaffenberger, C.D. and Horning, E.C. (1975) J. Chromatogr. 112, 581.
297 Völlmin, J.A. (1971) Clin. Chim. Acta 34, 207.
298 Luyten, J.A. and Rutten, G.A.F.M. (1974) J. Chromatogr. 91, 393.
299 Spiteller, G. (1978) Pure Appl. Chem. 50, 205.
300 Pfaffenberger, C.D., Malinak, L.R. and Horning, E.C. (1978) J. Chromatogr. 158, 313.
301 Smith, L.L., Smart, V.V. and Ansari, G.A.S. (1979) Mutat. Res. 68, 23.
302 Korahani, V., Bascoul, J., Crastes de Paulet, A. (1981) J. Chromatogr. 211, 392.
303 Schomburg, G. and Husmann, H. (1975) Chromatographia 8, 517.
304 Knights, B.A. (1967) J. Gas Chromatogr. 5, 273.
305 Novotny, M. and Zlatkis, A. (1971) J. Chromatogr. 56, 353.
306 Novotny, M., Lee, M.L., Low, C.-E. and Maskarinec, M.P. (1976) Steroids 27, 665.
307 Ballantine, J.A. and Williams, K. (1978) J. Chromatogr. 148, 504.
308 Takagi, T., Sakai, A., Hayashi, K. and Itabashi, Y. (1979) J. Chromatogr. Sci. 17, 212.
309 Brooks, C.J.W. and Smith, A.G. (1975) J. Chromatogr. 112, 499.
310 Edmonds, C.G., Smith, A.G. and Brooks, C.J.W. (1977) J. Chromatogr. 133, 372.
311 Homberg, E. (1977) J. Chromatogr. 139, 77.
312 Homberg, E. and Bielefeld, B. (1979) J. Chromatogr. 180, 83.
313 Kuksis, A. (1966) in Methods of Biochemical Analysis, (Glick, D., ed.) Vol. XIV, Interscience Publishers, New York.
314 Sjövall, J. (1967) in The Gas-Liquid Chromatography of Steroids (Grant, J.K., ed.) p. 243, Mem. Soc. Endocrinol. No. 16, Cambridge University Press, London.
315 Sjövall, J. (1964) in New Biochemical Separations (James, A.T. and Morris, L.T., eds.) p. 65, VanNostrand, London.
316 Sjövall, J., Eneroth, P. and Ryhage, R. (1971) in The Bile Acids (Nair, P.P. and Kritchevsky, D., eds) p. 209, Vol. 1, Plenum Press, New York.
317 Hoffman, A.F. (1967) J. Lipid Res. 8, 55.
318 Sjövall, J., Sjövall, K. and Turner, D.A. (1965) J. Lipid Res. 6, 182.
319 Makino, I. and Sjövall, J. (1972) Anal. Lett. 5, 341.
320 Almé, B., Bremmelgaard, A., Sjövall, J. and Thomassen, P. (1977) J. Lipid. Res. 18, 339.
321 Karlaganis, G. and Paumgartner, G. (1978) J. High Resolut. Chromatogr. 1, 54.
322 Fukunaga, A., Hatta, Y., Ishibashi, M. and Miyazaki, H. (1980) J. Chromatogr. 190, 339.
323 Campbell, R.L., Gantt, J.S. and Nigro, N.D. (1978) J. Chromatogr. 155, 427.

324 Szczepanik, P.A., Hachey, D.L. and Klein, P.D. (1976) J. Lipid Res. 17, 314.
325 Murata, T., Takahashi, S., Ohnishi, S., Hosoi, K., Nakashima, T., Ban, Y. and Kuriyama, K. (1982) J. Chromatogr. 239, 571.
326 Sjövall, K. and Sjövall, J. (1966) Clin. Chim. Acta 13, 207.
327 Nigro, N.D., Campbell, R.L., Gantt, J.S., Lin, Y.N. and Singh, D.V. (1977) Cancer Res. 37, 3198.
328 Summerfield, J.A., Billing, B.H. and Shackleton, C.H.L. (1976) Biochem. J. 154, 507.
329 Marinetti, G.V. (ed.) (1976) Lipid Chromatographic Analysis, Vol. 3, Marcel Dekker, New York.
330 Kuksis, A. (1977) J. Chromatogr. 143, 3.
331 Vaver, V.A. and Ushakov, A.N. (1980) in The Methods of Biochemical Analysis, Vol. 26 (Glick, D., ed.) p. 26, Wiley Interscience, New York.
332 Smith, S.L., Jorgenson, J.W. and Novotny, M. (1979) in Trace Organic Analysis: A New Frontier in Analytical Chemistry (Hertz, H.S. and Chesler, S.N., eds), p. 487, National Bureau of Standards Special Publication 519, U.S. Government Printing Office, Washington, D.C.
333 Smith, S.L., Jorgenson, J.W. and Novotny, M. (1980) J. Chromatogr. 187, 111.
334 Grob, Jr., K. (1979) J. Chromatogr. 178, 387.
335 Grob, Jr., K. (1981) J. Chromatogr. 205, 289.
336 Aitzetmüller, K. (1975) J. Chromatogr. 113, 231.
337 Kuksis, A., Hyher, J.J., Marai, L. and Geher, K. (1976) Anal. Biochem. 70, 302.
338 Skořepa, J., Mareš, P., Rublicová, J. and Vinogradov, S. (1979) J. Chromatogr. 162, 177.
339 Horning, M.G. in New Biochemical Separations (James, A.T. and Morris, L.J., eds) p. 59, Van Nostrand Company, Ltd., London.
340 Watts, R.B., Dils, R. and Wehr, H. (1972) J. Chromatogr. 66, 239.
341 Lin, S. and Horning, E.C. (1975) J. Chromatogr. 112, 465.
342 Hammarström, S. (1970) J. Lipid Res. 11, 175.
343 Viswanathan, C.V. (1974) J. Chromatogr. 98, 105.
344 Gaskell, S.J., Edmonds, C.G. and Brooks, C.J.W. (1976) J. Chromatogr. 126, 591.
345 Samuelsson, B. and Samuelsson, K. (1969) J. Lipid Res. 10, 41.
346 Gaskell, S.J. and Brooks, C.J.W. (1977) J. Chromatogr. 142, 469.
347 Kuksis, A., Stachnyk, O. and Holub, B.J. (1971) J. Lipid Res. 12, 257.
348 Kuksis, A., Marai, L. and Gornall, D.A. (1967) J. Lipid Res. 8, 352.
349 Horning, M.G., Gasparini, G. and Horning, E.C. (1967) J. Chromatogr. Sci. 7, 267.
350 Krupčík, J., Hrivňák, J. and Janák, J. (1976) J. Chromatogr. Sci. 14, 4.
351 Chalmers, R.A., Bickle, S. and Watts, R.W.E. (1974) Clin. Chim. Acta 52, 31.
352 Weatherston, J., MacDonald, L.M., Blake, T., Benn, M.H. and Huang, Y.Y. (1978) J. Chromatogr. 161, 347.
353 Zoccolillo, L., Liberti, A. and Goretti, G.G. (1969) J. Chromatogr. 43, 497.
354 Hrivňák, J., Soják, L., Beška, E. and Janák, J. (1972) J. Chromatogr. 68, 55.
355 Lipsky, S.R., Lovelock, J.E. and Landowne, R.A. (1959) J. Am. Chem. Soc. 81, 1010.
356 Ottenstein, D.M., Bartley, D.A. and Supina, W.R. (1976) J. Chromatogr. 119, 401.
357 Ackman, R.J. and Hooper, S.N. (1974) J. Chromatogr. Sci. 12, 131.
358 Jaeger, H., Kloer, H.V. and Ditschuneit, H. (1976) J. Lipid Res. 17, 185.
359 Ackman, R.J. and Sebedio, J.L. (1978) J. Chromatogr. Sci. 16, 204.
360 Klenk, E. and Kahlke, W. (1963) Hoppe-Seiler's Z. Physiol. Chem. 333, 133.
361 Gerstl, B., Eng, L.F., Tavastjerna, M., Smith, J.K. and Kruse, S.L. (1970) J. Neurochem. 17, 677.
362 Eto, Y., Suzuki, K. and Suzuki, K. (1970) J. Lipid Res. 11, 473.
363 Pullarkat, R.K. and Reha, H. (1976) J. Chromatogr. Sci. 14, 25.
364 Tichý, J., Skorkovská, I. and Baše, J. (1979) J. Chromatogr. 162, 185.
365 Lin, S.-N. and Horning, E.C. (1975) J. Chromatogr. 112, 483.
366 Gerhard, K.O. and Gehrke, C.W. (1977) 143, 335.
367 Rockerbie, R.A., Dobson, R.D. and Frohlich, J. (1979) Clin. Chem. 25, 1411.
368 Haan, G.J., Van der Heide, S. and Wolthers, B.G. (1979) J. Chromatogr. 162, 261.
369 O'Neill, H.J. and Gershbein, L.L. (1976) J. Chromatogr. Sci. 14, 28.

370 Farshy, D.C. and Moss, C.W. (1970) Appl. Microbiol. 20, 78.
371 Moss, C.W. and Dees, S.B. (1975) J. Chromatogr. 112, 595.
372 Mitruka, B.M. (1975) Gas Chromatographic Applications in Microbiology and Medicine, Wiley Biomedical-Health Publication, New York.
373 Tanaka, K. and Hine, D.G. (1982) J. Chromatogr. 239, 301.
374 Williams, V.P., Ching, D.K. and Cederbaum, S.D. (1979) Clin. Chem. 25, 1814.
375 Horning, E.C. and Horning, M.G. (1971) Clin. Chem. 17, 802.
376 Thompson, J.A. and Markey, S.P. (1975) Anal. Chem. 57, 1313.
377 Gates, S.C., Demdramis, N. and Sweeley, C.C. (1978) Clin. Chem. 24, 1674.
378 Fitch, W.L., Anderson, P.J. and Smith, D.H. (1979) J. Chromatogr. 162, 249.
379 Sims, P., Truscott, R. and Halpern, B. (1981) J. Chromatogr. 222, 337.
380 Horning, M.G., Boucher, E.A., Moss, A.M. and Horning, E.C. (1968) Anal. Lett. 1, 713.
381 Rocchiccioli, F., Leroux, J.P. and Cartier, P. (1981) Biomed. Mass Spectr. 8, 160.
382 Grupe, A. and Spiteller, G. (1981) J. Chromatogr. 226, 301.
383 Holland, M.L., Yancey, M. and Novotny, M., unpublished experiments.
384 Holland, M.L. (1982) Doctoral Thesis, Department of Chemistry, Indiana University, Bloomington, Indiana.
385 Nakamura, E., Rosenberg, L.E. and Tanaka, K. (1976) Clin. Chim. Acta 68, 127.
386 Jacobs, C., Solem, E., Ek, J., Halvorsen, K. and Jellum, E. (1977) J. Chromatogr. 143, 31.
387 Liebich, H.M., Pickert, A., Stierle, U. and Wöll, J. (1980) J. Chromatogr. 199, 181.
388 Liebich, H.M., Pickert, A. and Wöll, J. (1981) J. Chromatogr. 217, 255.
389 Niwa, T., Maeda, K., Ohki, T., Saito, A. and Tsuchida, I. (1981) J. Chromatogr. 225, 1.
390 Niwa, T., Maeda, K., Okhi, T. and Sakakibara, J. (1982) J. Chromatogr. 228, 59.
391 Kamerling, J.P., Duran, M., Gerwig, G.J., Ketting, D., Bruinvis, L., Vliegenthart, J.F.G. and Wadman, S.K. (1981) J. Chromatogr. 222, 276.
392 Wadman, S.K., Duran, M., Ketting, D., Bruinvis, L., de Bree, P.K., Kamerling, J.P., Gerwig, G.J., Vliegenthart, J.F.G., Przyrembel, H., Becker, K. and Bremer, J.H. (1976) Clin. Chim. Acta 71, 477.
393 Duran, M., van Biervliet, J.P.G.M., Kamerling, J.P. and Wadman, S.K. (1977) Clin. Chim. Acta 74, 297.
394 Kamerling, J.P., Gerwig, G.J., Duran, M., Ketting, D. and Wadman, S.K. (1978) Clin. Chim. Acta 88, 183.
395 Kaji, H., Saito, N., Murao, M., Ishimoto, M., Kondo, H., Gasa, H. and Saito, K. (1980) J. Chromatogr. 221, 145.
396 McInnes, A.G., Ball, D.H., Cooper, F.P. and Bishop, C.T. (1958) J. Chromatogr. 1, 556.
397 Bentley, R., Sweeley, C.C., Makita, M. and Wells, W.W. (1963) Biochem. Biophys. Res. Commun. 11, 14.
398 Horning, E.C. and VandenHeuvel, W.J.A. (1964) in New Biochemical Separations (James, A.T. and Morris, L.J., eds.) p. 25, Van Nostrand Company, Ltd., London.
399 Hanessian, S. (1971) in The Methods of Biochemical Analysis (Glick, D., ed.) Vol. 19, p. 105, Interscience Publishers, New York.
400 Clamp, J.R., Bhatti, T. and Chambers, R.E. (1971) in The Methods of Biochemical Analysis (Glick, D., ed.), Vol. 19, p. 229, Interscience Publishers, New York.
401 Ferrier, R.J. and Collins, P.M. (1972) Monosaccharide Chemistry, Penguin, Harmondsworth.
402 Eklund, G., Josefsson, B. and Roos, C. (1977) J. Chromatogr. 142, 575.
403 Szafranek, J., Pfaffenberger, C.D. and Horning, E.C. (1973) Anal. Lett. 6, 479.
404 Pfaffenberger, C.D., Szafranek, J. and Horning, E.C. (1976) J. Chromatogr. 126, 535.
405 Laine, R.A. and Sweeley, C.C. (1971) Anal. Biochem. 43, 533.
406 Pelletier, O. and Cadieux, S. (1982) J. Chromatogr. 231, 225.
407 Tesařík, K. (1972) J. Chromatogr. 65, 295.
408 Pfaffenberger, C.D., Szafranek, J., Horning, M.G. and Horning, E.C. (1975) Anal. Biochem. 63, 501.
409 Honda, S., Suzuki, S., Kakehi, K., Honda, A. and Takai, T. (1981) J. Chromatogr. 226, 341.
410 Anastassiades, T., Puzic, P. and Puzic, O. (1981) J. Chromatogr. 225, 309.

411 Gehrke, C.C., Waalkes, R.P., Borek, E., Swartz, W.F., Cole, T.F., Kuo, K.C., Abeloff, M., Ettinger, D., Rosenheim, N. and Young, R.C. (1979) J. Chromatogr. 162, 507.
412 Servo, C., Palo, J. and Pitkänen, E. (1977) Acta Neurol. Scand. 56, 104.
413 Laker, M.F. (1979) J. Chromatogr. 163, 9.
414 Pitkänen, E. (1973) Clin. Chim. Acta 48, 159.
415 Smith, S.L., Novotny, M. and Weber, E.L. (1978) Clin. Chem. 24, 545.
416 Novotny, M., Smith, S.L. and Karmen, A., submitted for publication.
417 Kissinger, P.T., Bruntlett, C.S. and Shoup, R.E. (1981) Life Sci. 28, 455.
418 Brooks, C.J.W. and Horning, E.C. (1964) Anal. Chem. 36, 1540.
419 Hiemke, C., Kauert, G. and Kalbhen, D.A. (1978) J. Chromatogr. 153, 451.
420 Anggard, E. and Sedvall, G. (1969) Anal. Chem. 41, 1250.
421 Karoum, F., Cattabeni, F., Costa, E., Ruthwer, C.R.J. and Sandler, M. (1972) Anal. Biochem. 47, 550.
422 Sunõl, C. and Gelpi, E. (1977) J. Chromatogr. 142, 559.
423 Holdiness, M.R., Rosen, M.T., Justice, J.B. and Neill, D.B. (1980) J. Chromatogr. 198, 329.
424 Sen, N.P. and McGeer, P.L. (1963) Biochem. Biophys. Res. Commun. 13, 390.
425 Horning, M.G., Moss, A.M. and Horning, E.C. (1967) Biochim. Biophys. Acta 148, 597.
426 Hattox, S.E. and Murphy, R.C. (1978) Biomed. Mass Spectrom. 5, 338.
427 Martinez, E. and Gelpi, E. (1978) J. Chromatogr. 167, 77.
428 Creveling, C.R., Kondo. K. and Daly, J.W. (1968) Clin. Chem. 14, 302.
429 Narasimhachari, N. and Vouros, P. (1972) Anal. Biochem. 45, 154.
430 Brandenberger, H. and Schnyder, D. (1972) Z. Anal. Chem. 261, 291.
431 Capella, P. and Horning, E.C. (1966) Anal. Chem. 38, 316.
432 Donike, M. (1975) J. Chromatogr. 103, 91.
433 Mosnaim, A.D., Callaghan, O.H. and Wolf, M.E. (1981) J. Chromatogr. 224, 481.
434 Yamamoto, S., Kakuno, K., Okahara, S., Kataoka, H. and Makita, M. (1980) J. Chromatogr. 194, 399.
435 Doshi, P.S. and Edwards, D.J. (1981) J. Chromatogr. 210, 505.
436 Bock, U.E.G. and Waser, P.G. (1981) J. Chromatogr. 213, 413.
437 van de Calseyde, J.F., Scholtis, R.J., Schmidt, R.J. and Leyton, C.J. (1971) Clin. Chim. Acta 32, 361.
438 Chattoraj, S.C. (1976) in Fundamentals of Clinical Chemistry (Tietz, N.W., ed.) W.B. Saunders Company, Philadelphia, 1976.
439 Mita, H., Yasueda, H. and Shida, T. (1980) J. Chromatogr. 221, 1.
440 LeGatt, D.F., Baker, G.B. and Coutts, R.T. (1981) J. Chromatogr. 225, 301.
441 Koslow, S.H., Cattabeni, F. and Costa, E. (1972) Science 176, 177.
442 Curtius, H.C., Wolfensberger, M., Steinmann, B., Redweik, U. and Siefgried, J. (1974) J. Chromatogr. 99, 529.
443 Miyazaki, H., Hashimoto, Y., Iwanaga, M. and Kubodera, T. (1974) J. Chromatogr. 99, 575.
444 Curtius, H.C., Baerlocher, K. and Völlmin, J.A. (1972) Clin. Chim. Acta 42, 235.
445 Cattabeni, F., Koslow, G.H. and Costa, E. (1972) Science 178, 166.
446 Gelpi, E., Peralta, E. and Segura, J. (1974) J. Chromatogr. Sci. 12, 701.
447 Curtius, H.C., Farner, H. and Rey, F. (1980) J. Chromatogr. 199, 171.
448 Fitzpatrick, F.A. and Wynalda, M.A. (1976) Anal. Biochem. 73, 198.
449 Vane, F. and Horning, M.G. (1969) Anal. Lett. 2, 357.
450 Rosello, J., Mahy, N., Tusell, J. and Gelpi, E. (1976) J. Chromatogr. 122, 469.
451 Rosello, J., Tusell, J. and Gelpi, E. (1977) J. Chromatogr. 130, 65.
452 Jonsson, H.T., Middleditch, B.S., Schexnayder, M.A. and Desiderio, D.M. (1976) J. Lipid Res. 17, 1.
453 Hubbard, W.C. and Watson, J.T. (1976) Prostaglandins 12, 21.
454 Rosello, J. and Gelpi, E. (1978) J. Chromatogr. Sci. 16, 177.
455 Tusell, J.M. and Gelpi, E. (1980) J. Chromatogr. 181, 295.
456 Gréen, K. (1969) Chem. Phys. Lipids 3, 254.
457 Miyazaki, H., Ishibashi, M., Yamashita, K. and Katori, M. (1978) J. Chromatogr. 153, 83.

458 Miyazaki, H., Ishibashi, M., Yamashita, K., Ohguchi, I., Saitoh, H., Kurono, H., Shimono, M. and Katori, M. (1982) J. Chromatogr. 239, 595.
459 Erlenmaier, T., Müller, H. and Seyberth, H.W. (1979) J. Chromatogr. 163, 289.
460 Sjöquist, B., Olin, E., Lunden, I. and Anggard, E. (1979) J. Chromatogr. 163, 1.
461 Jouvenaz, G.H., Nugteren, D.H., Beerthius, R.K. and van Dorp, D.A. (1970) Biochim. Biophys. Acta 202, 231.
462 Middleditch, B.S. and Desiderio, D.M. (1972) Prostaglandins 2, 115.
463 Fitzpatrick, F.A., Wynalda, M.A. and Kaiser, D.G. (1977) Anal. Chem. 49, 1032.
464 Fitzpatrick, F.A., Stringfellow, D.A., Maclouf, J. and Rigaud, M. (1979) J. Chromatogr. 177, 51.
465 Min, B.H., Pao, J., Garland, W.A., DeSilva, J.A.F. and Parsonnet, M. (1980) J. Chromatogr. 183, 411.
466 Barrow, S.E., Waddell, K.A., Ennis, M., Dollery, C.T. and Blair, I.A. (1982) J. Chromatogr. 239, 71.
467 deDeckere, E.A.M., Nugteren, D.H. and Tenhoor, F. (1977) Nature (London) 268, 160.
468 Samuelsson, B., Hamberg, M. and Sweeley, C.C. (1970) Anal. Biochem. 38, 301.
469 Rigaud, M., Chebroux, P., Soustre, A., Durand, J., Rabinowitch, H. and Breton, J.C. (1979) in Advances in Chromatography 1979 (Zlatkis, A., Ettre, L.S. and Kováts, E. Sz., eds.) p. 615, Chromatography Symposium, Houston.
470 Smith, A.G., Harland, W.A. and Brooks, C.J.W. (1977) J. Chromatogr. 142, 533.
471 Suzuki, M., Morita, I., Kawamura, M., Murota, S.-I., Nishizawa, M., Miyatake, J., Nagase, H., Ohno, K. and Shimizu, H. (1980) J. Chromatogr. 221, 361.
472 Fitzpatrick, F.A. (1978) Anal. Chem. 50, 47.
473 Moore, S. and Stein, W.H. (1951) J. Biol. Chem. 192, 663.
474 Bayer, E. (1958) in Gas Chromatography 1958 (Desty, D.H., ed.) p. 333, Butterworths, London.
475 Gehrke, C.W., Roach, D. and Zumwalt, R.W. (1969) J. Chromatogr. 43, 311.
476 Makisumi, S. and Saroff, H.A. (1965) J. Gas Chromatogr. 3, 21.
477 Moodie, I.M. (1974) J. Chromatogr. 99, 495.
478 Husek, P. and Macek, K. (1979) J. Chromatogr. 113, 139.
479 MacKenzie, S.L. (1981) in Methods of Biochemical Analysis (Glick, D., ed.) Vol. 27, Wiley Interscience, New York.
480 Jaeger, H., Kloer, H.U., Ditschuneit, H. and Frank, H. (1981) in Applications of Glass Capillary Gas Chromatography (Jennings, W.G., ed.), p. 331, Marcel Dekker, New York.
481 Pettitt, B.C. and Stouffer, J.E. (1970) J. Chromatogr. Sci. 8, 735.
482 Thenot, J.-P. and Horning, E.C. (1972) Anal. Lett. 5, 519.
483 Gehrke, C.W. and Leimer, K.R. (1971) J. Chromatogr. 57, 219.
484 Jönsson, J., Eyem, J. and Sjöquist, J. (1973) Anal. Biochem. 51, 204.
485 Poole, C. and Verzele, M. (1978) J. Chromatogr. 150, 439.
486 Pearce, R.C. (1977) J. Chromatogr. 136, 113.
487 Adams, R.F., Vandemark, F.L. and Schmidt, G.J. (1977) J. Chromatogr. Sci. 15, 63.
488 Leighton, W.P., Rosenblatt, S. and Chanley, J.D. (1979) J. Chromatogr. 164, 427.
489 Siezen, R.J. and Mague, T.H. (1977) J. Chromatogr. 130, 151.
490 Pisano, J.J., Bronzert, T.J. and Brewer, Jr., H.B. (1972) Anal. Biochem. 45, 43.
491 Eyem, J. and Sjöquist, J. (1973) Anal. Biochem. 52, 255.
492 Lamkin, W.M., Weatherford, J.W., Jones, N.S., Pan, T. and Ward, D.N. (1974) Anal. Biochem. 58, 422.
493 Inglis, A.S. and Nicholls, P.W. (1973) J. Chromatogr. 86, 117.
494 Cavadore, J.C., Nota, G., Prota, G. and Previero, A. (1974) Anal. Biochem. 60, 608.
495 Brock, D.J.H. and Mayo, O. (1978) (eds.), The Biochemical Genetics of Man, Academic Press, London.
496 Davis, L.G., Sass, N.L., Manna, B. and Nusynowitz, M.L. (1979) Clin. Chem. 25, 218.
497 Cooper, A.J.L., Dhar, A.K., Kutt, H. and Duffy, T.E. (1980) Anal. Biochem. 103, 118.
498 Ramsdell, H.S. and Tanaka, K. (1980) J. Chromatogr. 181, 90.
499 Mussini, E., Cotellessa, L., Colombo, L., Cani, D., Sfondrini, P., Marcucci, F. and Poy, F. (1981) J. Chromatogr. 224, 94.

500 Desgres, J., Boisson, D. and Padieu, P. (1979) J. Chromatogr. 162, 133.
501 Clay, K.L. and Murphy, R.C. (1979) J. Chromatogr. 164, 417.
502 Chauhan, J., Darbre, A. and Carlyle, R.F. (1982) J. Chromatogr. 227, 305.
503 Bengtsson, G. and Odham, G. (1979) Anal. Biochem. 92, 426.
504 Petersen, B.A. and Vouros, P. (1977) Anal. Chem. 49, 1304.
505 Halpern, B., Ricks, J. and Westley, J.W. (1966) Anal. Biochem. 14, 159.
506 Wegmann, H., Curtius, H.-Ch. and Redweik, U. (1978) J. Chromatogr. 158, 305.
507 Holdiness, M.R., Justice, J.B., Salamone, J.D. and Neill, D.B. (1981) J. Chromatogr. 225, 283.
508 Stenhagen, E. (1961) Z. Anal. Chem. 181, 462.
509 Biemann, K. and Vetter, W. (1960) Biochem. Biophys. Res. Commun. 3, 578.
510 Thomas, D.W., Das, B.C., Géro, S.D. and Lederer, E. (1968) Biochem. Biophys. Res. Commun. 32, 199.
511 Nau, H. and Biemann, K. (1976) Anal. Biochem. 73, 139, 154, 186.
512 Seifert, W.E., McKee, R.E., Beckner, C.F. and Caprioli, R.M. (1978) Anal. Biochem. 88, 149.

Note

A wealth of references regarding applications as well as instrumentation, detection procedures, derivatization and other practical aspects of GC separations can be obtained from the respective part of the Bibliography section published regularly in the *Journal of Chromatography*.

CHAPTER 4

Liquid column chromatography

4.1 Types of liquid column chromatography
by S.H. Hansen, P. Helboe and U. Lund

4.2 Instrumentation
by S.H. Hansen, P. Helboe and U. Lund

4.3 Detection
by S.H. Hansen, P. Helboe and U. Lund

4.4 Adsorption and partition chromatography
by S.H. Hansen, P. Helboe and U. Lund

4.5 Ion exchange chromatography
by O. Mikes

4.6 Gel chromatography
by D. Berek and M. Marcinka

4.7 Bioaffinity chromatography
by J. Turkovà

Deyl (ed.) Separation Methods

CHAPTER 4.1

Types of liquid column chromatography

STEEN H. HANSEN [a], PER HELBOE [b] and ULLA LUND [c]

[a] Royal Danish School of Pharmacy, Department of Chemistry BC, 2 Universitetsparken, DK 2100, Copenhagen, [b] National Board of Health, Drug Standardization Laboratory, 378 Frederikssundsvej, DK 2700, Brønshøj, and [c] Water Quality Institute (VKI), 11 Agern Allé, DK 2970, Hørsholm, Denmark

4.1.1 Introduction

The birth of column liquid chromatography is ascribed to the work of the Russian botanist Tswett who, in 1906 [1,2], published two papers in which he described a method to separate plant pigments contained in a petroleum ether extract of green leaves, by percolating it through a vertical glass column containing fine grains of calcium carbonate.

Despite this early description of adsorption chromatography, it was not until the 1930's that development of chromatography really began.

With the technological evolution during the last 15 years high efficiency column materials have appeared, and with them the development of high-performance liquid chromatography.

Several types of liquid column chromatography have been developed and the most important of these are described in the following.

4.1.2 Adsorption

In adsorption chromatography the retention of the solute is a consequence of the interaction with the surface of the solid adsorbent. The adsorbent surface has a rigid structure making this type of chromatography uniquely useful for separations of geometric and structural isomers with molecular weights up to about 1000.

4.1.3 Partition

Liquid-liquid partition chromatography was first described by Martin and Synge [3,4] in 1941. The distribution of solutes takes place between two immiscible solvents. In normal phase (straight phase) chromatography the more polar liquid —

Deyl (ed.) Separation Methods

CHAPTER 4.2

Instrumentation

STEEN H. HANSEN [a], PER HELBOE [b] and ULLA LUND [c]

[a] *Royal Danish School of Pharmacy, Department of Chemistry BC, 2 Universitetsparken, DK 2100, Copenhagen,* [b] *National Board of Health, Drug Standardization Laboratory, 378 Frederikssundsvej, DK 2700, Brønshøj, and* [c] *Water Quality Institute (VKI), 11 Agern Allé, DK 2970, Hørsholm, Denmark*

4.2.1 Introduction

The equipment needed for liquid column chromatography (LC) ranges from a simple and inexpensive glass column eluted by gravity flow to a sophisticated, computerized high-performance liquid chromatograph.

In any column chromatographic system the column is the most important part, as it is in the column the separation takes place. However, the systems for solvent delivery and for detection are absolutely necessary.

Some simple classic systems are shown in Fig. 4.2.1. These columns have to be repacked after being used only once or a few times. The application of the sample is made directly onto the column, whereupon the flow of the mobile phase is started. The mobile phase may flow either by gravity or may be delivered by a low pressure pump. Detection and quantitation are achieved by photometry (flow-through cell) or by discrete analysis of the individual fractions after fraction collection.

Classic liquid column chromatography is rather time and solvent consuming, but is still used for preparative work because of the simple and low priced apparatus and column materials.

Several monographs [1–3] on the classic LC techniques may be consulted.

From the late sixties until now (1982) the high-performance LC procedure has evolved to be the method of choice within liquid chromatography. The instrument set-up is shown schematically in Fig. 4.2.2.

It is seen that the system in principle is built from the same three parts as the classic one: the column or column system, the solvent delivery system, and the detection device. The components of the total system will be discussed in the following.

constant flow pumps. Pumps in the first category are generally the cheapest, but they have one drawback: if the back pressure of the column is altered the flow will change and hence the measurements are compromized.

It is important to achieve a constant flow, and therefore most pumps are delivered with a feedback mechanism to ensure a constant flow even if the back pressure of the column changes.

Changes in temperature also change the viscosity of the solvent, and thereby change the flow if no feedback control is used. This, together with the fact that all phase equilibria are dependent on the temperature, necessitates thermostating of the columns if accurate measurements are to be performed.

If the sample to be chromatographed contains compounds of very different polarities gradient elution may be of great help in giving a better overall separation in a shorter time. Gradient elution is also convenient when 'scouting' for a new chromatographic system.

4.2.5 Detectors

There seems to be no limitation on what kind of detectors may be used for HPLC. Even detectors designed for use in gas phase analyses have been applied to HPLC. The principles of the detectors used for routine analysis are based on absorption of light in the ultraviolet and visible spectrum, refractive index or fluorescence. Detection problems are discussed below.

4.2.6 Technical optimization of the LC system

In order to get the highest possible efficiency of the chromatographic column the connection lines between the different parts of the chromatographic equipment should be considered to avoid extra-column effects (band broadening). Any extra-column volume between the point of sample injection and sample detection tend to spoil the separation which may be obtained on the column, and should therefore be minimized. The items to be considered are the volume of the connection tubes, the column end-fittings and the detector cell. The significance of these extra-column effects are increased with decreasing column length.

4.2.7 Conclusion

In this survey of LC equipment attention has been drawn to the more important parts of the chromatographic system. When more detailed information is required special literature may be consulted, and a list of books on LC equipment is given below [8–12].

References

1 Lederer, E. and Lederer, M. (1962) in Comprehensive Biochemistry, pp. 32–268, Vol. 4, American Elsevier, New York.
2 Bobbitt, J.M., Schwarting, A.E. and Gritter, R.J. (1968) Introduction to Chromatography, pp. 84–105, Reinhold Science Studies.
3 Deyl, Z., Macek, K. and Janak, J. (eds.) (1975) Liquid Column Chromatography, Elsevier, New York.
4 Martin, M., Eon, C. and Guiochon, G. (1975) J. Chromatogr. 110, 213.
5 Halasz, I., Schmidt, H. and Vogtel, P. (1976) J. Chromatogr. 126, 19.
6 Knox, J.H. (1977) J. Chromatogr. Sci. 15, 352.
7 Brown, J.N., Hewins, M., van der Linden, J.H.M. and Lynch, R.J. (1981) J. Chromatogr. 204, 115.
8 Snyder, L.R. and Kirkland, J.J. (1979) Introduction to Modern Liquid Chromatography, 2nd Edn. Wiley-Interscience, New York
9 Huber, J.F.K. (1978) Instrumentation for High-Performance Liquid Chromatography, Elsevier, Amsterdam.
10 Scott, R.P.W. (1977) Liquid Chromatography Detectors. Elsevier, Amsterdam.
11 Parris, N.A. (1976) Instrumental Liquid Chromatography. Elsevier, Amsterdam.
12 Knox, J.H. (1978) High-Performance Liquid Chromatography. Edinburgh University Press.

CHAPTER 4.3

Detection

STEEN H. HANSEN [a], PER HELBOE [b] and ULLA LUND [c]

[a] *Royal Danish School of Pharmacy, Department of Chemistry BC, Universitetsparken, DK 2100, Copenhagen,* [b] *National Board of Health, Drug Standardization Laboratory, 378 Frederikssundsvej, DK 2700, Brønshøj, and* [c] *Water Quality Institute (VKI), 11 Agern Allé, DK 2970, Hørsholm, Denmark*

4.3.1 Introduction

In biochemical analysis the determination of solutes at low concentration levels in complex matrices is a frequent task. Here the choice of an appropriate detector and the use of proper methods for the enhancement of detection are very important, and will be treated in more detail later in this chapter. It must, however, be kept in mind that the whole chromatographic system has an influence on the detectability that can be achieved with a given liquid chromatographic analysis.

In the design of the chromatograph dead volumes should be kept at an absolute minimum, since these volumes will result in a dilution of the sample, and thus cause

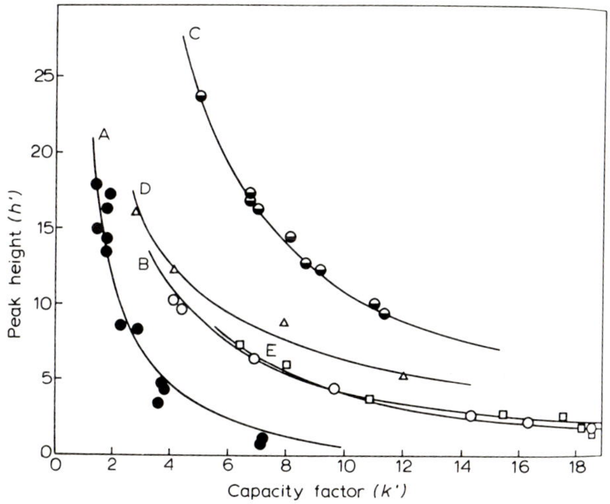

Fig. 4.3.1. Effect of capacity factor on peak height in a number (A–E) of chromatographic systems [1]. Reprinted with permission.

4.3.2.4 *The refractive index detector*

This detector is almost universal, though not very sensitive, and its lack of sensitivity in most cases restricts its use to situations where other detectors fail to detect the compounds of interest.

The refractive index detector is very sensitive to variations in temperature, but as long as this is realized and the detector kept properly thermostated, this is usually not a cause for concern.

4.3.2.5 *The radioactivity detector*

The radioactivity detector has an obvious application in metabolic studies. Unfortunately it is necessary when using this detector to trade resolution and speed for sensitivity. The response of the radioactivity detector is a function of the total amount of radioactivity in the detector cell, which means that the detector cell should be as large as possible. On the other hand a large cell volume will cause dispersion of the chromatographic peaks, and a compromise must therefore be found. The response is also a function of the residence time of the solute in the cell, calling for slow pumping velocity of the mobile phase, thus giving increased time of analysis.

4.3.2.6 *Liquid chromatography-mass spectrometry*

Detection by on-line coupled liquid chromatography-mass spectrometry (LC-MS) is at present at an experimental stage. The technique and instrumentation in LC-MS is improving rapidly at present, and recent reviews on the subject should be consulted for up-to-date information.

4.3.3 *Detection enhancement*

The detection properties of a solute can in many cases be enhanced, and for this purpose derivatization, either pre- or post-column, is often used. But non-derivatization techniques have also been described, e.g., continuous post-column ion-pair extraction [4], or post-column modification of pH [5] have been used to enhance detectability.

A schematic presentation of the equipment used for post-column reactions is shown in Fig. 4.3.2. Here a simple system is shown with only one reagent added to the effluent from the chromatographic column, and the reaction takes place at room temperature. The system is easily extended for several reagents and the reactors can be heated using standard thermostatic equipment. For details on reactor design and other factors of importance to the successful application of reaction detection specialized texts (e.g., Refs. 6, 7) should be consulted.

Pre-column derivatization procedures have the advantage, over post-column procedures, that long reaction times and radical reaction conditions can be used, and

TABLE 4.3.1

Typical properties of liquid chromatographic detectors for biochemical analysis

	Ultraviolet absorption	Fluorescence	Electrochemical detection	Refractive index	Radioactivity	Mass spectrometry
Sensitive to a favorable sample	10^{-10} g/ml	10^{-11} g/ml	10^{-12} g/ml	10^{-7} g/ml	100 cpm ^{14}C/ml	10^{-9} g/s
Selectivity	selective	selective	selective	universal	selective	universal/selective
Linear dynamic range	10^4–10^5	10^1–10^4	10^6	10^4	?	wide
Can be used with gradient elution	Yes	Yes	No	No	Yes	Yes
Cell volume	8–12 μl	8–20 μl	approx. 1 μl	approx. 10 μl	20–100 μl	—
Response time	Fast	Fast	Fast	Fast	Fast	Fast
Destructive	No	No	Yes	No	No	(Yes)

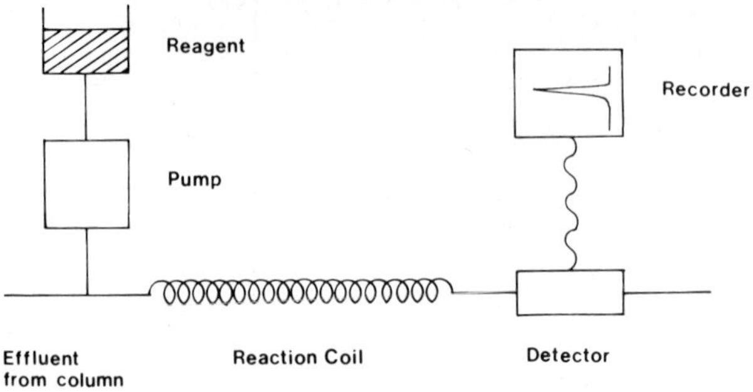

Fig. 4.3.2. Schematic representation of the equipment necessary for post-column reactions.

it is possible to use reagents that possess the same detection properties as the derivative. This is not possible with post-column reactions, since the reagent is fed with the effluent to the detector. Also long reaction times should be avoided with post-column reactions because excessive extra-column peak broadening will otherwise occur. A reaction time of about 20 min will normally be the maximum for high efficiency columns.

A disadvantage of the pre-column reaction technique, compared to post-column reaction, is that side products formed in the derivatization procedure may give rise to interferences in the chromatogram.

For a short series of analyses it is easier to use a pre-column procedure, since no instrumental modifications are required, but for longer series post-column procedures allow for a much higher degree of automatization.

References

1 Kirkland, J.J. (1974) Analyst 99, 859.
2 Snyder, L.R. and Kirkland, J.J. (1979) Introduction to Modern Liquid Chromatography, 2nd edn., John Wiley and Sons, New York.
3 Scott, R.P.W. (1977) Liquid Chromatography Detectors, Elsevier Scientific Publishing Company, Amsterdam.
4 Lawrence, J.F., Brinkman, U.A.Th. and Frei, R.W. (1979) J. Chromatogr. 185, 473.
5 Kissinger, P.T., Bratin, K., Davis, G.C. and Pachla, L.A. (1979) J. Chromatogr. Sci. 17, 137.
6 Lawrence, J.F. and Frei, R.W. (1976) Chemical Derivatization in Liquid Chromatography, Elsevier Scientific Publishing Company, Amsterdam.
7 Frei, R.W. (1979) J. Chromatogr. 165, 75.

Deyl (ed.) Separation Methods

CHAPTER 4.4

Adsorption and partition chromatography

STEEN H. HANSEN [a], PER HELBOE [b] and ULLA LUND [c]

[a] Royal Danish School of Pharmacy, Department of Chemistry BC, 2 Universitetsparken, DK 2100, Copenhagen, [b] National Board of Health, Drug Standardization Laboratory, 378 Frederikssundsvej, DK 2700, Brønshøj and [c] Water Quality Institute (VKI), 11 Agern Allé, DK 2970, Hørsholm, Denmark

4.4.1 Phase systems

4.4.1.1 General aspects

One of the principal reasons that liquid chromatography has become a most powerful tool in separations of mixtures of almost any origin, not least in the field of biochemistry, is the widespread possibilities in the choice of phase system for a given separation. At the same time, however, this has given rise to a situation where choosing the phase system may seem a little confusing for the inexperienced chromatographer.

In this section we discuss the choice of a chromatographic system on the basis of, firstly, the type of stationary phase and, secondly, the composition of the eluent. A division in subsections is made from the kinds of stationary phases available, bearing in mind that the more specialized phases for ion-exchange, gel permeation (size exclusion), and bioaffinity chromatography are treated in separate chapters (4.5 to 4.7). The following four types of chromatography are dealt with: adsorption; liquid-liquid partition; chemically bonded phases; and dynamically coated phases. It should be emphasized that the borders between the four modes are often rather fluid.

In the treatment of each type of chromatography importance is attached to two fundamental problems in the elaboration of separation methods: controlling the resolution by changing either the retention or the selectivity. In Fig. 4.4.1 from Snyder and Kirklands' fundamental book [1] on HPLC, the situation is illustrated for the separation of two compounds. When a suitable retention (k' between 1 and 5) has been reached without achieving sufficient separation there are three possible solutions. Two of these possibilities, further increase in retention or an enlargement of the number of theoretical plates, however, have implicit disadvantages, i.e., loss in speed of analysis, or, when replacing the column with one of higher efficiency, an increased back pressure. The remaining, and most appropriate, solution is to change

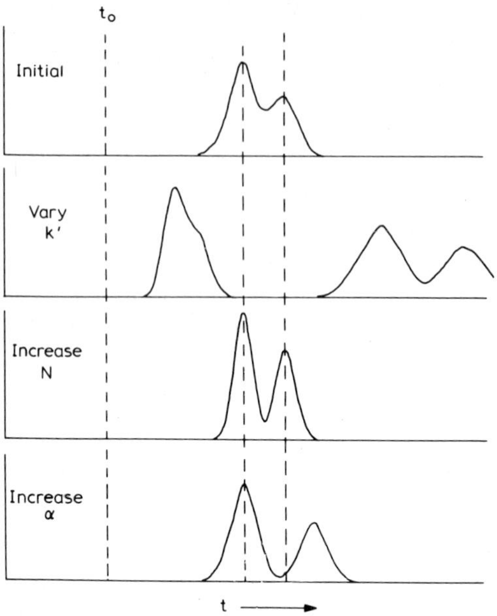

Fig. 4.4.1. The effect on sample resolution of change in k', N or α. Reprinted from Ref. 1 with permission.

the selectivity, α, by modifying the properties of the stationary or mobile phase or both.

4.4.1.2 Adsorption chromatography

Adsorption chromatography is based on liquid-solid partition, i.e., a partition of solute molecules between the liquid mobile phase and active sites on the surface of the solid stationary phase.

The solid support used is most often silica or alumina, the former being far more widely used. As the chromatographic mechanisms are largely the same for the two sorts of support only silica is discussed below. During the chromatographic process solute molecules are adsorbed to active sites on the support surface in competition with solvent molecules. The active sites consist of hydroxyl groups (for silica silanol groups, Si-OH) and the affinity to them is greater the more polar the molecules, and the greater their ability to form hydrogen bondings. The surface structures of different brands of silica are largely identical, and thus their selectivity towards various mixtures of solutes does not exhibit great variation. Variations due to differences in pore size, however, may occur.

As eluents, non-polar solvents (e.g., hexane and dichloromethane) are used with the addition of various amounts of polar solvents as so-called modifiers, e.g., water or low molecular weight alcohols. The polar modifiers are added in order to control the retention. The greater the concentration of modifier the greater the part of the active sites which have adsorbed modifier molecules, thus the retention decreases by increasing the concentration. The above process is also known as deactivation of the stationary phase.

It is possible to achieve the same degree of retention using different modifiers by adjusting their percentage. The selectivity of the chromatographic system, however, is strongly dependent on the nature of the modifier. When choosing the components of the eluent, Tables of solvents arranged according to their eluting strength (eluotropic series) are of great value. Table 4.4.1 shows a limited number of solvents, more detailed information may be found in chromatographic textbooks. The separation of vitamin D metabolites (Fig. 4.4.2) is shown as an example of a separation.

Separation by adsorption chromatography of ionic solutes is not immediately

TABLE 4.4.1
Properties of commonly used solvents

The order of solvents is according to increasing eluotropic strength in normal phase chromatography [2].

Solvents	R.I. ($n^{D}_{20°C}$)	UV cut off (nm) (T 10%)	B.P. (°C)	Viscosity cP (20°C)
Heptane — least polar	1.39	200	98	0.42
Hexane	1.38	200	69	0.32
Cyclohexane	1.43	202	81	0.93
Carbon disulphide	1.63	380	47	0.37
Carbon tetrachloride	1.46	265	76	0.97
Toluene	1.49	285	111	0.59
Propyl chloride	1.39	225	47	0.35
Benzene	1.50	280	80	0.65
1,2-Dichloroethane	1.44	230	83	0.79
Chloroform	1.45	245	61	0.57
Dichloromethane	1.42	233	40	0.44
Dibutyl ether *	1.40		143	
Nitromethane	1.38	380	101	0.67
n-Butyl acetate	1.40	255	125	
Diethyl ether	1.36	202	35	0.23
Ethyl acetate	1.37	260	77	0.45
n-Butanol	1.40		117	
Methyl ethyl ketone	1.38	330	80	
Tetrahydrofuran *	1.41	230	66	0.51
Dioxane *	1.42	215	101	1.54
Acetone	1.36	330	56	0.32
Isopropanol	1.38	207	82	2.30
Ethanol	1.36	205	78	1.20
Acetic acid	1.37	230	118	1.26
Methanol	1.33	208	65	0.60
Acetonitrile	1.34	212	82	0.37
Formamide	1.45	210	210	
Water — most polar	1.33	200	100	1.01

* These solvents often contain stabilizers which are strong UV absorbers.

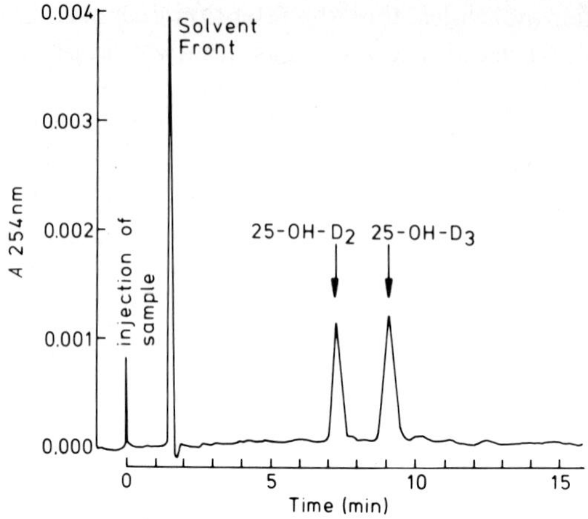

Fig. 4.4.2. Separation of 25-hydroxyvitamin D_2 (25-OHD_2) and 25-hydroxyvitamin D_3 (25-OHD_3) on a μPorasil column (0.4×30 cm). Eluent: 2.5% isopropanol in hexane. Reprinted from Ref. 3 with permission.

possible, these compounds being so polar that total retention may occur. Hence a considerable degree of deactivation of the stationary phase is needed for such separations. By the addition of an acidic or basic modifier to the eluent it may be possible to decrease the retention due to ion-suppression of the solute, or eventually to a blocking (deactivation) of the active site by strong adsorption of cationic compounds from the eluent; a special type of this technique is used in the 'dynamically coating' approach (cf., section 4.4.1.5). An example of chromatography on a strongly deactivated silica column is shown in Fig. 4.4.3.

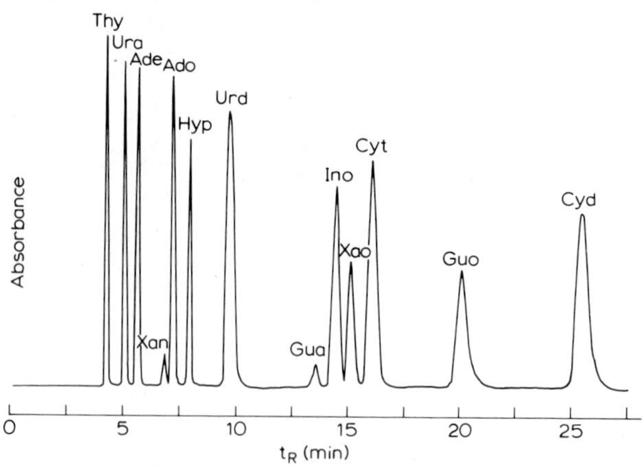

Fig. 4.4.3. Separation of nucleobases and nucleosides on a LiChrosorb SI 100 column (0.46×25 cm). Eluent: Dichloromethane/methanol/2.65 M formic acid (80:18:2). Peaks: Thy, thymine; Ura, uracil; Ade, adenine; Xan, xanthine; Ado, adenosine; Hyp, hypoxanthine; Urd, uridine; Gua, guanine; Ino, inosine; Xao, xanthosine; Cyt, cytosine; Guo, guanosine; Cyd, cytidine. Reprinted from Ref. 4 with permission.

4.4.1.3. Liquid-liquid partition chromatography

Separation by liquid-liquid partition chromatography is the result of the difference in the distribution between two immiscible liquid phases of the individual components of a mixture. The liquid stationary phase is coated on a solid support, which is ideally inert. As the support relatively large pore silica (10–50 nm pore diameter) is most often used.

Coating the stationary phase onto the support may be performed in various ways. The solvent evaporation technique (as in the preparation of GC column materials) may be used, provided that the stationary phase is non-volatile and that the particle size of the support exceeds 20 μm. When using a volatile stationary phase, or modern microparticular support materials (5–10 μm), the in situ coating technique must be used. This is done either by repeated injection of small volumes of pure stationary phase or by eluting the column, which has been packed with bare support material, with eluent saturated with the stationary phase until equilibrium has been established. The two-phase systems originally used consisted of two immiscible solvents, but several workers (e.g., Refs. 5, 6) have shown that excellent separations can be performed using two-phase ternary mixtures of solvents. Depending on the relative polarity of the stationary and the mobile phase, liquid-liquid partition chromatography is divided into normal (or straight) phase chromatography (polar stationary phase) and reversed phase chromatography (non-polar stationary phase). The preparation of reversed phase systems by the in situ technique presupposes the use of a non-polar support, e.g., silanized silica. However, most reversed phase chromatography is performed on bonded phase materials (cf., section 4.4.1.4). As liquid-liquid partition chromatographic separations basically depend on the two liquid phases, no solid phase induced selectivity changes are seen, implying the possibility of preparing chromatographic systems of high reproducibility with respect to selectivity.

In the choice of eluent/solvent mixtures, and in the adjustment of retention, the previously mentioned eluotropic series is valuable together with information on the solubilities of the solutes. To avoid loss of stationary phase it is important that the eluent is fully saturated with stationary phase and, furthermore, that the temperature of the system is well controlled. Equilibrium establishment for in situ coated columns often requires several hours of elution, for which reason the elaboration of separation methods by the use of this technique is time consuming. Typical examples of separations are shown in Figs. 4.4.4 and 5. Gradient elution for the separation of complex mixtures is very difficult to perform due to the necessity of saturating the eluent with stationary phase, and thus for this purpose a bonded phase system should be used.

In the separation of ionic solutes the possibility of ion-pair chromatography using liquid-liquid partition was developed by Schill and co-workers some ten years ago (e.g., Refs. 8, 9) as a useful alternative to ion-exchange chromatography. The technique is based on the fact that ionic solutes (which stay totally in the polar phase), on the addition of an appropriate counterion, may form outwardly un-

4.4.1.4 Bonded phase chromatography

From the previous sections it appears that polar, and in particular ionic, solutes are advantageously separated by partition rather than by adsorption chromatography. Conventional liquid-liquid partition chromatography, however, implies the previously stated inconvenient demands on temperature control, saturation of eluent with stationary phase, and often long equilibration times. All these problems were minimized when the so-called bonded phase materials were introduced in the early 1970s. The physical coating of the stationary phase onto the support was replaced by covalent bonding and thus no saturation of eluent with stationary phase was needed, temperature changes were no longer critical, and equilibration times were drastically reduced, often only 10 to 15 min are needed to equilibrate the column with a new eluent.

The materials introduced initially, and still used most often, were silicas with apolar long chain alkylgroups (e.g., octadecylsilyl (ODS) groups) chemically bonded to the surface. Thus, the materials were intended for reversed phase chromatography (cf., Fig. 4.4.7). Since then various materials have been introduced with bonded short chains carrying polar groups, e.g., nitril, amino, dimethylamino, dihydroxy, nitro,

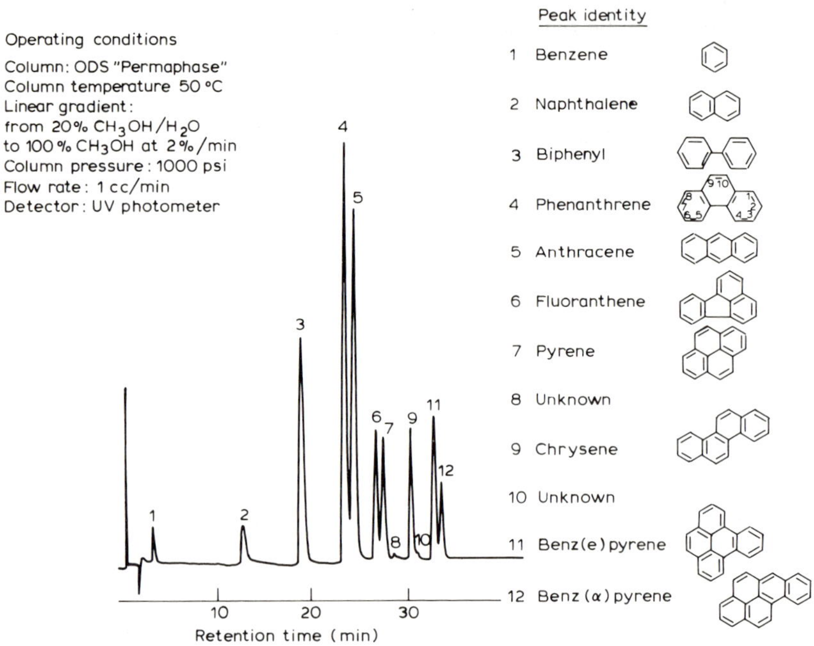

Fig. 4.4.7. Separation of a series of fused ring aromatics on chemically bonded ODS silica. Reprinted from Ref. 11 with permission.

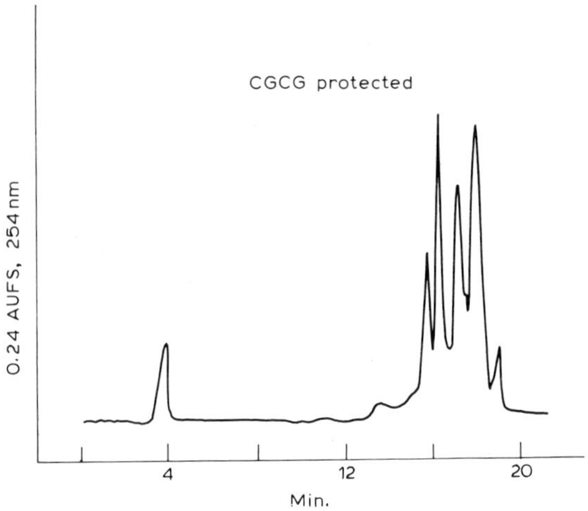

Fig. 4.4.8. Separation of protected oligonucleotide on a Partisil PAC column (Cyano-amino-bonded phase) (0.4 × 25 cm). The mixture chromatographed is the result of the synthesis of a tetramer of cytidine and guanine, protected by, e.g., dimethoxytrityl groups. Eluent: Gradient from methanol/dichloromethane (5 : 95) to (50 : 50). Reprinted from Ref. 12 with permission.

etc. These materials also offer several possibilities of performing straight phase separations on bonded phase materials (Fig. 4.4.8).

Supports for bonded phase chromatography are prepared from silica by derivatizing the surface silanol groups. Although several possibilities of performing this derivatization have been demonstrated, the majority of materials are now made by reaction with various sorts of alkylchlorosilanes, as the materials so obtained are largely hydrolytically stable. The reaction appears from the scheme below

$$\text{-Si-OH} + \text{Cl-}\overset{\overset{\displaystyle R_1}{|}}{\underset{\underset{\displaystyle R_2}{|}}{\text{Si}}}\text{-}R_3 \rightarrow \text{-Si-O-}\overset{\overset{\displaystyle R_1}{|}}{\underset{\underset{\displaystyle R_2}{|}}{\text{Si}}}\text{-R3} + \text{HCl}$$

R_1, R_2 = Cl or CH_3, R_3 = alkyl

Different kinds of layers may be obtained, depending upon the number of chloroatoms in the silane. If di- or trichloroalkylsilanes are used it is impossible to react all chloroatoms with silanol groups, and a final treatment is required. This treatment may be reaction with further di- or trichloroalkylsilanes in the presence of traces of water, resulting in a polymer layer, or by hydrolysis resulting in the formation of further silanol groups. Whatever the method used, it is not possible to derivatize the total number of silanol groups, and the residuals may cause tailing of peaks, in particular when chromatographing basic compounds. To reduce the

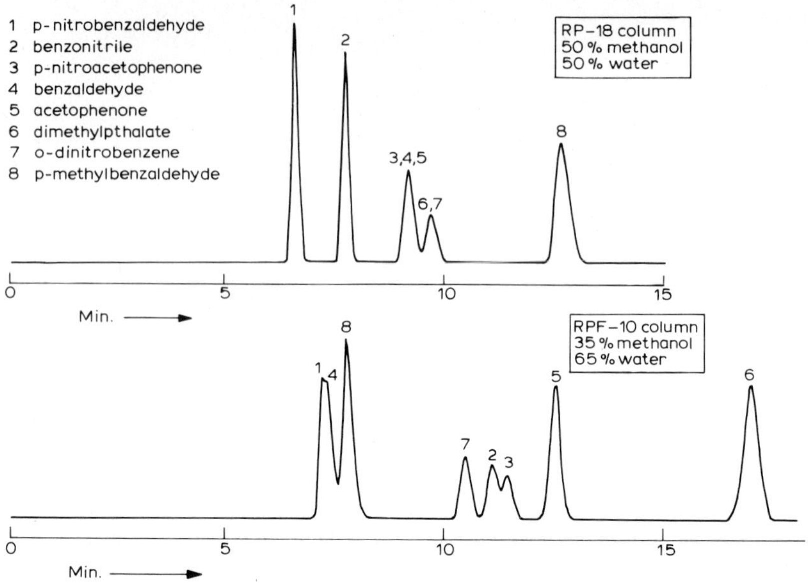

Fig. 4.4.9. Separation of aromatic solutes on two different bonded phase materials. Chemically bonded ODS silica and perfluorodecylsilyl silica respectively. Reprinted from Ref. 13 with permission.

number of residual silanol groups most materials are now prepared using alkyldimethylchlorosilanes, and furthermore several manufacturers retreat their materials with trimethylchlorosilane in order to mask silanol groups unreacted with the more bulky long chain alkylchlorosilane.

The most typical examples of the possibility of solid phase-induced selectivity changes are found in the area of bonded phase chromatography. Obviously, the possibility of using materials with a different structure in the bonded phase may cause changes in the selectivity as shown in Fig. 4.4.9. When using different brands of bonded phase materials which are claimed to be of the same type (e.g., different brands of ODS materials), however, large variations in the selectivity may also be seen [14–17]. In Fig. 4.4.10 is shown the separation of a mixture of polyaromatic hydrocarbons (PAHs) on columns made from five different brands, exhibiting widely varying selectivities towards the mixture. It appears that the most efficient materials for this type of separation are those of the largest carbon content, i.e., the most marked degree of polymer bonded layer. Even within the same brand variations may be seen due to either inadequate reproducibility in the production process or a change in the production technique, e.g., a change in the type of silane used. The insufficient batch-to-batch reproducibility was more pronounced in the early days of bonded phase chromatography, whereas changes in the manufacturing process have recently been observed for several brands.

Solid phase-induced selectivity changes can offer a great advantage when elaborating separation methods to be used in individual laboratories. The same

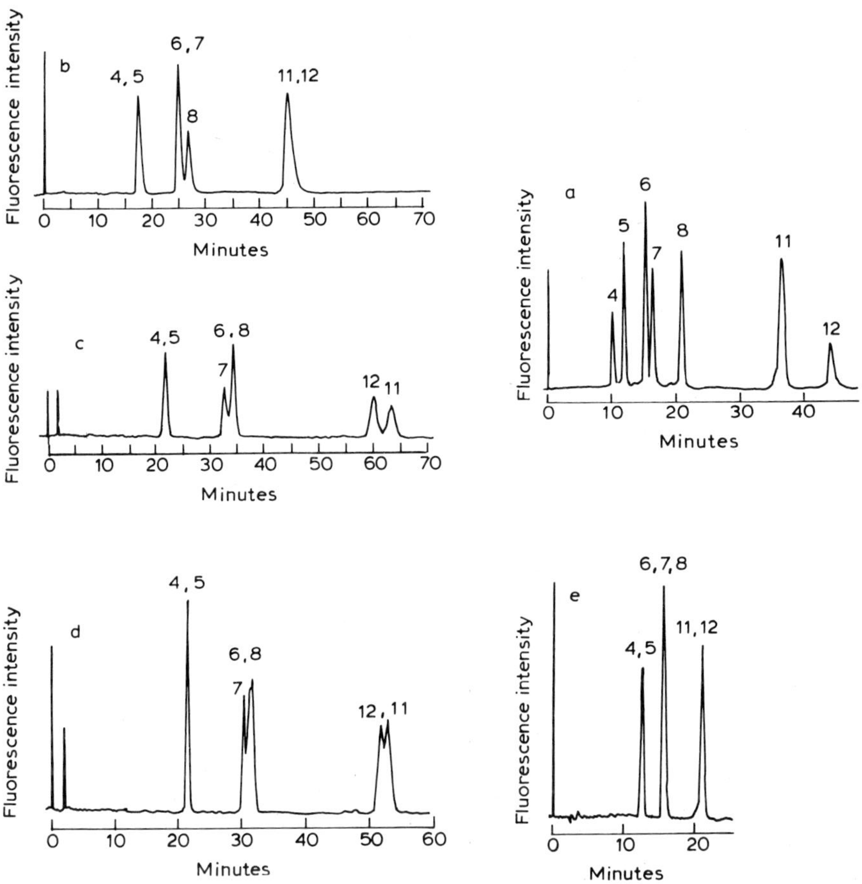

Fig. 4.4.10. Separation of a polyaromatic hydrocarbon mixture on 5 different brands of chemically bonded ODS-silica, the carbon content of the individual materials are given in brackets. a, HC-ODS (8.5%); b, LiChrosorb RP-18 (19.8%); c, Partisil-10 ODS-2 (16%); d, Zorbax ODS (10%); e, μBondapak C_{18} (10%). Peaks: 4, benz[a]anthracene; 5, chrysene; 6, benzo[e]pyrene; 7, benzo[b]fluoroanthrene; 8, benzo[k]fluoroanthrene; 11, benzo[ghi]perylene; 12, indeno[1,2,3-c,d]pyrene. Reprinted from Ref. 14 with permission.

phenomenon, however, can cause problems in the standardization of methods to be used in several laboratories, and even in different countries. In those cases it may be necessary to specify not only the type of material but also the brand, and even then there is the risk of a change in the production of the individual brand. When a high degree of standardization of methods is required it might be advantageous to use conventional liquid-liquid partition chromatography for straight phase separations, and the dynamically coated silica approach (cf., section 4.4.1.5) for reversed phase separations.

As mentioned previously residual silanol groups present on the surface of bonded phase materials affect the separation and the peak shape, in particular when using

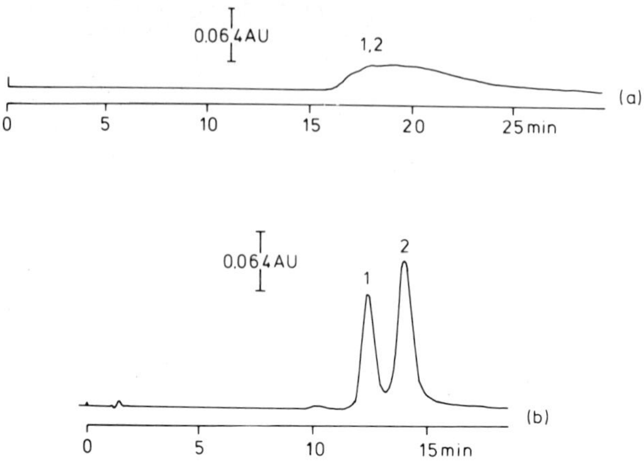

Fig. 4.4.11. Separation of *cis*- (1) and *trans*-clopenthixol (2) on a LiChrosorb RP-8 column (0.46 × 10 cm). Eluents: a, Methanol/0.05 M sodium acetate buffer (pH 4) (70 : 80); b, Methanol/0.05 M dimethyloctylammonium acetate buffer (pH 4) (50 : 50). Reprinted from Ref. 20.

eluents of high pH (> 6–7), and especially when chromatographing basic solutes. At increasing pH, the degree of ionization of silanol groups increases and cationic compounds exhibit a considerable degree of affinity, resulting in peak tailing. The tailing may be reduced by the addition to the eluent of compounds which also are able to be adsorbed to the silanols. As shown by Wahlund and Sokolowski [18,19] unsymmetrical tertiary amines or quaternary ammonium compounds can be used as tailing reducers. An example is shown in Fig. 4.4.11.

The considerations on the choice of eluents for straight phase separations on polar bonded phase materials do not differ much from those used in adsorption chromatography and in conventional liquid-liquid chromatography. Non-polar solvents with the addition of polar modifiers are used, and eluotropic series as in Table 4.4.1 are useful in the adjustment of eluting strength.

In reversed phase chromatography, which is almost exclusively carried out on bonded phases, some different reflections are required. The eluotropic series of Table 4.4.1 cannot simply be reversed, as only the most polar solvents are used, and hence a more differentiated version of this part of the table is needed. In Table 4.4.2 is shown a kind of eluotropic series suited for reversed phase chromatography. The basic solvent used is water, and the elution strength is controlled by the addition of a modifier, i.e., a stronger eluting organic solvent. The most frequently used modifiers are methanol and acetonitrile, alternatively, tetrahydrofuran or dioxan may be used. When elaborating reversed phase chromatographic methods for the separation of non-ionic compounds methanol-water mixtures are considered as first choice, starting with a high percentage of methanol or with pure methanol, and then adding water until sufficient retention is reached.

The separation of ionic solutes may be performed by ionic suppression. Thus, carboxylic acids can be separated as non-ionized molecules by using an acidified

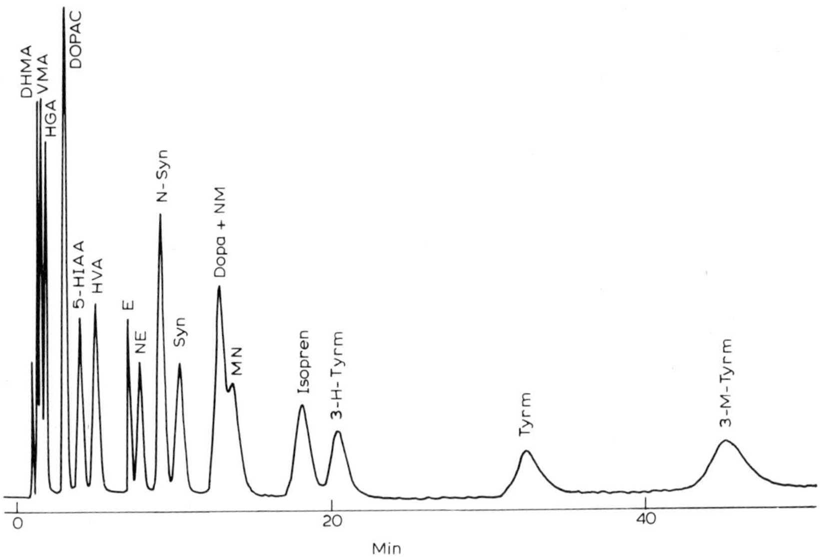

Fig. 4.4.12. Separation of a mixture of acidic and basic catecholamines on a LiChrosorb RP-8 column (0.28×25 cm). Eluent: Water containing 0.02 M citrate (pH 2.5)/1% propanol/$NaClO_4$ (0.08 M)/0.3% sodium dodecyl sulphate. Peaks: DHMA, 3,4-dihydroxymandelic acid; VMA, vanilmandelic acid; HGA, 2,5-dihydroxyphenylacetic acid; DOPAC, 3,4-dihydroxyphenylacetic acid; 5-HIAA, 5-hydroxyindole-3-acetic acid; HVA, homovanillylmandelic acid; E, epinephrine; NE, norepinephrine; N-Syn, norsynephrine; Syn, synephrine; Dopa, 3,4-dihydroxyphenylalanine; NM, normetanephrine; MN, metanephrine; Isopren, isoprenaline; 3-H-Tyrm, dopamine; Tyrm, tyramine; 3-M-Tyrm, 3-methoxytyramine. Reprinted from Ref. 21 with permission.

eluent, e.g., a methanol-water mixture with the addition of 0.01 M phosphoric acid. Another possibility is to carry out reversed phase ion-pair partition chromatography. The basic principles for ion-pair partition separations were outlined in section 4.4.1.3. The application of the reversed phase approach has, as a considerable advantage, that the counterion for the ion-pairing is added to the eluent, whereas in

TABLE 4.4.2
Solvents commonly used as modifiers in reversed phase chromatography

The order of solvents is according to increasing eluting strength.

Ethylene glycol
Methanol
Dimethylsulfoxide
Ethanol
Acetonitrile
Dioxane
Tetrahydrofuran
2-Propanol

straight phase separations the counterions are added to the stationary phase. The counterion in the eluent is easily replaced, thus facilitating the elaboration of methods.

As the counterions, quaternary ammonium compounds may be used for the separation of anions, whereas alkane sulphonates may be used for separation of cations. The concentrations of counterions are typically about 0.005 M.

Following the adjustment of pH of the eluent to achieve, if possible, total ionization of the solutes, a counterion is added, starting with one of low hydrophobicity, e.g., a short chain alkane sulphonate. The retention may be increased by increasing the alkane chain length or the concentration of the counterion. Fig. 4.4.12 shows a typical example of the separation of a mixture of catecholamines. It should be emphasized that the previously mentioned short equilibration times of bonded phase chromatography do not apply to ion-pair partition chromatography. The counterions used often exhibit a considerable affinity to the apolar stationary phase and also, eventually, to residual silanol groups, hence the column is able to adsorb a certain amount of counterions. Equilibrium is not established until the column is

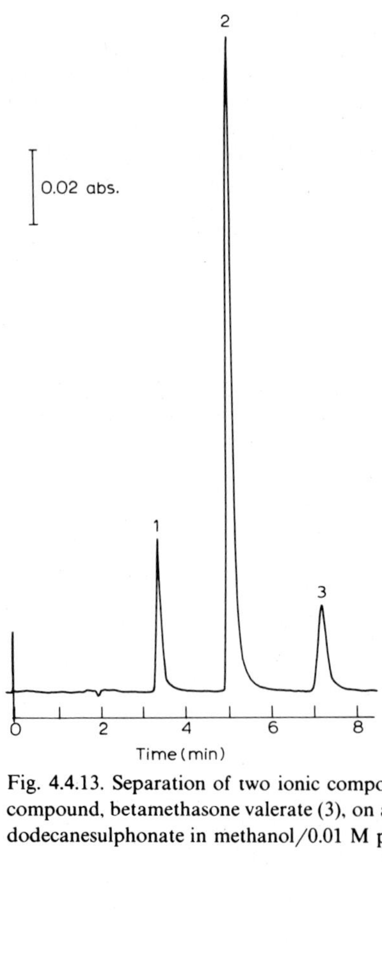

Fig. 4.4.13. Separation of two ionic compounds, phenylephrine (1) and lidocaine (2), and one non-ionic compound, betamethasone valerate (3), on a Nucleosil C8 column (0.46 × 15 cm). Eluent: 0.005 M sodium dodecanesulphonate in methanol/0.01 M phosphate buffer (pH 4.8) (7 : 3). Reprinted from Ref. 25.

completely saturated. The adsorption of counterions to the stationary phase has caused several chromatographers (e.g., Refs. 21–23) to discuss the possible characterization of the mechanism of these separations as a dynamic ion-exchange mechanism. As a practical consequence of the said adsorption it must be noted that, if exchanging a long chain counterion with one of shorter alkyl chain length, the column should be rinsed first (using e.g., 90% of methanol acidified with phosphoric acid), otherwise the long chain compound is difficult to replace due to its greater affinity to the column material.

For the separation of mixtures containing both ionic and non-ionic compounds the reversed phase ion-pairing approach has proven especially valuable [24,25]. When elaborating such methods the column material and the water/organic modifier percentage are first established because of the chromatographic properties of the non-ionic compounds. Afterwards the retention of the ionic compounds is controlled by proper adjustment of the three main parameters affecting the ion-pair partition process, i.e., pH, concentration and nature of counterion. Fig. 4.4.13 shows a separation elaborated according to the principles outlined above.

The selectivity of a reversed phase chromatographic system is markedly dependent on the nature of the organic modifier. If a sufficient separation of a mixture of solutes is not achieved by using methanol as the modifier, acetonitrile or tetrahydrofuran may be tried, bearing in mind that the concentrations should be decreased

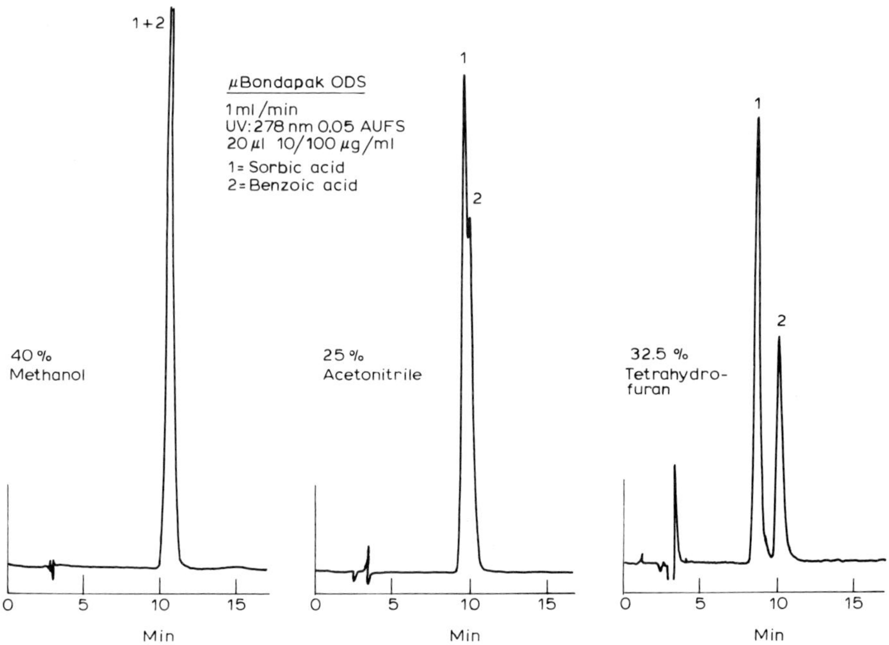

Fig. 4.4.14. Separation of sorbic acid (1) and benzoic acid (2) on a μBondapak C_{18} column (0.4 × 30 cm) using three various eluents as stated. Reprinted from Ref. 26 with permission.

relative to the methanol concentration due to the higher eluting strength of the said solvents (cf., Table 4.4.2). An example of this phenomenon is shown in Fig. 4.4.14. Supposing that the change of modifier cannot provide a proper separation either,

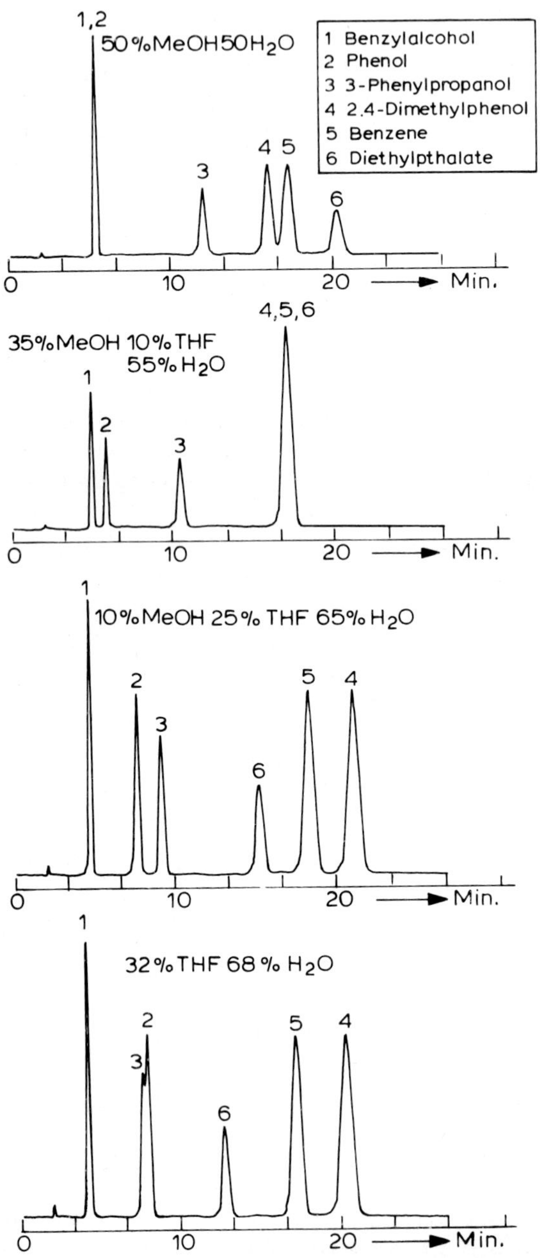

Fig. 4.4.15. Chromatograms illustrating the variations in selectivity obtained by eluting with some iso-eluotropic mixtures of methanol, tetrahydrofuran and water. Reprinted from Ref. 27 with permission.

there is still the possibility of improvement due to solvent-induced selectivity changes. It has been demonstrated [28,29] that the use of ternary solvent mixtures, i.e., water and two organic modifiers in various proportions, might cause considerable changes in selectivity relative to the use of the individual modifiers and water alone. When using this approach it may be possible to replace an eluent consisting of 50% of methanol in water, by a mixture containing 40% of methanol and 10% of acetonitrile. Further, one might substitute tetrahydrofuran, dioxan, etc., for acetonitrile. An example is shown in Fig. 4.4.15.

4.4.1.5 Dynamically coated phases

The designation 'dynamically coated phases', which covers a recently introduced technique in HPLC [30–32], requires a few explanatory remarks to establish its relation to the previously discussed, well-known, chromatographic modes. In this section the dynamic coating approach is used as the designation for columns of bare silica which, through a dynamic process, are coated with a stationary phase suited for reversed phase separations (cf., Fig. 4.4.16). In the section on liquid-liquid partition chromatography (4.4.1.3) it was mentioned that the stationary phases, when using modern microparticular support materials, were placed on the silica by in situ coating, i.e., by equilibrating with the eluent saturated with stationary phase. This method is most often used for straight phase chromatography, as only a polar stationary phase will exhibit a sufficient affinity to the silica surface. In the section

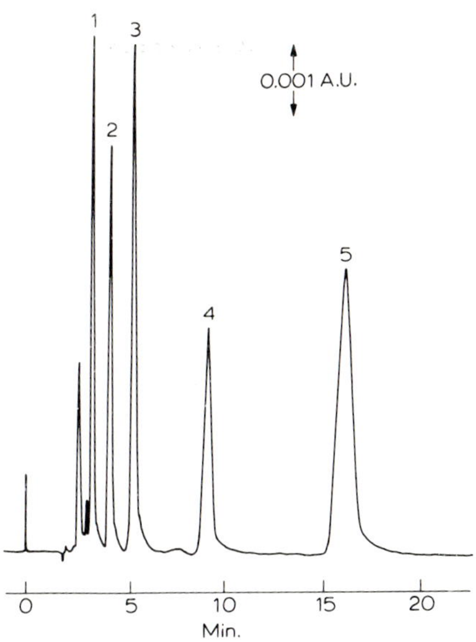

Fig. 4.4.16. Separation of aromatic hydrocarbons on a LiChrosorb SI 60 column (0.46 × 15 cm). Eluent: 0.08% CTMA in methanol/water/0.2 M potassium phosphate buffer (pH 8.0) (60:35:5). Peaks: 1, benzene; 2, toluene; 3, ethylbenzene; 4, 2-methylnaphthalene; 5, phenanthrene. Reprinted from Ref. 32.

for the confirmation of the identity of substances in a sample. A short introduction to preparative liquid chromatography is given.

4.4.3.2 Sample pre-treatment

The first requirement for a sample for liquid chromatography is that it must be a clear and homogeneous solution. Any particles present in the sample must be

TABLE 4.4.3
Derivatization reactions for biochemical substances

Substance	Reagent	Pre/post columns	Minimum detectable quantity	Refs.
Amino acids/peptides	phenylisothiocyanate	pre-column	10 pmol	38
				39
	ninhydrin	post-column	100 pmol	40
	dansyl chloride	pre-column		41
	o-phthalaldehyde	pre-column	50 fmol	42
			100 pmol	43
				44
		post-column	100 ng	45
				46
			10 pmol	40
	7-chloro-4-nitrobenzo-2-oxa-1,3-diazole	post-column	1–30 pmol	47
Amines	fluorescamine	pre-column		48
	o-phthalaldehyde	pre-column	5 pmol	49
				50
	dansyl chloride	pre-column	25 fmol	51
	trihydroxyindole reagents	post-column	50 fmol	52
	alkaline borate		0.25 ng	53
Carboxylic acids	5-chloromethylanthracene	pre-column	5 fmol	54
	4-bromomethyl-7-methoxycoumarin	pre-column	5 ng	55
	4-bromomethyl-7-acetoxycoumarin	pre-column		56
Steroids	sulphuric acid	pre-column	0.5 ng	57
	dansylhydrazine	pre-column		58
Saccharides	dansylhydrazine	pre-column	20 pmol	59
				60
Bile acids	ethylanthranilate azoderivatives	pre-column		61
	4-iodoaniline azoderivative	pre-column		62
Vanillylmandelic acid	dansyl chloride	pre-column		63
Guanidino compounds	phenanthrenequinone	post-column	5–50 ng	64

removed before injection either by filtering or by centrifugation, since particles will build up in the injector or on the top of the column and eventually block the flow.

Whenever possible, samples should be dissolved in the mobile phase to provide longest column life and maximum precision in quantitative analysis. The sample solvent must not have stronger eluting properties than the mobile phase, since this will result in wider peaks, and possibly in peak distortion.

Samples of biological origin must be deproteinized before injection since, otherwise, proteins will precipitate on top of the column resulting in rapid destruction of the column. Precipitation of proteins may in most cases be accomplished by simple methods, e.g., addition of methanol, acetonitrile or whichever modifier is used in the chromatographic system, followed by removal of the precipitated proteins by centrifugation. Using this simple technique serum or urine samples can be injected without further purification onto columns in reversed phase systems.

4.4.3.3 Choice of the chromatographic system

In biochemical analysis a reversed phase chromatographic system will often be the best choice, since such a system allows for the direct analysis of biological samples. If at all possible isocratic elution should be used, since isocratic systems are more stable than gradient elution systems, and give better reproducibility in quantitative analysis. Furthermore the necessity of solvent purity is not as strict in isocratic elution as in gradient elution, and less expensive equipment is needed.

In many cases, however, the sample composition makes it necessary to use gradient elution in order to elute all compounds of interest within a reasonable time and with sufficient resolution. In gradient elution it is important to use very pure solvents since, otherwise, impurities from the solvents may accumulate on top of the column as the gradient runs with low solvent strength, to be eluted as interfering peaks as the solvent strength is increased.

For reproducible retention times the gradient must be run in a reproducible fashion. The best way to accomplish this is to use a fixed delay between each gradient run. This does not necessarily mean that the column is in equilibrium with the mobile phase at the start of the run, and thus the chosen delay must be strictly adhered to. Another possibility is to ensure that the column is at equilibrium before each run, but this may often require a very long delay, perhaps hours, between each run. In order to obtain a reasonable throughput a short, fixed delay should thus be chosen.

4.4.3.4 Quantitative analysis

Quantitative analysis by liquid chromatography should preferably be performed by isocratic elution, with samples dissolved in the mobile phase. This is not always possible, and in this case a somewhat lower, but for most applications quite acceptable, degree of reproducibility must be accepted.

The valve injectors available for liquid chromatographs give very reproducible

4.4.3.6 Preparative liquid chromatography

In preparative liquid chromatography the technique is dependent on the amount of substance to be processed. In biochemical analysis preparative chromatography is most often used in order to collect, for spectroscopic identification, a sample of unknown substances. This can be done with quite small amounts of substance, and thus repeated sample collection from analytical columns is sufficient. If preparative

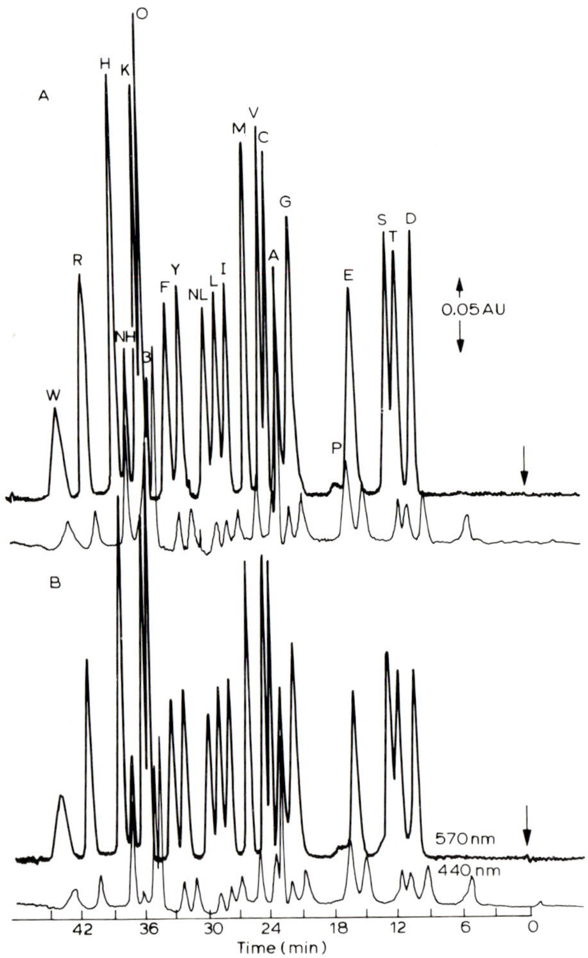

Fig. 4.4.19. Separation of amino acids on a cation-exchange resin and with post-column derivation with ninhydrin. B was obtained 50 analyses after A. Peak identification: D, aspartic acid; T, threonine; S, serine; E, glutamic acid; P, proline; G, glycine; A, alanine; C, cysteine; V, valine; M, methionine; I, isoleucine; L, leucine; NL, norleucine; F, phenylalanine; O, ornithine; K, lysine; NH_3, ammonia; H, histidine; R, arginine; W, tryptophan. Reprinted from Ref. 96 with permission.

chromatography is used to obtain pure analytical standards or material for testing, etc., preparative size columns have to be used. In the following some guidelines for the preparative collection of sample for identification are given.

The mobile phase used for preparative separations must not react with the solutes, neither during chromatography, nor during removal of the solvent. The solvents used in the mobile phase must be very pure, redistilled or distilled in glass, in order not to leave any residue upon removal of the solvent. The mobile phase should preferably be totally volatile, or if residue is left it must not interfere with the proposed identification technique.

Liquid-solid chromatography will often be the most convenient technique, but bonded phase chromatography using volatile buffers (e.g., ammonium formate) can also be used.

Samples should be collected in clean vials and identification performed as soon as possible after collection. Solvents are most conveniently removed by evaporation under a gentle stream of nitrogen with moderate heating of the vial. When volatile buffers are used elevated temperatures are required and rotary evaporation under reduced pressure is preferred. A third, very gentle, method for the removal of solvent is freeze drying.

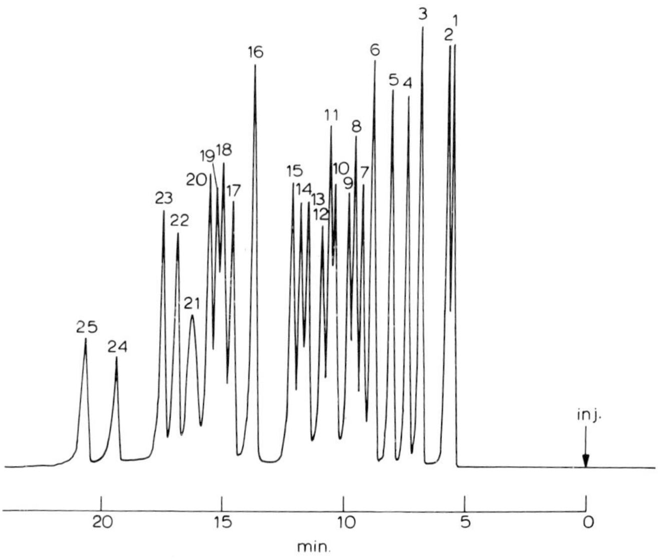

Fig. 4.4.20. Separation of amino acids after pre-column derivation with *o*-phthalaldehyde. Fluorescence detection at em.: 330 nm; ex.: 418 nm. Peaks: 1, cysteic acid; 2, Asp; 3, Glu; 4, *S*-carboxymethyl cysteine; 5, Ans; 6, Ser; 7, Gln; 8, His; 9, methionine sulfone; 10, Thr; 11, Gly; 12, Arg; 13, β-Ala; 14, Tyr; 15 Ala; 16, α-aminobutyric acid; 17, Trp; 18, Met; 19, Val; 20, Phe; 21, NH_4^+; 22, Ile; 23, Leu; 24, Orn; 25, Lys. Reprinted from Ref. 99 with permission.

TABLE 4.4.4

Nucleotides

Sample	Column material	Eluent	Detection	Refs.
Nucleosides and bases in plasma, etc.	Hypersil ODS	1% Methanol in 0.004 M KH_2PO_4	UV 254 and 280 nm	65
Nucleosides and bases	LiChrosorb Si 100	Dichloromethane-methanol-formate buffer (pH 2.5) (80 : 18 : 2)	UV 254 nm	66
2′ and 3′-Nucleotide monophosphates in cell extract	Radialpak C_{18}	0.2 M KH_2PO_4 (pH 4.4)–0.006 M KH_2PO_4 (pH 4.4)	UV 254 nm	67
Uridine in human and animal serum and plasma	Whatman PXS ODS	0.01 M acetate buffer (pH 4.5)	UV 254 and 280 nm	68
Nucleosides and bases	μBondapak C_{18}	0 → 10% methanol in 0.005 M KH_2PO_4 (pH 5.0)	UV 254 nm	69
Cyclic ribonucleotides and deoxyribonucleotides in biological fluids	Spherisorb 10 ODS	0.05 M phosphate buffer (pH 5.6) –same + 25% methanol and 25% water	UV 254 and 280 nm	70
Nucleosides and bases in hydrolyzed DNA	Zorbax SIL a.o.	Dichloromethane-methanol-water (835 : 150 : 15) + 0.01 M butansulphonate	UV 254 nm	71
Nucleotides	Hypersil ODS	88% methanol with zwitterionic pairing agent added	UV 254 nm	72
Nucleotides and nucleosides	μBondapak C_{18}	Phosphate buffer (pH 6.0 or 6.4) with or without Mg^{2+} added	UV 254 nm	73
Nucleosides in hydrolyzed tRNA	μBondapak C_{18}	2.5%–10% methanol in 0.01 M $NH_4H_2PO_4$ (pH 5.10)	UV 254 nm	74
3′,5′-Cyclic ribonucleotides in rat brain extracts	μBondapak C_{18}	0–25% B in A. A, 20 nmol KH_2PO_4/l (pH 3.7); B, 60% methanol	UV 254 nm	75
Nucleosides and bases in human urine	μBondapak C_{18}	0.1 M ammonium formiate buffer (pH 4.5)	UV 254 nm	76
Nucleotides in extract from human erythrocytes	μBondapak C_{18}	0–25% methanol in 0.1 M phosphate buffer (pH 6.0)	UV 254 nm	77
Nucleosides in hydrolyzed plant DNA	Nucleosil 5 C_{18}	0.1% phosphoric acid	UV 273 nm	78
Nucleosides in urine	μBondapak C_{18}	6% methanol in 0.01 M $NH_4H_2PO_4$ (pH 5.1)	UV 254 nm	79
Adenosine in cell extracts	Partisil ODS-1	10% methanol in 0.01 M KH_2PO_4	UV 260 nm	80
Adenosine phosphates, NAD, and NADP in cell extracts	μBondapak C_{18}	0–5% methanol in 0.1 M phosphate buffer (pH 6.0)	UV 260 and 340 nm	81
Adenosine phosphates in heart tissue	LiChrosorb RP-8	Acetonitrile-0.3% TBA and 0.65% KH_2PO_4 (pH 5.8) (24 : 86}	UV 254 nm	82

TABLE 4.4.5
Carbohydrates

Sample	Column material	Eluent	Detection	Refs.
Peracetylated oligosaccharides from hydrolyzed amylose	μBondapak C_{18}	10–70% acetonitrile	Moving wire FID	83
Mono- and disaccharides in hydrolyzed cellulose	Micro-Pak NH_2	80% acetonitrile	RI and UV 192 nm	84
Cyclodextrins	μBondapak Carbohydrate	70% acetonitrile	RI	85
Malto-oligosaccharides in hydrolyzed amylose	LiChrosorb NH_2	75% acetonitrile	Fluorescence. Post-column reaction with ethanolamine and boric acid	86
Mono- and disaccharides from, e.g., soybean extract	Spherisorb S5 NH_2	Acetonitrile/water	Mass-detector	87
Mono- and disaccharides in urine	LiChrosorb NH_2	80% acetonitrile	UV 510 nm. Post-column reaction with blue tetrazolium	88
Mono- and disaccharides in polyols	Radial Pak Silica	81% acetonitrile and 0.02% tetraethylenpentamine	RI	89
Mono- and disaccharides in hydrolyzed cellulose	μBondapak C_{18}	Water	RI	90
Benzyloxime-perbenzoyl derivatives of mono- and disaccharides in serum	μPorasil	Hexane-dioxane (80 : 20)	UV 230 or 254 nm	91
Mono- and disaccharides	Aminopropyl-silica	75% acetonitrile	RI or UV 188 nm	92
Dansylhydrazones of mono- and disaccharides	LiChrosorb Si 60	Chloroform/ethylacetate/methanol/2-propanol/acetic acid (30 : 50 : 10 : 10 : 1)	Fluorescence	93

TABLE 4.4.6
Amino acids

Sample	Column material	Eluent	Detection	Refs.
Amino acids	LiChrosorb RP-18	0.01 M sodium acetate (pH 5.2) + acetonitrile + dichloroethane (68.5 + 31.5 + 0.5) at 62°C	As phenylthiohydantoins at 254 nm	94
Amino acids	Ultrasphere ODS	0.065 M ammoniumacetate (pH 4.54) + acetonitrile (52 + 48) at 55°C with flow-programming	As phenylthiohydantoins at 269 nm	95
Amino acids	Cation exchangers DC-4A, DC-6A and Aminex A-9	Buffers	Post-column derivation with *o*-phthalaldehyde or ninhydrin; fluorescence	96
Amino acids	LiChrosorb RP-2, RP-8 and RP-18	0.01 M citrate (pH 2.75) + 1% *tert*-pentanol and 0.3% sodium dodecylsulphate	As ninhydrin derivatives at 400 and 570 nm	97
Amino acids	LiChrosorb NH_2	A: 0.01 M KH_2PO_4 (pH 4.3) B: Acetonitrile + water (500 + 70) 95%–50%B	At 200 nm	98
Amino acids in urine	Nucleosil C-18	A: 0.1 M phosphate (pH 7.5) B: Methanol 40–75%B	Pre-column derivation with *o*-phthalaldehyde Fluorescence at 330/418 nm	99
Amino acid in urine	μBondapak C_{18}	Ethanol + 0.02 M sodium acetate (pH 4.0) (4 + 6)	As dansyl derivatives at 425 nm	100

Taurine in CSF	LiChrosorb RP-18	Acetonitrile + 0.015 M phosphate buffer (pH 2.7) (28 + 72)	As fluorescamine derivative Fluorescence 390/450 nm	101
Tryptophan and its metabolites	Partisil 10/25 C-8	A: 0.1 M KH_2PO_4 (pH 5.50) B: Methanol + water (3 + 2) 0%B–80%B	Electrochemical (+0.70 V) and at 254 nm	102
3-Hydroxyproline in urine	Dowex 50-M82	10% methanol + 0.2 M sodium citrate (pH 3.0)	As ninhydrine derivative at 440 nm and 570 nm	103
D- and L-amino acids	ODS-HC SIL-X-1	20% acetonitrile + 5×10^{-3} M L-proline and ammonium acetate (pH 7.0) with 2.5×10^{-3} M $CuSO_4 \cdot 5H_2O$	As dansyl derivatives Fluorescence 340/480 nm	104
D- and L-amino acids	Supelcosil LC-8 and LC-18 Hypersil C_8	Methanol + water (60 + 40) with 4×10^{-3} M L-prolyl-*n*-octylamide-Ni(II) and 8.75×10^{-2} M ammonium acetate (pH 9.0)	As dansyl derivatives at 254 nm or by fluorescence	105
D- and L-amino acids	L-proline chemically bonded to LiChrosorb SI 100	0.05 M KH_2PO_4 (pH 4.6)	At 220 nm and by polarimetry	106
D- and L-amino acids	Supelcosil LC-18	0.008 M copper acetate and 0.017 M L-proline (pH 5)	Post-column derivation with *o*-phthalaldehyde Fluorescence	107
D- and L-amino acids	LiChrosorb RP-18 coated with C_{16}-L-hydroxyproline	15% Methanol with 10^{-3} M $CuAc_2$ (pH 5.0)	At 254 nm	108

TABLE 4.4.7
Peptides and proteins

Sample	Column material	Eluent	Detection	Refs.
Pharmaceutically important peptides	Reversed-phase materials	Various	Various	109
Tryptic digests	μBondapak alkylphenyl μBondapak C-18	Gradients	At 210 nm	110
Peptides and proteins	Reversed phase, size exclusion and ion exchange	Various	Various	111
Insulin	Bondagel E-125, Toya-Soda G2000 SW and G3000 SW and Waters I-125 protein column	Various buffers	At 280 nm	112
Preparative chrom. of peptides	Radial Pak A/C_{18}	A: 0.1 M ammonium bicarbonate B: Acetonitrile + 0.1 M ammonium-bicarbonate (50 + 50) 0%B–100%B	At 210–225 nm or by scintillation	113
Enkephalines Human chorionic gonadotropin	μBondapak C_{18}	A: 0.05 ml TFA in water B: 0.05 ml TFA in acetonitrile 0%B–60%B	At 206 nm	114
Cholecystokinin peptides	μBondapak C_{18}	0.25 N triethylammonium phosphate (pH 3.5) and acetonitrile	At 210 nm	115
Insulin	μBondapak C_{18} and Waters I-125 protein column	0.01 M Tris-HCl/0.05 M NaCl in 7 M urea (pH 7.0)	At 254 nm	116
Proteins	Partisil-10 ODS LiChrosorb RP-18 Supelcosil LC-18	Phosphate-buffer (pH 2.1) + various amounts of propanol-2 or ethanol	At 254 nm or by fluorescence 280/370 nm	117
Proteins	LiChrosorb DIOL	0.1 M phosphate (pH 2.1)	At 200 nm	118
Urinary proteins	TSK G 3000 SW	0.15 M phosphate (pH 6.8) with 0.1 M NaCl and 6×10^{-4} M NaN_3	At 280 nm	119
Peptides	Chemically bonded L-Val-L-Ala-L-Pro	Citrate buffers at various pH values	At 254 and 280 nm	120
Proteins	Ultrasphere SAC	A: 0.155 M NaCl (pH 2.1) B: Acetonitrile 0%B–75%B	At 215 nm	121

TABLE 4.4.8
Lipids and steroids

Sample	Column material	Eluent	Detection	Refs.
Phospholipids from microsomes	Hewlett Packard SI 100 10 μm	Linear gradient from hexane/propanol/water (6 : 8 : 0.75) to hexane/propanol/water (6 : 8 : 1.5) in 5 min	UV 206 nm	122
Lipoproteins in plasma	Supelcosil LC-18, 5 μm	Methanol/chloroform/acetonitrile (1 : 1 : 1)	RI	123
Lipoproteins in serum	TSK-GEL, G 5000 PW, G 4000 SW, G 3000 SW, Toyo Soda	0.15 M sodium chloride	Upon post column enzymatic reaction UV 500 nm	124
Prostaglandins in seminal fluid	LiChrosorb RP-18, 5 μm	Linear gradient from 30 to 90% acetonitrile in water in 64 min	As derivatives with 4-bromomethyl-7-acetoxy-coumarin Fluorescence 365/460 nm	125
Prostaglandins	Ultrasphere ODS	0.017 M phosphoric acid + acetonitrile (67.2 : 32.8)	UV 192.5 nm	126
Cortison, cortisol in plasma	Silica, 5 μm (Altex)	Chloroform/water saturated chloroform/methanol/tetra-hydrofuran/acetic acid (668.5 : 300 : 21 : 10 : 0.5)	UV 254 nm	127
Corticosteroids in adrenal extracts or plasma	LiChrosorb SI 100, 5 μm	Linear gradient from 1 to 10% methanol in water saturated chloroform in 30 min	UV 254 nm	128
17α-Hydroxyprogesterone, cortisol, 11-deoxycortisol in serum	μBondapak C_{18}	Methanol/water (55 : 45)	UV 254 nm	129

Continued on p. 198.

TABLE 4.4.8 (continued)

Sample	Column material	Eluent	Detection	Refs.
Oestrogens in urine	Pyrrolidone coated on silica	Methanol/water (1:1)	UV 280 nm	130
6β-Hydroxycortisol in urine	Perkin Elmer RP8, 10 μm	0.01 M potassium dihydrogen phosphate/acetonitrile/trichloroacetic acid (90:9.95:0.05)	UV 243.5 nm	131
Oestrogens in pregnancy urine	Perkin Elmer C_{18}, 10 μm	Methanol/water (60:40) to pH 3.9 with acetic acid	As dansyl derivatives Fluorescence 350/540 nm	132
Oestrogens in pregnancy urine	Partisil 5	Heptane/ethanol (95:5)	UV 280 nm	133
	Partisil 10 ODS	0.1% Ammonium carbonate in water/methanol (45:55)		

TABLE 4.4.9
Porphyrins and bile acids

Sample	Column material	Eluent	Detection	Refs.
Protoporphyrin and zinc protoporphyrin in blood	μBondapak C_{18}	Methanol/acetic acid/water (39:4:7)	Fluorescence 365/595 nm	134
Porphyrins in erythrocytes	μBondapak C_{18}	Linear gradient from methanol/water/acetic acid (6:4:1) to methanol/acetic acid (10:1) in 10 min	Fluorescence 396/500 nm	135
Porphyrins in erythrocytes	μBondapak C_{18}	Methanol/0.02 M phosphate buffer (pH 3.4) (92:8)	Fluorescence 404/550 nm	136
Porphyrins in plasma	Perkin Elmer Silica ^{A}, 10 μm	Linear gradient from 2 to 90% acetone in 0.25 M acetic acid	Fluorescence 400/600 nm	137
Porphyrins in urine	μBondapak Phenyl	15.6 mM pentane sulphonate and 0.1 mM EDTA in methanol: water (pH 2.1) with sulphuric acid (60:40)	Fluorescence 403/600 nm	138
Porphyrins in urine	μBondapak C_{18}	Linear gradient from 50% methanol in phosphate buffer (pH 3.5) to 100% methanol in 10 min	Fluorescence 420/510 nm	139
Porphyrins in faeces and urine	μ-Porasil	*n*-Heptane/methylacetate (3:2)	UV 404 nm	140
Bilirubin in serum	Hypersil ODS	Acetonitrile/dimethylsulphoxide/water (40:40:20)	UV 450 nm	141
Bile pigments in bile	LiChrosorb RP-8, 5 μm	Acetonitrile/ethyl acetate/methanol/water (30:33:45:50) with varying pH and concentration of tetrabutyl ammonium ions	As 4-iodoaniline azo-derivatives UV 546 nm or 365 nm	142
Conjugated bile acids in human bile	μBondapak fatty acid	8.8 mM phosphate buffer (pH 2.5)/2-propanol (68:32)	UV 193 nm	143
Conjugated bilirubin in rat serum and human amniotic fluid	LiChrosorb SI 60, 5 μm	Gradient from 0 to 1% methanol in chloroform/acetic acid (99.5:0.5) in 6 min	UV 430 nm	144
Conjugated bilirubin in bile	Shimadzu RP-material, 5 μm	Linear gradient from 20 to 60% acetonitrile in 0.1 M acetate buffer pH 4.0 in 80 min	As ethylanthranilate azoderivatives UV 530 nm	145

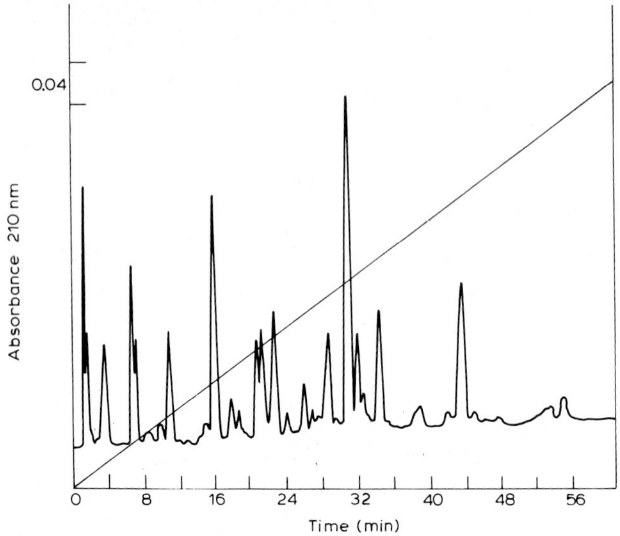

Fig. 4.4.21. Separation of a tryptic digest of chick lysozyme on a reversed phase column with gradient elution. Reprinted from Ref. 113 with permission.

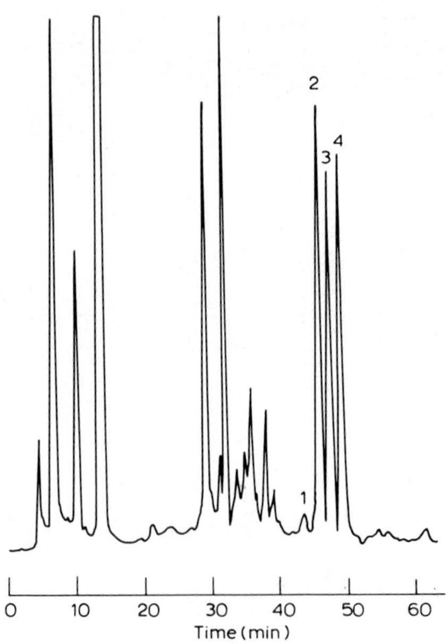

Fig. 4.4.22. High-performance liquid chromatogram obtained from human seminal fluid sample. Peaks: 1, $PGF_{2\alpha}$; 2, PGE_2; 3, PGE_1; 4, 16-methyl-$PGF_{1\alpha}$ (internal standard). Reprinted from Ref. 125 with permission.

4.4.4 *Applications* (The relevant Figures & Tables for this section can be found on pp. 188–200.)

The literature on HPLC, and especially on HPLC-applications, is growing extremely fast, and it is therefore impossible and useless to make a complete survey of the applications even within the field of biochemistry.

In this section a number of applications have been collected within some specific areas. The survey is far from complete, but should give an idea of what it is possible to separate by the use of modern HPLC techniques (Tables 4.4.4–9, Figs. 4.4.17–22). The references in the cited applications will give the key to most of the literature in that specific area. Further information about the available literature can be obtained from the Bibliography Section of the Journal of Chromatography.

References

1 Snyder, L.R. and Kirkland, J.J. (1979) Introduction to Modern Liquid Chromatography, 2nd edn., John Wiley and Sons, New York.
2 Parris, N.A. (1976) Instrumental Liquid Chromatography, Elsevier, Amsterdam.
3 Eisman, J.A., Shepard, R.M. and DeLuca, H.F. (1977) Anal. Biochem. 80, 298.
4 Ryba, M. and Beranek, J. (1981) J. Chromatogr. 211, 337.
5 Huber, J.F.K., Meijers, C.A.M. and Hulsman, J.A.R.J. (1972) Anal. Chem. 44, 111.
6 Parris, N.A. (1974) J. Chromatogr. Sci. 12, 753.
7 Done, J.N., Kennedy, G.J. and Knox, J.H. (1972) Nature 237, 77.
8 Schill, G., Modin, R. and Persson, B.A. (1965) Acta Pharm. Suec. 2, 119.
9 Eksborg, S. and Schill, G. (1973) Anal. Chem. 45, 2092.
10 Persson, B.A. and Karger, B.L. (1974) J. Chromatogr. Sci. 12, 521.
11 Schmit, J.A., Henry, R.A., Williams, R.C. and Dieckman, J.F. (1971) J. Chromatogr. Sci. 9, 645.
12 Crowther, J.B., Jones, R. and Hartwick, R.A. (1981) J. Chromatogr. 217, 479.
13 Billiet, H.A., Schoenmakers, P.J. and de Galan, L. (1981) J. Chromatogr. 218, 443.
14 Ogan, K. and Katz, E. (1980) J. Chromatogr. 188, 115.
15 Hansen, S.H. and Thomsen, M. (1981) J. Chromatogr. 209, 77.
16 Engelhardt, H. and Ahr, G. (1981) Chromatographia 14, 227.
17 Goldberg, A.P. (1982) Anal. Chem. 54, 342.
18 Wahlund, K.G. and Sokolowski, A. (1978) J. Chromatogr. 151, 299.
19 Sokolowski, A. and Wahlund, K.G. (1980) J. Chromatogr. 189, 299.
20 Lund, U., Thesis, Copenhagen 1979.
21 Crombeen, J.B., Kraak, J.C. and Poppe, H. (1978) J. Chromatogr. 167, 219.
22 Kraak, J.C., Jonker, K.M. and Huber, J.F.K. (1977) J. Chromatogr. 142, 671.
23 Horvath, C., Melander, W., Molnar, I. and Molnar, P. (1977) Anal. Chem. 49, 2295.
24 Korpi, J., Wittmer, D.P., Sandemann, B.J. and Haney, W.G. (1976) J. Pharm. Sci. 65, 1087.
25 Helboe, P. and Thomsen, M. (1979) Int. J. Pharm. 2, 317.
26 Thomsen, M., Unpublished results.
27 Schoenmakers, P.J., Billiet, H.A.H. and de Galan, L. (1981) J. Chromatogr. 218, 261.
28 Bakalyar, S.R., McIlwrick, R. and Roggendorf, E. (1977) J. Chromatogr. 142, 353.
29 Tanaka, Y., Goodell, H. and Karger, B.L. (1978) J. Chromatogr. 158, 233.
30 Ghaemi, Y. and Wall, R.A. (1979) J. Chromatogr. 174, 51.
31 Wall, R.A. (1980) J. Chromatogr. 194, 353.
32 Hansen, S.H. (1981) J. Chromatogr. 209, 203.
33 Hansen, S.H., Helboe, P., Thomsen, M. and Lund, U. (1981) J. Chromatogr. 210, 453.
34 Hansen, S.H., Helboe, P. and Lund, U. (1982) J. Chromatogr. 240, 319.

35 Ghaemi, Y., Knox, J.H. and Wall, R. (1981) J. Chromatogr. 209, 191.
36 Helboe, P. (1982) J. Chromatogr. 245, 229.
37 Hansen, S.H. and Døssing, M. (1982) J. Chromatogr. 229, 141.
38 DiMari, S., Robinson, J.P. and Hash, J.H. (1981) J. Chromatogr. 213, 91.
39 Tarr, G.E. (1981) Anal. Biochem. 111, 27.
40 Hughes, G.J., Winterhalter, K.H., Boller, E. and Wilson, K.J. (1982) J. Chromatogr. 235, 417.
41 Weiner, S. and Tishbee, A. (1981) J. Chromatogr. 213, 501.
42 Lindroth, P. and Mopper, K. (1979) Anal. Chem. 51, 1667.
43 Cooper, J.D.H. and Turnell, D.C. (1982) J. Chromatogr. 227, 158.
44 Larsen, B.R. and West, F.G. (1981) J. Chromatogr. Sci. 19, 259.
45 Friedman, Z. and Smith, H.W. (1980) J. Chromatogr. 182, 414.
46 Neckes, L.M., Delisi, L.E. and Wyatt, R.J. (1980) Clin. Chem. 27, 146.
47 Yoshida, H., Sumida, T., Masujina, T. and Imai, H. (1982) J. High Resolut. Chromatogr. Chromatogr. Commun. 5, 509.
48 Imai, K. and Tamura, Z. (1978) Clin. Chim. Acta 85, 1.
49 Tsuruta, Y., Kohashi, K. and Ohkura, Y. (1981) J. Chromatogr. 224, 105.
50 Davis, T.P., Gehrke, C.W., Gehrke, C.W., Jr., Cunningham, T.D., Kuo, K.C., Gerhardt, K.O., Johnson, H.D. and Williams, C.H. (1979) J. Chromatogr. 162, 293.
51 Brown, N.D., Strickler, M.P. and Whaun, J.M. (1982) J. Chromatogr. 245, 101.
52 Yui, Y., Kimura, M., Itokawa, Y. and Kawai, C. (1979) J. Chromatogr. 177, 376.
53 Nimura, N., Ishida, K. and Kinoshita, T. (1980) J. Chromatogr. 221, 249.
54 Korte, W.D. (1982) J. Chromatogr. 243, 153.
55 Voelter, W., Huber, R. and Zech, K. (1981) J. Chromatogr. 217, 491.
56 Tsuchiya, H., Hayashi, T., Nanuse, H. and Takagi, N. (1982) J. Chromatogr. 231, 247.
57 Gotelli, G.R., Wall, J.H., Kabra, P.M. and Marton, L.J. (1981) Clin. Chem. 27, 441.
58 Kawasaki, T., Maeda, M. and Tsuji, A. (1981) J. Chromatogr. 226, 1.
59 Takeda, M., Maeda, M. and Tsuji, A. (1982) J. Chromatogr. 244, 347.
60 Alpenfels, W. (1981) Anal. Biochem. 114, 153.
61 Onishi, S., Itoh, S., Kawade, N., Isobe, K. and Sugiyama, S. (1980) J. Chromatogr. 182, 105.
62 Cole, K.D. and Little, G.H. (1982) J. Chromatogr. 227, 503.
63 Yamada, K., Kayama, E. and Aizawa, Y. (1981) J. Chromatogr. 223, 176.
64 Baker, M.D., Mohammed, H.Y. and Veening, H. (1981) Anal. Chem. 53, 1658.
65 Simmonds, R.J. and Harkness, R.A. (1981) J. Chromatogr. 226, 369.
66 Ryba, M. and Beranek, J. (1981) J. Chromatogr. 211, 337.
67 Kessler, M.J. (1982) J. Liq. Chromatogr. 5, 111.
68 Karle, J.M., Anderson, L.W., Dietrick, D.D. and Cysyk, R.L. (1980) Anal. Biochem. 109, 41.
69 Hartwick, R.A., Grill, C.M. and Brown, P.R. (1979) Anal. Chem. 51, 34.
70 de Abreu, R.A., van Baal, J.M., de Bruyn, C.H.M.M., Bakkeren, J.A.J.M. and Schretlen, E.D.A.M. (1982) J. Chromatogr. 229, 67.
71 Brugman, W.J.T., Heemstra, S. and Kraak, J.C. (1982) Chromatographia 15, 282.
72 Knox, J.H. and Jurand, J. (1981) J. Chromatogr. 218, 341.
73 Chow, F.K. and Grushka, E. (1979) J. Chromatogr. 185, 361.
74 Davis, G.E., Gehrke, C.W., Kuo, K.C. and Agris, P.F. (1979) J. Chromatogr. 173, 281.
75 Krstulovic, A.M., Hartwick, R.A. and Brown, P.R. (1979) Clin. Chem. 25, 235.
76 Schöch, G., Thomale, J., Lorenz, H., Suberg, H. and Karsten, U. (1980) Clin. Chim. Acta 108, 247.
77 Schweinsberg, P.D. and Loo, T.L. (1980) J. Chromatogr. 181, 103.
78 Wagner, I. and Capesius, I. (1981) Biochim. Biophys. Acta 654, 52.
79 Gehrke, C.W., Kuo, K.C., Davis, G.E., Suits, R.D., Waalhes, T.P. and Borek, E. (1978) J. Chromatogr. 150, 455.
80 Henderson, R.J. and Griffin, C.A. (1981) J. Chromatogr. 226, 202.
81 Jones, D.P. (1981) J. Chromatogr. 225, 446.
82 Juengling, E. and Kammermeier, H. (1980) Anal. Biochem. 102, 358.

83 Wells, G.B. and Lester, R.L. (1979) Anal. Biochem. 97, 184.
84 Yang, M.T., Milligan, L.P. and Mathison, G.W. (1981) J. Chromatogr. 209, 316.
85 Zsadon, B., Otta, K.H., Tüdös, F. and Szejtli, J. (1979) J. Chromatogr. 172, 490.
86 Kato, T. and Kinoshita, T. (1980) Anal. Biochem. 106, 238.
87 Macrae, R. and Dich, J. (1981) J. Chromatogr. 210, 138.
88 Nnor, S.K. (1979) Analusis 7, 381.
89 Hendrix, D.L., Lee, R.E., Baust, J.G. and James, H. (1981) J. Chromatogr. 210, 45.
90 Heyraud, A. and Rinaudo, M. (1980) J. Liq. Chromatogr. 3, 721.
91 Thompson, R.M. (1978) J. Chromatogr. 166, 201.
92 Binder, H. (1980) J. Chromatogr. 189, 414.
93 Dutot, G. (1981) Fenill. Biol. 22, 101.
94 Lottspeich, F., (1980) Hoppe-Seyler's Z. Physiol. Chem. 361, 1829.
95 Tarr, G.E. (1981) Anal. Biochem. 111, 27.
96 Hughes, G.J., Winterhalter, K.H., Boller, E. and Wilson, K.J. (1982) J. Chromatogr. 235, 417.
97 Kraak, J.C., Jonker, K.-M. and Huber, J.F.K. (1977) J. Chromatogr. 142, 671.
98 Schuster, R. (1980) Anal. Chem. 52, 617.
99 Lindroth, P. and Mopper, K. (1979) Anal. Chem. 51, 1667.
100 Lin, J.-K. and Wang, C.-H. (1980) Clin. Chem. 26, 579.
101 Shihabi, Z.K. and White, J.P. (1979) Clin. Chem. 25, 1368.
102 Krstulovic, A.M., Friedman, M.J., Sinclair, P.R. and Felice, J. (1981) Clin. Chem. 27, 1291.
103 Sźymanovicz, A., Poulin, G., Randoux, A. and Borel, J.P. (1979) Clin. Chim. Acta 91, 141.
104 Lam, S., Chow, F. and Karmen, A. (1980) J. Chromatogr. 199, 295.
105 Tapuhi, Y., Miller, N. and Karger, B.L. (1981) J. Chromatogr. 205, 325.
106 Gübitz, G., Jellenz, W. and Santi, W. (1981) J. Chromatogr. 203, 377.
107 Gil-Av, E., Tishbee, A. and Harl, P.E. (1980) J. Am. Chem. Soc. 102, 5115.
108 Davankov, V.A., Bochkov, A.S., Kurganov, A.A., Roumeliotis, P. and Unger, K.K. (1980) Chromatographia 13, 677.
109 Krummen, K. (1980) J. Liq. Chromatogr. 3, 1243.
110 Hearn, M.T.W. (1980) J. Liq. Chromatogr. 3, 1255.
111 Richmond, W. (1980) in Current Developments in the Clinical Applications of HPLC, GC and MS. (Lawson, A.M., Lim, C.K. and Richmond, W., eds.) Academic Press, New York, London.
112 Welinder, B.S. (1980) J. Liq. Chromatogr. 3, 1399.
113 Hearn, M.T.W., Grego, B. and Bishop, C.A. (1981) J. Liq. Chromatogr. 4, 1725.
114 Putterman, G.J., Spear, M.B., Meade-Cobun, K.S., Vidra, M. and Hixson, C.V. (1982) J. Liq. Chromatogr. 5, 715.
115 Faurmy, D., Pradayrol, L., Antoniotti, H., Esteve, J.P. and Ribet, A. (1982) J. Liq. Chromatogr. 5, 757.
116 Pocker, Y. and Biswas, S.B. (1982) J. Liq. Chromatogr. 5, 1.
117 Barfod, R.A., Sliwinski, B.J., Breyer, A.C. and Rothbart, H.L. (1982) J. Chromatogr. 235, 281.
118 Buchholz, K., Gödelmann, B. and Molnar, I. (1982) J. Chromatogr. 238, 193.
119 Ratge, D. and Wisser, H. (1982) J. Chromatogr. 230, 47.
120 Fong, G.W.-K. and Grushka, E. (1978) Anal. Chem. 50, 1154.
121 Nice, E.C., Capp, M.W., Cooke, N. and O'Hare, M.J. (1981) J. Chromatogr. 218, 569.
122 James, J.L., Clawson, G.A., Chan, C.H. and Smuckler, E.A. (1981) Lipids 16, 541.
123 Perkins, E.G. (1981) Lipids 16, 609.
124 Okazaki, M., Hagiwara, N. and Hara, I. (1982) J. Chromatogr. 231, 13.
125 Tsuchiya, H., Hayashi, T., Naruse, H. and Takagi, N. (1982) J. Chromatogr. 231, 247.
126 Terragno, A., Rydzik, R. and Terragno, N.A. (1982) Prostaglandins 21, 101.
127 Frey, F.J., Frey, B.M. and Benet, L.Z. (1979) Clin. Chem. 25, 1944.
128 Cavina, G., Moretti, G., Alimenti, R. and Gallinella, B. (1979) J. Chromatogr. 175, 125.
129 Canalis, E., Caldarella, A.M. and Reardon, G.E. (1981) Clin. Chem. 27, 1241.
130 Mourey, T.H. and Siggia, S. (1980) Anal. Chem. 52, 881.

131 Lodovici, M., Dolara, P., Bavazzano, P., Colzi, A. and Pistolesi, V. (1981) Clin. Chim. Acta 114, 107.
132 Schmidt, G.F., Vandemark, F.I. and Slavin, W. (1978) Anal. Biochem. 91, 636.
133 Dolphin, R.J. and Pergande, P.J. (1977) J. Chromatogr. 143, 267.
134 Smith, R.M., Doran, D., Mazur, M. and Bush, B. (1980) J. Chromatogr. 181, 319.
135 Salmi, M. and Teukunen, R. (1980) Clin. Chem. 26, 1832.
136 Scoble, H.A., McKeag, M., Brown, P.R. and Kavarnos, G.J. (1981) Clin. Chim. Acta 113, 253.
137 Longas, M.O. and Poh-Fitzpatrick, M.B. (1980) Anal. Biochem. 104, 268.
138 Hill, R.H., Jr., Bailey, S.L. and Needham, L.L. (1982) J. Chromatogr. 232, 251.
139 Ford, R.E., Ou, C.-N. and Ellejson, R.D. (1981) Clin. Chem. 27, 397.
140 Gray, C.H., Lim, C.K. and Nicholson, D.C. (1977) Clin. Chim. Acta 77, 167.
141 Lim, C.K., Bull, R.V.A. and Rideout, J.H. (1981) J. Chromatogr. 204, 219.
142 Cole, K.D. and Little, G.H. (1982) J. Chromatogr. 227, 503.
143 Sian, M.S. and Rains, A.J.H. (1979) Clin. Chim. Acta 98, 243.
144 Rosenthal, R., Blackaert, N., Kabra, P.M. and Thaler, M.M. (1981) Clin. Chem. 27, 1704.
145 Onishi, S., Itoh, S., Kawade, N., Isobe, K. and Sugiyama, S. (1980) J. Chromatogr. 182, 105.

Deyl (ed.) Separation Methods

CHAPTER 4.5

Ion exchange chromatography

OTAKAR MIKEŠ

Institute of Organic Chemistry and Biochemistry, Czechoslovak Academy of Sciences, 166 10 Prague 6, Czechoslovakia

4.5.1 Ion exchange in biochemistry

Ion exchange in the form of batch processes and chromatography represents one of the most important separation methods in biochemistry. The chief advantage of this technique, compared with others, e.g., gel-permeation (size-exclusion), partition and affinity chromatography and electrophoretic column methods, is much higher sorption or separation capacity. This permits considerably higher loads to be applied to columns of equal bed size. The separation possibilities are multiplied by the fact that gradients of both ionic strength and pH can be used.

This chapter summarizes the majority of the literature on ion exchange chromatography in biochemistry over the five years preceding 1981. Occasionally some important older works are cited (especially reviews or monographs). It can be seen from the literature that even today many authors are satisfied with the use of older, classically proven, but time consuming, low pressure separation methods and develop them further. On the contrary, at the same time a broad shift to modern trends can be observed, which is represented by numerous applications of the rapid medium and high pressure liquid chromatography in various fields of biochemistry. Therefore both approaches must be discussed here.

4.5.1.1 Classic methods

Several monographs or monographic essays have been written on ion exchange resins and their use in classic column chromatography [1–6]. Many reviews treat this subject in detail [7–10]. All papers on ion exchange and liquid column chromatography are summarized every second year by Walton [11]. The essays mentioned present the necessary information, not only regarding the theoretical approach for the use of ion exchange chromatography, but also information on instrumentation, experimental techniques and various fields of application including biochemistry. The separation of low and middle molecular weight substances have usually never presented a great problem, because the majority of ion exchange resins produced have been suitable for this purpose. Classic liquid column chromatography of biopolymers (especially of proteins and nucleic acids) used to be problematic since suitable chromatographic supports were not available. Various inorganic compounds showed strong irreversible sorption and organic ion exchange resins with aromatic matrices often denatured the proteins by strong hydrophobic interactions. Only weakly acidic acrylate and methacrylate cation exchangers (of the Amberlite IRC 50 type) served this purpose. One of the earliest pioneering examples of this type of separation is the paper on isolation of cytochrome *c* by Paleus and Neilands [12]. These packings were, however, microporous and only the ionogenic groups, localized on the particle surface, were in function.

Sober and Peterson [13–15] developing ion exchange cellulose derivatives, Porath and Flodin [16] crosslinked polydextran, and Porath and Lindner [17] polydextran ion exchange derivatives, prepared hydrophilic and macroporous supports suitable for classic chromatography of biopolymers. These supports have successfully been used in thousands of studies, and it is the opinion of the author that the contribution of these workers to the development of modern biochemistry, molecular biology and related areas has not been fully appreciated as yet. Recently, new supports based on cross-linked agarose described by Porath et al. [18] and its ion exchange derivatives were introduced.

These hydrophilic macroporous packings for classic ion exchange chromatography of biopolymers, and methods of their application are described in detail in commercial brochures available on request [19,20].

4.5.1.2 Modern trends

Newer approaches to the development of ion exchange separation methods in biochemistry are represented by two general trends: (a) medium and high pressure liquid chromatography (MPLC, HPLC) *, (b) chromatofocusing; both trends are briefly discussed here.

The main difference between MPLC and HPLC is not only in height of the performance, speed of the separation process, and the pressure required, but also in the instrumentation. A home made MPLC apparatus (Fig. 4.5.1) can be improvised in nearly every laboratory using, e.g., spare parts of amino acid analyzers or sugar analyzers. However, for the HPLC methods it is usually necessary to use commercial equipment or parts, because here the requirements of construction are much more rigid. For the separation of low and middle molecular weight substances HPLC methods are mostly used.

---

* The author uses the classification of column chromatographic methods according the pressure applied: (1) Low pressure liquid chromatography (LPLC) methods, where the overpressure (often only hydrostatic or developed by a peristaltic pump) is usually expressed in cm of water column. (2) Medium pressure liquid chromatography (MPLC) with the overpressure usually expressed in MPa (atmospheres) up to 3–4 MPa (30–40 at), allowing the application of glass columns and plastic tubings and fittings. (3) High-pressure liquid chromatography (HPLC) with higher overpressures (up to hundreds of at) and with metal equipment.

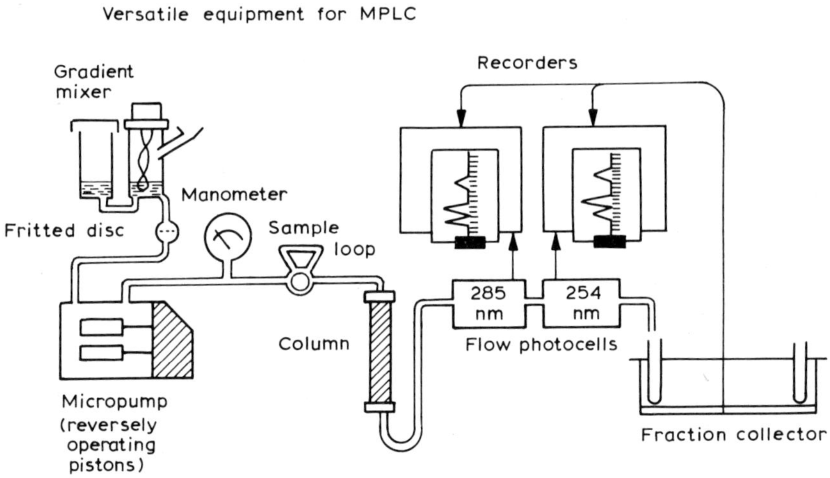

Fig. 4.5.1. Medium pressure liquid chromatograph for rapid separation of biopolymers [37,83,313]. The home-made equipment was constructed using spare parts (Glass columns, plastic tubings, fittings) for amino acid or sugar analyzers. A gradient mixer provided with a side funnel made possible the continuous linking of linear gradients without interruption of the operation. Reversibly operating pistons of the programmable micropump adjusted to the same output power contributed to reduction of pressure pulses. Fraction collector yielded electrical step-impulses to be recorded with both recorders. Column 20×0.8 cm packed with Spheron ion exchange derivatives, 20–40 μm, flow-rate up to 6 ml/cm^2 column cross-section per min, pressure up to 3 MPa (30 atm).

On the contrary in the field of biopolymers both MPLC and HPLC methods are being applied successfully.

Several monographs describing HPLC procedures have been issued. After the famous book by Kirkland [21] on rapid column chromatography other monographs on HPLC appeared (Hamilton and Sewell [22], Pryde and Gilbert [23], Huber [24] (describing instrumentation), Engelhard [25], Knox [26] and Brown [27] (specializing in biochemical and biomedical applications of HPLC methods)). The subject has also been treated in many reviews; some of them general [28–34] and others specialized, e.g., on HPLC of peptides [35], proteins [36–38], enzymes and enzymic reactions [37–40], biopolymers [41], biogenic amines [42]. Some other specialized reviews will be mentioned in sections 4.5.2.3 and 4.5.5. The essays cited present broad essential information on one of the modern trends in column chromatographic separations of biochemically important substances. This trend is characterized by the use of very fine packings, which require only short diffusion time to reach the equilibrium, so that the chromatographic separation process is very rapid.

The other new trend in ion exchange column separation methods is ampholyte displacement chromatography (ADC) and chromatofocusing (CF). Leaback and Robinson [43] — who first published the ADC method — used conventional ion exchangers for the separation of proteins, and carrier ampholytes for the elution. Using this approach the authors succeeded in resolution of isoenzymes unresolvable by conventional ion exchange chromatography or by isoelectric focusing. The method was tested also by other authors — e.g., by Young and Webb [44] — for the separation of serum proteins.

Sluyterman and Wijdenes [45] presented another idea which they designated chromatofocusing (CF): they prepared a slowly moving pH gradient inside the column of a conventional ion exchanger by pumping an influent buffer of different pH from the equilibration buffer. Proteins passing slowly through the column separated and emerged in order of their isoelectric points. Focusing effects occurred. In subsequent papers Sluyterman and co-workers elaborated the principle theoretically [46] and verified their results experimentally [47]. They also tested various ion exchangers [48,49] for their suitability for chromatofocusing of proteins using amphoteric buffers, thereby preparing both a theoretical and practical base for further development of CF methods in biochemistry. Several other authors (e.g., Young and Webb [50]) tested this method and compared CF with ADC experimentally.

The Swedish firm Pharmacia contributed to the development of CF by producing special amphoteric buffers (cf., 4.5.3.3) and ion exchangers (cf., 4.5.2.4). These products are described in detail in a brochure [51] available on request, which serves as a laboratory manual for CF. There is no doubt that chromatofocusing is an effective method for ion exchange separation of proteins. However, because it is a new method, we cannot yet evaluate its contribution to the progress in the field of biochemistry. The author hopes that this method will be found as useful as the electrofocusing method.

4.5.2 Ion exchangers

Ion exchangers are insoluble substances liberating ions by electrolytic dissociation. The majority of the most important for biochemistry are solid substances, but liquid ion exchangers for extraction chromatography have been also described (cf., review by Ghersini [52]). The survey and description of types of ion exchangers can be found in the cited monographs [1–6] and reviews [7–11]. From the chemical point of view the ion exchangers are for the most part resins. However, for separations of biopolymers, the application of ion exchangers with polysaccharide matrices is typical [13–20,53]. Modern high-performance liquid chromatography usually applies beads of inorganic macroporous matrices (variously modified); This will be discussed in section 4.5.2.3. In addition to the essays mentioned the survey of ion exchange material is presented also in reviews by Kressman [54,55], Helfferich et al. [56] and Horwath [57], the latest specializing on pellicular ion exchangers.

4.5.2.1 Classification and fundamental properties of ion exchangers

According to the literature cited [1–11,53–57] the ion exchangers can be classified, from the chemical point of view, as *inorganic* (very seldom used in biochemistry) and *organic* (the most important for biochemical applications). The latter are further classified according the their functional groups. *Cation exchangers* (or *catexes*) liberate and exchange cations, and *anion exchangers* (*anexes*) liberate and exchange anions. *Heteroionic ion exchangers* contain several cationic (or several anionic) functional groups. *Homoionic ion exchangers* are materials which contain only one type of cationic or anionic functional group; they show the sharpest separation effects in chromatography. *Amphoteric ion exchangers* contain both cation and anion exchanging groups in the same matrix. They dissociate in contact with electrolyte and bind both its components. This is the reason why electrolytes move only slowly down the column and can easily be separated from non-electrolytes; therefore these packings are also called ion retardation resins. When they are free from separated substances, the opposite charged groups form internal salts and this is why the amphoteric ion exchangers can be simply regenerated by washing with water only. Their special form are *dipolar ion exchangers* the functional groups of which form dipoles in aqueous solutions. They were prepared for biochemical purposes by Porath and Fryklund [53] (for other citations and examples of applications see Ref. 6). It is a pity that these ion exchangers, which have been proved to be effective for the séparation of biopolymers are not yet produced commercially.

Chelating ion exchangers carry functional groups capable of forming complexes with metal ions. *Selective ion exchangers* have a limited binding ability, binding certain group of ions only. *Specific ion exchangers* contain a special functional group which reacts selectively with one type of ion only. The principle of specific ion exchangers resembles that of affinity chromatography. Other specialized ion exchangers may be prepared, e.g., *oleophilic, decolorizing* and *redox exchangers* (transferring electrons instead of ions). Conventional ion exchange resins characterized by crosslinking X2–X8 are classified as *microporous* and are penetrable by low and medium molecular weight substances (e.g., by amino acids, peptides). *Isoporous ion exchangers* are characterized by such regular distribution of meshes in the network as possible. *Macroporous ion exchangers* are penetrable by biopolymers. *Pellicular ion exchangers* have the form of a very thin porous surface ionogenic layer on microscopic, inert non-porous spherical beads (these packings are also designated porous layer beads, PLB). *Mixed bed resins* consist of equivalent amounts of catex (H^+) and anex (OH^-). They are used for deionization of water. However, classification of ion exchangers into cation and anion exchangers is of the most general importance.

Ion exchange [5,6] is the process in which the ion A (the so called *counter-ion*), electrolytically dissociated from a cation or anion exchanger, is substituted by another ion, C, from the solution possessing a charge of the same sign, and greater affinity towards the ion exchanger I. Such an ion exchange can be expressed by the equation

$$I - A + C \rightleftarrows I - C + A \tag{1}$$

It is a reversible process and the direction of the reaction is affected not only by the affinity, but mainly by the concentration of the ions. Therefore ion C may easily be substituted reversibly by ion A, if its concentration in the solution is substantially higher. This is the main principle of sorption and desorption of ionogenic substances by means of ion exchangers.

The degree of such an exchange under static conditions (Eq. 1), in a grain of ion exchange resin in contact with an electrolyte solution, is governed both by Donnan equilibria (which also control the swelling of the particle) and by the affinity of ions to the exchanger. It can be expressed quantitatively by Samuelson's equation [58]

$$k_{\mathrm{A}}^{\mathrm{C}} = \frac{[\mathrm{C}]_{\mathrm{r}}^{a} \cdot [\mathrm{A}]_{\mathrm{s}}^{c}}{[\mathrm{A}]_{\mathrm{r}}^{c} \cdot [\mathrm{C}]_{\mathrm{s}}^{a}} \tag{2}$$

where $k_{\mathrm{A}}^{\mathrm{C}}$ is the selectivity coefficient, a, c are the absolute values of charges of ions A,C to be exchanged, and brackets represent the concentration in the resin phase ($_{\mathrm{r}}$) or in the external solution ($_{\mathrm{s}}$).

The *rate* at which equilibrium [grain ⇄ external solution] in mixed system is reached, is very important for the study of a chromatographic process. It has been found in *kinetic studies*, that the process can be divided into five individual steps: (1) the transport of ion C from the surrounding moving solution to the surface of the bead through the stable thin layer of the solution attached to the particle (so-called Nernst film); (2) the diffusion of C through the matrix to functional groups; (3) the proper chemical exchange expressed by Eq. 1; (4) the diffusion of exchanged ion A from the functional groups to the surface of the bead; (5) the transport of A through the Nernst film into the external solution. The slowest of these five processes should be rate controlling. In the majority of cases it is the 'particle diffusion', however, under certain conditions 'film diffusion' is rate controlling in ion exchange kinetics. Film diffusion is inversely proportional to the diameter of the particle, whereas particle diffusion is inversely proportional to the square of the diameter. An increased flow rate through the chromatographic column will decrease the thickness of the Nernst film and thus will favour mass transfer control in the particle.

Martin and Synge [59] were the first to apply the idea of a theoretical plate to column chromatography, and Mayer and Tompkins [60] were the first to extend it to ion exchange columns. It has been found that the particle size has the most important influence on the efficiency of the chromatographic process. Therefore, particles as small as possible in a given case are usually applied. If the elution is slow enough (allowing to reach equilibrium), then the height equivalent of a theoretical plate (HETP) is approximately equal to the diameter of the particle. Of course, such low ideal values are seldom obtained in practice.

Channelling in the columns due to irregularities in the packing causes the HETP to be several times higher. Glueckauf [61] derived formulae for calculation of the plate height under non-equilibrium conditions usually found in practical ion exchange chromatography (cf. also Ref. 5). The final term H for HETP is divided into three partial individual terms

$$H = H_{\mathrm{ps}} + H_{\mathrm{pd}} + H_{\mathrm{fd}} \tag{3}$$

where ps is particle size, pd particle diffusion, and fd film diffusion. $H_{ps} = 1.64\ r$ (where r = particle radius), is roughly valid for equilibrium conditions, and increases to higher values of H depending on speed of the chromatography process. For formulae regarding H_{pd} and H_{fd} see, e.g., Ref. 5.

Giddings [62] generally defines H, in liquid chromatography, as the function of linear flow rate u of the mobile phase in this form

$$H = C_s u + C_m u + (1/A + 1/C_m u)^{-1} \tag{4}$$

where C_s is the coefficient of the mass transfer resistance in the stationary phase, C_m in the mobile phase, and A is the coefficient of eddy diffusion. According to Eq. 4 the limiting factors in liquid chromatography are mass transfer resistance in the stationary and mobile phases.

Ion exchange chromatography can be realized by three methods [6]: (1) *frontal analysis*; (2) *displacement chromatography*; and (3) *elution chromatography*. The first method is obsolete and not used today. In displacement chromatography the separated substances originally sorbed at the top are displaced from their positions in the column by some other substances acting as a piston with greater affinity to the ion exchanger. While they are pressed down they separate and displace each other; this is the reason why they must remain in close contact and usually do not separate perfectly. The latter method is most important for biochemistry and is briefly described. The load of material to be separated (represented here by ions A, C), which exhausts only a small part of the column capacity, is first sorbed at the top of the column and then eluted down using a buffer (M) with the ions of a lower affinity to the ion exchanger I than the separated ions A and C. The movement of ions A,C is then governed solely by equilibria

$$\begin{aligned} &\mathrm{IA + M \rightleftarrows IM + A} \\ &\mathrm{IC + M \rightleftarrows IM + C} \end{aligned} \tag{6}$$

depending on the different affinities of the separated ions to the exchanger only. Therefore their mobilities are mutually independent, and the ions eluted may appear in the form of individual well separated peaks. This is the greatest advantage of elution chromatography.

4.5.2.2 Materials for batch processes and packings for low-pressure liquid column chromatography (LPLC)

The older classic low speed LPLC methods are often used for biochemical separations, because they are very simple, modest in instrumentation, and therefore cheap. They can be realized in every laboratory without special equipment. Also, ion exchange materials for these purposes need not be so finely and carefully prepared nor stand high pressure, and are usually inexpensive. The most important materials for batch processes and column packings can be classified according to composition of the matrix into two groups: (1) resins, and (2) polysaccharide materials.

The resins [1–11,54–57] are usually bead-like styrene-divinylbenzene (DVB) copolymers, containing 2–8% DVB, which is equivalent to X2–X8 crosslinking of the network. Pearl or ground (meth)acrylate-divinylbenzene or (meth)acrylate-ethylene dimethacrylate copolymers are also used. The most usual products of these types

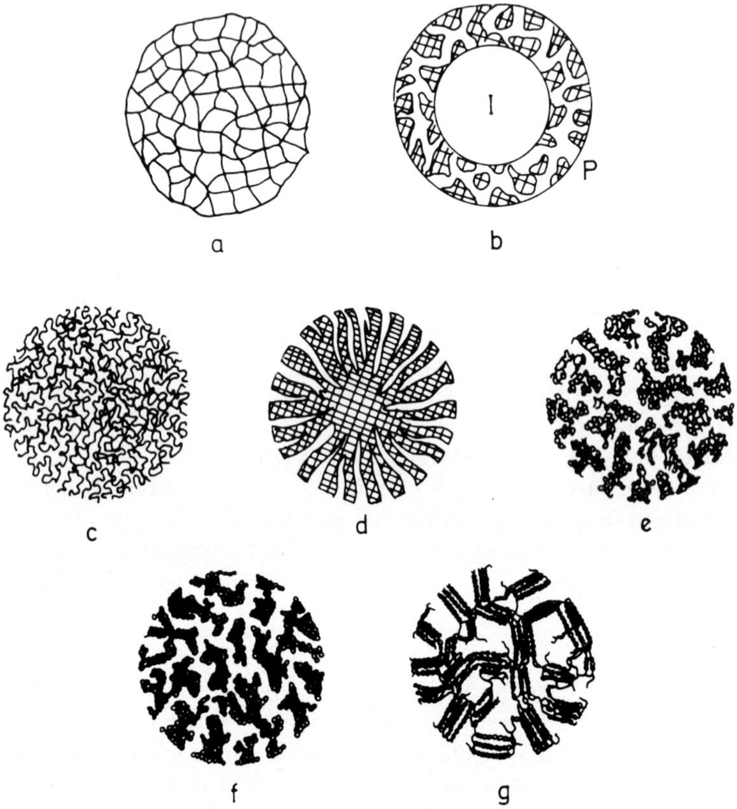

Fig. 4.5.2. Schematic representation of internal bead macrostructures [6,421]. (a) Microporous resin (xerogel); ion exchange functional groups (cf., Table 4.5.1) are attached to the swelling network. (b) Pellicular ion exchanger (porous layer bead, PLB); functional groups are located within the pores P; I = inert nucleus, which ions cannot penetrate. (c) Crosslinked polydextran (classical xerogel); functional groups are attached to polysaccharide chains of the swelling matrix. (d) Porous glass and (e) porous silica gel (both are classic aerogels); OH groups of glycophase covering large inner and outer surface are the sites of ionogenic substitution (however, ionogenic groups can also be bound in other ways). (f) Organic macroreticular polymer (xerogel-aerogel hybrid); its macrostructure (macrosphere = bead) is composed of many submicroscopic densely crosslinked xerogel microspheres with macroporous channels between their clusters; macropores on the surface of the bead can be of aerogel type; functional groups are attached to the large inner and outer surfaces of the macrostructure. (g) Crosslinked agarose (xerogel-aerogel hybrid); functional groups substitute H in some OH groups of the polysaccharide chain similar to cellulose ion exchange derivatives.

(Fig. 4.5.2) are known under the original trade names Amberlite, Dowex, Duolite, Lewatit, Wofatit, ZeoKarb, etc. The microreticular matrix of conventional ion exchange resins usually has the form of *xerogel* network (changing dimensions of pores with the degree of swelling). Only low and middle molecular weight substances can penetrate these microporous matrices. However, newer ion exchangers, so-called *macroreticular resins* have also been developed. Their beads are aggregates consisting of tiny granules (with diameters of a few hundred Ångstroms) with large pores among them (up to 1000 Å in diameter). They are prepared by a special suspension copolymerization, have large internal surface area (tens of m^2/g), allow the penetration of large molecules, and can also be used in non-aqueous solutions. They were first commercially introduced under the name Amberlyst. In spite of their macroporosity they are not suitable for the separation of biopolymers because of their strong hydrophobicity.

The most frequent types of functional groups of ion exchangers used for biochemical purposes are summarized in Table 4.5.1. In addition, ion exchangers have been prepared with other functional groups, but their applications are not as common. The detailed tables of ion exchangers commercially available can be found, e.g., in monographs [5,6]. In catalogues of several firms supplying chemicals a series of relatively cheap ion exchange resins of *commercial* (or *practical grade*) are advertised. They may be used for common laboratory applications (both for batch processes and column chromatography) if they are of suitable particle size, and if they are extracted and recycled. Some other firms producing packings for chromatography do this work themselves and offer pure ion exchangers labeled AG (Analytical Grade). In addition, some firms also sieve the ion exchangers to be as size homogeneous as possible, and thus prepare ion exchangers labeled CG (Chromatographic Grade). Such firms usually issue catalogues and brochures available on request with full description of the ion exchangers offered. RG (Reactor Grade) ion exchangers are also produced for work with radioactive ions. The most usual types of ion exchangers prepared specially for chromatographic applications in biochemistry are known under the trade names Bio-Rad, Bio-Rex or XE and others. Many producers of chromatographic equipment (e.g., of amino acid or sugar analyzers) as a rule recommend to their own special chromatographic packings, tested and delivered by themselves (e.g., Beckman or Durrum resins). The various Aminex ion exchangers are also specialized for biochemical purposes.

The polysaccharide materials [13–20,53,63] used as matrices for ion exchangers are cellulose, crosslinked polydextran and crosslinked agarose. The ion exchangers with the matrices of this type are macroporous and hydrophilic; large macromolecules of biopolymers penetrate them without danger of denaturation. Therefore they are very often used for biochemical separations both for batch and column processes. The *advanced cellulose ion exchangers* are now produced. They are characterized by shorter and more defined length of the fibres which results in very good chromatographic separations. The cellulose ion exchangers are prepared in the following functional groups (Table 4.5.1): CM, DEAE and SE [19]; sometimes also AE, ECTEOLA, GE, P, PEI and TEAE. The DEA- and CM-derivatives are most often used for separation of biopolymers. Besides the fibre form [19], cellulose ion exchangers (DEAE) are also now obtainable in the form of beads [20] which are advantageous both for column chromatography and batch processes (cf., review by Štamberg et al. [64]). *Polydextran ion exchangers* are available [20] with functional groups (Table 4.5.1): BD, CM, DEAE, QAE, SE or SP, and with two types of macroporosity of matrix. The BD derivative is used for chromatography of tRNAs.

TABLE 4.5.1.

Functional groups of main organic ion exchangers for LPLC

No.	Abbreviations *	Formulae in ionized form	Types
(a) *Cation exchangers*			
1	S-, SM-, SE-, SP-, SB-	$-SO_3^-$, $-CH_2\cdot SO_3^-$, $-(CH_2)_2\cdot SO_3^-$, $-(CH_2)_3\cdot SO_3^-$, $-(CH_2)_4\cdot SO_3^-$	Strongly acid
2	P-	$-PO_3^{2-}$	Medium acid
3	C-, CM-	$-COO^-$, $-CH_2\cdot COO^-$	Weakly acid
(b) *Anion exchangers*			
4	TMAM-; TEAE-	$-CH_2\cdot \overset{+}{N}(CH_3)_3$; $-(CH_2)_2\cdot \overset{+}{N}(C_2H_5)_3$	
5	HEDMAM-	$-CH_2\cdot \overset{+}{N}(CH_3)_2\cdot C_2H_4OH$	
6	QAE-	$-CH_2\cdot \overset{+}{N}(C_2H_5)_2\cdot CH_2\cdot CH(OH)\cdot CH_3$	Strongly basic
7	GE-	$-C_2H_4\cdot C(NH_2)\!=\!\overset{+}{N}H_2$	
8	MP-	$-C_5H_5\overset{+}{N}\cdot CH_3$	
9	DEAE-	$-(CH_2)_2\cdot \overset{+}{N}H\cdot (C_2H_5)_2$	
10	DEAHP-	$-CH_2\cdot CH(OH)\cdot CH_2\cdot \overset{+}{N}H(C_2H_5)_2$	
11	ECTEOLA-	(undefined mixture of amines)	Medium basic
12	AE-	$-(C_2H_4)\overset{+}{N}H_3$	
13	PEI-	$-(C_2H_4NH_2)_n\cdot C_2H_4\overset{+}{N}H_3$	
14	BD-	(benzoyl- and DEAE-derivative of polydextran)	Medium basic and hydrophobic
15	PAB-	$-CH_2\cdot C_6H_4\cdot \overset{+}{N}H_3$	Weakly basic
(c) *Dipolar ion exchangers [53,63]*			
16	BA	$=\overset{+}{N}H\cdot (CH_2)_2\cdot COO^-$	Neutral **
17	CAI-	$-\overset{+}{N}H_2\cdot (CH_2)_n\cdot COO^-$	Neutral **
18	A-	$=N\cdot CH\cdot COO^-$ | $(CH_2)_3$ | $NH\cdot C(NH_2)\!=\!\overset{+}{N}H_2$	Medium basic **
19	SAHP-	$-O\cdot CH_2\cdot CH(OH)\cdot CH_2\cdot \overset{+}{N}H_2\cdot C_6H_4\cdot SO_3^-$	Medium acid **

* Explanations to abbreviations derived from trivial terms for ion exchange functional groups: (1) Sulfo-, sulfomethyl-, sulfoethyl-, sulfopropyl-, sulfobutyl-; (2) Phosphonic-; (3) Carboxyl-, carboxymethyl-; (4) Trimethylaminomethyl-, Triethylaminoethyl-; (5) Hydroxyethyldimethylaminomethyl-; (6) Quaternary aminoethyl-; (7) Guanidinoethyl-; (8) Methylpyridinium-; (9) Diethylaminoethyl-; (10) Diethylaminohydroxypropyl-; (11) Arbitrary term; (12) Aminoethyl-; (13) Polyethyleneimine-; (14) Benzoyl DEAE-; (15) Paraaminobenzyl-; (16) β-alanyl-; (17) Carboxyalkylimino-; (18) Arginyl-; (19) Sulfonic acid-hydroxypropyl-;

** Ionized form (type of exchanger), depends on the pH used.

Sepharose ion exchangers differ from polydextran derivatives in the type of macromolecular structure, greater rigidity, better flow properties, low swelling, and resistance to organic solvents. They can be obtained [20] with DEAE- or CM-functional groups under the name Sepharose CL (CL = cross-linked), and are used for the separation of biopolymers up to $M_r \sim 1\,000\,000$. Other packings of similar type are DEAE Bio-Gel A and CM Bio-Gel A, described in catalogues of the American firm Bio-Rad Laboratories [158].

In addition to the products mentioned other materials have also been prepared. The present status of the development of ion exchangers has been reviewed by Bogoczek and Suruwiec [65]. Rahm [66] prepared a cation exchanger crosslinked with *m*-divinylbenzene, he showed that it had better properties for the separation of amino acids and reviewed this problem [67]. Selezneva et al. [68] described the cation exchanger KMT (carboxylated methacrylate-triazine copolymer) for the sorption of proteins. Kohn et al. [69] prepared a cation exchanger with the matrix of crosslinked pectic acid, and Popova and Krachanov [70] used sunflower pectic acid as an ion exchanger for the resolution of enantiomers. Antal and Toman [71] described DEAHP-starch and DEAHP-cellulose for fractionating polysaccharides and other biopolymers. Smith and Billett [72] prepared ECTHAM-cellulose by coupling THAM (Tris) to cellulose with epichlorohydrin and used this ion exchanger for the fractionation of chromatin.

4.5.2.3 Packings for medium- and high-pressure liquid chromatography

How to accelerate ion exchange chromatography? From the considerations in section 4.5.2.1. we know that mass transfer in the particle is one of the limiting factors for speed of the chromatography. There are two principal ways to lower its influence: (1) use of thin porous ion exchange layers spread on the surface of beads with inert non-porous rigid core; (2) use of very fine porous ion exchange particles. In both cases the diffusion is rapid, the functional groups are reached quickly, and the equilibrium can be attained in a short time. The former method used in application of pellicular ion exchangers has been described and discussed [21,57]. The general disadvantage of this approach is the low ion exchange capacity in relation to the volume of the bead. Therefore this method is not suitable for preparations. In spite of that, very good and rapid analytical separation was achieved in numerous biochemical applications.

In the latter method the lowering of bead diameter is important. The smaller the particles, the higher the pressure which must be used for pumping the influent into the column, thus approaching HPLC. The particles should possess sufficient rigidity to pressure deformation and, moreover, the differences in bead diameter should be as small as possible, as they negatively influence the uniformity in attainment of the equilibrium, and thus the chromatographic process. The HPLC methods are described in monographs [22–27] and reviews [28,29,31–34]. In addition, Majors [73] summarized in detail recent advances in HPLC packings and columns. Care and use of ion exchange packings for HPLC-separations has been reviewed by Rabel [74].

TABLE 4.5.2

Ion exchangers for HPLC and MPLC

Types	Functional groups	Trade names
(A) Packings for separation of low and middle molecular weight substances *		
(a) *Pelliculars*		
Strongly acidic cation exchangers (SCX)	$-SO_3^-$	BondaPak CX/Corasil; HC-Pellionex SAX; HC-Pellionex SCX; HS-Pellionex SAX; Perisorb KAT; Vydac SC Cation; Zipax SCX
Strongly basic anion exchangers (SAX)	$-\overset{+}{N}R_3$	AE- or AS-Pellionex-SAX; BondaPak AX/Corasil; Perisorb AN; Permaphase-AAX; Permaphase-ABX; Vydac SC Anion; Zipax-SAX
Weakly basic anion exchangers (WAX)	$-NH_2$ or $-NR_2$	AL-Pellionex-WAX; Zipax WAX
(b) *Silica gel derivatives*		
Strongly acidic cation exchangers (SCX)	$-SO_3$	Chromegabond SCX; LiChrosorb KAT; Microsil SCX; Nucleosil SA; Partisil-10-SCX; RSIL-CAT; Vydac TP Cation; Vydac 401TP; Zorbax SCX
Strongly basic anion exchangers (SAX)	$-\overset{+}{N}(CH_3)_3$ or $-\overset{+}{N}R_3$	Chromegabond SAX; LiChrosorb-AN; MicroPak SAX; Microsil SAX; Nucleosil SB; Partisil-SAX; Partisil 10 SAX; RSIL AN; RSL Anion; Vydac 301 TP; Zorbax SAX
Medium basic anion exchanger (MAX)	$-\overset{+}{N}H(CH_3)_2$	Nucleosil NMe_2
Weakly basic anion exchangers (WAX) (High polarity packings for 'normal phase' or partition chromatography)	$-NH_2$	Alltech NH_2; Amino Sil-X-1; Apex Amino; APS-Hypersil; Chromegabond Diamine; Chromegabond NH_2; Chromosorb LC-9; Finepak SIL NH; HiChrom NH_2; Hypersil APS; LiChrosorb NH_2; MicroPak AX; MicroPak-NH_2; Microsil NH_2; Nucleosil NH_2; Polygosil NH_2; RSIL NH_2; Separon Si NH_2; Sil 60 DNH_2; Spheri-5 Amino; Spherisorb NH_2; Spherisorb S5 NH; Techsil Amino; Zorbax-NH_2; μBondapak NH_2; μBondapak-Carbohydrate
(c) *Resins*		
Strongly acidic cation exchangers	$-SO_3^-$	Aminex A4–A7; Aminex HPX-87; Beckman AA; Benson BC–X; Chromex Cation; Dionex DC; Durrum DC; Hamilton HC; Hamilton HPX; Hitachi Gel 3011 S; Ionex SA; Shodex Cxpak
Weakly acidic cation exchangers	$-COO^-$	Benson BCOOH; Hitachi Gel 3011-C
Strongly basic anion exchangers	$-\overset{+}{N}(CH_3)_3$ or $-\overset{+}{N}R_3$	Aminex A14 to A28; AN-X; Benson BA-X; Chromex; Hamilton HA; Hitachi Gel 3011 N; Ionex SB; Shodex Axpak

TABLE 4.5.2 (continued)

Types	Functional groups	Trade names
Medium basic anion exchanger	$-\overset{+}{N}HR_2$	Benson BWA
(B) Packings for separation of biopolymers **		
Strongly acidic cation exchangers	$-SO_3^-$	Mono S, SP-Glycophase/CPG; Spheron S
Medium acidic cation exchanger	$-PO(O^-)_2$	Spheron-Phosphate
Weakly acidic cation exchangers	$-COO^-$	CM-Glycophase/CPG; CM TSK; Spheron C, Syn-Chropak CM300
Strongly basic anion exchangers	$-\overset{+}{N}R_3$	Mono Q, QAE-Glycophase/CPG; Spheron TEAE
Medium basic anion exchangers	$-\overset{+}{N}HR_2$	DEAE-Glycophase/CPG; DEAE TSK; MicroPak MAX; Spheron DEAE
Medium basic anion exchanger	Polymeric imine	SynChropak AX 300
Strongly and medium basic	Special ***	Mono P ***

* Detail lists and commentary on ion exchangers of this type summarized up to 1980 can be found in comprehensive reviews by Majors [73].

** A short commentary on these ion exchangers can be found in the text of section 4.5.2.3; for more detailed data see Refs. 6, and 37–39.

*** For chromatofocusing (cf. section 4.5.2.4).

Packings for HPLC of biopolymers are described or discussed in reviews [37–39] and in a lecture [41]. Scott [75] deals with the HPLC separation of complex biochemical mixtures.

Ion exchangers produced commercially and suitable for MPLC or HPLC of low molecular weight substances are summarized in the upper part of Table 4.5.2. *Pellicular ion exchangers* have very low pressure drop. This is an advantage from the point of view of instrumentation. They are prepared in two types: (1) glass beads coated with resins, or resins polymerized into the pores of silica gel; and (2) bonded to silica gel by means of silylation reaction. Exceptionally they have an alumina core. Type (2) is less chemically stable in alkaline media, is resistant only up to pH 7.5 (type (1) is stable up to pH 12). Neither can be used below pH 2. *Silica ion exchangers* possess low chemical stability (pH 2–7.5, sometimes 9; with 'presaturator columns' they can be used up to pH 10.5) and there are problems with purification and regeneration; no cycling is possible. Therefore they have short life-time. In addition to the packings mentioned, some highly crosslinked conventional microparticular *ion exchange resins* can be used for MPLC or HPLC of biochemically important substances. However, they have relatively high pressure drop and low degree of rigidity. The advantage of polystyrene-divinylbenzene ion exchangers is their chemical stability and high exchange capacity. Therefore they are suitable for rapid preparative applications, and can also be used in some automated analysers (e.g., for amino acids, peptides and sugars).

The ion exchange packings for HPLC of biopolymers can be classified into three main groups: (1) resins, (2) porous glass, and (3) porous silica.

Resins

In 1973 Čoupek et al. [76] prepared a macroreticular glycol-methacrylate copolymer, which was later described by Janák et al. [77] under the name Spheron (cf., also Borák et al. [78]), and also recently by Čoupek [79] under the name Separon HEMA. Mikeš et al. [80] have shown that Spheron can be used with advantage for the preparation of ion exchangers suitable for the rapid separation of biopolymers. The authors characterized the matrix [81], described preparation and properties of its DEAE-derivative [82], and of the carboxylic cation exchanger [83]. The applications of these Spheron ion exchangers have been reviewed [37,39]. The great advantage of these packings is their extreme stability (2 M sodium hydroxide or 2 M hydrochloric acid may be used for cycling, and organic solvents for washing), macroporosity (exclusion limit 10^6 Da), great capacity (1.5–3.5 meq/g), hydrophobicity (they do not denature biopolymers and do not manifest irreversible sorption), rigidity (beads can withstand more than 100 atmospheres) and they are not attacked by any known enzyme. They can be used not only for the separation of proteins and polypeptides, but also for oligonucleotides. For the separation of proteins Sain-Blancard et al. [84] described the application of Trisacryl gel (*N*-[tris(hydroxymethyl)-methyl]-acrylamide), and also of macroporous weakly cationic and anionic ion exchangers of this type.

Porous glass

Controlled pore glass (CPG, with defined macroporosity) was developed in 1965 by Haller [85], and is produced commercially in bead form. It has very large inner surface (tens to hundreds m^2/g). However, macroporous glass manifests irreversible sorption of some proteins. In spite of this Mizutani and Mizutani [86,87] have shown that in some cases it can act as a mild cation exchanger in the process of chromatography of proteins. Chang et al. [88,89] developed suitable bound phases on the outer and inner surface of porous glass (and also of porous silica) preventing the irreversible sorption of biopolymers. The most important is the hydrophilic layer called the 'glycophase'. The latter is essentially glycerol, covalently linked through one hydroxy group via propyl silane to the total bead surface

$$-\overset{|}{\underset{|}{S}}i-O-(CH_3)_3-O-CH_2-CH(OH)-CH_2OH$$

The silicious macroporous supports — glass and silica gel — modified in this way are free from irreversible sorption of proteins and can be used for size (steric) exclusion chromatography (SEC) under conditions required for HPLC. Moreover, one hydroxy group can be ionogenically modified, and ion exchangers are prepared in such a way. Ionogenic supports of this kind (and also others) have been reviewed by Regnier and Gooding [38].

Porous silica

Porous silica gel manifests irreversible sorption of some proteins in the same way as porous glass. Therefore Shechter [90] in 1974, in his paper on HPLC of proteins, described desactivation of Porous Silica Gel 1000 and Porasil DX by Carbowax before chromatographic application. In 1973 Eltekov et al. [91] first prepared ion exchange derivatives of macroporous silica for chromatography of proteins without irreversible sorption; they modified Silichrom C-80 by treatment with γ-aminopropyl triethoxysilane. Tayot et al. [92] prepared composite column packings, that combine the mechanical properties of Spherosil porous beads and the ion exchange properties of DEAE-dextran; they impregnated columns packed with Spherosil with DEAE-dextran, and in some cases fixed the impregnation by crosslinkage with 1,4-butanedioldiglycidyl ether, or with other bifunctional agents. The application of DEAE-silica (in addition to DEAE-CPG) for chromatography of glycolipids has been described [93], as has the separation of oligonucleotides [94]. Alpert and Regnier [95] developed a microparticulate anion exchange support, SynChropack AX 300, for the separation of proteins by coating the surface of silica beads with a continuous layer of crosslinked polyethylene imine; cf. also Vaněček and Regnier [96]. This packing has high capacity and has been found useful for the separation of enzymes and other proteins. The ion exchangers suitable for the separation of biopolymers are summarized in the lower part of Table 4.5.2.

4.5.2.4 Packings for ampholyte displacement and chromatofocusing

Ampholyte displacement chromatography [43] does not require special packings to be applied. Conventional ion exchangers are sufficient. The main principle is the elution of the column with carrier ampholytes. Leaback and Robinson [43] used CM-cellulose for separation of acetyl hexosamidase isoenzymes (Fig. 4.5.3), Young and Webb [44,50] separated serum proteins on DEAE-cellulose, and Chapuis-Cellier et al. [97] described ampholyte displacement chromatography of hemoglobins on DEAE-cellulose.

The method of chromatofocusing originated [45] from the application of an internal pH-gradient moving down the column and also in the bed of a conventional ion exchanger. In a later paper Sluyterman and Vijdenes [47] used low capacity DEAE Bio-Gel A and CM Bio-Gel A, and the proteins were eluted with Ampholine buffers. The described experiments led to the necessity of using ion exchangers with elevated nominal capacity, and therefore in the next communication [48] the maximum capacity DEAE-Sepharose CL was applied. Better results were achieved with this approach, but some negative aspects influenced the development of a new type of ion exchanger, described in the last paper [49] of this series. A special ion exchanger was synthesized: polyethylenimine attached to epoxide-activated Sepharose. The resolving power of this new ion exchanger PI-Sepharose was further improved, when some of the imino groups were converted into guanidino groups by modification with *O*-methylisourea. The ion exchanger PI-gu-Sepharose was suitable for separation at pH $\leq$ 9.2. The authors state [49] that a new ion exchanger with

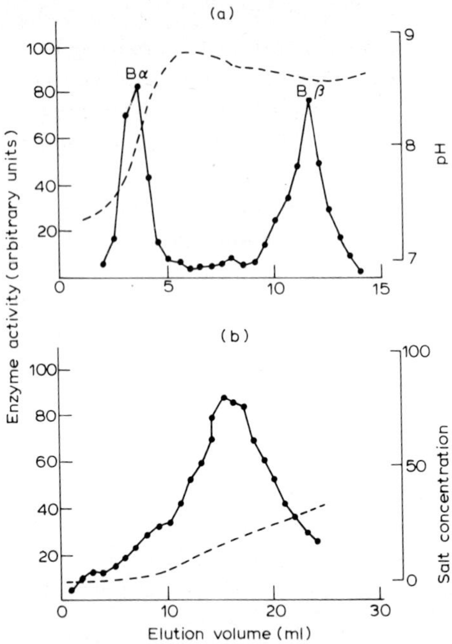

Fig. 4.5.3. An example of efficiency of ampholyte displacement chromatography (a) in comparison with conventional ion exchange chromatography (b) [43]. (a) β-*N*-acetyl-D-hexosaminidase activity) ●———●) and pH(- - - - - -) profiles in effluent from a column (45 × 11 mm) of CM-cellulose after loading with partially purified epididymal enzyme and eluting with 4% (w/v) (pH 8–10) Ampholine carrier ampholytes. (b) β-*N*-acetyl-D-hexosamidinase activity (●———●) and NaCl concentration (- - - - - -) profiles from a column loaded as above, but eluted with 0.02 M Tris/HCl buffer (pH 7.6) and an increasing concentration of NaCl.

higher buffer capacity should be devised for a higher pH range (9–10). Pharmacia [51] has tested two special ion exchangers for chromatofocusing and supplies them under the name ‘Polybuffer exchangers’: PBE 94 for the pH range 4–9, and PBE 118 for pH 7–10.5. The firm also offers special amphoteric buffers for this purpose (cf. section 4.5.3.3).

4.5.3 Mobile phase systems

4.5.3.1 Aqueous solutions and organic solvents

Ion exchange chromatography is usually carried out in *aqueous solutions*, because electrolytic dissociation in water splits the dissolved substances to be chromatographed into ions, which can then interact with the exchanger. Water also contributes to the dissociation of the functional groups and to the swelling of the exchanger, and in this way influences positively the rapid penetration of ions to the functional groups. After the admixture of miscible organic solvent (e.g., of alcohol), the swelling and the useful exchange capacity of resins is decreased. The attainment of equilibrium then takes much longer. In pure organic solvents the rate of exchange can be about three orders of magnitude less than in water. This is the reason why, for applications in organic solvents, macroreticular ion exchangers with large pores filled only with the relevant solvent proved best. On the contrary, organic solvents affect, to different degrees, the solvation of various ions to be separated and thus the selectivity coefficients and contribute in such a way in some cases to better quality of chromatographic separations. Therefore mixed or non-aqueous phases are sometimes tested. The most commonly used organic solvents for this purpose are alcohols (methanol, ethanol, propanol, ethylenglycol, glycerol), ketones (acetone, methylethyl ketone) and ethers (dioxane). Organic solvents are also used for partition chromatography on ion exchange columns, where two-phase systems are formed. They are sometimes even applied for the elution of substances which have lost their electrical charge, are not soluble in aqueous solutions, and remain sorbed in the bed of the ion exchanger. Generally, chromatographic methods in non-aqueous or mixed solutions are seldom used compared with chromatography in aqueous solutions.

There are two factors that must be considered carefully when using ion exchange chromatography in aqueous solutions: pH and ionic strength. Very acidic solutions regenerate the H^+ form of cation exchangers and completely convert anion exchangers to the corresponding salt forms. The same is valid with very alkaline solutions for the OH^- form of anion exchangers, and the salt form of cation exchangers. Under these extreme conditions no, or very limited sorption of other substances is possible. Therefore ion exchange chromatography is usually effective only in a certain range of pH values. Apart from this range, and for the reasons mentioned above, *diluted strong acids* can be used in special cases for displacement chromatography on anion exchangers, and for elution chromatography on cation exchangers. On the contrary *diluted strong bases* can be applied for displacement chromatography using cation exchanger columns, and for elution chromatography using anion exchangers. The top limit for true ion exchange on all cation exchangers is pH 12, the bottom limit for all anion exchangers is pH 2. Strongly acidic cation exchangers can operate from pH 2, moderately acidic from pH 4, and weakly acidic from pH 5. Strongly basic anion exchangers have the top limit roughly at pH 12, moderately basic at pH 8, and weakly basic at pH 6. This approach considered only the ion exchange and not the stability of the matrix. Therefore recommendation of procedures must be followed. In general, resins and cellulose are usually stable through the whole ion exchange range at laboratory or lower temperatures, polydextran derivatives are stable in the range pH 2–10, and agarose between pH 3–10. The stability of silica supports is discussed in section 4.5.2.3.

Ionic strength

Ionic strength (μ, a measure of the intensity of the electrical field due to the ions in the solution) strongly influences the nominal capacity of ion exchangers. High ionic strength (high concentration of buffers or salts) generally decreases the sorption of sorbed ions — very often zwitter-ions — also at non-extreme values of pH. Therefore the elevation of ionic strength is a common characteristic for method of elution. The most frequently used method of sorption and desorption when operating with buffered ion exchangers is the following: sorption at very low ionic strength $\mu = 0.05$ or 0.1 (which corresponds roughly to 0.05 or 0.1 M buffers with univalent ions), and a stepwise or gradient elution using buffers up to $\mu = 0.5$; higher values, e.g. $\mu \geqq 1.0$ are used only in special cases. At $\mu = 2$–3 regeneration usually proceeds. The pH is kept constant throughout the whole chromatographic process or increases with use of cation exchangers and decreases with anion exchangers.

The third factor of importance for chromatography is the *temperature of the mobile phase*. Most chromatographic experiments are carried out at room temperature. Higher temperature usually accelerates the speed of the chromatography, because it diminishes the viscosity of the solution and elevates the diffusion. As a rule, higher temperature does not change the exchange potential substantially, but diminishes the sorption due to Van der Waals forces. Therefore changes in temperature, without any other change in chromatographic conditions, can enable the separation in cases when the substances are bound to an ion exchanger in a more complicated way. When stable substances are chromatographed temperatures up to 100°C may be used. On the contrary many biochemical separations must be carried out at 1–4°C, which is usual when enzymes or other sensitive substances are chromatographed.

Mobile phase systems in LC were described and discussed by Mikeš and Vespalec [98], and solvent interactions in ion-exchange were reviewed by Pietrzyk [99]. Robinson [100] described thoroughly buffers for ion exchange columns used in amino acid analyses. Theory and practice of LC on ion exchangers using water as eluent and applied in saccharide analysis can be found in papers by Ladisch and Tsao [101] and Ladisch et al. [102]. The elution of cation exchange columns with the solutions of hydrochloric acid for the separation of amines has been described by Sharpe and Gray [103], and for citric acid cycle acids by Turkeson and Richards [104]. Thomasz [105] used two-dimensional cation-exchange thin layers (I, 1N HCl; II, phosphate buffer (pH 6.5)) for chromatography of nucleic acid bases. Replacement of similar types of ions in elution buffers sometimes contributes to better separations. A well known example of this effect is the sodium-lithium replacement described by Benson et al. [106] for buffers in amino acid analyzers in order to achieve better separation of glutamine and asparagine. Lithium-citrate buffers were used also by Otterburn et al. [107] for separation of isopeptides. A mixed counter-ion elution (sodium citrate and kalium perchlorate) from an anion exchanger has been found to be effective for sharp separation of cyclic ribonucleotides [108] and amino acid buffers have been used to prevent adsorption of proteins on controlled-pore glass [109].

Organic buffers

Chen et al. [110] applied heterocyclic cations (imidazole) and aromatic anions (*p*-hydroxybenzoic acid) in buffers for determination of biogenic amines using amino acid analyser. Also, Joy et al. [111] found organic buffers suitable for amino acid analysis. Ethanol suppresses non-ionic interactions and therefore was used by Asteriadis et al. [112] in solvent systems for separation of nucleotides and nucleosides on anion exchangers. Ethanol partly diluted with water served as eluting solvent in analysis of carbohydrates by HPLC; e.g., Lawrence [113] described *influence of counter-ion* on separation using this solvent and strongly acidic cation exchangers: a column in the trimethylammonium form was used for the separation of monosaccharides, the lithium form was the most suitable for disaccharides. Mopper [114] improved separation of sugars on anion exchanges using partition chromatography in ethanol. Alcohol-induced improvement of the separation of amino acids in automatic analyser was published by Warchalewski et al. [115]. Bruinsma and Le Tourneau [116] have shown that ethanol-formic acid-water eluents also proved effective in separation of the Krebs cycle acids with beaded diethylaminoethyl cellulose.

4.5.3.2 Volatile or complex-forming buffers (special additives)

Collected fractions from ion exchange chromatography separations often contain many ballast substances originated from inorganic and non-volatile organic buffers. Desalting is an additional process and sometimes presents problems. Therefore *volatile buffers* were sought, which decompose and escape in a rotary evaporator or by lyophilization. Thus, the pure solid substances of chromatographed peaks are obtained in a simple way. The most usual volatile buffers contain pyridine, picoline, *N*-substituted morpholine, aliphatic amines or amino alcohols as a basic component, and volatile organic acids (acetic acid or formic acid) as an acidic component. In some cases semi-volatile buffers are used, e.g., ammonium acetate or ammonium carbonate, which can be removed by sublimation (they are decomposed in a high vacuum and at elevated temperatures). Volatile buffers for preparative chromatography of peptides were described by Schroeder [117]. They can be used also in other fields of biochemistry. Some examples of their use in peptide separations have been given [98]. Pyridine-acetate buffers have been used for elution of purine organic acids quantitatively retained by DEAE-Sephadex [118], for effective peptide fractionation using an ion exchange resin for amino acid analyzer [119], for HPLC of thyrotropin-releasing hormone analogs [120], and when testing peptide separations on substituted polystyrene resins [121]. Breter [122] described an improved separation of various compounds related to purine and thiopurine metabolism using HPLC on strongly acidic cation exchanger and ammonium carbonate buffer.

Various organic *complexing agents* (polyvalent acids, hydroxy-, keto- and imino-derivatives of acids, derivatives of ketones) proved effective for the selective desorption of many inorganic ions. Examples of excellent successes in this field are ion exchange separations of rare earths or uranium fission products. On the contrary, some inorganic complexing agents were used with great advantage for separations in biochemistry. One of the most frequently used reagents is *boric acid* (or *borate salts*) reacting with vicinal hydroxy groups of organic substances, e.g., glycols. Complex anions formed in this way are much stronger acids than the less dissociated boric acid, and can be separated on anion exchangers with great success. Borate complexes are very often used for the separation of saccharides; for a review of this subject see Čapek and Staněk [123]. Novel applications have been described by Mopper [124], Torii et al. [125] and Wagner et al. [126]. Borate complex ion exchange chromatography in the presence of ethylenediamine for fluorimetric detection of saccharides was described by Mopper et al. [127]. The majority of the separations of sugars in borate buffers has been realized using microporous, strongly basic, anion exchange xerogel resins. On the contrary, Chytilová-Hostomská et al. [128] and Vrátný et al. [129] described the advantageous application of medium basic, rigid and macroreticular DEAE-Spheron for this purpose. Novel applications of sugar-borate complexes for separation of ribo-, deoxyribo- and arabino-nucleosides were published by Pal [130] and Moran and Werkheiser [131]; nucleosides were converted to corresponding anions by complexation with borate and then separated on strongly acidic cation exchangers (Aminex A6 or Dowex 50 eluted with ammonium borate). Complexation of boric acid with vicinal hydroxy groups has often been used also for ion exchange chromatography separations of pyrocatechine derivatives. Catecholamines (epinephrine derivatives, dopamine, serotonine, isoproterenol, etc.) have been separated on weakly acidic cation exchangers using phosphate buffers and borate or boric acid containing eluents [132–134], and a similar method has been used for analysis of catecholamines in urine [135].

A very interesting utilization of *cyclodextrin complexation* for separation of E, A and B prostaglandins was published by Uekama et al. [136]. Cyclodextrin, which forms an inclusion complex with prostaglandins, was added to the mobile phase (phosphate buffer), and significantly reduced retention times of prostaglandins when they were chromatographed on strong anion-exchange pellicular support. A new approach to the *resolution of optical isomers* of amino acids was published by Hare and Gil-Av [137]: the mobile phase contained L- or D-proline in acetate buffer with the admixture of cupric sulphate. The chiral eluent formed diastereoisomeric complexes with Cu and the unresolved amino acids. The isomers were then readily separable with a separation factor > 4 in some cases. The *copper complexes* of amino acids in human serum were used by Kahn and VanLoon [138] for atomic absorption spectrophotometry after ion exchange chromatography.

Special additives are used to improve properties of mobile phases for chromatography in biochemical separations. Antimicrobial reagents are the most often applied; a list of their characteristics is given in Ref. 98 (Table 10.8) and in Ref. 6 (Table 5.10). Commercial brochures also recommend special agents to keep the products offered free from microbial contamination. Detergents (e.g., Brij 25) are sometimes used for improving the wetting of the resins. Anti-oxidants (e.g., thiodiglycol, β-mercaptoethanol or dithiothreitol) are added to protect sensitive solutes against air oxidation. On the contrary, Repaske [139] presented *anaerobic column chromatography* systems for application where the reducing agents in elution buffers are not sufficient (cf., also Mikeš [140]). Poston et al. [141] described in detail the entire, strictly anaerobic, laboratory room used both for biochemical and for microbiological procedures. Another application of anaerobic ion exchange chromatography to the separation of hemoglobin conformers was published by Shih and Jones [142]. Urea up to 8 M concentration in buffers is often used to dissolve less soluble modified proteins or their large fragments and to keep them in solution. All of these special additives should be chosen for their ability not to firmly bind to the exchangers. The admixtures of alcohols or other organic solvents to mobile phases have already been discussed in section 4.5.3.1.

4.5.3.3 Amphoteric buffers for ampholyte displacement chromatography and chromatofocusing

In the process of ampholyte displacement chromatography [43] LKB 'Ampholine' carrier ampholytes were used. These amphoteric buffers are well known through their use in 'isoelectric focusing' or 'electrofocusing' [143,144]. Leaback and Robinson [43] eluted acetylhexosaminidase isomers from CM-cellulose using 4% (w/v) Ampholine buffers (pH 8–10). Young and Webb [44] applied Ampholine carrier ampholytes (pH 4–6 and 3.5–10) for the separation of serum proteins on DEAE-cellulose (cf., also [50]). Chapuis-Cellier et al. [97] fractionated hemoglobin variants on a DEAE-cellulose column equilibrated with 0.05 M Tris-HCl (pH 8.9); after the application of Ampholine buffer solution proteins were eluted in the order of decreasing isoelectric points. LKB 'Ampholine' carrier ampholytes are mixtures of a large number of amphoteric polyamino-polycarboxylic acids with different isoelectric points in the pH range 2.5–11 (for synthesis see Vesterberg [145]). After the synthesis they are fractionated by electrophoresis. They should possess the important characteristics of high buffering capacity, high conductivity, low UV-absorbance, they should not be biologically damaging, and should be easily removable from separated proteins. They are available in several different overlapping pH ranges, and still other ranges can be prepared by simple mixing. Similar to 'Ampholines' are other commercially available products designated 'Servalyte' and 'Bio-Lyte'. 'Pharmalyte®' are ampholytes synthesized by another method: they are prepared in a separate batch for each pH interval by condensation of glycine, or glycylglycine, with amines of defined p*K* values and with epichlorohydrine.

In spite of the fact that chromatofocusing originated [45] from the formation of 'internal pH-gradient' in a column of a conventional ion exchanger using conventional buffers, for its further development [46–49] not only were special ion exchangers synthesized (cf., section 4.5.2.4), but special amphoteric buffers were also applied. LKB 'Ampholines' were used for experiments with isoelectric elution [47,48], but in the last paper of this series [49] Pharmacia 'Pharmalyte' buffers (pH 8–10.5 and 6.5–9) were tested in addition to ethanolamine-propionic acid and Tris-propionic acid buffers for the elution from PI-Sepharose and PI-gu-Sepharose columns. The results obtained led the firm Pharmacia to the production of special Polybuffers® for chromatofocusing. Polybuffer 96 and Polybuffer 74 are recommended for ranges pH 9–6 and 7–4, respectively, together with Polybuffer exchanger PBE 94. For chromatofocusing exceeding pH 9 (more specifically for the range pH 8–11) Pharmalyte pH 8–10.5, and the Polybuffer exchanger PBE 118, should be used. The Polybuffers contain a large number of buffering substances and therefore show sufficient buffering capacity in the given interval. They have low absorption at 280 nm, and when applied with the exchangers recommended they allow to prepare nearly linear pH elution gradient without application of any gradient mixer. The focusing effect is efficient and the proteins can be eluted with peak widths as narrow as 0.05 pH units. More detailed information can be found in the cited papers and in the commercial brochure [51]. A great advantage of 'Polybuffers' is relatively low cost of application. They need not be used in high concentrations as is the case with the carrier ampholytes for ampholyte displacement. Because of these points chromatofocusing can now compete with isoelectric focusing, both for analytical and preparative purposes, as no special equipment is necessary.

4.5.4 Experimental techniques

Ion exchange experimental techniques have been described in monographs [1–6] and reviews [7–14]. Information on high-speed ion exchange chromatography techniques can be found in books [21–26] and assays [27,29,32,34,57,73–75]. For instrumentation of LC see also chapters 4.1–4.4 of this volume.

In principle biochemical substances can be separated by two different methods using ion exchangers: (a) by the application of *batch processes*, and (b) by *ion exchange chromatography*. The first is very simple from the experimental point of view — the solution of the substances to be sorbed is mixed slowly with the exchanger in a suitable form for a suitable time interval (let us say 30 min) and then the suspension is centrifuged, or the ion exchanger with the sorbed substance is filtered off. Necessary conditions for successful application are that the ionic strength (μ) of the solution should not be high (e.g., $\mu \leqq 0.1$ or $\leqq 0.05$), and the proper pH is chosen. Desorption is realized simply by washing the ion exchanger with a solution of high ionic strength (i.e., of $\mu \geqq 1$–2) and relevant pH. In the following sections some general approaches to chromatographic separation techniques are discussed, the possibilities of which are more varied than the batch process.

4.5.4.1 Principles of chromatographic separation procedures

True ion exchange is the most usual process in which one type of ions, attracted by electrostatic forces to functional groups of the ion exchanger, is interchanged by another type of ion with the same charge. Ion exchange is mostly combined with elution chromatography. Details of these processes are discussed in section 4.5.2.1.

Ion exclusion is the chromatographic separation of non-electrolytes (which do not interact with functional groups and penetrate the ion exchangers freely) or ions of weak electrolytes (which enter the grains of ion exchangers of the same charge relatively easily) from ions of strong electrolytes (which, having the same charge are repulsed, excluded and pass the grains without being retained). A typical classic example is the separation of hydrochloric and acetic acid on the sulphonic cation exchanger, Dowex 50 X8, in acid form (charges: R-SO_3^-; Cl^-; $CH_3COO^- \rightleftarrows CH_3COOH$, p$K = 4.5$), where Cl^- are excluded first.

Ion retardation is a process opposite in its result to ion exclusion. It is effected on special amphoteric 'snake-cage' resins (cf., section 4.5.2.1). These ion exchangers will bind electrolytes but not non-electrolytes. One of the first typical examples described by Hatch et al. [146] is the chromatographic separation of glycerol (which leaves the column first) from sodium chloride using Retardion 11A8. The salt retained can be washed from the column by elution with water only.

Ion sieving is a combination of molecular sieving and ion exchange. Sterically well defined inorganic ion exchangers have been used with advantage for this purpose. Organic resins, of suitable degree of crosslinking, can be used for sorption using this principle only if the ions to be separated are sufficiently different in size. Small ions can then diffuse into the grains to functional groups and large ones are excluded. Because there are functional groups also on the surface of ion exchanger beads, it is best to use large pearls, since they have higher volume-to-surface ratios and give better separation. This approach is now only seldom used for separation purposes, because *size exclusion chromatography* (SEC) and *gel permeation chromatography* (GPC) are better developed on the non-ionic principle. *Partition chromatography on ion exchangers* is often used for separation in biochemistry, especially in the field of saccharides. The swollen non-soluble matrix in contact with aqueous solution can serve as stationary phase for this purpose [101,102]. However, this two-phase system can be extended to many other forms using organic or mixed aqueous-organic solvents [114]. The affinity of non-ionic substances to be separated to different phases is governed in principle by the partition coefficients, thereby creating the conditions for partition chromatography. When modern packings are used (cf., section 4.5.2.3), such a system is usually called *normal phase chromatography* as opposed to *reversed phase chromatography*, and slightly basic anion exchangers of silica matrix (containing NH_2 groups) are often applied in the function of the phase with enhanced polarity.

Ligand exchange or *ligand chromatography* (Fig. 4.5.4) is a method using metal ions capable of forming complexes (e.g., Au^+, Cu^{2+}, Fe^{3+} or Ni^{3+}) attached firmly to exchanger for the chromatographic separation of so-called ligands. These substances, molecules of which coordinate metal ions with different affinities, can be exchanged and well separated. *Argentation chromatography* of this type is used for special separations in organic chemistry; silver forms complexes due to the interactions with π-electrons of double bonds of separated substances. For biochemical purposes the experiments of Seigel and Degens [147] are important, because these authors have shown, that Cu^{2+} bound to chelating resin, Chelex 100, can be used for the isolation of amino acids from sea water. These authors have thus extended the use of ion exchange chromatography to an area where this method could not previously be applied — to solutions of high ionic strength (e.g., the brines and

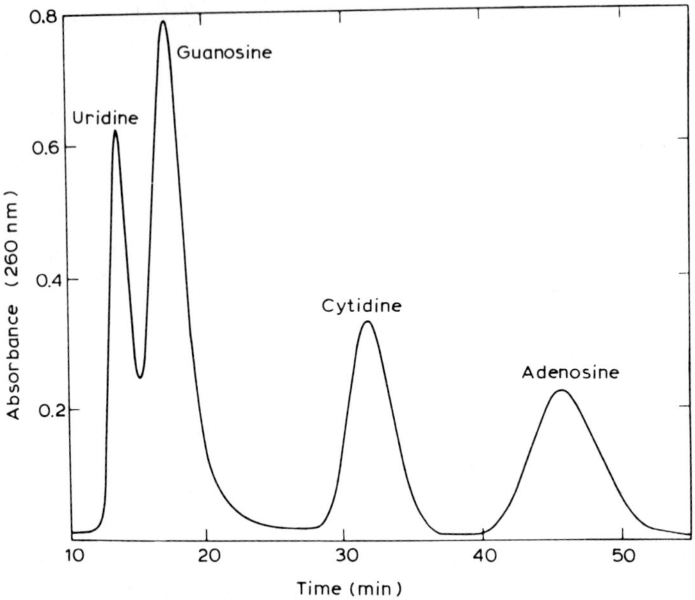

Fig. 4.5.4. An example of separation of uridine, guanosine, cytidine and adenosine by ligand exchange chromatography [151]. Column 33 × 0.9 cm, Chelex 100 resin in Cu^{2+} form, 200–400 mesh, elution with 1 M ammonium hydroxide, flow rate 0.77 ml/min.

saline). Buist and O'Brien [148] separated peptides from amino acids in urine by the same procedure. Ligand exchange separation of amino acids was systematically studied by Doury-Berthod et al. [149] using acrylic type cation exchanger BioRex 70 in the Cu^{2+} or Zn^{2+} form. Divan and Joshi [150] fractionated histones on the Al^{3+}-equilibrated cation exchanger Amberlite IR 120. Goldstein [151] isolated nucleic acid bases, nucleosides and nucleotides by ligand exchange chromatography. Shankar and Joshi [152] extended the studies to fractionation of RNA on Amberlite IR 120 in the Al^{3+}-form. Ion exchange separation of copper complexes of amino acids and peptides on DEAE-Sephadex was described by Sampson and Barlow [153]. Ivanova et al. [154] studied the sugar-borate complexes mentioned in section 4.5.3.2 in a novel form of binding: the boric acid was not present as a component in the buffer but it was covalently attached to the exchanger. Monosaccharides were chromatographed through a column containing *N,N*-diethyl-*N*-(*p*-methylphenylboryl) aminoethyl-Sephadex and anionic complexes formed were indicated by their relative retention volumes.

Bioaffinity elution (also known as *substrate* or *biospecific* elution) was described by Pogell [155], and later by Haar [156], and has recently been reviewed by Scopes [157]. In this process conventional ion exchangers are used for the sorption of proteins. The specific elution of a certain protein is enabled by its conformational or other stereochemical change in the ion exchange bed caused by a *biospecific interaction*. For instance, an ion exchange column is completely saturated with a raw enzyme preparation and successively eluted using an ionic strength gradient up to the moment in which the desired enzyme should be eluted. Before this stage the gradient is stopped and replaced by isocratic elution with a buffer, of such a composition that the desired enzyme remains on the column and other weakly bound impurities are completely washed out. Then a suitable enzyme substrate or a competitive inhibitor is added to the same eluting buffer. In many cases the desired enzyme is washed out in pure form. This method can compete with bioaffinity chromatography, because it does not require expensive specific packings and covalent immobilization of affinity ligands (affinants). Conventional ion exchangers are sufficient and universally applicable, because they can be well regenerated. *Ampholyte displacement chromatography* [43,44,50] is a special form of displacement chromatography of proteins on conventional ion exchangers in which amphoteric buffers are used. This process (cf., section 4.5.3.4) which, in some cases enables better separation in comparison with usual methods of elution, has not yet been fully theoretically explained. *Chromatofocusing* [45] is a method in which 'moving internal pH gradient' is formed in an ion exchange column by a proper combination of elution buffers so that proteins are arranged, focused and eluted in the order of their isoelectric points. In the final form of the method [46–49] special ion exchangers [51] (cf., section 4.5.2.4) and special amphoteric buffers [51] (cf., section 4.5.3.3) are used.

4.5.4.2 Choice of a suitable ion exchanger

Ion exchange cellulose, polydextran and agarose derivatives are used for the LC separations of biopolymers [19,20]. In some cases (meth)acrylate derivatives can also be applied for such a purpose. LC of other organic substances (various bases and acids, amino acids and peptides, nucleosides and nucleotides, sugars, etc.) can be realized using ion exchange resins (cf., Ref. 5, Tables 13.1–13.9, and Ref. 6, Tables 5.6–5.7). The degree of crosslinking is important; e.g., for the separation of amino acids X8 or X4 are suitable, but for larger peptides X2 should be used. For HPLC of biopolymers, ion exchange derivatives with Glycophase spread on controlled macropore glass or porous silica, or derivatives of macroreticular glycolmethacrylate polymers or microparticulate polyethylenimine derivatives of porous silica should be chosen (cf., Table 4.5.2B). For the selection of ion exchangers for HPLC of other substances the list in Table 4.5.2A should be consulted. A good survey of ion exchangers of all types suitable for chromatographic application in biochemistry can also be found in commercial catalogues, e.g., Ref. 158. For all types of chromatography, beaded monofunctional ion exchangers are preferred, as homogeneous in particle size as possible, with the lowest acceptable particle diameter from the point of view of pressure limitations. Basic substances are chromatographed in the form of cations on cation exchangers, acid substances in the form of anions on anion exchangers. Amphoteric substances in weakly acid medium behave like cations and can be separated on cation exchangers, whereas in weakly basic medium they form anions, and anion exchangers should be used for their separation. For the chromatography of proteins their isoelectric points (p*I*) are important. Acid proteins with low p*I* are preferentially sorbed on anion exchangers, the opposite is true for basic proteins of high p*I*. Of course, the sorption depends on pH used (cf., section 4.5.5.4). Generally, for most separations strongly acidic or basic ion exchangers are preferred. Weakly acidic or basic ion exchangers are used with success only for the chromatography of amphoteric biopolymers. Choice of suitable ion exchangers for biochemical applications has also been discussed by Neumann and Wallenborg [159].

4.5.4.3 Preliminary operations, equilibration (buffering) of ion exchangers, and filling or packing of chromatographic columns

Ion exchange resins of all types, macroporous ion exchangers of aerogel or hybrid type (e.g., Spheron derivatives), dry polydextran ion exchangers of xerogel type (Sephadex derivatives), and dry cellulose ion exchangers require some preliminary operations before chromatographic application. *Deaeration* of aerogels and cellulose derivatives can be simply realized by mixing the ion exchanger in a suitable salt form with the buffer solution containing the same counter-ion, which should coincide with the ions in the starting buffer for chromatography; the pH should be close to neutrality when cellulose is treated. The slurry is evacuated in a round-bottomed flask using a water pump and water bath with elevated temperature (40–50°C). After several minutes the initial 'boiling' stops and appearance of 'bumping' indicates the end of deaeration. The following is the procedure for *accelerated swelling* of polydextran derivatives: the ion exchanger is warmed at atmospheric pressure in a boiling water bath for 1–2 h. *Decantation* separates particles which are too fine from the main part of the packing. The slurry of the ion exchanger is mixed about 1 : 10 with the starting buffer, and the suspension is allowed to sediment in a cylindrical vessel. In a certain time (depending on the type of exchanger and the granulation) most particles settle down. The supernatant containing the fine particles is decanted. Some cellulose derivatives require this treatment. The process of decantation is repeated several times (if necessary) until the supernatant remains clear. Decantation cannot be used for purifying very fine packings (e.g., for HPLC), because suspensions of such materials do not sediment in a reasonable time.

Some commercial ion exchangers do not require all these procedures. Agarose ion exchangers [20], beaded DEAE-Sephacel [20] and some forms of microgranular advanced ion exchange celluloses [19] are delivered already in the preswollen state. The first two ion exchangers mentioned also do not require decantation. However, dry ion exchange celluloses, and also new commercial-grade resins which have not yet been used, require *pre-cycling*. Anion exchangers are cycled at laboratory temperature, first with 0.5 M hydrochloric acid, and after 30 min the acid is washed with water and the second treatment using 0.5 M sodium hydroxide follows, which is also washed after 30 min. The treatment of cation exchangers starts with the alkaline solution and ends with the acid. Pre-cycling improves the separation quality of dry

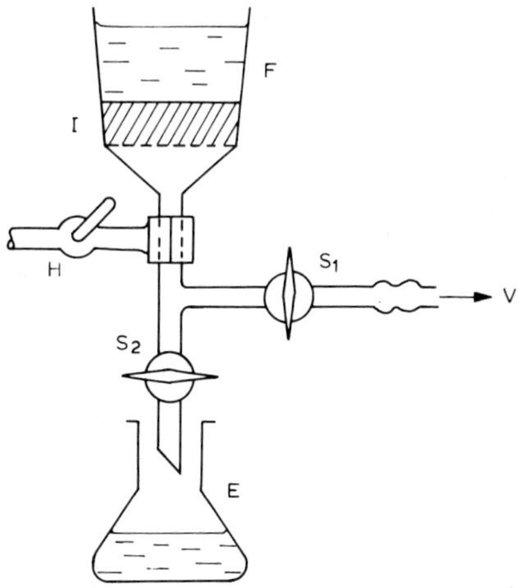

Fig. 4.5.5. Simple modification of sintered glass funnel for purposes of cycling, buffering (equilibration) and regeneration of ion exchangers. F, sintered glass funnel; I, ion exchanger; H, laboratory holder; S_1 and S_2, stop-cocks; V, vacuum water pump; E, conical flask for checking attainment of equilibrium. This arrangement proved to be successful for all the chemical treatments of ion exchangers.

ion exchange celluloses, purifies the new resins, and modifies their conformation in a proper way.

Buffering (equilibration) of ion exchangers can be realized by several different methods. The following procedure is often used in practice: the swollen and decanted ion exchanger in the form of the required counter-ion is stirred slowly with about a 5-fold amount of the starting buffer in a closed frita (Fig. 4.5.5). After 5 min the pH of the suspension is checked and, if it differs from the starting buffer, titration with an appropriate base or acid component of the buffer follows. After brief mixing and being left to stand for 5 min the supernatant liquor is filtered off and the ion exchanger is slowly mixed with the second portion of the starting buffer, which is also filtered off after 5 min. On addition of the third portion of the buffer the pH of the suspension is re-checked, and the process is repeated until the original pH and electrical conductivity of the starting buffer is reached.

The best method of *column filling for LC* is the connection of a chromatographic column with an extension tube or with a reservoir, which enables the ion exchanger suspension to be poured all at once. The column containing a little buffer should be fixed vertically in a place out of draught, direct sunlight or thermal radiation. The degassed suspension (slurry) is poured into the extension tube down a glass rod (or down the inner wall of the tube) to avoid formation of bubbles. The filling of the column and particle settling must be rapid enough to avoid separation of particles of different size; this is important especially with cellulose derivatives. The settling of ion exchange resins, of cellulose ion exchangers, and of relatively rigid gel derivatives (Sephadex A25 or C25, crosslinked Sepharose) can be accelerated by careful pumping of the starting buffer by a peristaltic pump through the top connector, using the same speed as in the chromatography or a little higher. After the column has been filled, care must be taken to prevent drying up (air must not enter the formed column from above). A layer of buffer should always be kept in the column above the top of the ion exchanger. The column should be washed before chromatography with the starting buffer used for equilibration and the pH, conductivity and UV-absorbance of the effluent should be compared with the influent. Small differences can be corrected by continuation in the washing.

Column packing for MPLC and HPLC cannot usually be brought into effect using the methods described above. The settling time for small particles is very long. Pellicular and other ion exchangers which do not swell can be filled into columns by a dry packing technique. The column is packed completely with dry material using gentle tapping and the air remaining in the bed is displaced by the pumped liquid influent. The other dynamic method called 'slurry technique' has several variants. In one of them thick slurry is displaced from a chamber mounted above the bed by the pumped buffer. In another method the ion exchanger is thrust into the column from a stainless-steel cartridge of larger diameter (where it is prepared in the form of a fixed bed) by the pressure of the influent. The slurry technique was modified by *stepwise pulse-packing* of Spheron ion exchangers into glass columns (of amino acid analyzers) used for MPLC of biopolymers [37]; a portion of the slurry in the starting buffer is introduced into a column and overlayered with the buffer, the head connector is mounted on the top of the column and settling is accelerated under pressure, using a flow rate approximately equal to that in chromatography. Occasionally the outlet from the column is stopped briefly and when the pressure has quickly risen to 1.5–2.5 MPa, it is suddenly released, and a pressed layer is formed. The supernatant is sucked off, a new portion of the slurry is added, and the process is repeated. These pulses led to a successive compact packing of the column. The highest pulse-packing pressure was 3 MPa. Values above this limit are dangerous for the glass column. The packing of stainless-steel HPLC columns specified for very high pressures cannot be recommended as a normal technique for biochemical laboratories, unless the operator is a specialist in the field. Such work requires not only special experience but also special equipment. The firms producing HPLC equipment deliver packed high performance columns with certificates of their tested efficiency.

4.5.4.4 Application of samples and methods of elution

In many applications of ion exchange chromatography *sorption* of sample components and their *desorption* are separate processes. In such a case the sample solution may be diluted and can be simply poured or pumped into the column. The components are sorbed in the form of a narrow zone in the upper part of the column. This is the case when the elution power of the sorption buffer is very low in comparison with the first elution buffer. If there is no guarantee that any of the sample components are eluted during the sample application, that is, in the process of *starting condition chromatography*, more concentrated sample solutions should be used, otherwise the eluted peaks are too wide. Two methods are used for preparative purposes in such a case.

(1) In the process of *overlayering* the sample solution is carefully applied, by means of a bent pipette above the top of an ion exchanger covered with a wet filter paper disc, to the empty space of the upper part of the column. When it is soaked — and this moment is the start of the chromatography — the walls are washed several times and after soaking the column is filled with the first buffer.

(2) In *underlayering*, the upper part of the column, above the top of ion exchanger (also covered with a paper disc), is full of the first buffer. The concentrated, cooled (and therefore heavy) sample solution is carefully introduced by means of syringe with a long needle at the bottom of the eluent just above the disc. In this way a layer of sample is formed above the ion exchanger and under the excess of elution buffer. After the head connector is mounted, the chromatography can start immediately. The time necessary for washing of inner walls is saved in this case.

In another technique of sample application a *three-way valve* is used, which enables the injection of the sample solution to the feed line. Automation of successive sample application at high pressures has been developed for analytical purposes on the basis of this procedure. Some commercial columns have a *septum* built into the head, which permits the application of the sample with a hypodermic syringe into the closed system just above the top of the bed, while the buffer is flowing through the column. This method is effective for high-speed analysis but it is not suitable for extremely high pressures. Generally applicable is the use of a *sample loop*. Usually a six-port valve and volume-calibrated capillary loops are used for this purpose. In modern amino acid analyzers, and in some liquid chromatographs, a special system has been developed, allowing automatic stepwise sample application of large series at suitable time intervals; the samples are stored in an attached refrigerator in the meantime.

Generally the elution can be realized in three different ways.

(1) *Isocratic* or *simple elution* (also called 'starting condition procedure') is often used for high-speed analytical separations. Only one buffer is applied in this case, and long columns are used in order to achieve a satisfactory separation. The great advantage of the isocratic elution method is that regeneration of the column is not necessary.

(2) For *stepwise elution* several buffers are applied with increasing pH (for cation exchange chromatography) or decreasing pH (for anion exchange columns). The pH need not change in particular cases, but the ionic strength increases in all instances.

(3) In *gradient elution* the composition of eluting buffers is changed continually. Additional 'stronger' buffer flowing from the reservoir is continuously mixed into the preceding 'weaker' buffer in a mixing chamber, from which the solution of continuously changing composition is pumped to the column. Special pumping systems can be used instead of mixing chambers for preparing the gradient of influent buffer, the composition of which can be changed according to a concave or linear shape (which is the most usual), or according to a convex or various

compound shapes (applied in specific cases). Because gradient elution is the most effective for ion exchange separations in biochemistry, it will be discussed here in a little more detail.

The method was proposed in 1952 by Alm et al. [160]. General description and calculation of gradients can be found in Ref. 98, and for a short review of gradient methods in ion exchange see Ref. 159. Further details can be found in the book specializing on gradient liquid chromatography by Liteau et al. [161]. Voelter and Bauer [162] described a special gradient-generating system for a high-resolution carbohydrate analyzer, and Robinson [163] published details of a gradient system with nine chambers for improved resolution of amino acids. An automatic gradient generator and gradient terminator controller were developed by Koester and Frank [164] for chromatography of protected oligonucleotides. A computer assisted buffer gradient system was used by Otterburn et al. [107] for separation of isopeptides. Davies et al. [165] described a microcomputer-controlled buffer gradient generator for ion exchange chromatography. Other elution methods were also studied. A discontinuous gradient elution from a miniature ion exchange column for the separation of isoenzymes of serum lactate dehydrogenase was used by Hsu et al. [166]. Porath and Fornstedt [63], when studying applications of dipolar ion exchangers, used '*pulse elution*', i.e., a short zone of eluting buffer moving down the column in the form of peak displacing the sorbed proteins. Pulse elution is quick and saves buffers. However, it can be applied with amphoteric ion exchangers only.

4.5.4.5 Evaluation of fractions

The evaluation of fractions can be realized immediately after the effluent leaves the chromatography column. Various detector systems are used for this purpose (cf., Krejčí et al. [167] and Meloun [168]). In some instances fractions are collected and tested only after the whole chromatographic procedure is finished. In such experiments the fractions should be stored in hermetically sealed containers in darkness and cold to prevent undesirable reactions and contaminations. For control of the course of ion exchange chromatography it is very important to measure *pH* and *electric conductivity* of every fifth or tenth fraction. Gradient elution defined by these parameters is well reproducible. In every case methods chosen for evaluation of fractions should be familiar, or tested, before the beginning of chromatography. *Detection methods* can be classified into the *destructive* and *non-destructive* group. Continuous monitoring of fractions by measuring UV absorbance at 280 nm (for detection of proteins), 220 nm (peptides) and 254–260 nm (nucleic acid constituents), as well as measuring refractive index, or absorbance or transmittance in the visible and IR regions, are examples of the second group. On the other hand applications of all chemical reactions in order to obtain coloured or fluorescent products of fraction components, or application of pyrolytic reactions, etc., and also bioassays and additional flat bed chromatographic or electrophoretic analyses of fractions, are considered destructive methods.

A short survey of various more recent evaluation methods is now given. In the field of amino acid analysis (cf., Zmrhal et al. [169]) new detection systems reviewed by Robinson [170] contributed to the high analytical standard. One of the substantial innovations in *ninhydrin colorimetry* is replacement of the capillary reactor by a glass bead packed bed reactor, leading to shorter reaction time and also peak narrowing (cf., Jonker et al. [171]). However, the most important contribution to amino acid analysis from the point of view of sensitivity is *fluorescence detection*. Roth [172] and Benson and Hare [173] described the application of *o*-phthalaldehyde (phthaldialdehyde) in the presence of 2-mercaptoethanol for this purpose. The method has been extended to the detection of peptides [174], and further improved [175]. Villanueva et al. [176] applied the *o*-phthalaldehyde method for the detection of mono- and polyamines in platelets, plasma and urine. Stein et al. [177] and Böhlen et al. [178] used fluorescamine (Fluoram®) for fluorimetric detection, and increased the sensitivity of the amino acid determination to picomole level. For a comparison of the *o*-phthalaldehyde, fluorescamine and ninhydrin methods of detection see Ref. 173. Rubinstein et al. [179] detected opioid peptides from rat pituitary using fluorescamine. Fluorimetric assays are often used for detection of many biochemically important substances after ion exchange chromatography. Measurement of biogenic amines and polyamines [180–182], catecholamines and tryptophan derivatives [183], and pyruvic and α-ketoglutaric acid [184] have been described using the sensitive fluorescence method. In the latter case the acids separated by anion exchange chromatography reacted with N^1-methylnicotinamide chloride and, after thermal treatment, fluorophores were formed. Kato and Kinoshita [185] described fluorimetric detection of carbohydrates using ethanolamine. The application of various aliphatic amines for automatic fluorimetric analysis of carbohydrates was studied by Honda et al. [186], and on-line post column fluorimetric detection of borate complexes of saccharides using ethylenediamine was used by Mopper et al. [127]. Toktamysova and Buzurbaev [187] applied a bioluminescence method for the determination of adenosine triphosphate in ascite tumor cells.

A survey of various destructive and non-destructive methods for the automated detection of carbohydrates was presented by Čapek and Staněk [123]. The destructive orcinol-sulphuric acid method was first used by Kesler [188] for the determination of free saccharides, and Kennedy and Fox [189] described newer fully automatic carbohydrate analysis using this type of detection. *p*-Hydroxybenzoic acid hydrazide was successfully applied as a colorimetric detection reagent for reducing sugars by Davies et al. [190], in their accelerated ion exchange chromatography analysis. A new detection principle for liquid chromatography of non-reducing oligosaccharides was described by Vrátný et al. [191]: the oligosaccharides are hydrolyzed by passing the effluent from chromatography separation through a small reaction column packed with a strongly acidic cation exchanger as a catalyst, and then detected as reducing saccharides using *p*-hydroxybenzoic acid hydrazide. A specific colour reaction for automated determination of carboxylic acids after anion exchange chromatography was tested by Kasai et al. [192]. When substances labelled with radioactive isotopes are separated, the continuous measurement of radioactivity is

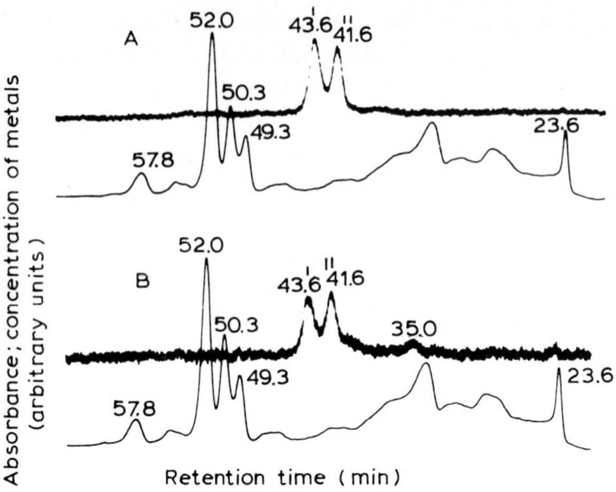

Fig. 4.5.6. An example of the combination of gel permeation (ion exchange) chromatography of metalloproteins with atomic absorption spectrometry for evaluation of fractions [196]. Rat liver supernatant (0.2 ml) obtained after injection of cadmium chloride was applied to a TSK Gel SW 3000 column 600×21.5 mm, and eluted with 50 mM Tris-HCl buffer solution (pH 8.6 at 25°C). Absorbance at 280 nm (lower curves) and concentration of cadmium (A) or zinc (B) (upper curves) were continuously monitored. I and II indicate metallothionein-I and -II, respectively. This chromatogram indicates the contribution of ion exchange to TSK Gel SW gel-permeation chromatography, because both separated proteins have the same molecular weight, and the sequence of peaks emerging corresponds to the ion exchange chromatography separation.

used. For the survey of detectors for this purpose see the reviews by Hais and Dršata [193] and Dybczynski [194]. Fenn et al. [195] described a liquid chromatography detector based on single and twin electrode thin-layer electrochemistry, and applied it in the determination of catecholamines in blood plasma after cation exchange chromatography. Atomic absorption spectrophotometry was used by Kahn and Loon [138] for the detection of Cu-amino acid complexes in human serum after their separation on ion exchange columns. Suzuki [196] also used this method for the detection of Cd and Zn in the effluent from gel permeation chromatography of metallothioneins to demonstrate the participation of ion exchange chromatography in the separation obtained (Fig. 4.5.6). This method can also be applied for the detection of other metalloproteins.

The simultaneous spectrophotometric detection at two different wavelengths can sometimes be useful. In the analysis of technical enzyme preparations Mikeš et al. [197,198] consistently used a series of two through-flow photocells, measuring at 285 and 254 nm, for the detection of the emergence of the effluent components after ion exchange chromatography on Spheron derivatives. In some cases this tandem system was able to distinguish the protein peaks from coloured contaminants at first sight. Similar systems can be used for evaluations of the first fractionation procedures of various biological extracts and fluids, because they can distinguish the nucleic acid fractions or their fragments from proteins. Munktell [199] described a new evaluation principle for liquid column chromatography of proteins, which consists of simultaneous monitoring of ion exchange column effluent at high and low concentrations, using two independent measuring systems at two wavelengths simultaneously. Schlabach and Abbott [200] followed HPLC-profiles of proteins also using detection methods at multiple wavelengths. Katz et al. [201] combined UV absorption and fluorescence in their 'dual detector' developed for comparative serum and urine analyses after cation exchange chromatography separation. Lou et al. [202] described a split-stream ion exchange chromatographic monitoring system used for isolating glycopeptides from biological fluids, allowing to obtain both ninhydrin and orcinol reaction patterns simultaneously.

The *determination of enzymes* in the chromatography effluents can be solved by routine analysis of the fractions collected. Technicon autoanalyzer and similar equipment (cf., short survey in [140], using sample splitters and full automation of various enzyme assays, are developed to a high degree of precision. For example Raschbaum and Everse [203] described an automatic detection instrument based on similar principles. Recently, a step forwards in biochemical detection systems was made by the authors, who developed post-column on-line enzyme detectors for evaluation of one type of enzyme activity only. Detectors of this type operate in the following manner: the column effluent is mixed with an appropriate substrate, the mixture is passed through a thermostated reactor, and the amount of the product is either measured in the ultraviolet region at a wavelength where proteins do not absorb, or else it is detected by fluorescence technique. Such detectors have undergone an intensive development and have been used mainly in clinical diagnostics for specific detection of groups of isoenzymes. There are two systems for utilization of these detectors. One system is based on a temperature controlled column packed with very fine non-porous hydrophilized glass beads functioning as the reactor [204–207]. The other system is characterized by the flow of the column effluent, together with the substrate, through a thermostated reaction capillary [208–212]. The influence of the background was eliminated by subtraction of the absorbance of the unreacted mixture from that of reaction mixture [209,210].

Nakatsu [213] described a fraction collector controller for separate collecting of liquid chromatography peaks.

4.5.4.6 Regeneration and storage of ion exchangers

Ion exchangers should never be stored for long periods in columns or in aqueous suspensions, as there is a risk of microbial infection. After the chromatography has been completed the ion exchangers should be taken out of the column, regenerated and stored. If there are some impurities in the top part of the bed, the coloured or contaminated thin layer of the ion exchanger should be discarded. The remaining part can then be extruded, washed thoroughly with 2 M sodium chloride and water. After this procedure cation exchangers remain in Na^+-form, anion exchangers in Cl^--form. If the last washings contained a conservation agent (cf., section 4.5.3.2), then all ion exchangers may be stored wet for short periods (e.g., for a month in cold); for this the surface water is sucked off using a sintered-glass funnel, and the wet ion exchanger is then stored in a hermetically sealed flask. Resins can be kept in this form for a longer time — a year and more. They should never be allowed to dry up completely. When they are to be used again, it is advisable to regenerate them before application. Polydextran derivatives cannot be stored for long in the wet form. They should be washed successively with gradually increasing concentrations of alcohol (up to 98%) on a sintered glass funnel, and the last traces of alcohol are eliminated by drying in the air. Attached particles which sometimes form break up again when the ion exchanger is next used. The cellulose ion exchangers can be stored in the wet form in cold provided that preservative agents are used. If cellulose derivatives were completely dried for very long storage, then cycling must be used before further chromatographic application (cf., section 4.5.2.2).

Labelling of stored ion exchangers should be as complete as possible. The following information should be given on the label: (1) type of exchanger; (2) name; (3) degree of crosslinking; (4) granulation or particle size; (5) ionic form; (6) data of regeneration; (7) name of the experimenter; and (8) number of experiment in which the ion exchanger was last used. This information secures repeatable utilization of the often expensive ion exchange materials and prevents useless waste.

4.5.5 Areas of application

In this section ion exchange chromatography applications in biochemistry are documented according to the substances to be separated, isolated or studied. A broader and more detailed survey of many substances of biochemical interest separated by chromatographic methods can be found in a comprehensive book on LC edited by Deyl et al. [214]. Nevertheless, some reviews and novel original references are given here. (See Note following References, p. 269.)

4.5.5.1 Biochemically important bases and acids

Separation of amines [215,216], biogenic amines [42], lower carboxylic acids [217] and higher carboxylic acids [218] has been previously described. The concise list of bases and acids separated by ion exchange chromatography is given in Table 4.5.3. This section does not consider amino sugars, sugar acids and sugar phosphates (cf., section 4.5.5.2).

4.5.5.2 Saccharides and their derivatives

A general essay on column chromatography of carbohydrates was written by Čapek and Staněk [123], the same authors [224] reviewed polysaccharides, and Juřicová and Deyl [225] summarized chromatographic separation of polysaccharide-protein complexes. Reviews specialized on ion exchange chromatography of oligosaccharides

TABLE 4.5.3

Ion exchange chromatography of biochemically important bases and acids

Compounds	Methodical notes	References
Bases		
Primary amines (peptides, pituitary extract neurophysin)	Preparative; automatic fluorescamine monitoring	Böhlen et al. [178]
Biogenic amines	Heterocyclic cation and aromatic anion elution buffers	Chen et al. [110]
	Cation exchange chromatography; fluorimetry	Atack [180]
Amines (separation from amino acids)	CM-cellulose; acid elution	Sharpe and Gray [103]
Amines (mono-, di-, poly-)	*o*-Phthalaldehyde fluorimetric detection; picomol level	Villanueva et al. [176]
Amines (poly-, histamine, 1-methylhistamine)	Automated cation exchange chromatography; fluorimetric detection	Perini et al. [181]
Amines (poly-, polypeptides)	Cation exchange chromatography; nanomole range	Radhakrishnan et al. [219]
Amines (poly-, amino acids, histidine dipeptides)	Cation exchange chromatography; *o*-phthaldehyde fluorimetric detecttion in picomol range	Nakamura et al. [220]
Natural polyamines (norspermidine, norspermine, putrescine, spermidine)	Cation exchange chromatography; fluorimetry; integrator	Adlakha et al. [182]
Catecholamines (dopamine, deoxyepinepherine, epinephrine, isoproterenol, norepinephrine, serotonine)	Sulfonic cation exchanger; fluorimetric detection	Marini et al. [183]
	Weak cation exchange resin; fluorimetric detection	Cebecauer et al. [135] Hamaji and Seki [133] Ogasahara et al. [134]
	Weak cation exchange resin	Seki [132]
Catecholamines	Strong cation exchanger; electrochemical detector	Fenn et al. [195]
	Strong cation exchanger	Atack and Magnusson [221]

TABLE 4.5.3 (continued)

Compounds	Methodical notes	References
Acids		
Carboxylic acids (of white wine)	Anion exchange chromatography; specific hydroxamate-ferric ion colour detection	Kasai et al. [192]
Organic acids (from urine and other fluids)	Anion exchange chromatography; volatile buffers elution	Chalmers [118]
Dicarboxylic acids	Anion exchange chromatography	Keier et al. [222] Lefevre et al. [223]
Pyruvic and α-ketoglutaric acid	Fluorimetric detection, fluorophores made by *N*-methylnicotinamide thermal reaction	Nakamura and Tamura [184]
Citric acid cycle acids	Strong cation exchange resin; acid elution; no regeneration	Turkelson and Richards [104]
	Anion exchange chromatography (beaded DEAE--cellulose); volatile ethanol-formic acid eluents	Bruinsma and Le Tourneau [116]

were written by Samuelson [226] and Kennedy and Fox [227]. Scott [75] dealt with HPLC of complex biochemical mixtures including sugars, and several papers were published describing methodical improvements in this field. Ladish et al. [101,102] discussed rapid LC of sugars using water as eluent, and Lawrence [113] HPLC (partition chromatography) of carbohydrates on a cation exchanger using various counter-ions. Improved partition chromatography on anion exchangers allowing separation of mixture of 18 monosaccharides was published by Mopper [114], while Palu and Samuelson [228] noticed the negative influence of some aromatic impurities from anion exchange resins interfering with sugar determination. Other contributions to saccharide analyses have already been cited, e.g., application of boroarylated DEAE-Sephadex [154], medium basic DEAE-Spheron [128] and DEAHP-starch [71]. Dawson and Mopper [229] developed a hybrid method using conventional amino acid analyzers and sugar analyzers for automatic specific determination of aminosugars. Kennedy and Fox [189] modified an amino acid analyzer for carbohydrate analysis. Fluorimetric [127,185,186], orcinol-sulphuric acid [188,189], *p*-hydroxybenzoic acid hydrazide [190] and hydrolytic [191] detection methods have been discussed in Section 4.5.4.5. Dawson and Mopper [229] applied a noncorrosive detection reagent containing $CuSO_4$, aspartic acid, Na_2CO_3 and Na-bicinchoninate

TABLE 4.5.4

Ion exchange chromatography of mono- and oligosaccharides and their derivatives

Compounds	Methodical notes	References
Mono- and oligosaccharides		
Reducing sugars	Accelerated analysis; anion exchanger (SO_4^{2-}); parahydroxybenzoic acid hydrazide	Davies et al. [190]
Sugars in body fluids; 16 carbohydrates in < 4h	Anion exchanger; orcinol-H_2SO_4 detection	Voelter and Bauer [162]
Hexoses and pentoses	Anion exchange chromatography; 10^{-10} mol sensitivity	Morrison et al. [231]
Reducing sugars and sugar alcohols	Simultaneous determination	Simutupang et al. [230]
Sugars and sugar alcohols from isomerization and hydrogenation of glucose	Anion exchange chromatography in borate buffers	Verhaar and Dirkx [232]
6-Deoxyhexoses	Anion exchange chromatography; microscale; single borate buffer	Wagner et al. [126]
Neutral monosaccharides and oligosaccharides	Combination of IEC and GPC	Kennedy and Fox [189]
Monosaccharides	DEAE-macroreticular polymer; borate buffers; orcinol-H_2SO_4 detection	Chytilová et al. [128]
Monosaccharides and oligosaccharides, analyses of molasses	DEAE-macroreticular polymer; borate buffers; orcinol-H_2SO_4 detection	Vrátný et al. [129]
Oligosaccharides	Anion exchange chromatography; borate buffers	Derevitskaya et al. [233]
Gluco-oligosaccharides	Preparative application of sugar autoanalyzer; anion exchange in borate buffers	Torii et al. [125]
Nonreducing oligosaccharides	Countinuous hydrolysis; *p*-hydroxybenzoic acid hydrazide	Vrátný et al. [191]
Starch hydrolysate	HPLC; methodical study	Richter and Woelk [234]
Cellodextrins, cellulose hydrolysate	TEAE- and DEAE-macroreticular polymer; borate buffers; orcinol-H_2SO_4 detection	Hostomská-Chytilová et al. [235]

TABLE 4.5.4 (continued)

Compounds	Methodical notes	References
Cyclodextrins	HPLC; strongly acidic cation exchanger (Ca^{2+}), refractometric detection	Hokse [236]
Amino sugars and amino alcohols		
Amino sugars	Hybrid analyzer; non-corrosive detection (cf., text)	Dawson and Mopper [229]
Galactosamine, glucosamine, mannosamine	Sulfonic acid cation exchanger	Chocha [237]
Anomeric aminodeoxyglycofuranosides and -glycopyranosides	Strongly acidic cation exchange resin; boric acid containing buffer	Hirano and Ishigami [238]
Hexosamines, hexosaminitols and amine-containing di- and trisaccharides	Reduction $NaBH_4$, *N*-deacetylation; anion exchange chromatography, borate-citrate buffers	Donald [239]
Amino sugars, amino alcohols and amino acids from bacterial cell-wall peptidoglycans	Reduction $NaBH_4$; amino acid analyzer	Hadzija and Keglevic [240]
Sugar acids		
Oligomeric aldonic acids	Strongly basic anion exchanger (CH_3COO^-)	Havlicek and Samuelson [241]
Oligomeric aldonic acids and oligomeric sugars	Strongly basic anion exchanger (CH_3COO^-)	Larsson and Samuelson [242]
Anionic oligosaccharide species (carboxylic and phosphate derivatives)	HPLC; weakly basic amino phase bound to silica gel; phosphate buffer	Baenziger and Natowicz [243]
Inositol phosphate intermediates (scyllo-inositol phosphates)	Anion exchange chromatography	Cosgrove [244]
Sugar phosphates and nucleotides from tissue extracts	Anion exchange chromatography; orcinol-H_2SO_4 detection	Blanshard [245]
Disaccharides from enzymic degradation of isomeric chondroitin sulphates	HPLC; weakly basic amino phase bound to silica gel; methanol ammonium formate eluents	Lee and Tieckelman [246]
Sulfated disaccharides of heparin	HPLC; weakly basic aminophase bound to silica gel; phosphate buffers	Delaney et al. [247]

for amino sugars, Simatupang et al. [230] used Neocuproin for ketosugars, an orcinol-sulphuric acid method for sucrose, and a periodate method for detection of alcoholic sugars. A concise list of mono- and oligosaccharides and their derivatives is given in Table 4.5.4. This survey considers amino sugars, amino alcohols, and sugar acids including sugar phosphates; however it does not contain separations of glycopeptides (cf., section 4.5.5.3) or glycoproteins (cf., 4.5.5.4), nucleosides and nucleotides (cf., 4.5.5.6), but sugar analyses of glycopeptides or glycoproteins are included.

4.5.5.3 Amino acids and lower peptides

Column chromatographic separation of amino acids have been reviewed by Zmrhal et al. [169], Davies et al. [248], Bober [249], Khan [250], Ersser [251], Robinson [100,170,252] and Dilley [253]. Scott [75] dealt with separations of complex biochemical mixtures including amino acids and Deyl and Juřicová [254] with amino acid derivatives. Peptides have been reviewed by Kluh [255], Stein [256] (essay specialized on neurobiology), Smith and McWilliams [35] (essay specialized on IE-HPLC), and Chin and Wood [257]. Various improvements have been published concerning both the resins used [66,67,121] and the composition of buffers [106,111,115]. Robinson [163] and Murren et al. [258,259] modified and improved single column analysis using gradient elution. Mal'tsev et al. [260] studied mathematical optimalization of automatic amino acid analysis, Charpentier [261] described rapid analysis (80 samples/24 h) of selected amino acids for studies of inborn errors of metabolism. Lee et al. [262] proposed a highly selective and inexpensive amino acid analyzer.

Detection methods both for amino acids and peptides [171,179] have already been mentioned (section 4.5.4.5), as have ligand exchange [147–149] (4.5.4.1), and separation of copper complexes of amino acids [138,153] (4.5.3.2). Mitchell et al. [263] described quantitative determination of D- and L-amino acids by the reaction with *tert*-butyloxycarbonyl-L-leucine *N*-hydroxysuccinimide ester and chromatographic separation as L,D- and L,L-dipeptides. Application of chiral eluents for separation of D- and L-enantiomers of amino acids [137] has been explained in 4.5.3.2. Hurlbut and Balka [264], and Heathcote et al. [265], used a combination of ion exchange and thin layer chromatography for separation of amino acids and short peptides. Volatile buffers for peptide fractionation were described in essays [117,119,121]. Separation of enzymic digests of proteins on microbore columns was studied by Machleidt et al. [266]. Table 4.5.5 summarizes other examples of ion exchange chromatography of amino acids and lower peptides including glycopeptides.

4.5.5.4 Proteins and their high molecular weight fragments

Column-chromatographic separation of proteins has been reviewed by Prusík [298] and Fasold [299], large scale chromatography by Janson [300] and Curling [301], separation of polysaccharide-protein complexes by Juřicová and Deyl [225], and fractionation of histones by Spring and Cole [302]. Chromatography of hemoglobin

TABLE 4.5.5

Ion exchange chromatography analysis or preparation of amino acids and peptides

Compounds	Methodical notes	References
Amino acids		
Separation of amino acids from amines	CM-cellulose; elution with 2 M HCl	Sharpe and Gray [103]
Separation of amino acids, amino sugars and amino alcohols (peptidoglycan components)	Amino acid analyzer; 3-buffer elution program	Hadzija and Keglevic [240]
Amino acid analyses	Conditions for high speed; single column	Fuzita et al. [267]
Labelled glutamine and other amino acids in whole blood	Amino acid analyzer; single column	Joo et al. [268]
Methionine and homocystine determination in deproteinized blood plasma or serum	Fast method using amino acid analyzer	Pospíšil [269]
Tryptophan and its metabolites in body fluids	HPLC; sulfonated cation exchanger	Saito et al. [270]
Halogen- and nitro-derivatives of aromatic amino acids	Amino acid analyzer; single column method	Ohtsuki and Hatano [271]
Thyroid hormones and their metabolites	Strongly acidic cation exchanger	Sorimachi [272]
γ-Carboxyglutamic acid determination in proteins via amino acid analysis	Alkaline hydrolysis; amino acid analyzer	Madar et al. [273]
Arginine-related metabolites in tissues	Strongly acidic and strongly basic ion exchangers; various detections	Gopalakrishna and Nagarajan [274]
Preparation of radioactive ε-*N*-methylated lysine via microbial methylation	Acid hydrolysis of protein; strongly acid cation exchanger	Paik et al. [275]
Preparation of amino acids with the chromatofuge	Strongly acid cation exchanger in the basket	Finley et al. [276]
Amino acids from extracts of tissues and body fluids	Simple semiautomatic small quantities preparation	Blanshard et al. [277]

TABLE 4.5.5 (continued)

Compounds	Methodical notes	References
Determination of amino acid composition of plant samples	Hydrolysates of TCA-precipitated proteins and of non-soluble proteins	Krishchenko [278]
Analysis of amino acids in biological material (blood, urine)	Methodical recommendations; 2-column amino acid analyzer; 55-ninhydrin-positive compounds were determined	Murayama and Shindo [279]
Free amino acids in biological fluids (diagnostic significance, metabolic disorders)	Amino acid analyzer	Gianek [280]
Free amino acids in the follicular fluids and serum of normally cycling and synchronized cows	Amino acid analyzer	Dave and Graves [281]
Amino acids in physiological fluids	Short programs for amino acid analyzers; various resins; economical aspects	Olek et al. [282]
Free amino acids in deproteinized blood plasma	A collaborative study	Williams et al. [283]
Peptides		
Polypeptides and polyamines	Cation-exchange-HPLC; nanomole range	Radhakrishnan et al. [219]
Separation of peptides from amino acids in urine	Ligand exchange; chelating ion exchanger in Cu^{2+} form	Buist and O'Brien [148]
Underivatized peptides; separation with quantitative yields	HPLC; SEC, RPC, IEC	Hollaway et al. [284]
Conditions for separation of di- to tetra--peptides	Amino acid analyzer	Nys et al. [285]
Analysis of products of abiogenic synthesis of peptides	GPC, IEC; separation on SP-polydextran; amino acid analyzer	Zayats et al. [286]
Digest of human collagen by Clostridiopeptidase A	HPLC-finger-prints on amino acid analyzer (cation exchanger resin)	Van der Rest et al. [287]

TABLE 4.5.5 (continued)

Compounds	Methodical notes	References
Tryptic digests of proteins formed in a cell free system	Doubly labelled (^3H; ^{14}C) protein hydrolysates; strongly acidic cation exchanger	Skarlat [288]
L,L- and L,D-dipeptides (cf., text of 4.5.5.3)	Amino acid analyzer; short column, basic buffers	Mitchell [263]
Isopeptides from heated proteins	Amino acid analyzer (computer assisted buffer gradient system)	Otterburn et al. [107]
Homo-oligopeptides (composed of alanine and glycine)	Amino acid analyzer	Oshima et al. [289]
Glycopeptides from biological fluids	Split-stream cation exchange chromatography; ninhydrin and orcinol detection	Lou et al. [202]
Peptides of cystine and glycine	Amino acid analyzer	Armstrong [290]
γ-Glutamyl peptides	Anion and cation exchange chromatography	James [291]
Histidine containing dipeptides, polyamines, amino acids	Cation exchange chromatography; fluorimetric detection in picomole range	Nakamura et al. [220]
Methionine containing peptides	Located modification by [^{14}C]iodoacetamine; cation exchange chromatography; thiolysis to eliminate the extra positive charge; second cation exchange chromatography yields pure peptides	Degen and Kyte [292]
Opioid peptides from rat pituitary (β-lipotropin)	HPLC; strong cation exchange groups bound to silica gel; fluorimetric detection	Rubinstein et al. [179]
Enkephalins	HPLC; strong cation exchange resin; electrochemical detection	Meek and Bohan [293]
Thyrotropin-releasing hormone analogs	Preparative HPLC; strongly acidic cation exchange resin; volatile buffers; ninhydrin detection	Voelter et al. [120]

TABLE 4.5.5 (continued)

Compounds	Methodical notes	References
Glutathione automated assay	Anion exchange resin in amino acid analyzer	Tabor and Tabor [294]
Glutathione in blood	HPLC; pellicular strong cation exchanger; mercury based electrochemical detector	Rabenstein and Saetre [295]
Aspartyl-phenylalanine methylester (sweetener) in food products	HPLC; pellicular strong cation exchanger	Fox et al. [296]
Carcinine (β-alanyl-histamine)	Determination by cation exchange chromatography	Arnould and Tankosic [297]

has been dealt with in the book by Schroeder and Huisman [303] and in the review by Atwater and Schwartz [304], and antigens involved in anaerobic infections in the review by Golshmid [305]. Applications of HPLC or MPLC in the study of separation of proteins were summarized in the essays [36–39] mentioned in 4.5.1.2.

Principles of chromatographic separation of proteins on ion exchangers were reviewed and discussed by Tiselius [306] in 1954. In addition to the literature cited above a concise recent approach can be found in commercial brochures [19,20]. Proteins are amphoteric substances with many positively and negatively charged groups on their surface. Their net charge changes with the changing pH of the solution, and at a certain pH-value (i.e., at their isoelectric point pI) they have zero net charge. At this point they should not be sorbed by any type of hydrophilic ion exchanger. The pI-values are therefore important characteristics for ion exchange chromatography of proteins, and about one thousand data can be found in *lists of isoelectric points of proteins* compiled by Malamud and Drysdale [307] and by Righetti et al. [308,309]. Below the pI proteins gain a positive net charge and can be chromatographed on cation exchangers, above the pI they have a negative net charge and anion exchangers should be used. The recommended pH for chromatography is a value minimally 1 pH-unit from the pI. For the selection of conditions for chromatography the range of stability of the particular protein must also be considered.

Detailed studies and methodical improvements have been published in this field. Torres and Peterson [310] have shown that separation of protein mixtures on ion exchangers (e.g., carboxymethyl dextrans) is fundamentally a displacement process. Dmitrenko et al. [311] studied equilibrium and kinetics of the sorption of proteins on macroporous ion exchangers, and Selezneva et al. [68] the distribution of proteins in the sorbed state inside grains of a cation exchanger. High-speed ion exchange chromatography of proteins has been developed and described in the papers already

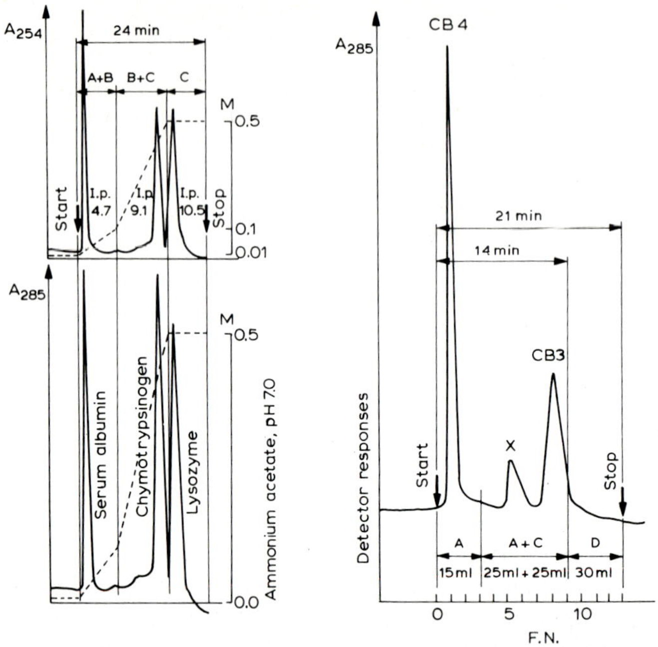

Fig. 4.5.7. An example of medium pressure rapid ion exchange chromatography of 16 mg of a protein mixture (left-hand figure) and of 18 mg of a mixture containing protein fragments (right-hand chromatogram) on 20×0.8 cm columns of CM-Spheron 20–40 μm [37]. Buffers: A, 0.01 ammonia and acetic acid (pH 7); B, 0.1 M ammonia+acetic acid (pH 7); C, 0.5 M buffer of the same composition was 1 M in NaCl; D, 2 M NaCl. The sequence of protein peaks corresponds to their isoelectric points p*I*. Cyanogen bromine fragments of human serum albumin CB4 and CB3 contain 31 and 185 amino acids.

cited using macroreticular glycolmethacrylate polymers [37,80] (Fig. 4.5.7), non-substituted controlled-pore glass [86,87,109], and bonded phase supports on pore glass and silica [88,89,95,96] (these results were briefly summarized by Chang [312]; cf., also Ref. 38 and a lecture by Mikeš [313]). Barford et al. [314] studied interactions of proteins with bonded phase ion exchangers. Porath and Fryklund [53] applied dipolar ion exchangers for the separation of proteins.

Ampholyte displacement chromatography and chromatofocusing [44–51] were discussed in detail in sections 4.5.1.2, 4.5.2.4 and 4.5.3.3. The column effluent monitoring for the detection of proteins [196–200] was described in 4.5.4.5. Tomana [315] developed microdetermination of monosaccharides in glycoproteins. Hashizume [316], Edwards and Wellington [317] and Horwath and Meretey [318] substantially improved methods for chromatographic separation of serum proteins. Jones and Phillips [319] demonstrated the possibilities of separations of egg white and bovine serum proteins. Examples of separation of other particular proteins or their groups including glycoproteins are listed in Table 4.5.6. (cf. Fig. 4.5.8).

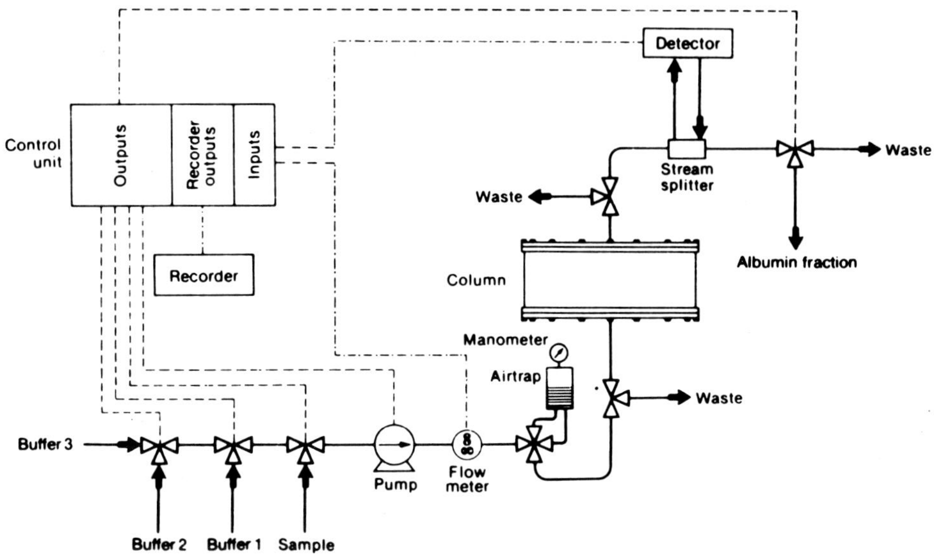

Fig. 4.5.8. An automated set-up for large scale fractionation of plasma proteins for production of serum albumin by ion exchange chromatography on DEAE-Sepharose CL-65 [422]. Stainless steel column details: Sephamatic® Gel Filter (Pharmacia Fine Chemicals AB, Sweden) 80×30 cm, bed volume 150 l, upward flow rate 100 l/h. The automation programme is volume based, but the conditions are automatically corrected according to the results of running UV-checking of the effluent, which overrides the volume programme.

4.5.5.5 Enzymes

Column separations of enzymes have been reviewed by Mikeš [140], analyses of enzymic reactions using HPLC by Studebaker [40], and an anaerobic enzyme laboratory was described by Poston et al. [141], an anaerobic column system by Repaske [139], and substrate-, biospecific- or affinity-elution of enzymes were reviewed by Pogel [155] and Von der Haar [156]. Guilbault [342] has written a handbook of enzymatic methods of analysis, dealing with assays for many enzymes. Raschbaum and Everse [203] described an instrument for automatic detection of enzyme activity in column chromatography effluents. Lim [343] developed an HPLC method for the determination of enzymes of the heme biosynthetic pathway. Post column enzyme detectors [204–212] discussed in section 4.5.4.5 were used for automated HPLC-analysis of serum lactate dehydrogenase isoenzymes [204–208,210,211], isoenzymes of creatine phosphokinase [204–207], alkaline phosphatase [206], hexokinase [206], aryl sulfatase (described by Bostik et al. [344]), and for detection of tryptic activity and of -SH proteins [205,206]. These detectors enable evaluation of isoenzyme peaks in the column effluent independently of the presence of other nonenzymic protein peaks. Examples of separations or chromatographic studies of particular enzymes or their groups are collected in Table 4.5.7.

TABLE 4.5.6

Ion exchange chromatography of proteins and their fragments

Compounds	Methodical notes	References
Various proteins		
Cytochrome *c* (example of one pioneer application of ion exchange resins for protein separation)	Mild acidic methacrylate microporous cation exchanger	Paléus and Neilands [12]
Plant proteins	Fractional separation with buffer solutions	Krishchenko [278]
Chick intestinal calcium-binding proteins	Anion exchange chromatography	Friedlander and Norman [320]
Sendai virus glycoprotein	Carboxymethyl-agarose cation exchange chromatography	Fukami et al. [321]
Chromosomal proteins	Analytical microscale cation exchanger procedure	Hesslinger and Pollow [322]
Fractionation of histones	Aluminium equilibrated strongly acidic cation exchanger	Divan and Joshi [150]
Microdetermination of the protein content of mucous glycoproteins	Modification of amino acid analysis	Carrol et al. [323]
Large scale purification of muscular parvalbumins	Anion exchange chromatography	Haiech et al. [324]
Smooth muscle proteins purification	Ion exchange column methods	Feramisco and Burridge [325]
Brain specific protein α-albumin (GFA)	Anion exchange and affinity chromatography	Noppe et al. [326]
Serum albumin (heterogeneity)	Anion exchange chromatography (both analytical and preparative procedures)	Janatová et al. [327]
BrCN-fragments of human serum albumin	MPCL; carboxymethyl derivative of hydrophilic macroreticular resin	Mikeš [37]
Fibrinogen and its degradation products	Anion exchange chromatography; purification and separation in the presence of calcium ions	Lawrie et al. [328]
Human globin chains	Carboxymethyl derivative of crosslinked agarose	Sparham et al. [329]

TABLE 4.5.6 (continued)

Compounds	Methodical notes	References
Haptoglobin and α_1-glycoproteins	Anion exchange chromatography	Schoessler et al. [330]
Hemoglobins		
Hemoglobin of *Moina macrocopa* (oxyHbs from metHbs)	GPC and anion exchange chromatography	Kobayashi [331]
Abnormal adults and newborn	Anion exchange chromatography using glycine-KCN-KCl developers	Abraham et al. [332]
A_{1a+b} and A_{1c}	Microchromatographic cation exchange column	Friedman and Humbert [333]
A_2 and Hb-variants	HPLC; anion exchange chromatography; microparticulate polyethylenimine derivative	Gooding et al. [334]
Separation of HBs in R from T conformational state	Anaerobic cation exchange chromatography	Shih and Jones [142]
Hb variants	Ampholyte displacement chromatography (using anion exchanger and carrier ampholytes)	Chapuis-Cellier et al. [97]
Glycosylated Hbs	Cation exchange chromatography, various improvements	Alboni et al. [335], Canham and Lockett [336]; Mosca et al. [337,338]
Large scale (or industrial) protein preparations		
Placental globulins and serum proteins	Composite column packings (anion exchange polydextran and porous silica gel)	Tayot et al. [92]
Purification of albumin	Anion and cation exchange chromatography	Curling and Bergloef [339]
Production of pyrogene free human albumin	Anion and cation exchange chromatography; aseptic conditions	Cunningham and O'Sullivan [340]
Proteins obtained in the manufacture of cheese and casein	Silica gel ion exchangers; economical considerations	Mirabel [341]
Immunoglobulin G and albumin from human plasma	Trisacryl derivatives: gel, anion exchanger and affinity sorbent	Saint-Blancard et al. [84]

TABLE 4.5.7

Ion exchange chromatography of enzymes

Compounds	Methodical notes	References
Various enzymes		
Endopolygalacturonanase	Crosslinked pectic acid cation exchanger	Kohn et al. [69]
Pectic enzymes	Crosslinked polyuronides	Rombouts et al. [345]
Human pepsins	Repetitive anion exchange chromatography	Roberts and Taylor [346]
Technical enzymes (protease, glucosa oxidase, pectic enzyme)	MPLC; ion exchange derivatives of hydrophilic macroreticular resin	Mikeš et al. [197,198]
Pectolytic enzymes	MPLC; ion exchange derivatives of hydrophilic macroreticular resin	Rexová-Benková et al. [347]
Cellulolytic enzymes (*Trichoderma viride-reesei*)	MPLC; ion exchange derivatives of hydrophilic macroreticular resin	Hostomská and Mikeš [348]
Cellulolytic enzymes (*Trichoderma viride--reesei*)	HPLC; weakly anion exchange groups bound to silica gel	Montenecourt et al. [349]
Pyruvate kinase	Weakly cation exchanger chromatography	Scopes [157]
Isoenzyme separations		
β-N-acetyl-D-hexosaminidase	Ampholyte displacement chromatography (weakly acidic cation exchanger, carrier ampholytes)	Leaback and Robinson [43]
Creatine kinase (tissue and serum)	Anion exchange chromatography	Mercer [350]
Creatine kinase	HPLC; pellicular anion exchanger	Kudirka et al. [351]
Creatine kinase	Medium basic anion exchanger chromatography	Klein et al. [352]
	Anion exchange chromatography and centrifugal analyzer	Elkins [353]
	Anion exchange chromatography and evaluation of all separation methods	Morin [354]
Lactate dehydrogenase	HPLC; anion exchange glycophase bound to porous glass	Kudirka et al. [355]
Multiple forms of soybean lipoxygenase-1	HPLC; cation exchange chromatography on modified silica gel	Aoshima [356]
Tissue and serum acid phosphatase	Medium basic anion exchange chromatography	Mercer [357]

TABLE 4.5.8

Ion exchange chromatography of nucleic acids and their constituents

Compounds	Methodical notes	References
Nucleic acid bases and their derivatives		
Electro-chemical oxidation products of biologically important purines	GPC; strongly basic anion exchanger	Cleary and Dryhurst [368]
Purine and 6-thiopurine metabolic products (22 compounds)	Strongly acidic cation exchanger, semivolatile buffers	Breter [122]
Nucleic acid bases in photographic gelatins	HPLC; cation exchange resins	Ohno et al. [369]
Nucleosides		
Ribo-, deoxyribo- and arabinonucleosides	Cation exchange chromatography after complexation with borate	Pal [130]
Ribo- and deoxyribonucleosides	Cation exchange chromatography after complexation with borate	Moran and Werkheiser [131]
Nucleotides		
Sugars, sugar phosphates and nucleotides present in tissue extracts	Anion exchange chromatography of borate complexes; orcinol-H_2SO_4 detection	Blanshard et al. [245]
Cyclic ribonucleotides, rapid sharp separation	Conventional anion exchanger with mixed counter-ion elution	Khym [108]
Nucleotide analysis using microcolumns	Strongly basic anion and strongly acidic cation exchangers	Vlasov et al. [370]
Pyrimidine nucleotides	Medium basic polydextran anion exchanger	Tonooka et al. [371]
Deoxyriboadenylic acids from hydrolysates of oxidized herring sperm DNA, preparative isolation	Strongly basic polydextran anion exchanger	Schott and Schwarz [372]
Nucleoside monophosphates, preparation of radioisotope labelled materials	Medium basic macroreticular glycolmethacrylate copolymer	Svoboda and Kleinmann [373]
Nucleoside mono-, di- and triphosphates	Strongly acidic cation exchanger, diluted HCl and NaCl elution ion-exclusion/exchange	Leigh and Cashion [374]

TABLE 4.5.8 (continued)

Compounds	Methodical notes	References
Individual nucleoside monophosphates, preparative isolation	Strongly acidic polydextran cation exchanger	Koshkina et al. [375]
Ribo- and deoxyribo-, mono-, di- and triphosphates	Rapid resolution using conventional anion exchange chromatography	Khym [376]
Adenine nucleotides, simple determination	Strong basic macroporous anion exchanger, mild HCl gradient	Hsu and Chen [377]
Adenosine- and riboflavin nucleotides (including cyclic AMP)	Strongly acidic cation exchange resin; semivolatile buffer	Brunius [378]
Phosphoric esters of guanosine, quantitative separation	Medium basic anion exchange column and thin layer chromatography	Alvarez Pez et al. [379]
Oligonucleotides		
Di- and trinucleotides in DNAse digest of DNA	HPLC; weakly basic anion exchange column	Wakizaka et al. [380]
DNA- and RNA-oligonucleotides	HPLC; anion exchange chromatography	Dizdaroglu and Hermes [381]
Pyrimidine oligonucleotides; specific partial hydrolysate of *Bacillus subtilis* DNA	Anion exchange derivatives of macroporous glycol-methacrylate copolymer	Šatava et al. [382]
Oligo- and monoribonucleotides, RNA-constituents	HPLC; weakly basic pellicular anion exchanger	Vandenberghe et al. [383,384]
tRNA oligonucleotides after RNAse digestion	HPLC; aminopropyl silica	Mc Laughlin et al. [385]
Synthetic oligonucleotide, analysis	HPLC; amino-silica column	Aukaty et al. [386]
Chemical synthesis of oligonucleotides, separation of trisubstituted pyrophosphates-active dinucleotide derivatives	Medium basic anion exchanger chromatography	Zarytova and Lebedev [387]
Enzymic synthesis of oligoribonucleotides, simple large scale methods	RP-anion exchange chromatographic system	Shum and Crothes [388]
Nucleic acid constituents		
Bases, nucleosides, nucleotides	Ligand exchange; chelating resin in Cu^{2+} form	Goldstein [151]

TABLE 4.5.8 (continued)

Compounds	Methodical notes	References
Bases, nucleosides, nucleotides (continued)	Thin layer ion exchange chromatography	Tomasz [105]
Oligonucleotides, nucleotides and nucleosides	Anion exchange resins, ethanol containing eluents	Asteriadis et al. [112]
Bases, nucleosides, nucleotides; nucleic acid hydrolysates	HPLC; strongly acidic cation exchange resin for bases and strongly basic anion exchange resin for other constituents	Demushkin and Plyashkevich [389]
Nucleic acid constituents	HPLC; silica based weak anion exchange column	Edelson et al. [390]
Nucleic acids		
Ribonucleic acids, fractionation study	Strongly acidic cation exchange resin in Al^{3+} form	Shankar and Joshi [152]
Nucleic acid related substances, analysis of extracts of leafy vegetables	Cation exchange chromatography	Tashiro et al. [391]
tRNAs	Comparison of RP and strongly basic anion exchange resin chromatography	Singhal et al. [392]
tRNAs (recovery study)	Medium basic cellulose anion exchanger; sorption in 2 M ammonium sulfate, elution with 1 M sodium chloride	Gillam and Tener [393]
High molecular weight DNAs fractionation	Benzoylated medium basic cellulose anion exchanger	Popov [394]
tRNAs	Medium basic anion exchange phase bound to silica gel	Wehr and Abbott [395]
Automated sequence analysis of synthetic DNA fragment	HPLC; strong anion exchange resin	De Rooij et al. [396]

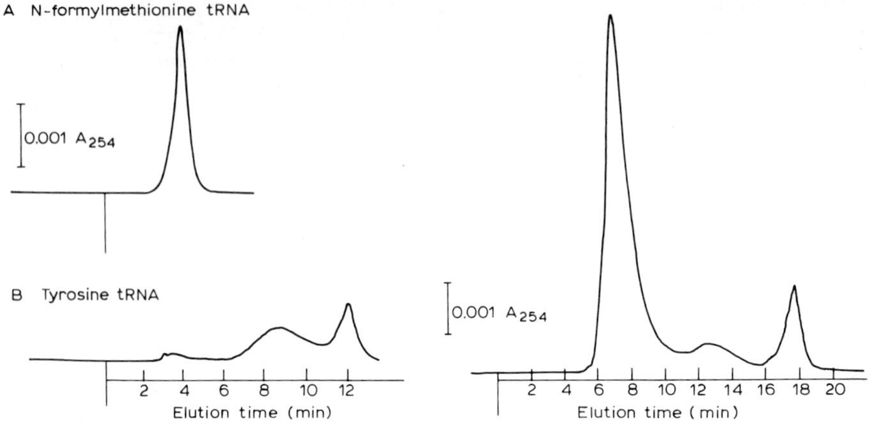

Fig. 4.5.9. Rapid separation of transfer RNA on MicroPak MAX-500 [395]. Left: selected tRNAs; right: rabbit liver tRNA extract. Conditions: Solvent A, 0.1 M Tris (pH 6.8); solvent B, 0.1 M Tris (pH 6.8)+1 M NACl. Gradient: 0% B 5 min, then 0–45% B in 15 min. Flow rate 1.0 ml/min (left); 0.5 ml/min (right), temperature 30°C.

4.5.5.6 Nucleic acids and their constituents

LC-separation of low molecular-weight constituents of nucleic acids and intact nucleic acids was reviewed by Zadražil [358,359]. Brown [360] described an 'enzyme peak shift method' verifying peak identities of nucleotides. Singhal [361] reviewed separation and analysis of nucleic acids and their constituents by ion-exclusion and ion exchange column chromatography, and Brown [31,362] summarized the latest developments and state-of-art in HPLC of nucleic acid constituents. Plunkett [363] dealt with the use of HPLC in research of purine nucleoside analogs.

Advances in development and application of IEC in this field have been published. Koester and Frank [164] described automated chromatography of protected nucleotides in connection with total chemical synthesis of a structural gene for human angiotensin II, and Leutzinger [364] also applied HPLC techniques in the synthesis of oligonucleotides. Perret [365] constructed a low pressure high-resolution nucleotide analyzer (~ 10 pmol sensitivity). Jost et al. [94] and Yanagawa [366] separated oligonucleotides and oligodeoxynucleotides using medium basic ion exchange phases bonded on silica gel. Toktamysova and Buzurbaev [187] applied a bioluminescence method for detection of ATP after anion exchange chromatography. Bujarski and Augustyniak [367] dealt with large-scale preparation of oligonucleotides from commercial sodium ribonucleate (from yeast RNA hydrolysates with pancreatic RNAse) using anion exchange chromatography. Other examples of separations of nucleic acids and their constituents are summarized in Table 4.5.8 (cf., also Fig. 4.5.9).

TABLE 4.5.9

Ion exchange chromatography of miscellaneous biochemically important substances

Compounds	Methodical notes	References
Glycolipids		
Ganglioside separation and mapping	Medium and strongly basic polydextran and agarose anion exchangers; silica gel thin layer chromatography	Iwamori and Nagai [407]
Gangliosides, neutral glycosphingolipids	Medium basic polydextran or silica gel anion exchangers, comparision	Kundu et al. [93]
Mono- to pentasialoganglioside fractions from human infant forebrain and cerebellum	Porous silica or glass covered with medium basic derivative of cross-linked polydextran	Fredman et al. [408]
Miscellaneous substances		
Antibiotics; lincomycin, streptomycin, kinamycin, monomycin	Sorption on weakly acidic cellulose cation exchanger	Kaputskii et al. [409]
Corticosteroid binder IB, evidence for physiological role	Medium basic polydextran anion exchanger minicolumns	Markovic et al. [410]
B-vitamins, automatic analysis	Strongly acidic cation exchange resins	Noe and Psallidi [411]
Prostaglandins A,B,E	Strongly basic pellicular anion exchanger; cyclodextrin complexation	Uekama et al. [136]
Mannosylretinyl phosphate from dolichylmannosil phosphate separation	Beaded medium basic cellulose anion exchanger	Sasak et al. [412]
Cytokinin glucosides from Ginkgo biloba leaves (zeatin glucoside, zeatin riboside)	Strongly acidic cation exchange resin	Van Staden [413]
Endogenous cytokinins from *Mercurialis annua* shoots	Medium acidic polydextran cation exchanger	Dauphin et al. [414]
Gibberellins, their conjugates and other phytohormones, abscisic acid	Medium basic polydextran anion exchanger	Graebner et al. [415]

TABLE 4.5.9 (continued)

Compounds	Methodical notes	References
Plant polyphenols	Weakly and medium basic anion exchanger celluloses and polydextran	Nagels et al. [416]
Neurochemicals in tissue samples (subpicomole levels)	HPLC; strong cation exchangers; thin layer electrochemical detector	Christensen and Le Roy Blank [417]
Cytostatic drug methotrexate	RPC and IEC; chemically bound anion exchanger to silica	Lankelma and Poppe [418]
Porphyrin analysis in urine	Strongly basic anion exchange resin	With and Pedersen [419]
Urinary acids from patients with amino acid metabolic errors	Strongly basic anion exchange resin extraction; combination with GC after methylation	Guarino et al. [420]

4.5.5.7 Other biochemically important substances

The separations of substances reviewed in section 4.5.5.1–4.5.5.6 are mostly realized using ion exchange chromatographic methods. However, for fractionation of other biochemically important substances ion exchange techniques are not usually the main separation principle, and therefore all examples are collected in this last common section. LC separations of some groups of such substances have been reviewed by the following authors: lipids by Pokorný [397], antibiotics by Betina [398] and Anhalt [399], ion exchange chromatography of antibiotics by Umezawa and Kondo [400] and Lotscher et al. [401], steroids by Procházka [402], vitamins by Davidek [403], alkaloids by Macek [404], alcohols and polyols by Churáček [405] and HPLC of complex biochemical mixtures, including indoles, by Scott [75].

Some methodical improvements have also been published in this field. Saito et al. [270] described an automated analysis of tryptophan and its metabolites (including β-indolepropionic and β-indoleacetic acid) using HPLC on a sulfonated cation exchanger. Marini et al. [183] studied simultaneous assays for 5-hydroxyindoleacetic acid, catecholamines and tryptophan in brain using strongly acidic cation exchange resin and fluorimetric detection. Katz et al. [201] used dual-detector anion exchange chromatography (by UV absorption and fluorescence monitoring) for comparative analyses of a large number of serum and urine constituents. Equipment for kinetic studies of photosynthesis based on anion exchange chromatography was developed by Laisk et al. [406]. Popova and Krachanov [70] resolved racemates of erhythro-3-

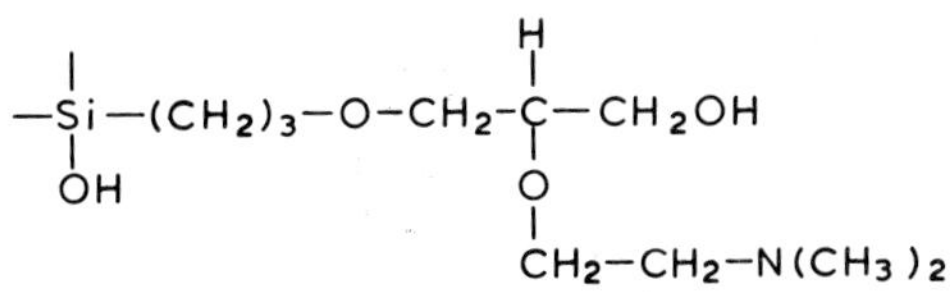

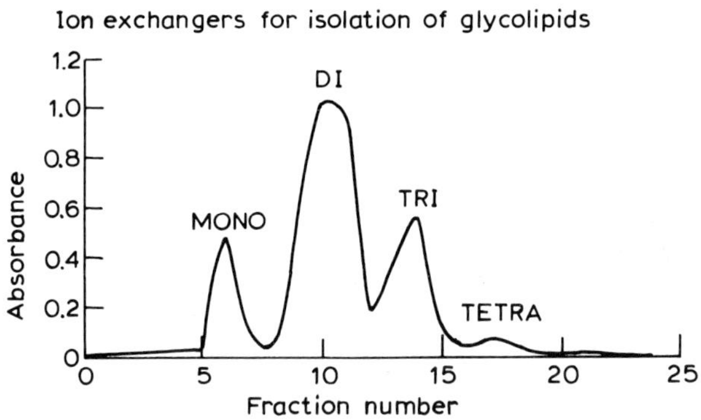

Fig. 4.5.10. Elution profile of total gangliosides from beef brain on DEAE-silica gel (acetate form) [93]. Bed dimensions 60×2 cm. Sample applied, 40 mg ganglioside mixture. Flow-rate 1.5 ml/min. Eluents: methanol, 0.2 M ammonium acetate in methanol and 0.5 M ammonium acetate in methanol; gradient elution. Fractions of 15 ml were collected. Resorcinol assay method for detection of sialic acid. Mono- to tetra- denote the numbers of sialic acid residues in the ganglioside fractions. The gangliosides were isolated after removal of ammonium acetate by dialysis against cold water.

amino-2,3-diphenylpropanoic acid methyl ester using a natural optically active ion exchanger. Other examples were collected into Table 4.5.9 (cf., also Fig. 4.5.10).

References

For papers published in a language other than English, and for journals not easily available in the majority of libraries, Chemical Abstracts (C.A.) citations are also added.

1 Helfferich, F.G. (1962) Ion Exchange, McGraw-Hill, New York.
2 Marinski, J.A. (ed.) (1966) Ion Exchange, M. Dekker, New York.
3 Walton, H.F. (1975) in Chromatography, 3rd Edn. (Heftman, E., ed.) pp. 312–43, 344–61, Van Nostrand-Reinhold, New York.
4 Walton, H.F. (1976) Ion Exchange Chromatography, Halsted Press, Stroudsburg, USA.
5 Mikeš, O. (1975) in Liquid Column Chromatography (Deyl, Z., Macek, K. and Janák, J., eds.) pp. 69–87, 325–68, Elsevier, Amsterdam.
6 Mikeš, O., Štamberg, J., Hejtmánek, M. and Šebesta, K. (1979) in Laboratory Handbook of Chromatographic and Allied Methods (Mikeš, O., ed.) pp. 218–333, Ellis Horwood Limited, Chicester, England.
7 Helfferich, F.G. (1965) Adv. Chromatogr. 1, 3–60.

8 Clayton, R.C. (1974) Rep. Prog. Appl. Chem. (Pub. 1875) 59, 253–61, C.A. 86, 78297x.
9 Brown, P.R. and Krstulovic, A.M. (1978) Tech. Chem. (N.Y.) 12, 197–255. C.A. 91, 79254w.
10 Gold, H. and Calmon, C. (1980) AIChE Symp. Ser. 76, 60–67. C.A. 92, 149028x.
11 Walton, H.F. (1976–1980) Anal. Chem. 48, 52R–66R; 50, 36R–50R; 52, 12R–27R.
12 Paléus, S. and Neilands, J.B. (1950) Acta Chem. Scand. 4, 1024–30.
13 Sober, H.A. and Peterson, E.A. (1954) J. Am. Chem. Soc. 76, 1711–12.
14 Peterson, E.A. and Sober, H.A. (1956) J. Am. Chem. Soc. 78, 751–55.
15 Peterson, E.A. (1970) Cellulosic Ion Exchangers, Elsevier, Amsterdam.
16 Porath, J. and Flodin, P. (1959) Nature 183, 1657–59.
17 Porath, J. and Lindner, E.B. (1961) Nature 191, 69–70.
18 Porath, J., Låås, T. and Janson, J.Ch. (1975) J. Chromatogr. 103, 49–62.
19 Anonymous (1981) Advanced Ion Exchange Celluloses; Laboratory Manual, Whatman Chemical Separation Ltd., Springfield Mill, Maidstone, Kent, England.
20 Anonymous (1980) Ion Exchange Chromatography-Principles and Methods, Pharmacia Fine Chemicals, Uppsala.
21 Kirkland, J.J. (ed.) (1971) Modern Practice of Liquid Chromatography, Wiley, New York.
22 Hamilton, R.J. and Sewell, P.A. (1977) Introduction to High Performance Liquid Chromatography, Chapman and Hall, London.
23 Pryde, A. and Gilbert, M.T. (1978) Application of High Performance Liquid Chromatography, Chapman and Hall, London.
24 Huber, J.F.K. (ed.) (1978) Instrumentation for High Performance Liquid Chromatography, Elsevier, Amsterdam.
25 Engelhardt, H. (1979) High Performance Liquid Chromatography. Chemical Laboratory Practice (in German), Springer-Verlag, Berlin.
26 Knox, J.H. (ed.) (1979) High-Performance Liquid Chromatography, Edinburgh University Press, Edinburgh, Scotland.
27 Brown, P.R. (1973) High Pressure Liquid Chromatography (Biochemical and Biomedical Applications), Acad. Press, New York.
28 Scott, C.D. (1974) Science 186, 226–33.
29 Rosset, R., Caude, M. and Jardy, A. (1976) Pract. High Perform. Liq. Chromatogr. 1976, 133–52.
30 Kremmer, T. (1977) Magy. Kem. Lapja 32, 71–75 (Hung.); C.A. 88, 2502f.
31 Brown, P. (1978) Chem. Aust. 45, 257–64. C.A. 90, 50699p.
32 Schmid, J. and Beschke, K. (1978) Arzneim. Forsch. 28, 1969–74 (Ger.). C.A. 90, 33660b.
33 Schaettle, E. (1978) LaborPraxis 1978, 36–38, 40 (Ger.). C.A. 91, 35023r.
34 Meyer, V. (1979) SLZ, Schweiz. Lab. Z. 36, 319–25 (Ger.). C.A. 92, 121268b.
35 Smith, J.A. and McWilliams, R.A. (1980) Int. Lab. 1980, 29–32.
36 Hollaway, W.I., Bhown, A.S., Mole, J.E. and Bennett, J.C. (1979) Chromatogr. Sci. Biolog. Biomed. Appl. 10, 163–76.
37 Mikeš, O. (1979) Int. J. Peptide Protein Res. 14, 393–401.
38 Regnier, F.E. and Gooding, K.M. (1980) Anal. Biochem. 103, 1–25.
39 Mikeš, O. (1981) Ernährung/Nutr. 5, 88–98 (Engl.); Chemické Listy 76, 59–79 (Czech.).
40 Studebaker, J.F. (1979) Chromatogr. Sci., Biol. Biomed. Appl. 10, 261–81.
41 Unger, K.K. and Roumeliotis, P. (1982) in Affinity Chromatography and Related Techniques (Gribnau, T.C.J., Visser, J. and Nivard, R.J.F., eds.) Analytical Chemistry Symposia Series, Vol. 9., pp. 471–82, Elsevier, Amsterdam.
42 Seiler, N. (1977) J. Chromatogr. 143, 221–46.
43 Leaback, D.H. and Robinson, H.K. (1975) Biochem. Biophys. Res. Commun. 67, 248–54.
44 Young, J.L. and Webb, B.A. (1978) Sci. Tools 25, 54–56.
45 Sluyterman, L.A.Ae. and Wijdenes, J. (1976) in Electrofocusing Isotachophoresis, Proc. Int. Symp. (Radola, B.J. and Graesslin, D., eds.) pp. 463–66, DeGruyter, Berlin (1977).
46 Sluyterman, L.A.Ae. and Elgersma, O. (1978) J. Chromatogr. 150, 17–30.
47 Sluyterman, L.A.Ae. and Wijdenes, J. (1978) J. Chromatogr. 150, 31–44.

48 Sluyterman, L.A.Ae. and Wijdenes, J. (1981) J. Chromatogr. 206, 429–40.
49 Sluyterman, L.A.Ae. and Wijdenes, J. (1981) J. Chromatogr. 206, 441–47.
50 Young, J.L. and Webb, B.A. (1979) in Protides of Biological Fluids, Proceedings of the 27th Colloquium (Peeters, M., ed.) pp. 739–42, Bruges.
51 Anonymous (1980) Chromatofocusing (with Polybuffer™ and PBE™), Pharmacia Fine Chemicals, Uppsala.
52 Ghersini, G. (1974) J. Chromatogr. 102, 299–307.
53 Porath, J. and Fryklund L. (1970) Nature 226, 1169.
54 Kressman, T.R.E. (1974–5) Rep. Prog. Appl. Chem. 59, 235–41. C.A. 85, 178221r.
55 Kressman, T.R.E. (1975) Inf. Chim. 142, 133–37 (Fr.). C.A. 83, 11237x.
56 Helfferich, F., Schmidt, E., Lieser, K.H., Naumann, G., Martinola, F. and Ladendorf, P. (1977) in Ullmanns Encykl. Tech. Chem. (Bartholome, E., Bicker, E. and Hellmann, H., eds.) Vol. 13, pp. 279–346, Verlag Chemie, Weinheim (Ger.). C.A. 89, 199578k.
57 Horvath, C. (1974) in Bonded Stationary Phases in Chromatography (Grushka, E., ed.) pp. 59–91, Ann Arbor Sci., Ann Arbor, Mich., USA. C.A. 82, 59192k.
58 Samuelson, O. (1963) Ion Exchange Separation in Analytical Chemistry, Wiley, New York.
59 Martin, A.J.P. and Synge, R.L.M. (1941) Biochem. J. 35, 1358–68.
60 Mayer, S.W. and Tompkins E.R. (1947) J. Am. Chem. Soc. 69, 2866–74.
61 Glueckauf, E. (1955) Trans. Faraday Soc. 51, 34–44.
62 Giddings, J.C. (1965) Dynamics of Chromatography, Marcel Dekker, New York.
63 Porath, J. and Fornstedt, N. (1970) J. Chromatogr. 51, 479–89.
64 Štamberg, J., Peška, J., Dautzenberg, H. and Philipp, B. (1982) in Affinity Chromatography and Related Techniques (Gribnau, T.C.J., Visser, J. and Nivard, R.J.F. eds.) pp. 131–41, Elsevier, Amsterdam.
65 Bogoczek, R. and Surowiec, J. (1978) Przem. Chem. 57, 51–56 (Pol.). C.A. 88, 177600p.
66 Rahm, J. (1975) J. Chromatogr. 115, 455–60.
67 Rahm, J. (1978) Chem. Listy 72, 381–96 (Czech.). C.A. 89, 19339b.
68 Selezneva, A.A., Dubinina, N.I., Fluknitskaya, O.F. and Samsonov, G.V. (1976) Kolloidn. Zh. 38, 1194–98 (Russ.). C.A. 86, 85564k.
69 Kohn, R., Rexová-Benková, L', Luknár, O. and Kuniak, L'. (1976) Coll. Czech. Chem. Commun. 41, 1879–93.
70 Popova, M. and Krachanov, K. (1979) J. Chromatogr. 175, 332–34.
71 Antal, M. and Toman, R. (1976) J. Chromatogr. 123, 434–38.
72 Smith, A.J. and Billett, M.A. (1979) Biochem. Soc. Trans. 7, 37980. C.A. 91, 136383p.
73 Majors, R.S. (1977, 1980) J. Chromatogr. Sci. 15, 334–51; 18, 488–511.
74 Rabel, F.M. (1979) Adv. Chromatogr. 17, 53–100.
75 Scott, Ch.D. (1974) Sep. Purif. Methods 3, 263–97.
76 Čoupek, J., Křiváková, M. and Pokorný, S. (1973) J. Polymer. Sci. 42. (part C) 185–90.
77 Janák, J., Čoupek, J., Krejčí, M., Mikeš, O. and Turková, J. (1975) Sorbents, in monograph cited in 214.
78 Borák, J., Čaderský, I., Kiss, F., Smrž, M. and Víška, J. (1978) in Chromatography of Synthetic and Biological Polymers, Vol. 1., Column Packings, GPC, GF and Gradient Elution (Epton, R., ed.) pp. 91–108, Ellis Horwood Ltd., Chichester, England.
79 Čoupek, J. (1982) in Affinity Chromatography and Related Techniques (Gribnau, T.C.J., Visser, J. Nivard, R.J.F., eds.) pp. 165–79, Elsevier, Amsterdam.
80 Mikeš, O. Štrop, P., Zbrožek, J. and Čoupek, J. (1976) J. Chromatogr. 119, 339–54.
81 Mikeš, O., Štrop, P. and Čoupek, J. (1978) J. Chromatogr. 153, 23–36.
82 Mikeš, O., Štrop, P., Zbrožek, J. and Čoupek, J. (1979) J. Chromatogr. 180, 17–30.
83 Mikeš, O., Štrop, P., Smrž, M. and Čoupek, J. (1980) J. Chromatogr. 159–72.
84 Saint-Blancard, J., Kirzin, J.M., Riberon, P., Petit, F., Fourcart, J., Girot, P. and Boschetti, E. (1982) in Affinity Chromatography and Related Techniques (Gribnau, T.C.J., Visser, J. and Nivard, R.J.F., eds.) pp. 305–12, Elsevier, Amsterdam.

85 Haller, W. (1965) Nature 206, 693.
86 Mizutani, T. and Mizutani, A. (1976) J. Chromatogr. 120, 206–10.
87 Mizutani, T. and Mizutani, A. (1979) J. Chromatogr. 168, 143–50.
88 Chang, S.H., Gooding, K.M. and Regnier, F.E. (1976) J. Chromatogr. 120, 321–33; 125, 103–14.
89 Chang, S.H., Noel, R. and Regnier, F.E. (1976) Anal. Chem. 48, 1839–45.
90 Shechter, I. (1974) Anal. Biochem. 58, 30–38.
91 Eltekov, Yu.A., Kiselev, A.V., Khokhlova, T.D. and Nikitin, Yu.S. (1973) Chromatographia 6, 187–89.
92 Tayot, J.K., Tardy, M., Gattel, P. Plan, R. and Roumiantzeff, M. (1976) in Chromatography of Synthetic and Biological Polymers, Vol. 2., (Epton, R., ed.) pp. 95–110, Ellis Horwood Ltd., Chichester, England.
93 Kundu, S.K., Chakravarty, S.K., Roy, S.K. and Roy, A.K. (1979) J. Chromatogr. 170, 65–72.
94 Jost, W., Unger, K.K., Lipecky, R. and Gassen, H.G. (1979) J. Chromatogr. 185, 403–12.
95 Alpert, A.J. and Regnier, F.E. (1979) J. Chromatogr. 185,375–92.
96 Vaněček, G. and Regnier, F.E. (1980) Anal. Biochem. 109, 345–53.
97 Chapuis-Cellier, C., Francina, A. and Arnaud, P. (1979) Protides Biol. Fluids (27th, Pub. 1980), 743-46. C.A. 92, 211107e.
98 Mikeš, O. and Vespalec, R. (1975) Mobile phases, pp. 233–80 in monograph cited in 214.
99 Pietrzyk, D.J. (1976) CRC Crit. Rev. Anal. Chem. 6, 131–70. C.A. 85, 178233w.
100 Robinson, G.W. (1978) in Amino Acid Determination, 2nd. Edn., (Blackburn, S., ed.) pp. 69–88 Dekker, New York.
101 Ladisch, M.R. and Tsao, G.T. (1978) J. Chromatogr. 166, 85–100.
102 Ladisch, M.R., Huebner, A.L. and Tsao, G.T. (1978) J. Chromatogr. 147, 185–93.
103 Sharpe, S. and Gray, D.O. (1976) J. Chromatogr. 120, 473–76.
104 Turkelson, V.T. and Richards, M. (1978) Anal. Chem. 50, 1420–23.
105 Tomasz, J. (1976) J. Chromatogr. 128, 304–8.
106 Benson, J.V., Jr., Gordon, M.J. and Patterson, J.A. (1967) Anal. Biochem. 18, 228–40.
107 Otterburn, M., Healy, M. and Sinclair, W. (1977) Adv. Exp. Med. Biol. 86B, 239-62. C.A. 88, 18473h.
108 Kym, J.X. (1978) J. Chromatogr. 151, 421–23.
109 Mituzani, T. and Mituzani, A. (1975) J. Chromatogr. 111, 214–16.
110 Chen, K.C.S., Buchanan, T.M., Davick P.R. and Holmes K.K. (1979) Anal. Biochem. 97, 282–85.
111 Joy, K.W., Shay, C. and McLimont, M.J. (1980) J. Chromatogr. 194, 76–79.
112 Asteriadis, G.T., Armbruster, M.A. and Gilham, P.T. (1976) Anal. Biochem. 70, 64–74.
113 Lawrence, J.G. (1974) Scan 5, 19–24. C.A. 85, 71766v.
114 Mopper, K. (1978) Anal. Biochem. 85, 528–33.
115 Warchalewski, J.R., Waliszewski, K. and Skupin, J. (1976) Bull. Acad. Pol. Sci., Ser. Sci. Biol. 24, 67–71. C.A. 85, 1983q.
116 Bruinsma, B.L. and Le Tourneau, D. (1978) J. Assoc. Off. Anal. Chem. 61, 751–52. C.A. 89, 38918f.
117 Schroeder, W.A. (1967) Methods Enzymol. 11, 351–61, 361–69.
118 Chalmers, R.A. (1976) Methodol. Dev. Biochem. 5, 135–40. C.A. 86, 67724h.
119 Johnson, P. (1979) J. Chromatogr. Sci. 17, 406–9.
120 Voelter, W., Bauer, H., Fuchs, S. and Pietrzik, E. (1978) J. Chromatogr. 153, 433–42.
121 Bradshaw, R.A., Bates, O.J. and Benson, J.R. (1980) J. Chromatogr. 187, 27–33.
122 Breter, H.J. (1977) Z. Naturforsch., C. Biosci. 32C, 905–7. C.A. 88, 34011m.
123 Čapek, K. and Staněk, J., Jr. (1975) pp. 465–522 in monograph cited in 214.
124 Mopper, K. (1978) Anal. Biochem. 87, 162–68.
125 Torii, M., Alberto, B.P., Tanaka, S., Tsumuraya, Y., Misaki, A. and Sawai, T. (1979) J. Biochem. (Tokyo) 85, 883–86. C.A. 90, 164157r.
126 Wagner, H., Habermeier, H. and Wegener, G. (1980) Planta Med. 39, 135–39 (Ger.). C.A. 93, 163946x.
127 Mopper, K., Dawson, R., Liebezeit, G. and Hansen, H.P. (1980) Anal. Chem. 52, 2018–22.
128 Chytilová, Z., Mikeš, O., Farkaš, J. and Štrop, P. (1978) J. Chromatogr. 153, 37–48.

129 Vrátný, P., Mikeš, O., Farkaš, J., Štrop, P., Čopíková, J. and Nejepínská, K. (1979) J. Chromatogr. 180, 39–44.
130 Pal, B.C. (1978) J. Chromatogr. 148, 545–48.
131 Moran, R.G. and Werkheiser, W.C. (1978) Anal. Biochem. 88, 668–74.
132 Seki, T. (1976) J. Chromatogr. 124, 411–14.
133 Hamaji, M. and Seki, T. (1979) J. Chromatogr. 163, 329–36.
134 Ogasahara, S., Mandai, T., Yamatodani, A., Watanabe, T., Wada, H. and Seki, T. (1979) J. Chromatogr. 180, 119–26.
135 Cebecauer, L., Svobodová, E. and Sebo, M. (1979) Biochem. Clin. Bohemoslov. 8, 109–13 (Slo.). C.A. 91, 170998c.
136 Uekama, K., Hirayama, F., Ikeda, K. and Inaba, K. (1977) J. Pharm. Sci. 66, 706–10. C.A. 87, 18337w.
137 Hare, P.E. and Gil-Av, E. (1978) Year book-Carnegie Inst. Washington 1978 (Pub. 1979) 655–58. C.A. 93, 21810z.
138 Kahn, N. and Van Loon, J.C. (1979) J. Liq. Chromatogr. 2, 23–36. C.A. 90, 134670k.
139 Repaske, R. (1971) Methods Enzymol. 22, 322–25.
140 Mikeš, O. (1975) Enzymes, pp. 807–30 in monograph cited in 214.
141 Poston, J.M., Stadtman, T.C. and Stadtman, E.R. (1971) Methods Enzymol. 22, 49–54.
142 Shih, T.-B. and Jones, R.T. (1980) Hemoglobin 4, 541–50. C.A. 93, 234240a.
143 Catsimpoolas, N., ed. (1976) Isoelectric Focusing, Academic Press, New York.
144 Anonymous (1979) Electrofocusing. Seminar Notes (Parts 1–3). LKB-Produkter AB, Bromma, Sweden.
145 Vesterberg, O. (1976) in Isoelectric Focusing (Catsimpoolas, N., ed.) pp. 53–76, Academic Press, New York.
146 Hatch, M.J., Dillon, J.A. and Smith, H.B. (1957) Ind. Eng. Chem. 49, 1812–19.
147 Seigel, A. and Degens, E.T. (1966) Science 151, 1098–1101.
148 Buist, N.R.M. and O'Brien, D. (1967) J. Chromatogr. 29, 398–402.
149 Doury-Berthod, M., Poitrenaud, C. and Tremillon, B. (1979) J. Chromatogr. 179, 37–61.
150 Divan, A.M. and Joshi, P.N. (1979) J. Chromatogr. 173, 373–78.
151 Goldstein, G. (1967) Anal. Biochem. 20, 477–83.
152 Shankar, V. and Joshi, P.N. (1974) J. Chromatogr. 95, 65–74.
153 Sampson, B. and Barlow, G.B. (1980) J. Chromatogr. 183, 9–15.
154 Ivanova, E.A., Panchenko, S.I., Kolodkina, I.I. and Yurkevich, A.M. (1975) Zh. Obshch. Khim. 45, 208–12 (Russ.). C.A. 82, 156599f.
155 Pogell, B.M. (1966) Methods Enzymol. 9, 9–15.
156 Von der Haar, F. (1974) Methods Enzymol. 34, 163–71.
157 Scopes, R.K. (1982) in Affinity Chromatography and Related Techniques (Gribnau, T.C.J., Visser, J. and Nivard, R.J.F., eds.) pp. 333–39, Elsevier, Amsterdam.
158 Anonymous (1981) Chromatography, Electrophoresis, Immunochemistry and HPLC. Bio-Rad Laboratories, Richmond, California.
159 Neumann, G. and Wallenborg, B. (1978) GIT Fachz. Lab. 22, 101-5 (Ger.). C.A. 88, 148318x.
160 Alm, R.S., Williams, R.J.P. and Tiselius, A. (1952) Acta Chem. Scand. 6, 826–36.
161 Liteanu, C., Gocan, S., Hodisan, T. and Nascu, H. (1974) Gradient Liquid Chromatography, Horwood, Chicester, England.
162 Voelter, W. and Bauer, H. (1975) Clin. Chem. (Winston-Salem, N.C.) 21, 1882–86. C.A. 84, 40360h.
163 Robinson, G.W. (1975) J. Chromatogr. 111, 416–18.
164 Koester, H. and Frank, R. (1976) Chromatographia 9, 497–501.
165 Davies, A.M.C., Prescott, E.H.A. and Stansfield, R. (1979) J. Chromatogr. 171, 117–33.
166 Hsu, M.-Y., Kohler, M.M., Barolia, L. and Bondar, R.J.L. (1979) Clin. Chem. 25, 1453–58.
167 Krejčí, M., Pechan, Z. and Deyl, Z. (1975) Instrumentation for Liquid Chromatography, pp. 101–68, in monograph cited in 214.
168 Meloun, B. (1979) in Chromatographic and Allied Methods (Mikeš, O., ed.) pp. 424–61, Horwood, Chicester, England.

169 Zmrhal, Z.J., Heathcote, J.G. and Washington, R.J. (1975) Amino acids, pp. 665–711, in monograph cited in 214.
170 Robinson, G.W. (1978) in Amino Acid Determination, 2nd Edn. (Blackburn, S., ed.) pp. 89–100, Dekker, New York.
171 Jonker, K.M., Poppe, H. and Huber, J.F.K. (1978) Chromatographia 11, 123–27.
172 Roth, M. (1971) Anal. Chem. 43, 880–82.
173 Benson, J.R. and Hare, P.E. (1975) Proc. Natl. Acad. Sci. USA 72, 619–22.
174 Craser, E.H. and Hughes, G.J. (1977) J. Chromatogr. 144, 69–75.
175 Lee, K.S. and Dresher, D.G. (1978) Int. J. Biochem. 9, 457–67.
176 Villanueva, V.R., Adlakha, R.C. and Giret, M. (1979) Feuill. Biol. 106, 141–46 (Fr.). C.A. 91, 2014k.
177 Stein, S., Böhlen, P., Stone, K., Dairman, W. and Udenfriend, S. (1973) Biochem. Biophys. 155, 203–12.
178 Böhlen, P. Stein, S. Stone, J. and Udenfriend, S. (1975) Anal. Biochem. 67, 438–45.
179 Rubinstein, M., Stein, S., Gerber, L.D. and Udenfriend, S. (1977) Proc. Natl. Acad. Sci. USA 74, 3052–55.
180 Atack, C. (1977) Acta Physiol. Scand. Suppl. 451, 1977 (Almqvist and Wikset, Goeteborg. Swed.) 1977, 99p. C.A. 88, 34210a.
181 Perini, F., Sadow, J.B., and Hixson, C.V. (1979) Anal. Biochem. 94, 431–39.
182 Adlakha, R.C. and Villanueva, V.R. (1980) J. Chromatogr. 187, 442–46.
183 Marini, J.L., Williams, S.P. and Sheard, M.H. (1979) Pharmacol., Biochem. Behav. 11, 183–87. C.A. 91, 171039c.
184 Nakamura, H. and Tamura, Z. (1979) Anal. Chem. 51, 1679–83.
185 Kato, T. and Kinoshita, T. (1980) Anal. Biochem. 106, 238–43.
186 Honda, S., Matsuda, Y., Takahashi, M., Kakehi, K., Honda, A., Ganno, S. and Ito, M. (1980) J. Pharmacobio-Dyn. 3, S-11, S-31. C.A. 93, 163927s.
187 Toktamysova, Z.S. and Buzurbaev, G.G. (1980) Izv. Akad. Nauk Kaz. SSR, Ser. Biol. 1980, p. 24–27 (Russ.). C.A. 93, 109906w.
188 Kesler, R.B. (1967) Anal. Chem. 39, 1416–22.
189 Kennedy, J.F. and Fox, J.E. (1977) Carbohydr. Res. 54, 13–21.
190 Davies, A.M.C., Robinson, D.S. and Couchman, R. (1974) J. Chromatogr. 101, 307–14.
191 Vrátný, P., Ouhrabková, J. and Čopíková, J. (1980) J. Chromatogr. 191, 313–17.
192 Kasai, Y., Tanimura, T., Tamura, Z. and Ozawa, Y. (1977) Anal. Chem. 49, 655–58.
193 Hais, I.M. and Dršata, J. (1975) Radiochromatographic Techniques, pp. 403–14, in monograph cited in 214.
194 Dybczynski, R. (1976) Nukleonika 21, 547-64. C.A. 86, 25433g.
195 Fenn, R.J., Siggia, S. and Curran, D.J. (1978) Anal. Chem. 50, 1067–73.
196 Suzuki, K.T. (1980) Anal. Biochem. 102, 31–34.
197 Mikeš, O., Štrop, P. and Sedláčková, J. (1978) J. Chromatogr. 148, 237–45.
198 Mikeš, O., Sedláčková, J., Rexová-Benková, L.' and Omelková, J. (1981) J. Chromatogr. 207, 99–114.
199 Munktell, G. (1979) Protides Biol. Fluids 1979 (Publ. 1980) 27th, 735–37. C.A. 92, 211105a.
200 Schlabach, T.D. and Abbott, S.R. (1980) Clin. Chem. 26, 1504–8.
201 Katz, S., Pitt, W.W., Jr. and Mrochek, J.E. (1975) J. Chromatogr. 104, 303–10.
202 Lou, M.F., Morrison, W.H. and Dueber, Ch. (1980) Biochem. Med. 23, 47–54. C.A. 93, 3219j.
203 Raschbaum, G.R. and Everse, J. (1978) Anal. Biochem. 90, 146–54.
204 Chang, S.H., Gooding, K.M. and Regnier, F.E. (1976) J. Chromatogr. 125, 103–14.
205 Schlabach, T.D., Chang, S.H., Gooding, K.M. and Regnier, F. (1977) J. Chromatogr. 134, 91–106.
206 Schlabach, T.D. and Regnier, F.E., (1978) J. Chromatogr. 158, 349–64.
207 Schlabach, T.D., Alpert, A.J. and Regnier, F.E. (1978) Clin. Chem. 24, 1351–60.
208 Schroeder, R.R., Kudirka, P.J. and Toren, E.C., Jr. (1977) J. Chromatogr. 134, 83–90.
209 Fulton, J.A., Schlabach, T.D., Kerl, J.E., Toren, E.C., Jr. and Miller, A.R. (1979) J. Chromatogr. 175, 269–81.
210 Fulton, J.A., Schlabach, T.D., Kerl, J.E. and Toren, C., Jr. (1979) J. Chromatogr. 175, 283–91.

211 Schlabach, T.D., Fulton, J.A., Mockridge, P.B. and Toren, E.C., Jr. (1979) Clin. Chem. 25, 1600–8.
212 Schlabach, T.D., Fulton, J.A., Mockridge, P.B. and Toren, E.C., Jr. (1980) Anal. Chem. 52, 729–33.
213 Nakatsu, K. (1981) J. Chromatogr. 212, 105–7.
214 Deyl, Z., Macek, K. and Janák, J., eds. (1975) Liquid Column Chromatography. J. of Chromatography Library, Vol. 3. Elsevier, Amsterdam.
215 Deyl, Z. (1975) Amines, pp. 637–56 in monograph cited in 214.
216 Churáček, J. (1975) Other Non-heterocyclic Nitrogen Compounds, pp. 657–64 in monograph cited in 214.
217 Churáček, J. and Jandera, P. (1975) Lower Carboxylic Acids, pp. 543–73 in monograph cited in 214.
218 Pokorný, J. (1975) Higher Carboxylic Acids, pp. 575–80 in monograph cited in 214.
219 Radhakrishnan, A.N., Stein, S., Licht, A., Gruber, K.A. and Udenfriend, S. (1977) J. Chromatogr. 132, 552–55.
220 Nakamura, H., Zimmerman, C.L. and Pisano, J. (1979) Anal. Biochem. 93, 423–29.
221 Atack, C. and Magnusson, T. (1978) Acta Pharmacol. Toxicol. 42, 35–57. C.A. 89, 175864k.
222 Keier, B.R., Lanin, E.P., Cherepov, A.G. and Plachenov, T.G. (1975) Sb. Tr. Leningr. Tekhnol. Inst. Im. Lensoveta 1975, 112–17 (Russ.) C.A. 85, 130962c.
223 Lefevre, J.P., Caude, M. and Rosset, R. (1976) Analusis 4, 16–24 (Fr.). C.A. 85, 13490q.
224 Čapek, K. and Staněk, J., Jr. (1975) Polysaccharides, pp. 523–28, in monograph cited in 214.
225 Juřicová, M. and Deyl, Z. (1975) Polysaccharide-Protein Complexes, pp. 529–42, in monograph cited in 214.
226 Samuelson, O. (1978) Adv. Chromatogr. 16, 113–50.
227 Kennedy, J.F. and Fox, J.E. (1980) Methods Carbohydr. Chem. 8, 3–12.
228 Palu, M. and Samuelson, O. (1977) Talanta 24, 264–66.
229 Dawson, R. and Mopper, K. (1978) Anal. Biochem. 84, 191–95.
230 Simatupang, M.H., Sinner, M. and Dietrichs, H.H. (1978) J. Chromatogr. 155, 446–49 (Ger.). C.A. 89, 175863y.
231 Morrison, W.H., Lou, M.F. and Hamilton, P.B. (1976) Anal. Biochem. 71, 415–25.
232 Verhaar, L.A.T. and Dirkx, J.M.H. (1977) Carbohydr. Res. 59, 1–10.
233 Derevitskaya, V.A., Arbatskii, N.P. and Kochetkov, N.K. (1975) Dokl. Akad. Nauk. SSSR 223, 1137–39 (Russ.). C.A. 84, 31320t.
234 Richter, K. and Woelk, H.U. (1977) Staerke 29, 273–77. C.A. 87, 153670q.
235 Hostomská-Chytilová, Z., Mikeš, O., Vrátný, P. and Smrž, M. (1982) J. Chromatogr. 235, 229–36.
236 Hokse, H. (1980) J. Chromatogr. 189, 98–100.
237 Chocha, A. (1977) Diagn. Lab. 13, 373–78 (Pol.). C.A. 89, 215682x.
238 Hirano, S. and Ishigami, M. (1977) Carbohydr. Res. 54, 139–41.
239 Donald, A.S.R. (1977) J. Chromatogr. 134, 199–203.
240 Hadzija, O. and Keglevic, D. (1977) J. Chromatogr. 138, 458–60.
241 Havlicek, J. and Samuelson, P. (1975) J. Chromatogr. 114, 383–88.
242 Larsson, K. and Samuelson, O. (1977) J. Chromatogr. 134, 195–98.
243 Baenziger, J.U. and Natowicz, M. (1981) Anal. Biochem. 112, 357–61.
244 Cosgrove, D.J. (1977) Cyclitols Phosphoinositides (Proc. Symp.) (Pub. 1978) (Wells, W.W. and Eisenberg, F., Jr., eds.) pp. 23–33, Academic Press, New York, C.A. 91, 206557z.
245 Blanshard, K.C., Das, I. and Thomas, A.J. (1977) Anal. Biochem 83, 1–6.
246 Lee, G.J.-L. and Tieckelmann, H. (1979) Anal. Biochem. 94, 231–36.
247 Delaney, S.R., Lener, M. and Conrad, H.E. (1980) Anal. Biochem. 106, 253–61.
248 Davies, M.G., Thomas, A.J. and Blanshard, K.C. (1974) Process Biochem. 9, 9–12.
249 Bober, H. (1976) GIT Fachz. Lab. 20, 299–302, 304–5, 643–44, 647–50, 653–54, 657–59 (Ger.). C.A. 85, 40396v, 188430e.
250 Khan, A. (1977) Pak, J. Sci. 29, 1–5, C.A. 92, 36965g.
251 Ersser, R.S. (1979) Lab. Equip. Dig. 17, 61, 63, 65. C.A. 93, 142323b.
252 Robinson, G.W. (1978) in Amino Acid Determination, 2nd Edn. (Blackburn, S., ed.) pp. 39–45, 47–67, 101–108, Dekker, New York. C.A. 90, 50701h, 50702j, 50705n.

253 Dilley, K.J. (1980) Am. Lab. (Fairfield, CT) 12, 113–14, 116. C.A. 93, 64939r.
254 Deyl, Z. and Juřicová, M. (1975) Amino Acid Derivatives, pp. 713–39, in monograph cited in 214.
255 Kluh, I. (1975) Peptides, pp. 741–72, in monograph cited in 214.
256 Stein, S. (1977) in Peptides in Neurobiology (Gaiger, H., ed.) p. 9, Plenum Press, New York.
257 Chin, C.H.C.Q. and Wold, F. (1977) Methods Enzymol. 47, 204–10.
258 Murren, C., Stelling, D. and Felstead, G. (1975) J. Chromatogr. 115, 236–39.
259 Murren, C.K. (1979) in Anal. Biol. Mater., Proc. Conf. 1977 (Pub. 1979) (Butler, L.R.P., ed.) pp. 57–62, Pergamon, Oxford, England. C.A. 91, 119704b.
260 Mal'tsev, V.G., Belen'kii, B.G. and Alexandrov, M.L. (1978) Zh. Anal. Khim. 33, 798–807 (Russ.). C.A. 89, 86609c.
261 Charpentier, Ch. (1976) Proc. Int. Symp. Lab. Screening Tech. Inborn Errors Metab. Newborn Sel. High-Risk Infants 1972 (Pub. 1976) pp. 50–57, NTIS, Springfield, VA. C.A. 90, 37241p.
262 Lee, H.-M., Bucher, D.J., and Seid, R.C., Jr. (1979) Ind. Eng. Chem. Prod. Res. Dev. 18, 122–26. C.A. 90, 199705v.
263 Mitchell, A.R., Kent, S.B.H., Chu, I.C. and Merrifield, R.B. (1978) Anal. Chem. 50, 637–40.
264 Hurlbut, J.A. and Balka, T.J. (1978) J. Chem. Educ. 55, 794. C.A. 90, 86019f.
265 Heathcote, J.G., Washington, R.J. and Keogh, B.J. (1975) J. Chromatogr. 104, 141–46.
266 Machleidt, W., Otto, J. and Wachter, E. (1977) Methods Enzymol. 47, 210–20.
267 Fuzita, K., Takeuchi, S. and Arikawa, Y. (1976) Nippon Kagaku Kaishi 1976, 623–26 (Jap.). C.A. 85, 71767w.
268 Joo, P.K., Heitmann, R.N., Reulein, S.S. and Bergman, E.N. (1976) J. Dairy Sci. 59, 309–14. C.A. 84, 86301t.
269 Pospíšil, R. (1978) Biochem. Clin. Bohemoslov. 7, 171–76 (Slo.). C.A. 91, 16053j.
270 Saito, A., Maeda, K., Kobayashi, K., Yazawa, F. and Yamamoto, Y. (1978) Igaku No Ayumi 106, 37–40 (Japan). C.A. 89, 142736f.
271 Ohtsuki, K. and Hatano, H. (1978) J. Chromatogr. 148, 536–38.
272 Sorimachi, K. (1979) Anal. Biochem. 93, 31–36.
273 Madar, D.A., Willis, R.A., Koehler, K.A. and Hiskey, R.G. (1979) Anal. Biochem. 92, 466–72.
274 Gopalakrishna, R. and Nagarajan, B. (1980) Anal. Biochem. 107, 318–23.
275 Paik, W.K., Dimaria, P., Pearson, E. and Kim, S. (1978) Anal. Biochem. 90, 262–72.
276 Finley, J.W., Krochta, J.M. and Heftmann, E. (1978) J. Chromatogr. 157, 435–39.
277 Blanshard, K.C., Bradford, H.F., Dodd, P.R. and Thomas, A.J. (1975) Anal. Biochem. 67, 233–44.
278 Krishchenko, V.P. (1978) Izv. Akad. Nauk SSSR, Ser. Biol. 1978 405–17 (Russ.). C.A. 89, 72590c.
279 Murayama, K. and Shindo, N. (1977) J. Chromatogr. 143, 137–52.
280 Gianek, J. (1978) Lab. Delo 1978, 28–31 (Russ.). C.A. 88, 85438p.
281 Dave, B.K. and Graves, C.N. (1978) Ind. J. Anim. Sci. 48, 257–61. C.A. 89, 86626f.
282 Olek, K., Uhlhaas, S. and Wardenbach, P. (1978) J. Clin. Chem. Clin. Biochem. 16, 283–88 (Ger.). C.A. 89, 55814m.
283 Williams, A.P., Hewitt, D., Cockburn, J.E., Harris, D.A., Moore, R.A. and Davies, M.G. (1980) J. Sci. Food. Agric. 31, 474–80. C.A. 93, 182133w.
284 Hollaway, W.L., Bhown, A.S., Mole, J.E. and Benett, J.C. (1978) J. High Resolut. Chromatogr., Chromatogr. Commun. 1, 177–78. C.A. 90, 50753b.
285 Nys, P.S., Petyushenko, R.M., and Savitskaya, E.M. (1975) Zh. Fiz. Khim. 49, 2330–34 (Russ.). C.A. 84, 71081x.
286 Zayats, R.F., Moiseeva, L.N. and Pavlovskaya, T.E. (1980) Prikl. Biochim. Mikrobiol. 16, 439–45 (Russ.). C.A. 93, 132797v.
287 Van Der Rest, M., Cole, W.G. and Glorieux, F.H. (1977) Biochem. J. 161, 527–34.
288 Skarlat, I.V. (1978) in Fiz.-Khim. Metody Mol. Biol. 1978 (Berezkin, I.V., ed.) pp. 119–26, Moskovskij Gos. Univ., Moscow, USSR (Russ.). C.A. 91, 206559b.
289 Oshima, G., Shimabukuro, H. and Nagasawa, K. (1978) J. Chromatogr. 152, 579–84.
290 Armstrong, M.D. (1979) J. Chromatogr. 175, 216–18.
291 James, L.B. (1979) J. Chromatogr. 172, 481–83.

292 Degen, J. and Kyte, J. (1978) Anal. Biochem. 89, 529–39.
293 Meek, J.L. and Bohan, T.P. (1978) in The Endorphins (Costa, E. and Trabucchi, M., eds.), serie Adv. Biochem. Psychopharmacol. 18, 141–47.
294 Tabor, C.W. and Tabor, H. (1977) Anal. Biochem. 78, 543–53.
295 Rabenstein, D.L. and Saetre, R. (1978) Clin. Chem. 1140–43.
296 Fox, L., Anthony, G.D. and Lau, E.P.K. (1976) J. Assoc. Offic. Anal. Chem. 59, 1048–50.
297 Arnould, J.M. and Tankosic, C. (1980) Arch. Int. Physiol. Biochim. 88, 293–302 (Fr.). C.A. 93, 217361y.
298 Prusík, Z. (1975) Proteins, pp. 773–806, in monograph cited in 214.
299 Fasold, H. (1975) in Chromatography, 3rd Edn. (Heftmann, E., ed.) pp. 466–526, Van Nostrand-Reinhold, New York.
300 Janson, J.Ch. (1977) DHEW Publ. (US) NIH-78-1422, Proc. Int. Workshop Technol. Protein Sep. Improv. Blood Plasma Fractionation, pp. 205–20. C.A. 91, 119667s.
301 Curling, J. (1978) Proc. FEBS Meet. 1978 (Pub. 1979). 52 (Protein: Struct., Funct. Ind. Appl.) 45–57. C.A. 91, 16031a.
302 Spring, T.G. and Cole, R.D. (1977) Methods Cell Biol. 16, 227–40. C.A. 87, 129714m.
303 Schroeder, W.A., and Huisman, T.H.J. (1980) The Chromatography of Hemoglobin, M. Dekker, New York.
304 Atwater, J. and Schwartz, E. (1977) in Hematology, 2nd Edn. (Williams, W.J., Beutler, E. and Erslev, A.J., eds.) pp. 1597–601, McGraw-Hill, New York.
305 Golshmid, C.K. (1975) Khim. Assocs. Vaktsiny 1975, 163–77 (Russ.). C.A. 84, 3170a.
306 Tiselius, A. (1954) Arkiv Kemi 7, 49.
307 Malamud, D. and Drysdale, J.V. (1978) Anal. Biochem. 86, 620–47.
308 Righetti, P.G. and Caravaggio, T. (1976) J. Chromatogr. 127, 1–28.
309 Righetti, P.G., Tudor, G. and Ek, K. (1981) J. Chromatogr. 220, 115–94.
310 Torres, A.R. and Peterson, E.A. (1979) J. Biochem. Biophys. Methods 1, 349–60.
311 Dmitrenko, L.V., Ostrovskii, D.I. and Samsonov, G.V. (1975) Rasshir. Terisy Dokl. -Vses. Simpl. Thermodin. Inoogo Obmena 1975, 42–44 (Russ.). C.A. 88, 132776j.
312 Chang, S.H. (1977) Diss. Abstr. Int. B 37, 3923–24.
313 Mikeš, O. (1981) in Recent Developments in Food Analysis (Proceedings of the First European Conference on Food Chemistry, Vienna 1981) Baltes, W., Czedik-Eysenberg, P.B. and Pfannhauser, W., eds.) pp. 306–21. Verlag Chemie, Weinheim 1982.
314 Barford, R.A., Sliwinski, B.J. and Rothbart, H.L. (1979) J. Chromatogr. 185, 393–402.
315 Tomana, M., Niedermeier, W., Spivey, C. and Gerard, J. (1978) Microchem J. 23, 93–103.
316 Hashizume, H. (1975) Rep. Fac. Sci., Shizuoka Univ. 10, 55–71. C.A. 85, 89503s.
317 Edwards, W.G. and Wellington, C.A. (1977) J. Chromatogr. 135, 463–69.
318 Horvath, M. and Meretey, K. (1978) Kiserl. Orvostud. 30, 289–94 (Hung.). C.A. 89, 159639z.
319 Jones, G. and Phillips, D. (1976) Educ, Chem. 13, 15–16. C.A. 84, 120560p.
320 Friedlander, E.J. and Norman, A.W. (1980) Methods Enzymol. 67, 504–8.
321 Fukami, Y., Hosaka, Y. and Yamamoto, K. (1980) FEBS Lett. 114, 342–46.
322 Hesslinger, H. and Pollow, K. (1978) Mol. Biol. Rep. 4, 171–75. C.A. 90, 99375u.
323 Carrol, M.A., Post, A., Griggs L.J. and Zarembo, J.E. (1979) Microchem. J. 24, 137–42.
324 Haiech, J., Derancourt, J., Pechere, J.F. and Demaille, J.G. (1979) Biochimie 61, 583–87.
325 Feramisco, J.R. and Burridge, K. (1980) J. Biol. Chem. 255, 1194–99.
326 Noppe, M., Gheuens, J., Karcher, D. and Lowenthal, A. (1979) Protides Biol. Fluids, 27th (Pub. 1980) 813-16. C.A. 92, 193707s.
327 Janatová, J., Crandall, R.E. and Andrade, J.D. (1980) Prep. Biochem. 10, 405–30.
328 Lawrie, J.S., Ross, J. and Kemp, G.D. (1979) Biochem. Soc. Trans 7, 693–94.
329 Sparham, S.J. and Huehns, E.R. (1979) Hemoglobin 3, 13–20. C.A. 91, 16064p.
330 Schoessler, W., Zorn, H. and Patzelt, D. (1979) Acta Biol. Med. Ger. 38, 669–72 (Ger.). C.A. 92, 18192y.
331 Kobayashi, M. (1977) Sci. Rep. Niigata Univ., Ser D 14, 1–5. C.A. 90, 18463w.

332 Abraham, E.C., Reese, A., Stallings, M. and Huisman, Th.J. (1977) Hemoglobin 1, 27–44. C.A. 86, 67710a.
333 Friedman, S. and Humbert, J.R. (1979) Hemoglobin 3, 411–28. C.A. 92, 142569r.
334 Gooding, K.M., Lu, K.-Ch. and Regnier, F. (1979) J. Chromatogr. 164, 506–9.
335 Alboni, P., Brini, M., Davoli, V., Manicardi, V., Modena, G. and Portioli, I. (1979) Perspect. Inherited Metab. Dis. (Pub. 1980). 3 (Fatty Acids Triglycerides: Biosynth. Transp. Norm. Pathol. Cond.) 175–79. C.A. 93, 182142y.
336 Canham, P.A.S. and Lockett, C.J. (1980) S. Afr. Med. J. 53, 133–35. C.A. 93, 217378j.
337 Mosca, A., Carenini, A., Samaja, M. and Saibene, V. (1980) Clin. Chem. 26, 1106–7.
338 Mosca, A., Carenini, A., Samaja, M. and Saibene, V. (1980) Ric. Clin. Lab. 10, 251–53. C.A. 93, 40688g.
339 Curling, J.M. and Bergloef, J.H. (1978) DHEW Publ. (US), NIH-78-1422, Proc. Int. Workshop Technol. Protein Sep. Improv. Blood Plasma Fractionation 1977, 269–75. C.A. 91, 181326h.
340 Cunningham, C.M. and O'Sullivan, S.A. (1978) DHEW Publ. (US), NIH-78-1422, Proc. Int. Workshop Technol. Protein Sep. Improv. Blood Plasma Fractionation 1977, 276–78. C.A. 91, 181327j.
341 Mirabel, B. (1980) Actual. Chim. 1980, 39–44 (Fr.). C.A. 93, 148214p.
342 Guilbault, G.G. (1976) Handbook of Enzymatic Methods of Analysis, Marcel Dekker, New York.
343 Lim, C.K. (1979) Proc. Anal. Div. Chem. Soc. 16, 305–7. C.A. 92, 71551r.
344 Bostik, W.D., Denton, M.S. and Dinsmore, S.R. (1980) Clin. Chem. 26, 712–17.
345 Rombouts, F.M., Geraeds, C.C.J.M., Visser, J. and Pilnik, W. (1982) in Affinity Chromatography and Related Techniques (Gribnau, T.C.J., Visser, J. and Nivard, R.J.F., eds.) pp. 255–60, Elsevier, Amsterdam.
346 Roberts, N.B. and Taylor, W.H. (1978) Biochem. J. 169, 607–15.
347 Rexová-Benková, L'., Omelková, J., Mikeš, O., and Sedláčková, J. (1982) J. Chromatogr. 238, 183–92.
348 Hostomská, Z. and Mikeš, O. (1983) J. Chromatogr. 267, 355–66.
349 Montenecourt, B.S., Kelleher, T.J., Eveleigh, D. and Pettersson, L.G. (1980) Biotechnol. Bioeng. Symp. No 10, 15–26.
350 Mercer, D.W. (1974) Clin. Chem. 20, 36–40.
351 Kudirka, P.J., Busby, M.G., Carey, R.N. and Torren, E.C., Jr. (1975) Clin. Chem. 21, 450–52.
352 Klein, B., Foreman, J.A., Jeunelot, C.J. and Sheehan, J.E. (1977) Clin. Chem. 23, 504–10.
353 Elkins, B.N. (1977) Clin. Chem. 23, 1510.
354 Morin, L.G. (1977) Clin. Chem. 23, 205–10.
355 Kudirka, P.J., Schroeder, R.R., Hewitt, T.E. and Toren, E.C., Jr. (1976) Clin. Chem. 22, 471–74.
356 Aoshima, H. (1979) Anal. Biochem. 95, 371–76.
357 Mercer, D.W. (1977) Clin. Chem. 23, 653–58.
358 Zadražil, Z. (1975) Low-molecular weight constituents of nucleic acids. Nucleosides, nucleotides and their analogues, pp. 831–57 in monograph cited in 214.
359 Zadražil, S. (1975) Nucleic Acids, pp. 859–86 in monograph cited in 214.
360 Brown, P.R. (1973) Enzyme Peak Shift, pp. 101–111, in monograph cited in 27.
361 Singhal, R.P. (1974) Sep. Purif. Methods 3, 339–98. C.A. 82, 166788f.
362 Brown, P.R., Krstulovic, A.M. and Hartwick, R.A. (1980) Adv. Chromatogr. 18, 101–38.
363 Plunkett, W. (1978) Adv. Chromatogr. 16, 211–48.
364 Leutzinger, E.E., Miller, P.S. and Ts'o, P.O.P. (1978) Nucleic Acid Chem. 2, 1037–43.
365 Perrett, D. (1976) J. Chromatogr. 124, 187–96.
366 Yanagawa, H. (1978) Nucleic Acids Res., Spec. Publ., 5, 461–64. C.A. 90, 83055d.
367 Bujarski, J., and Augustyniak J. (1977) Rocz. Chem. 51, 2273–78. C.A. 88, 85406b.
368 Cleary, M.T. and Dryhurst, G. (1977) Anal. Chim. Acta 94, 343–50.
369 Ohno, T., Morikawa, M., Irie, H. and Muto, G. (1980) J. Photogr. Sci. 28, 23–27. C.A. 92, 208577a.
370 Vlasov, V.V., Melamed, N.V., Chizhikov, V.E. and Tukalo, M.A. (1976) Bioorg. Khim. 2, 892–97 (Russ.). C.A. 85, 106152w.

371 Tonooka, S., Sasaki, A. and Kakimoto, S. (1978) Hokkaido Daigaku Meneki Kagaku Kenkyusho Kiyo, 38, 46–52 (Japan). C.A. 90, 134611s.
372 Schott, H. and Schwarz, M. (1978) J. Chromatogr. 147, 197–215 (Ger.). C.A. 90, 35746h.
373 Svoboda, V. and Kleinmann, I. (1979) J. Chromatogr. 176, 65–73.
374 Leigh, C.P.H. and Cashion, P.J. (1980) J. Chromatogr. 192, 490–93.
375 Koshkina, I.M., Remizova, L.A. and Favorskaya, I.A. (1976) Vestn. Leningr. Univ., Fiz. Khim., 1976, 135–39 (Russ.). C.A. 85, 193007u.
376 Khym, J.X. (1976) J. Chromatogr. 124, 415–17.
377 Hsu, D.-S. and Chen, S.C. (1980) J. Chromatogr. 192, 193–98.
378 Brunius, G. (1979) J. Chromatogr. 170, 486–89.
379 Alvarez Pez, J.M. and Garcia Fernandez, A. (1978) Ars. Pharm. 19, 189–95 (Span.). C.A. 91, 16058q.
380 Wakizaka, A., Kurosaka, K. and Okuhara, E. (1978) IRCS Med. Sci. Libr. Compend. 6, 485. C.A. 90, 50797u.
381 Dizdaroglu, M. and Hermes, W. (1979) J. Chromatogr. 171, 321–30.
382 Šatava, J., Mikeš, O. and Štrop, P. (1979) J. Chromatogr. 180, 31–37.
383 Vandenberghe, A., Van Broeckhoven, C. and De Wachter, R. (1979) Arch. Int. Physiol. Biochim. 87, 848–49.C.A. 92, 106680m.
384 Vandenberghe, A., Nelles, L. and De Wachter, R. (1980) Anal. Biochem. 107, 369–76.
385 McLaughlin, L.W., Cramer, F. and Sprinzl, M. (1981) Anal. Biochem. 112, 60–69.
386 Aukaty, M.F., Bubenschikova, S.N., Kagrammanova, V.K. and Baratova, L.A. (1977) J. Chromatogr. 137, 351–56.
387 Zarytova, V.F. and Lebedev, A.V. (1977) Bioorg. Khim. 3, 1211–18 (Russ.). C.A. 88, 38097s.
388 Shum, B.W.-K. and Crothers, D.M. (1978) Nucleic Acids Res. 5, 2297–311.
389 Demushkin, V.P. and Plyashkevich, Yu.G. (1976) Bioorg. Khim. 2, 1652–59 (Russ.). C.A. 86, 52303w.
390 Edelson, E.H., Lawless J.G., Wehr, C.T. and Abott, S.R. (1979) J. Chromatogr. 174, 409–419.
391 Tashiro, T., Fujita, E., Miyamoto, M. and Yoshi, J. (1980) Nippon Shokuhin Kogyo Gakkaishi 27, 131–35 (Japan). C.A. 93, 91278f.
392 Singhal, R.P., Griffin, G.D. and Novelli, G.D. (1976) Biochemistry 15, 5083–87.
393 Gillam, I.C. and Tenner, G.M. (1980) Anal. Biochem. 105, 405–6.
394 Popov, L.S. (1978) Biokhimiya (Moscow) 43, 511–15 (Russ.) C.A. 89, 19408y.
395 Wehr, C.T. and Abbott, S.R. (1979) J. Chromatogr. 185, 453–62.
396 De Rooij, J.F.M., Bloemhoff, W. and Van Boom, J.H. (1979) J. Chromatogr. 177, 380–84.
397 Pokorný, J. (1975) Lipids, pp. 581–92, in monograph cited in 214.
398 Betina, V. (1975) Antibiotics, pp. 979–1008, in monograph cited in 214.
399 Anhalt, J.P. (1981) in Liquid Chromatography in Clinical Analysis (Marton, L.J. and Kabra, P.M., eds.) pp. 163–84, The Human Press.
400 Umezawa, H. and Kondo, S. (1975) Methods Enzymol. 43, 263–78.
401 Lotscher, K.M., Brander, B. and Kern, H. (1975) Varian Instrum. Appl. 9, 12–15. C.A. 86, 111210p.
402 Procházka, Ž. (1975) Steroids, pp. 593–622, in monograph cited in 214.
403 Davídek, J. (1975) Vitamins, pp. 953–978, in monograph cited in 214.
404 Macek, K. (1975) Alkaloids, pp. 887–94, in monograph cited in 214.
405 Churáček, J. (1975) Alcohols and Polyols, pp. 431–39, in monograph cited in 214.
406 Laisk, A., Laarin, P., Oja, V. and Sober, J. (1980) Fiziol. Rast. (Moscow) 27, 897–904 (Russ.). C.A. 93, 163721v.
407 Iwamori, M. and Nagai, Y. (1978) Biochim. Biophys. Acta 528, 257–67.
408 Fredman, P., Nilsson, O., Tayot, J.L. and Svennerholm, L. (1980) Biochim. Biophys. Acta 618, 42–52.
409 Kaputskii, F.N., Yurkshtovich, T.L., Kaputskii, V.E., Balabaeva, M.D. and Tikavaya, S.N. (1977) Zh. Prikl. Khim. (Leningrad) 50, 2583–86 (Russ.). C.A. 88, 65911j.
410 Markovic, R.D., Eisen, H.J., Parchman, L.G., Barnett, C.A. and Litwack, G. (1980) Biochemistry 19, 4556–64.

411 Noe, V. and Psallidi, M. (1976) Riv. Soc. Ital. Sci. Aliment. 5, 137–40 (Ital.). C.A. 85, 89502r.
412 Sasak, W., Silvermann-Jones, C.S. and De Luca, L.M. (1979) Anal. Biochem. 97, 298–301.
413 Van Staden, J. (1976) Physiol. Plant. 38, 240–42. CA 86, 84564y.
414 Dauphin, B., Teller, G. and Durand, B. (1977) Physiol. Veg. 15, 747–62 (Fr.). C.A. 89, 55782z.
415 Graebner, R., Schneider, G. and Sembdner, G. (1976) J. Chromatogr. 110–15 (Ger.); C.A. 85, 58994y.
416 Nagels, L., Van Dongen, W. and Parmentier, F. (1979) Arch. Int. Physiol. Biochim. 87, 585–91. C.A. 92, 72067z.
417 Christensen, H.D. and Le Roy Blank, C. (1979) Chromatogr. Sci. 12, 133–64.
418 Lankelma, J. and Poppe, H. (1978) J. Chromatogr. 149, 587–98.
419 With, T.K. and Pedersen, J.S. (1978) Scand. J. Clin. Lab. Invest. 38, 279–88. C.A. 89, 72594g.
420 Guarino, A., Boffa, G., Taccone, W., De Vito, B., Malorni, H., Sodano, G. and Sannolo N. (1979) G. Ital. Chim. Clin. 4, 461–77 (Ital.). C.A. 93, 91264y.
421 Epton, R. (ed.) (1978) Chromatography of Synthetic and Biological Polymers, Vol. 1., Column Packings, GPC, GF and Gradient Elution; p. 2, Ellis Horwood, Chicester, England.
422 Curling, J. (1978) in Chromatography of Synthetic and Biological Polymers, Vol. 2., Hydrophobic, Ion Exchange and Affinity Methods (Epton, R., ed.) pp. 75–87, Ellise Horwood, Chicester, England.

Note

For additional information see The Bibliography Section of the *Journal of Chromatography.*

Deyl (ed.) Separation Methods

CHAPTER 4.6

Gel chromatography

DUŠAN BEREK [a] and KAMIL MARCINKA [b]

[a] *Polymer Institute, Centre for Chemical Research, Slovak Academy of Sciences, 84236, Bratislava, and* [b] *Institute of Virology, Slovak Academy of Sciences, 81703, Bratislava, Czechoslovakia*

4.6.1 Introduction

In chapter 1, chromatography was presented as a phenomenon of differential migration of the sample components in the two-phase system composed of the stationary and mobile phases. In contrast to other modes of liquid chromatography, the differences in the migration velocities of sample components in gel chromatography are primarily due to the different *sizes* of particular species to be separated (molecules, associates, colloidal particles, etc.) and secondarily only by their attractive interactions with the stationary phases. In other words, the 'driving force' of separation in gel chromatography is the difference in the dimensions of elementary units of the sample components. Consequently, according to the type of column packing, gel chromatography allows the separation of the species ranging from the low-molecular substances to macromolecules with molar mass reaching millions of Daltons, and to colloidal particles and viruses with the size of several hundred nanometers. For this reason, this method is often also called steric exclusion chromatography, or size exclusion chromatography. However, besides steric exclusion, several other separation mechanisms are often involved in the overall process. In this chapter, the classical term *gel chromatography* [1] will be used; this originates from the designation of the system rather than referring to the process. Simultaneously, the abbreviation GPC is coined to distinguish gel chromatography from gas chromatography. Another one of the numerous terms proposed before, namely gel *filtration*, still often used in connection with the separations in aqueous eluents and with the soft gels, describes adequately neither the system nor the process.

Attempts to separate the substances on the basis of the exclusion or penetration effects within porous column packings were already published early in the fifties, but the actual founders of the method are regarded to be Porath and Flodin [2]. From the beginning of the sixties, gel chromatography underwent a vivid development which was strongly influenced by other methods of liquid chromatography: the theoretical principles were formulated for the description of the main processes taking place within the GPC columns, both the material basis and the instrumental technique of the method were substantially extended and improved, and the interpretation of experimental results was refined. Simultaneously, several new applications of the method were proposed. In spite of this progress, many theoretical, experimental and application problems still remain unsolved, so that the development of gel chromatography cannot, by far, be considered finished. In addition, the transfer of knowledge from the field of separation of synthetic substances into the area of biological systems and vice versa is insufficient. In this chapter, we have tried to bridge this 'information barrier' and, with the exception of the sections devoted to the conventional applications of gel chromatography on biological material, a unifying view of this efficient separation method is given.

4.6.2 General concepts and principles of theory

The majority of gel chromatographic separations are carried out by elution development (cf., Chapter 1) of the molecularly dispersed macromolecular solutes through the gel bed packed in a column; this case will be considered in the following discussions unless otherwise stated.

4.6.2.1 Mechanism of ideal gel chromatography

The term ideal gel chromatography denotes the process in which, exclusively, the differences in the dimensions of separated molecules are operative.

Let us consider the bed of randomly packed porous gel particles within the column that is percolated by the mobile phase (eluent) carrying the sample components. The sizes of gel pores and separated molecules have mutually matched distribution so that the largest sample molecules are totally excluded from all pores of the gel. These molecules will not be retained in the column and are eluted by the volume of the eluent equal to the interstitial volume of the column, V_o. This means that the retention volume, V_R, of excluded molecules equals V_o. On the other hand, the smallest sample molecules with the sizes that are comparable with the size of eluent molecules permeate all pores of gel and diffuse back into the mobile phase. The consequence of this repeated process is the retention of the small molecules in the column and, for their elution, such volume of eluent is necessary that equals the interstitial volume, V_o, plus pore volume, V_i. The resulting retention volume is $V_R = V_m = V_o + V_i$.

The molecules of intermediate dimensions are partly retained in the gel bed by their permeation into pores large enough to accomodate them. Such molecules are eluted from the column in retention volume

$$V_R = V_o + K_d V_i \tag{1}$$

where K_d is the distribution coefficient expressed as the ratio of the solute concentration in the gel to the concentration in the mobile phase. Evidently, K_d reaches values between zero and one.

Eqn. (1) is a formal analogy to equation (17) in Chapter 1 and shows important inherent restriction of gel chromatography: the space in which separation takes place is confined to the volume of pores and, consequently, the peak capacity of the method is fairly limited.

The above simple model of a steric exclusion mechanism was considered by several authors attempting to describe quantitatively the gel chromatographic separation process. Distribution coefficients were expressed on the basis of the model considerations of the dimensions of both the separated molecules and the pores of gel, as well as of the stochastic model approaches (for reviews see e.g., Refs. 1, 3–6), and also of the thermodynamic reasoning on the changes of conformational entropy of macromolecules due to their transfer from the interstitial volume into the pores in the course of separation [7]. However, besides the steric exclusion from the pores, at least two other size-based mechanisms are operative in the ideal gel chromatography:

(i) The restriction of diffusion of solute molecules within the gel particles that depends on both the geometry of gel network and the size of solute molecules [4,8];

(ii) The decrease of mobile phase flow rate in the immediate vicinity of the gel particles. For geometrical reasons the larger molecules cannot enter slower eluent streamlines, and the relative velocity of their motion through the column increases [9].

It is assumed that the steric exclusion from the gel pores is the decisive separation mechanism in the ideal gel chromatography, and both the latter mechanisms contribute positively (but within a limited extent only) to the overall separation effect.

Generally, three fundamental problems are encountered at a priori calculation of the distribution coefficients K_d.

(i) The complexity of processes taking place in gel chromatographic column.

(ii) The difficulty of exact expression of the dimensions of (macro)molecules that are operative during separation. The rigid spheres approximation can be used in a few cases only, and the randomly coiled macromolecules change steadily their conformation and dimensions in solutions.

(iii) The impossibility of an exact description of the gel structure: the pores of the real gel systems have rather irregular shape and, moreover, the existence of a layer of the immobilized eluent that may not be accessible for macromolecules is to be expected in the immediate vicinity of gel matrix, i.e., also within the pores.

Consequently, gel chromatography should be regarded primarily as a powerful separation procedure and not as an absolute method for measurement of the molecular characteristics of polymers. This limitation means that each particular chromatographic system must be carefully calibrated prior to quantitative interpretation of experimental data (cf., Sections 4.6.2.3 and 4.6.2.4) or the fractions obtained must be characterized by an independent method.

4.6.2.2 *Real gel chromatography*

In the preceding section, we briefly discussed the separation mechanisms that are exclusively and directly connected with the dimensions of the separated species. Let us designate these mechanisms *primary*. In the real gel chromatographic systems, however, several other processes are operative, affecting both the retention volumes and the widths of chromatographic zones. These *secondary processes* can be classified into:

(i) *Additional separation mechanisms,* i.e., the processes that influence directly the rate of migration of separated species, and are caused mainly by the thermodynamic interactions between different components of gel chromatographic systems. The most important additional separation mechanisms are: adsorption of separated substances on the gel matrix; thermodynamic partition of sample components between gel phase and mobile phase; incompatibility of macromolecules with the gel; as well as solute — gel and solute — solvent ionic interactions and Donnan effects causing ion exclusion, ion inclusion and ion exchange [48]. The adsorption and ionic interactions are especially important in the systems containing polar or charged biological material, while thermodynamic partition is to be expected mainly in the systems composed of soft gels and mixed, multicomponent eluents.

(ii) *Side effects,* i.e., the processes that influence the separation results indirectly, above all by changing the dimensions of separated species due to temperature variations, as well as due to association, aggregation, intra- and intermolecular ionic interactions and incompatibility of sample molecules, or due to deformations of colloidal particles and degradation of macromolecules in the flow shear. Among side effects, one should also consider secondary exclusion of larger solute molecules from the pores already occupied by smaller, faster diffusing separated molecules.

(iii) *Parasitic effects,* i.e., the processes that bring about the perturbations of flow, the spreading of chromatographic zones and their irregular movement through the column. Typical examples are osmotic, mixing and viscosity effects, parasitic diffusion, etc.

The secondary processes often exhibit dynamic character, i.e., their extent may vary in the course of elution (e.g., as a consequence of progressive dilution of the separated substances within the column).

In some cases, several secondary processes may manifest themselves simultaneously. A typical combined secondary process is the so-called concentration effect in gel chromatography, i.e., the often observed rise of the retention volumes of macromolecular substances with the increased sample concentration. The concentration effect is caused by a decrease in the macromolecular coils' dimensions in solution with growing concentration, further by viscosity and osmotic effects as well as by secondary exclusion — and in the case of soft gels also by altering the pore geometry due to deswelling of gel particles in the zone of the sample.

The nature of particular secondary processes and their influence on the results of gel chromatographic separation are discussed in detail, e.g., in Refs. 6,10,11.

The retention volume in real gel chromatography can be expressed thus

$$V_R = V_o + K_d V_i + \sum K_j V_{ij}, \qquad (2)$$

where K_j stands for the distribution coefficients of a particular secondary process, and V_{ij} denotes the corresponding effective values of the volume of pores.

The secondary processes usually disturb the gel chromatographic separation or complicate the processing of the chromatographic analytical data. That is why it is necessary to remove, or at least to suppress, the secondary processes in common experimental practice by the appropriate choice of the operational variables. For example, the effects of adsorption, thermodynamic partition and incompatibility can be diminished by the choice of gel and eluent, while the ionic effects are suppressed by adding a suitable salt into eluent, and the concentration effects are not important when working with very low sample concentration or when applying the thermodynamically poor solvent as mobile phase.

Other possibilities for obtaining precise and reliable analytical results are the computational procedures applied to the corrections of raw data for both the zone spreading and the concentration effects.

The existence of secondary processes brings about another limitation of gel chromatography: it is often rather difficult and sometimes even impossible to compare directly the data obtained with different chromatographic systems or under different operational variables. The same limitation is valid for the transfer of experimental results from one type of sample to another — even if an identical column system and operational variables are applied. It is advisable to check the possible influence of secondary processes when separating an unknown sample.

The special working procedures where some of the secondary processes are intentionally used to enhance the separation selectivity are discussed in section 4.6.3.11.

4.6.2.3 Resolution power and calibration in gel chromatography

The resolution power in gel chromatography is determined by both the efficiency and selectivity of the chromatographic system.

The efficiency of a real GPC system depends above all on the rate of mass transfer between mobile phase and gel phase, as well as on the extent of secondary processes. Quantitatively, the efficiency can be expressed by the terms like width (w) or deviation (σ) of the chromatographic peak, as well as by other terms of the theoretical plate concept. Since the diffusion rate of solute molecules decreases with an increase of their dimensions, one has to expect generally lower efficiency in gel chromatography of macromolecules in comparison with any other mode of liquid chromatographic separation of low molecular substances.

In ideal gel chromatography, the selectivity of separation and, consequently, the peak capacity is limited by the existence of the two boundary K_d values (0 and 1). For an identical sample-eluent system, the decisive variables are the effective pore size distribution, the pore volume and the interstitial volume of the gel bed, i.e., the parameters of both inner structure of the gel and the gel bed geometry. In other words, the selectivity of a GPC system is defined by the region of effective molecular dimensions (r), in which the system can separate, as well as by the region of retention volumes available. This can be elucidated by means of the $\log r - V_R$ plots (calibration curves) shown schematically in Fig. 4.6.1. Curve (a) belongs to nonselective, 'universal' gel with wide pore size distribution and relatively large pore volume. This gel separates in the wide range of r values. Curve (b) shows the selectively separating gel with large pore volume and narrow pore size distribution. A small variation in r brings about a large change in V_R. Curve (c) corresponds to the gel with narrow pore size distribution but small pore volume. Curve (d) belongs to a gel with unsuitable, bimodal pore size distribution, and curve (e) represents the gel bed with fairly narrow pore size distribution but with large interstitial volume V_o.

The courses of calibration curves can be expressed mathematically by the appropriate polynomes. The central part of calibration curves that is most important for practical use can, however, often be quite well approximated by the straight line (cf., Fig. 4.6.1).

$$\log r = A - BV_R \tag{3}$$

where the slope B is usually called the GPC separation selectivity. Evidently, the value B is related mainly to the pore size and distribution of the gel, and reflects in the second instance only the pore volume and gel bed geometry.

For the pilot separations of unknown samples, the universal gel as in (a) is

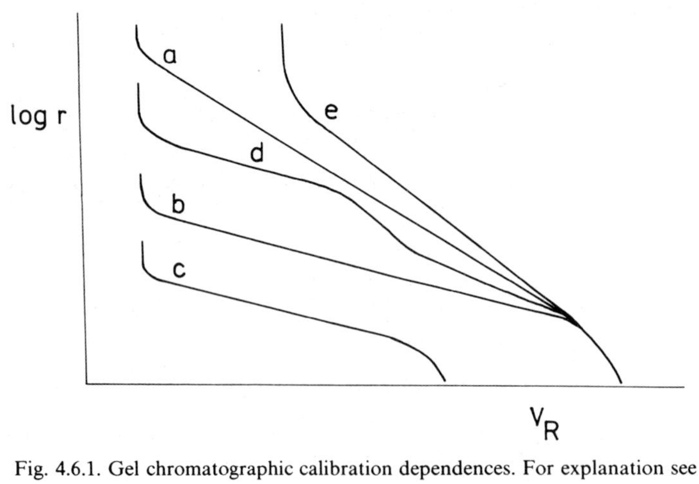

Fig. 4.6.1. Gel chromatographic calibration dependences. For explanation see text.

recommended while for highly selective separations the gel with narrow, sample-matched pore size distribution similar to that in (b) should be used.

In analytical practice, the logarithm of sample molar masses, M_M, or molar volumes, V_M, is plotted versus retention volumes in calibration dependences of low molecular substances while M_M values or effective hydrodynamic volumes, V_h, are used as size parameters in gel chromatography of macromolecules [12,13]. V_h is often called 'universal calibration parameter' because in ideal gel chromatography of randomly coiled macromolecules, it enables the transfer of data from one polymer to another regardless of both the physical (linearity, branching, tacticity, etc.) and the chemical (composition) structure of macromolecules [12]. The hydrodynamic volume of a particular polymer is proportional to the product of its molar mass and limiting viscosity number $[\eta]$, in the solvent that is used as mobile phase: $V_h \sim [\eta] M_M$.

Globular proteins are uniform (monodisperse) as to their molar mass and usually exhibit approximately spherical shape. In this case, the corresponding Stokes radius can be used as the calibration size parameter.

On the other hand, globular and helical macromolecules can be intentionally denaturated so that the samples are analyzed in the random coil conformation. The transition into random coil conformation can be caused by the elevation of temperature or by addition of structure-breaking substances to the eluent (cf., Section 4.6.4.2).

Various parameters are often plotted in the calibration dependences instead of retention volumes with the aim to exclude, or at least suppress, the influence of the gel bed dimensions. The most often used are the distribution coefficients, K_d, or the ratios V_R/V_o, V_R/V_t, and $(V_R - V_o)/(V_t - V_o)$, where V_t is the total volume of the gel bed [14].

For resolution, R, of a gel chromatographic system, one can write

$$R = \frac{2(V_{R,2} - V_{R,1})}{(w_1 + w_2)} \tag{4}$$

where subscripts denote samples 1 and 2.

Yau et al. [15] developed the concept of specific resolution factor, R_{sp}, that can be used for direct comparison of resolution power of unit length of various columns:

$$R_{sp} = 1/4(\sigma B' \sqrt{L}), \tag{5}$$

where L is the column length, B' is the slope of the linear part of calibration dependence log M_M vs. V_R, and σ is the deviation of peaks of narrow testing polymers, the molar masses of which differ by an order of magnitude.

4.6.2.4 Processing experimental data

The product of a gel chromatographic separation is either a fraction that can be utilized for further measurements or a chromatographic elution curve, most often the plot of sample concentration versus retention volume. The processing of chromatograms is usually preceded by the calibration of the chromatographic system, i.e., by the determination of dependences log M_M, log V_M, log V_h or log (Stokes radius) vs. V_R or K_d, etc. (cf., the preceding section). The necessary experimental data are usually obtained by successive measurement of solutions of reference materials (standards) with narrow molar mass distribution and different values of M_M. The retention volume at the peak maximum can be considered for the construction of the calibration dependence if the particular elution curves are symmetrical. Calibration can also be realized by means of well characterized polydisperse polymers [13].

The samples with lower molar masses exhibit large relative differences in molecular dimensions of the components, and their GPC chromatograms, obtained with highly selective column systems, often resemble the results of retentive liquid chromatography. A typical example of separation of oligomeric substances is shown in Fig. 4.6.2. In (a) the particular species are totally separated, and in (b) (i.e., by separation of a sample with higher mean molar mass) only components with lower molar masses were discriminated. The discrete chromatographic peaks are identified by means of calibration dependence or by the method of internal standards, and the relative concentration of sample components is determined from the heights or areas of corresponding peaks. Besides the above-mentioned secondary processes, especially adsorption effects, the dependence of the detector response on chemical composition, and sometimes on molar mass of separated substances, must often be considered when processing GPC data for lower molecular substances.

The chromatograms of well separated mixtures of monodisperse biopolymers with different molar masses contain discrete peaks for each component and can be evaluated similarly to the chromatograms of the low molecular samples.

On the other hand, the chromatograms of the polydisperse high molecular substances represent only the envelope curve, because the samples contain large

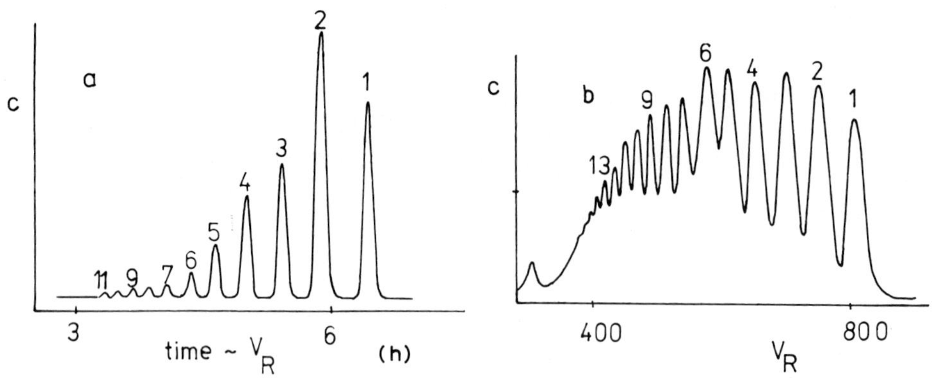

Fig. 4.6.2. Chromatograms of oligosaccharides. (a) Oligosaccharides synthesised by the action of *E. coli* ML 30 on maltose. Separation on Bio-Gel P-2 in pure water at 65°C; Gel bed 1.5 × 127 cm; flow rate 28 $cm^3\ h^{-1}$. 1–11 glucose to maltoundecaose (after John et al. [16]). (b) Malto-oligosaccharides separated on Bio-Gel P-2 in pure water at 55°C; Gel bed 2.65 × 200 cm; flow rate 15–32 $cm^3\ h^{-1}$. 1, glucose; 2, maltose; etc. (after Kainuma et al. [138]).

numbers of polymer homologues with slight differences in their molar masses (Fig. 4.6.3). In (a) the gel applied for separation has an unsuitable, too narrow pore size distribution and the largest macromolecules are already excluded from the gel pores, while the retention volumes of the lowest polymer fractions overlap the peaks of admixtures, impurities, air, etc. In (b) the gel has optimal pore size distribution. The chromatogram can be processed in the following manner:

(i) the base line is constructed (i.e., the straight line which connects detector responses before and after sample elution);

(ii) the chromatogram is divided by vertical lines into about 50 parts with identical width;

(iii) the height, h_i, of each vertical line is measured and corresponding retention volumes, V_R, are read;

(iv) the values M_{M_i} or $V_{h_i} = M_{M_i}[\eta]_i$ are read for each V_{R_i} from the calibration curve. In the latter case, the corresponding values M_{M_i} are calculated for each value V_{h_i} from the Mark-Houwink equation $[\eta]_i = KM^a_{M_i}$, where the corresponding constants $\underline{K}$ and $\underline{a}$ for each system consisting of (linear) polymer and solvent can be either found in the literature (e.g., Ref. 17) or determined by independent measurements;

(v) The total area below the chromatographic peak, P, is determined and chromatogram is normalized by division of each height h_i by P; $g_i = h_i/P$. Instead of area P, the sum $H = \Sigma h_i$ can be used.

The distribution curve of molar masses of analyzed macromolecular substance is constructed using g_i, M_{M_i} and V_{R_i} values, GPC calibration curve, and the equation for molar mass distribution:

$$f(M_M) = \frac{g(V_R)}{2.303 M_M (\mathrm{d} \log M_M / \mathrm{d}V_R)} \tag{6}$$

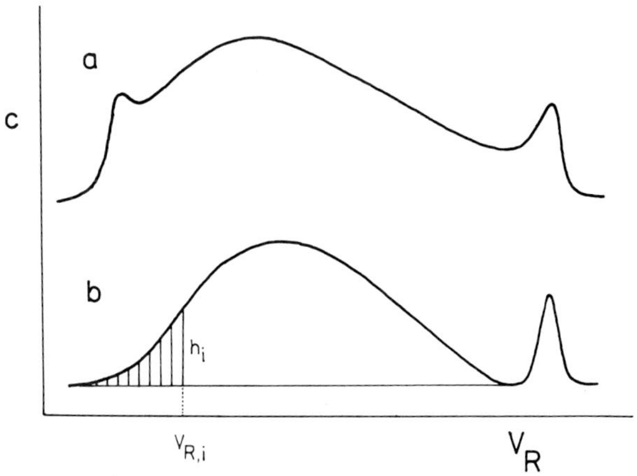

Fig. 4.6.3. Typical gel chromatograms of polydisperse polymers. For explanation see text.

The mean values of molar masses, number average, M_{M_n}, and mass average, M_{M_w}, can be calculated by means of equations

$$M_{M_n} = \left[\int_{V_0}^{V_m} \frac{g(V_R)\, dV_R}{10^{F(V_R)}} \right]^{-1} \tag{7}$$

$$M_{M_w} = \int_{V_0}^{V_m} 10^{F(V_R)} g(V_R)\, dV_R \tag{8}$$

The above described calculations are as a rule aided by a computer that can simultaneously correct the experimental data for zone spreading and the concentration effects. The modern gel chromatographs of the fourth generation (cf., Section 4.6.3) are equipped by computing units with the necessary software for on-line data acquisition.

4.6.3 Equipment and working procedures in gel chromatography

At the beginning of the development of gel chromatography, very simple equipment was used. The solvent flow was generated by gravitation and fractions were collected for further analysis. These devices are called the chromatographs of the first generation and corresponding separation procedures are denoted 'classical'. In the liquid chromatographs of the second generation, the column technology was improved, eluent was transported by means of a pumping system at the pressures of few mPa (tenths of bars), and detection was performed by sensitive flow detectors; termed conventional separations. Later, the effectivity of separation was substantially improved by application of the rigid microparticulate column-filling materials. Consequently, the working pressures and the linear velocities of the mobile phases could be increased. Presently, the microprocessor-controlled apparatus became available with fully automated sample application, processing of results and, if necessary, also fraction collection. Such instruments are now popular not only in analytical practice but also in the semipreparative and preparative separations. Gel chromatographic procedures using instruments of the third and fourth generations often bear the attributes high-efficient, high-performance, high-pressure or high-speed (HP GPC).

At present, the gel chromatographic equipment of all four generations, and GPC procedures of all three classes, are applied in both research and production aimed separations of biological material. In spite of the fact that HP GPC has been rapidly gaining importance in biology and biochemistry within last few years, the classical separations are still very often used for purification of biological samples. Because of this we shall discuss briefly the problematics of gel chromatographic equipment across the whole spectrum (including also the simplest laboratory improvisations).

4.6.3.1 Scheme of a gel chromatograph

A typical block scheme of gel chromatographic apparatus is shown in Fig. 4.6.4. The mobile phase flows from the solvent container, C, into degassing unit, D, and through filters, F, reaches the pumping system, P, which transports it via the pulse damper, PD, and the sample injecting system, I, into the column, CO. The effluent from the column enters the detector, DE, and flows through the volumeter, V, into the fraction collector, F.

The effluent may be returned through the pumping system back to the column by means of the special valve, RV. This procedure is called recycling and its most simple arrangement is distinguished in Fig. 4.6.4 by the dashed line. Recycling procedures will be discussed more in detail in Section 4.6.3.11.

The wavy line designates the electric circuits: the signal from the detector and the volumeter is recorded by the registration instrument, R.

The elementary apparatus shown in the above scheme can be either supplemented or simplified (e.g., the GPC devices of the first generation consist only of the eluent container and the column and, eventually, also of the fraction collector, while the instruments of the fourth generation include microprocessor-based automating and computing units that are integrated with the registration instruments and that serve for controlling the pumping system, the sample injector and, eventually, also the fraction collector, as well as for on-line data processing). Very often several different columns and detectors are connected in a series.

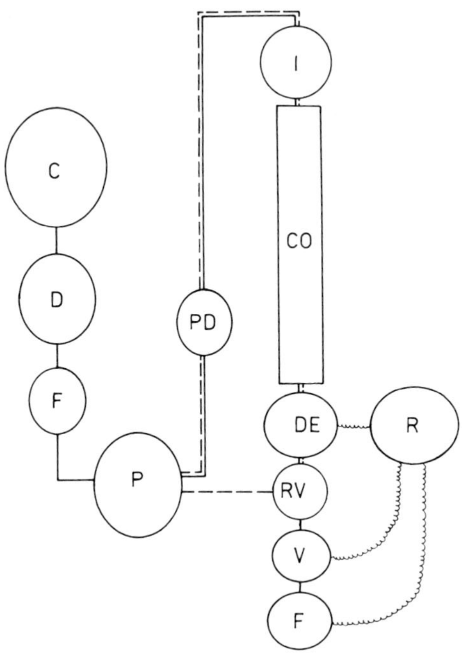

Fig. 4.6.4. Block scheme of gel chromatograph. For explanation see text.

A fundamental requirement for the efficient operation of a gel chromatographic instrument is that there should be no empty volumes where the sample components, which have already been separated, are again mixed. The equipment employed for the separation of biological materials must resist water and solutions of salts and be inert toward samples that can denaturate and otherwise change.

4.6.3.2 Transport of mobile phase

An *eluent container* can be a Mariotte flask that produces constant hydrostatic pressure, and thus it can serve for the generation of the gravitational eluent transport in the chromatographs of the first generation. Simple solvent tanks, or even open vessels, are used as eluent containers in the instruments provided with efficient pumping systems. The preferential evaporation effects, however, should be prevented when working with multicomponent, mixed eluents.

The *degassing unit* removes gases dissolved in the eluent. The gas bubbles released from the effluent may complicate the detection of sample. Aqueous eluents can be degassed by increased temperature, by application of vacuum or by displacing effect of helium.

Filters remove dust particles and other mechanical impurities from the eluents and solutions of samples. Filters made of sintered glass, porous polymers, and sometimes also of sintered stainless steel are used for aqueous mobile phases.

The *pumping system* transports eluent and sample through the column(s). Instead of the well known peristaltic pumps generating almost pulseless flow at low pressures, the pumps capable of producing high pressures are becoming more and more popular. These are the pneumatic pumps with amplifiers, the positive displacement-syringe type pumps, and above all, the reciprocating piston pumps that are sometimes provided with diaphragms (membrane pumps, cf., Fig. 4.6.5).

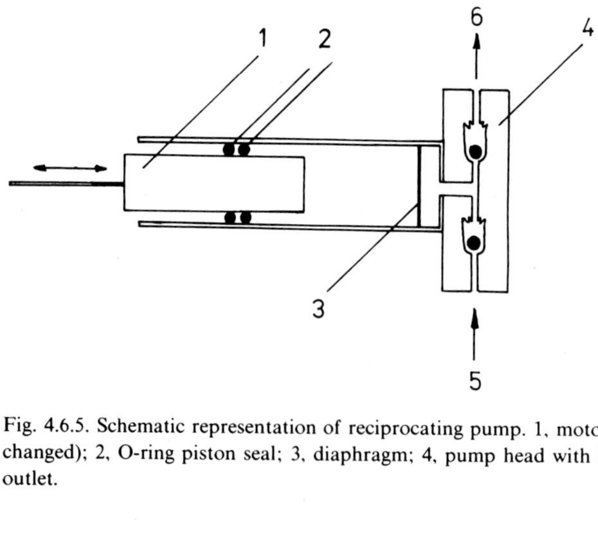

Fig. 4.6.5. Schematic representation of reciprocating pump. 1, motor driven piston (stroke height can be changed); 2, O-ring piston seal; 3, diaphragm; 4, pump head with back valves; 5, eluent inlet; 6, eluent outlet.

The modern reciprocating pumps with the extremely small head volume are able to produce very low and precise flow rate and are advantageously utilized for recycling. However, the piston seal of the reciprocating pump can be rapidly worn out when in direct contact with the aqueous mobile phases: the seals made most frequently of nonpolar, polytetrafluoroethylene based composites are not lubricated by water and the piston 'moves dry'. Further drawbacks of these widely used pumps are: (i) the sensitivity of their back valves to the mechanical impurities in the eluent (including the particles released from the seals); (ii) the inherent dependence of the flow rate on the hydrodynamic resistance of the chromatograph; and (iii) the pulsating flow they generate.

The problem of pulses is solved by means of damping devices that consist of a hydraulic resistor and a hydraulic capacitor. The modern reciprocating pumping systems are equipped with two or more pistons, their motion being mechanically or electronically controlled in such way that both short and long term flow rate variations are substantially reduced.

Connecting parts of particular components of the gel chromatographs are tubings made of metal, glass or plastics and the necessary unions, joining parts. Often, an excessive mixing occurs in the connecting tubings of large diameters and through untight joints, not only the eluent may leak out but also air can be sucked into the chromatographic system.

4.6.3.3 Sample preparation and application

The condition necessary for a successful GPC separation is the total solubility of the sample in the mobile phase at the temperature of experiment. If this condition is fulfilled, the sample can also be dissolved and applied into the column in the liquid different from the mobile phase, provided both solvents are fully miscible. Prior to application the sample solution must often be cleaned by filtration or centrifugation.

During filtration, the layer of very small, insoluble particles can also act as an ultrafilter trapping the largest macromolecules. In this case, either using filters with larger pores and larger surface areas, or centrifugation of samples, is recommended.

The application of the sample into the column can be carried out in different ways. In the GPC devices of the first generation, the sample is deposited on the gel bed surface after removing the excess of eluent by means of a syringe or pipette. In order to prevent the disturbance of the gel bed during the sample application, its surface should be protected by a filter or by a layer of an inert substance. In the chromatographs of higher generations, the samples are mixed with the stream of eluent just entering the column, or injected directly into the gel bed by means of special devices; most often by injection valves. The injecting valves enable the application of the sample without stopping the eluent flow and can be rather easily automated. The principle of a simple six-port three-way injection valve is shown schematically in Fig. 4.6.6. The valve loop can be changed and the volume of injected sample adjusted simply. Some injection valves are constructed so that a fraction of the loop volume can be reproducibly filled by sample. The injection

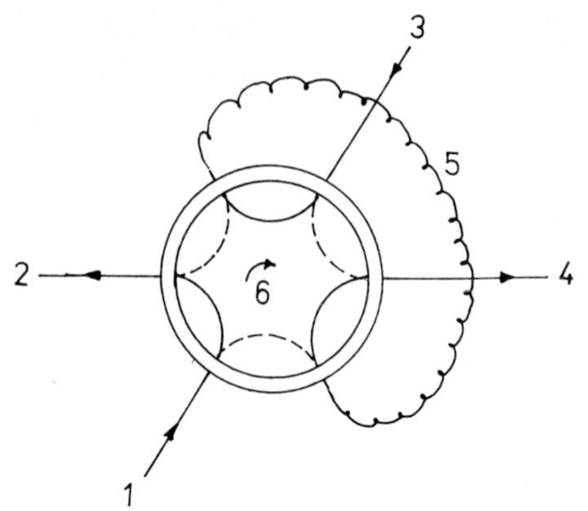

Fig. 4.6.6. Six-port three-way sample injection valve. 1, eluent inlet; 2, outlet to column; 3, sample inlet; 4, sample outlet; 5, sample loop; 6, rotor; ———, filling sample loop; ------, injecting sample.

valves can be further provided with a bypass capillary that allows the flow of the mobile phase into column, regardless of the position of the valve rotor.

4.6.3.4 Separation columns

Most often, the GPC separation process takes place in the gel bed packed into a straight column that is formed by a tube provided with column end fittings. Coiled columns show generally lower performances.

The columns are usually made of glass or stainless steel. The advantage of glass is the smoothness of its surface allowing higher efficiency and, above all, its inertness to the great majority of eluents and biological samples. The mechanical resistance of glass can be improved by special treatment with inorganic salts. The inner surface of metallic columns is usually smoothed by drilling and polishing, and sometimes is covered by a glass layer. For separations at low pressures, columns made of plastics, e.g., poly(methyl methacrylate) or polyolefins can also be used. Columns made of soft polymers can be cut and fractions extracted from gel ('drum chromatography').

The typical inner diameter of analytical gel chromatographic columns ranges from 2 to 8 mm. Recently, attempts were reported to apply packed capillary columns to gel chromatography. Preparative separations are performed in columns with diameters up to several decimeters.

The length of the gel chromatographic column covers the range from 100 to about 2000 mm, usually from 500 to 1250 mm in conventional GPC, and from 250 to 350 mm in HP GPC. Very often, several columns packed with different gels are combined forming the column system so that appropriate calibration curve is obtained (Fig. 4.6.1).

The most simple improvisation of the end part of a column of the first generation is shown in Fig. 4.6.7. In the modern column constructions, the gel is supported by special end fittings that can be divided into fixed (Fig. 4.6.8) and movable (piston-

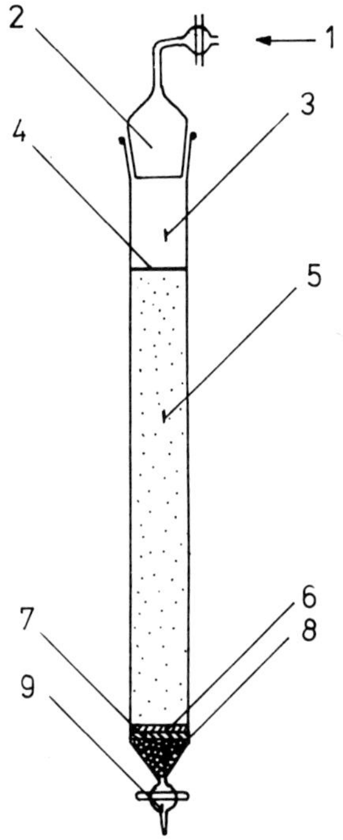

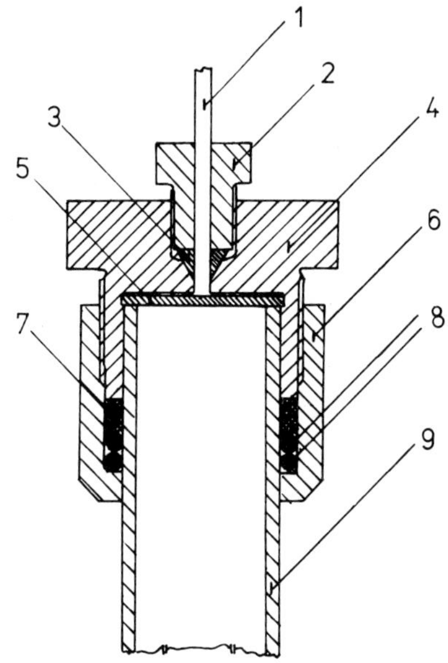

Fig. 4.6.7. (Left) Simple glass column for classical gel chromatography. 1, eluent inlet; 2, connecting adaptor; 3, mobile phase; 4, filter paper; 5, gel bed; 6, glass spheres 100–200 μm; 7, sintered glass plate; 8, glass spheres 0.2–2 mm; 9, stopcock.

Fig. 4.6.8. (Right) Example of the fixed column end fitting for glass columns, suitable for water and organic solvents (after Ref. 18). 1, inlet capillary; 2, thrust screw; 3, seal made of soft polytetrafluoroethylene; 4, end fitting body made of plastics or metal; 5, polyamide or polytetrafluoroethylene filter; 6, cap nut; 7, seal ring made of polytetrafluoroethylene powder or powdered mixture of graphite and polytetrafluoroethylene; 8, rubber rings; 9, column.

type, Fig. 4.6.9). The piston-type column end fittings are applied if the gel bed changes its volume in the course of use due to compression and deswelling of the gel particles.

Packing of columns belongs to the most important experimental procedures in gel chromatography, because it substantially influences the performance of the separation. Each type of gel needs a particular packing technique. Rigid inorganic gels with particle sizes over 30 μm can be dry-packed by means of tap-and-fill method. Semirigid and soft organic gels, as well as inorganic gels with particles smaller than 20 μm, must be packed by a slurry technique.

The dispersing medium for soft gels is the eluent, and both semirigid and rigid gels are packed from dispersing media chosen so that their density equals the density

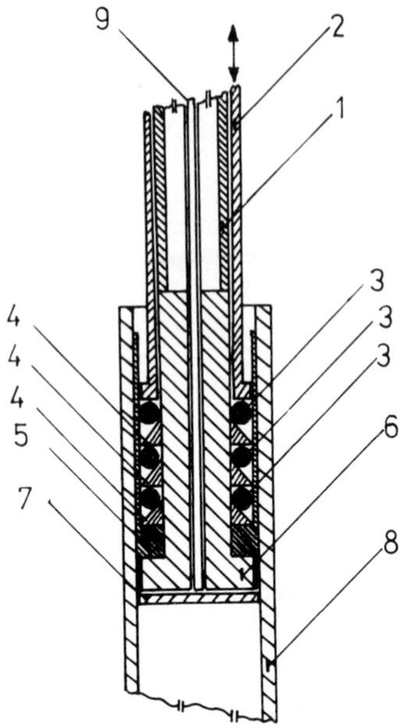

Fig. 4.6.9. Example of piston type column end fitting suitable for water and organic solvents (after Ref. 19). 1, fixed tube; 2, movable tube with pressure plate; 3, rubber O-rings; 4, rings; 5, polytetrafluoroethylene seal pressed toward column wall by rubber rings; 6, body of the end fitting made of plastics or inox steel; 7, filter; 8, column; 9, inlet capillary.

of gel matrix (*balanced slurry method*), or that their viscosity is rather high (*viscosity method*). Recently, a number of single and mixed dispersing media with matched polarity was reported to give good results. Further variables of the wet packing method are the slurry concentration, both packing pressure and rate, as well as the packing direction (from bottom up or vice versa). After packing is finished, the column is conditioned by exchanging the dispersing medium for eluent, if necessary in several steps. The packing procedures of major producers of high-efficient columns have not been published.

4.6.3.5 Operational variables

Within certain limits, the majority of operational variables do not determine the results of gel chromatographic separations in a decisive way. Nevertheless, it is advisable to test all variables for each particular system and application in order to obtain optimal and reproducible data. The most common operational variables in gel chromatography are the volume, v_i, and the concentration, c_i, of the sample applied, both influencing the retention volumes and the separation efficiency. In analytical separations, it is necessary to work with the lowest v_i and c_i allowed by the particular detection system (cf., section 4.6.3.6). The allowed sample volume depends primarily on the volume of the column and is usually several millilitres in the case of classical separations, 1–2 ml in the case of conventional separations and 5–100 μl for HP GPC. The concentration of applied solution usually ranges from 0.1 to 10 mg/ml.

In the preparative separations, c_i, is usually higher and v_i can be even substantially larger. The particular optimal operational variables are dictated by the required sample throughput on the one hand, and by the necessary separation efficiency on the other: it is well known that when certain limiting values of v_i and c_i are exceeded, the separation efficiency is completely lost. These limiting values depend on both the chromatographic system used and the sample separated: with increasing sample molar mass, the limiting c_i value may rapidly decrease.

The composition of the eluent, its pH and its ionic strength are very important operational variables for numerous gel chromatographic systems, mainly for the sensitive biological samples and for systems containing ionizable groups (cf., section 4.6.4.2) while temperature and mobile phase flow rate show less influence on the results of separation.

4.6.3.6 Detection

Detection is understood to be the determination of the concentration, and sometimes also the particular properties of the sample components in the effluent. The separated substances can be detected by discontinuous characterization of the collected fractions, by continuous measurement at the column outlet and — in special cases — also by direct column scanning in the course of elution. From the physical point of view, one can measure either the properties of the separated substances in the effluent (*solute property detectors* like photometers) or the properties of the effluent as a whole (*bulk property detectors,* e.g. differential refractometers). Many biological substances can be detected on the basis of measurement of their biological activity.

According to another classification, the detectors can be divided into specific, universal and mixed. For the choice of the detection method, the properties of sample and mobile phase as well as experimental requirements are decisive; however, sometimes several types of instruments based on the same detection principle are marketed. When judging the detectors one considers primarily: (i) *linearity, dynamic range* and *sensitivity*; (ii) erroneous responses like *noise* and *drift,* caused in the former by the instability of operational variables such as temperature variations, pulsating eluent flow, etc.; (iii) *response distortion* due to hydraulic broadening and skewing of the sample zone as well as *response delay.*

Presently, the most important detectors in gel chromatography are the photometers and the differential refractometers.

Photometers measure the changes in the intensity of light passing the measuring cell due to presence of the sample. The most frequently applied photometers determine the absorption of light in the ultraviolet range of spectrum. The necessary monochromatic light is obtained from the line spectra by means of filtration (e.g., the line at about 254 nm produced by the low pressure mercury arc is available in a great majority of monochromatic absorptiometers, while detectors using the band with a wavelength of about 215 nm, delivered by hollow cathode zinc lamp, have recently become available). The latter photometers may detect a rather wide spectrum of substances, e.g., saccharides. Another possibility is the use of a polychromatic source such as the deuterium lamp, and selection of the appropriate wavelength by a monochromator. The spectrophotometric detectors may also be utilized for scanning the absorption over the whole spectrum. This is important for peak identification in the case of complicated multicomponent samples and can be carried out either by applying the *stop-and-go technique,* i.e., by successive interrupting of the elution or by means of *rapidly scanning spectrophotometers,* even without flow interruption.

Infrared *spectrophotometers* and *fluorimetric* detectors are also very promising.

To the advantage of the photometric detectors are their generally very high sensitivity for samples intensively interacting with light and, if the measuring cell is properly built, also the independence of their response of temperature, flow rate and eluate refractive index variations. Evidently, the photometric detectors are applicable only to the substances containing corresponding chromophores.

Differential refractometers detect the changes in refractive index of eluate when compared with the refractive index of the mobile phase. Four principles are exploited in the construction of differential refractometers: deflection of light passing the boundary surface between the substances with different optical density (*deflexion refractometers*), and subsequent light scattering in case of multiple refraction (*refractometers based on Christiansen effect*), the Fresnel law of light reflection (*reflection refractometers*) and, finally, the interference of light (*interferometers*).

Differential refractometers are typical universal detectors. Their drawback is lower sensitivity in comparison with photometers, and the pronounced dependence of their response on the temperature and pulsating eluent flow, as well as on molar mass in the case of the lower members of polymer-homologous series.

Further important detectors utilize *electrochemical* as well as *conductivity* and *capacitance* measurements. Electrochemical detectors are absolute while conductimeters with extremely small measuring cells are very sensitive and rather specific detectors for gel chromatography of (poly)electrolytes.

A special group of detectors are those which continuously measure the molar mass of sample components in the eluate. Their application allows to omit the calibration of the gel chromatographic system. So far, *vapor phase osmometry* for solutes with lower molar masses, and *viscosity* and *light scattering measurements* for polymers have been applied.

An important and rapidly expanding detector group is that of the *derivatization* devices in which the sample components are reacted or interacted with appropriate substance so that the specific groups, e.g., chromophores, are introduced into their molecules and, consequently, they can be detected by the specific detectors (e.g. by photometers). *Turbidimeters,* in which the precipitate is formed by the controlled addition of a nonsolvent for the sample to the effluent, also belong to the group of derivatization detectors. Naturally, the nephelometers measuring turbidity can also be employed directly for detection of colloidal particles in the effluent, as well as of the aerosols formed by evaporation of mobile phase from the droplets of effluent containing macromolecules.

Several detectors connected in series can be used for simultaneous effluent detection, and there exist also the so-called combined detectors which measure the light absorption at two wave lengths, absorption plus fluorescence or absorption plus refractive index, in the same measuring cell.

The detailed discussion of detection in liquid chromatography can be found in the monograph by Scott [20].

4.6.3.7 Measurement of effluent volume

The retention volume of the sample is one of the most important data obtained by analytical gel chromatography.

In gel chromatographs of the first generation, the volume of each fraction collected is measured separately. In the instruments of the second generation, the retention volume is measured continuously, most frequently by means of a siphon. This device can easily be automated and utilized for controlling the injection system, and, eventually also the pumping system and fraction collector. The drawback of the siphon is its sensitivity to the variations of interfacial tension of the effluent and to the formation of the film of grease on its walls. Both effects are pronounced especially in the aqueous mobile phases. Smaller retention volumes can be measured by means of the drop counters or capillary volumeters. From the point of view of aqueous eluents, the drop counters exhibit the same limitations as siphons. The capillary volumeters are based upon counting how many times the effluent fills up the fixed volume of a capillary. The results of both, drop counters and capillary volumeters, depend on the flow rate and on the physical characteristics of the effluent.

According to the commercial literature, many modern pumping systems provide strictly constant flow rate so that it is sufficient to measure the retention time. It seems that this information must be taken into account with some precaution, especially in the cases of very low flow rates, long-term experiments and aqueous eluents. For measurements of very high precision, the retention volumes can be determined by automatic continuous weighing of the effluent [21].

4.6.3.8 Auxiliary equipment

Fraction collectors allow automatic collection of the effluent and are the first to be applied in preparative gel chromatography. The shift of the collecting vials is controlled by the impulses from the device responding to the change of their weight, from the volumeter, from the timer or from the peak-slope detector.

Thermostats control the temperature of the whole apparatus or its particular parts. Thermostats are used in GPC measurements of high precision, or in the case when it is necessary to perform the separation under reduced or elevated temperature due to sample sensitivity or low solubility, due to strong interactions between sample and gel.

Registration devices: for the registration of detector response in the instruments of the second and third generations, *potentiometric recorders* are most often used. The primary as well as integrated and otherwise processed data in the instruments of the fourth generation are registered by *printer-plotter devices*.

4.6.3.9 High speed separations

The acceleration of any chromatographic separation can generally be achieved by reducing the volume of the column, by elevating the linear velocity of mobile phase, or by the combination of both. In the former, one has to pay for the increased separation rate by the loss of selectivity, while the efficiency is sacrificed in the latter case. Since selectivity is principally rather limited in gel chromatography, the substantial increase in the general separation efficiency was the necessary condition for successful acceleration of gel chromatographic separations. In other words, high-speed GPC procedures can hardly be abstracted from high-efficient gel chromatography. The efficiency was increased by the application of microparticulate, rigid gels that are well packed in the column, by improving the sample application techniques and by the substantial reduction of all empty volumes in the system.

Presently, the high-speed GPC procedures are rapidly gaining importance also in the field of separation of biologic materials. Typically, one needs 10–20 min for a complete HP GPC analysis, compared with 2–5 h necessary for conventional separations and even several times longer classical separations. Semiquantitative HP GPC separations lasting only about 1–2 min were also reported using columns with very small volume and high flow rates.

Similar relative acceleration was achieved also in the modern preparative gel chromatographic separations.

Evidently, at high-speed GPC separations, very severe requirements are laid also on the other parts of the apparatus, in the first instance on the pumping system, injector and volumeters. The constancy of all operational variables must be more rigorously controlled, as well. On the other hand, the application of smaller columns means material savings, especially as the amount of both sample and eluent is substantially reduced.

4.6.3.10 Preparative separations

Under preparative gel chromatographic separation one understands the preparation of larger fractions that can be utilized either for further characterization by means of independent methods in clinical practice, for the calibration of the instruments, etc.

The sample throughput of a chromatograph can be increased by: (a) increasing the volume and concentration of applied sample; (b) repeating injection of the sample into a semipreparative apparatus and combining the corresponding fractions; (c) applying a continuous chromatograph; (d) utilizing the batch procedures; (e) applying a large preparative chromatograph.

The limitations of the first procedure were mentioned in section 4.6.3.5. The combination of fractions is highly exacting as to the separation reproducibility and repeatability, and exhibits rather low productivity. Continuous chromatography (cf., section 4.6.3.11) is a promising method for the preparative separation of simple mixtures. In the batch separations, the gel is brought into contact with the sample solution. After equilibrium has been reached, the gel that contains the smaller molecules is separated from the bulk solution by the filtration or centrifugation. Finally, the sample components that penetrated the gel are displaced by pure solvent. Evidently, this procedure can be considered chromatographic in the first approximation only. Batch procedures are suitable for the preseparation of large molecules, that are excluded from the gel pores, from the small molecules totally penetrating the gel. However, even in this advantageous case, the separation power is rather low. On the other hand, the method is rather simple and large quantities of the material can be rapidly processed.

Presently, preparative gel chromatographs are rapidly gaining wide acceptance. The volume of their columns can be as large as several litres, with the sample capacity reaching several hundred grams per run — depending on the type of gel and sample and on resolution desired. It is expedient to further increase the dimensions of the columns when using incompressible gels. Recently, high-efficient, preparative GPC columns became commercially available. They are packed with rigid gel particles of diameter about 15 μm which withstand the high pressures needed when the elevated mobile phase velocity is applied.

4.6.3.11 Special working procedures

The necessity to increase the resolution, speed and sample throughput, and to obtain some additional information about the sample, together with the effort to simplify or miniaturize the GPC apparatus on the other hand, gave rise to several special working procedures in gel chromatography [3,6,22]. These can be divided into:

(*i*) The procedures where the basic separation mechanism (steric exclusion) remains unchanged (recycling, continuous GPC, differential, vacancy and flat bed techniques).

(*ii*) The procedures where the steric exclusion mechanism is intentionally combined with another chromatographic separation mechanism such as adsorption or thermodynamic partition.

(*iii*) The procedures, where the gel chromatographic separation is combined with other, non-chromatographic, selective physico-chemical processes such as solvation, association, aggregation, precipitation, electrokinetic effects, etc.

The term *recycling* denotes the multiple elution of the sample through the same column or set of columns. By means of recycling, the gel bed available for separation

varied, is often called *multidimensional liquid chromatography*. In the first step, the samples are usually separated according to their size and the collected fractions are further separated in the columns packed by other gels or eluted by another eluent (sequential analysis). The effluent from the first column can be directly fed into various further columns by means of the special valve according to a present programme (*column switching*). The changes in the eluent composition can be made stepwise or in a gradient [3,29]. Modern liquid chromatographs are often provided with automated gradient forming devices.

Recently, the multistep combined procedures became fairly popular in the separations of complicated biological samples, especially in their purification, preseparation and preconcentration.

Large potentiality of the combination of different separation mechanisms within the same gel bed offer *bidimensional columns* [30], in which the bed is formed by a layer (as in TL GPC) in a sandwich arrangement. In one direction, the sample can be eluted in the steric exclusion mode and, using another eluent in the perpendicular direction, the elution is completed (as in the mode of sorption liquid chromatography).

Interesting results can also be obtained by the combination of steric exclusion with other physicochemical processes like selective complexation [31], selective precipitation [29], electrophoresis [32] or immunophoresis, as the selective complexing agent can serve the individually added substance and also the eluent itself [33]. The combination of gel chromatography with precipitation chromatography is called zone precipitation [29].

4.6.4 Materials for gel chromatography

Three fundamental kinds of materials used in gel chromatography will be discussed in this section, namely column fillings, mobile phases and reference materials.

4.6.4.1 Column filling materials — gels

The porous materials, gels applied in the gel chromatographic columns are almost exclusively in the particulate form.

In the beginnings of gel chromatography, the soft, swelling xerogels were used. The increasing elution rate and pressure drop in the columns required the application of semirigid and rigid aerogels.

The most important physical characteristics of the GPC gels are: pore shape, pore volume and pore diameters and their distribution, as well as particle size and shape.

Particle sizes of 100–200 μm are used in classical GPC, 35–80 μm particles are applied in conventional GPC, while HP GPC utilizes particles with diameters ranging from 3 to 15 μm. With decreasing particle sizes, their dimensional polydispersity gains in importance, and for obtaining high-efficient columns, the microparticles must be carefully sized and the dust removed. The most appropriate particle shape is spherical as it simplifies the column packing and decreases the hydrodynamic resistance of the gel bed.

The pore shape influences the mass transfer rate and thus the efficiency of separation. The effective diameter of pores determines the range of separated molar masses. The pore size distribution and the pore volume are decisive for selectivity of separation (section 4.6.2.3). The pore sizes of commercially available gels cover the region necessary for separation of the wide spectrum of substances — from low molecular samples to very high polymers, colloidal particles and viruses. The mean values of pore diameters range from few nanometers to about 2.5 μm. Gels with various pore sizes, but of the same type, can be combined within the same column.

Further important characteristics of the gels are: (i) their mechanical resistance toward pressure, abrasion and crushing; (ii) their chemical resistance toward eluents and samples; (iii) their biological resistance toward microorganisms that may grow in aqueous eluents and toward biologically active samples; (iv) their biological inertness — in order to avoid the changes and denaturation of samples; (v) their temperature resistance — if the separations at reduced or elevated temperatures are performed.

The above characteristics are connected in the first instance with the chemical structure of gels, and must be considered in the planning of the experiment. The bacterial resistance can be substituted by the addition of the bacteriostatic substance into eluent. The chemical composition of gels determines also their interactions with both the mobile phase and separated substances (cf., Section 4.6.2.2). Depending on the wettability of gel matrix with water, with organic solvents or with both, the column filling materials are classified as hydrophilic, lyophilic and universal. According to their nature, the gels are considered organic, inorganic and mixed.

The matrix of organic gels is formed by the organic macromolecular substances that are, with only few exceptions (agarose gels), crosslinked by chemical bonds. Polyreactions — polymerization and crosslinking, are usually performed in small droplets of dispersion and, consequently, spherical particles are formed. The physical structure of the gel depends largely on the course of the polyreactions and, according to this criterion, the gels are divided into homogeneously, semi-heterogeneously and heterogeneously crosslinked. In the first case, the polyreaction is carried out in the homogeneous system containing monomer(s) or polymer(s) and crosslinking agent, and the pore dimensions of the gel formed are determined by both the extent of the interactions between gel matrix and eluent and the crosslinking density (i.e., by the concentration of the crosslinking agent in the polymerization reaction mixture). Homogeneously crosslinked gels are soft and compressible. Before application, they must be swollen by water or the solvent that is subsequently used as eluent. Semi-heterogeneously crosslinked gels are formed by the copolymerization of the bifunctional and the multifunctional monomer(s) in the presence of an inert diluting agent that is the solvent for the linear form of the polymer formed. The resulting gels possess larger pores and exhibit higher rigidity in comparison with the homogeneously crosslinked gels. The heterogeneously crosslinked gels are formed in a similar way to the semi-heterogeneously crosslinked gels, but the inert diluting

TABLE 4.6.1

Commercially available gels for gel chromatography

Composition	Trade name	Supplier	Excluded M_M ($g \cdot mol^{-1}$)
1. *Dextran gels*			
a Dextran homogeneously crosslinked with epichlorhydrine	Sephadex, series G-10 to G-200	1	From 700 to 2×10^5 for dextrans or up to 8×10^5 for globular proteins
b Hydroxypropylated dextran gels	Sephadex LH-20 and LH-60	1	Depending on eluent up to 4×10^3 or 10^4
c Allyldextran crosslinked with *N,N*-methylene-bis-acrylamide	Sephacryl S-200 and S-300	1	2.5×10^5 or 1.5×10^6 for proteins and 8×10^4 or 7.5×10^5 for polysaccharides
2. *Agarose gels*			
a Agar cleared of polyelectrolytes, stabilized with H-bonds	Sepharose	1	
b Chemically crosslinked agar	Sepharose CL, 2B, 4B and 6B;	1	Typically up to 4×10^6 for proteins and 10^6 for polysaccharides for 6B or 1.5×10^8 for A-150m
	Bio-Gel A, series 0 to 150 m;	2	
	Super Ago Gel, several types;	3	
	Ultrogel A-2, A-4, A-6	4	
c Combined gels polyacrylamide-agarose	Ultrogel AcA, several types	3	From 1.5×10^4 to 1.4×10^7 for linear proteins
3. *Acrylate and methacrylate gels*			
a Copolymers of acrylamide and *N,N′*-methylene-bis acrylamide	Bio-Gel P, several types for oligomers and polymers	2	Up to 4×10^5 for globular proteins
b Copolymers of *N*-acryloylmorpholine with *N,N′*-methylene-bis acrylamide	Enzacryl Gel, Types K1 and K2	5	Up to 2×10^5 for dextrans
c Copolymers of hydroxyethylmethacrylate with ethyleneglycoldimethacrylate	Spheron, several types;	6	Up to about 10^8 for dextrans in water
	Separon HEMA, several types	7	
d Copolymers of oligoethyleneglycol dimethacrylate and ethyleneglycoldimethacrylate	Fractogel PGM	8	Up to about 2×10^3
e Copolymers of glycerolmethacrylate (?)	Shodex OH Pak B-804	9	About 4×10^5 for polysaccharides
4. *Polyvinylacetate gels*			
Copolymers of vinylacetate and butandiol-1,4-divinylether	Fractogel PVA, several types	8	Up to about 10^6 for polystyrene in tetrahydrofuran

Eluents	Notes
Water, water solutions of salts and very polar organic solvents	Soft gels. After swelling, lg of gel gives up to 40 cm^3 gel bed. Gels contain small amount of carboxyl groups. Strong adsorption of several compounds was observed. E.g. aromatics are generally retained.
Water and polar organic solvents	
Water and polar organic solvents	
Water and water solutions up to pH 9	Allowed temperature region 0–40°C. Gels contain small amount of charged groups
Water and water solutions at pH from 3 to 11 or from 4 to 10 according to the type	Temperature resistant up to 120°C. Contain small amount of charged groups
Water and water solutions	
Water and water solutions, pH from 2 to 10	Small amount of charged groups may be present
Water, water solutions and polar organic solvents	
Water, water solutions, polar and medium polar organic solvents	Fairly rigid gels. Separons are available also in glucose, or galactose modified forms
Medium polar organic solvents	Semirigid gel
Water and water solutions	Actual composition has not been published
Medium polar organic solvents	Polyvinyl acetate gels can be hydrolyzed to polyvinyl alcohol gels and used also in aqueous eluents

TABLE 4.6.1 (continued)

Composition	Trade name	Supplier	Excluded M_M ($g \cdot mol^{-1}$)
5. *Polystyrene gels*	*Produced by several companies:*		
a Copolymers of styrene and divinylbenzene	Styragel, several types; TSK gel H, HG and S, several types; Shodex A, several types; Chrompack Macrogel, several types;	10 11 9 12	Suitable for separation of oligomers and polymers up to about 10^8 for polystyrenes in tetrahydrofuran
	PL Gel, several types;	13	
	Hamilton neutral resins, several types	14	Suitable for oligomers
b Sulfonated copolymers of styrene and divinylbenze	Aquapak, several types; Shodex IonPak S, several types	10 9	For oligomers and polymers up to about 5×10^6 for polysaccharides
6. *Polyether gel* (?)	TSK gel PW, several types	11	Up to about 2×10^7 for polysaccharides
7. *SiO_2 based gels*	*Produced by several companies:*		
a Polymers of silicic acid, silica gels	Sphérosil, several types; Porasil; Lichrospher, several types; Zorbax PSM, several types; Vit-X, several types	15 10 8 16 17	Silica gels are generally not suitable for separation of oligomers based on steric exclusion. Polymers can be separated typically up to about 10^7 for polystyrenes in tetrahydrofuran
	Fractosil, several types;	8	Up to more than 10^8 for polystyrenes
	μ-Bondagel, several types; TSK gel SW, several types	10 11	Up to about 10^6 for proteins
b Porous glass	Controlled porosity glass, CPG-10, several types	18	Up to about 10^7 for polystyrene in tetrahydrofuran

Suppliers: 1, Pharmacia Fine Chemicals, Uppsala, Sweden; 2, Bio-Rad Laboratories, Richmond, CA, USA; 3, LKB Produkter, Bromma, Sweden; 4, Mann Research Laboratories, New York, NY, USA; 5, Koch-Light Laboratories, Colnbrook, UK; 6, Lachema n.e., Brno, Czechoslovakia; 7, Laboratorní Přístroje n.e., Prague, Czechoslovakia; 8, E. Merck, Darmstadt, FRG; 9, Showa-Denko Co., Tokyo, Japan; 10, Waters Associates Inc., Milford, MA, USA; 11, Toyo Soda Mfg.Co., Ltd., Tokyo, Japan; 12, Chrompack Nederland B.V., Middelburg, The Netherlands;

Eluents	Notes
Organic solvents of medium polarity	Polystyrene gels are either homogeneously crosslinked for oligomers, or heterogeneously crosslinked for high polymers. Majority of polystyrene gels is fairly rigid and can be used at elevated pressures. Several companies also produce microparticulate, 5–10 μm polystyrene gels for HP GPC. Some companies supply polystyrene gels exclusively packed in columns
Water and water solutions	
Water and water solutions	
Water and water solutions	Actual chemical structure has not been published
Typically water and water solutions at pH lower than 8 to 9 as well as organic solvents of different polarities	All SiO_2 based gels are rigid. Majority of silica gels is available in spherical form and several products also in microparticulate form for HP GPC Merck Co. also produces Lichrospher Diol for aqueous eluents
Water, water solutions and polar org. solv.	Irregular particles for classical and conventional GPC
	Chemically modified silica gel. Waters also sells Protein Columns I
Water, water solutions and polar org. solv.	Chemically modified silica gel. Sold also as Micro Pak TSK SW
Same as non modified silicagels	Irregular particles. Pierce Co. supplies also chemically modified CPG-10 porous glass, glyceryl phase for aqueous eluents

13, Polymer Laboratories Ltd., Salop, UK; 14, Hamilton-Micromesure, Bonaduz, Switzerland; 15, Rhône-Poulenc, Aubervilliers, France; 16, DuPont Co., Wilmington, DE, USA; 17, Perkin-Elmer Corp., Norwalk, CT, USA; 18, Electro Nucleonics Inc., Fairfield, NJ, USA.

Note: Only products known to the authors are given here and the list is necessarily incomplete and should in no case be considered as implying a recommendation of any company or product.

agent is a nonsolvent for the linear form of the polymer prepared. In such a system, the phase separation occurs in the course of the polyreaction that results in the formation of a rather rigid, corpuscular structure; the pores are formed by the voids within the clusters of primary, practically nonporous particles. Also interesting are the composite gels [34] in which the rigid matrix, prepared by the heterogeneous polyreaction, is in the second step filled by the soft, homogeneously crosslinked network. In this way, it is possible to combine to a certain extent the advantages of both types of gels.

Numerous materials were used for preparation of the GPC gels. The most important, commercially available column-filling materials are collected in Table 4.6.1 with the emphasis on the hydrophilic gels. The water-soluble substances were separated also in the columns packed by poly(vinyl alcohol) gels formed by the hydrolysis of the polyvinylacetate matrix, further by poly(*N*-acryloyl-L-prolyl morpholine), by cellulose and its derivatives, as well as by starch and lignin based porous materials.

The detailed discussion of the organic gels for chromatography can be found in the monographs [1,3,6,35–37], as well as in the reviews [38–41].

In the last period, the inorganic rigid aerogels are frequently used in gel chromatographic separations of biological samples. They are most frequently based on silicium dioxide [42] and are prepared by the controlled polymerization of silicic acid (silicagels) or by the selective leaching of borosilicate glasses in which phase separation had occurred (porous glasses).

Silicagel particles can be rather easily prepared in spherical shape, while porous glasses possess irregular shape.

The advantage of the SiO_2 based gels is their high mechanical, biological and thermal resistance. Their stability in the acidic milieu allows their purification by the concentrated mineral acids. Aerogels do not swell and therefore it is possible to change both mobile phase and temperature in the course of their use without the loss of separation efficiency. The drawback of silicagels and porous glasses is mainly their instability in the basic milieu, and slow dissolution of their matrix is observed even in pure water, and in aqueous solutions of many neutral salts. A further problem of application of SiO_2-based gels is their sorptive and ionic activity caused by the presence of the surface silanol groups that act as weak Lewis acids. The silanol groups may be shielded either by appropriate ionic or electroneutral polar substances added to the mobile phase, or by bonded organic groups of various polarities. The chemical modification of the silica surface can be performed, e.g., by its reaction with the appropriate chlorosilanes or oxosilanes. For separation in aqueous eluents, most important are diol and glylcerylpropyl phases. If the surface of the SiO_2-based gel is covered by nonpolar groups, it retains, selectively, the nonpolar sample components in the aqueous eluents due to strong hydrophobic interactions while the polar admixtures are eluted. The following elution by a nonpolar mobile phase results in the separation according to the steric exclusion mechanism (cf., Section 4.6.3.11). The problem is again the hydrolytic instability of the bonded organic groups.

The pores of silicagels can by filled by the homogeneously crosslinked organic gels [43].

Other materials that can be used for preparation of rigid aerogels are carbon [44] and aluminium oxide [45].

4.6.4.2 Mobile phases — eluents

Severe requirements are laid on the gel chromatographic mobile phases. They must dissolve the sample analyzed, must be compatible with both the detector and the gel, must be chemically and biologically inert toward the sample, and non-corrosive for the constructing materials of the instrument. Further on, the eluents should suppress the interactions between the sample and the gel, should exhibit low viscosity, low coefficient of compressibility, and low vapour tension. Finally, they should be minimally poisonous and explosive as well as cheap and readily available in rather high purity.

To fulfill simultaneously all the above requirements with a single solvent is practically impossible, and therefore multicomponent eluents are used rather extensively. However, the presence of the second component in the eluent can bring about several problems: the preferential evaporation of one of the components causes changes in the mobile phase composition. Due to the preferential solvation of the gel matrix [46], the composition of the gel phase will differ from the composition of the eluent. The sample may displace the preferentially sorbed eluent molecules from the gel and, consequently, the zones of the mobile phase with changed composition travel through the column. Similar effects, and sometimes also the detection problems, are brought about by the preferential solvation of the sample molecules [47]. The above phenomena cause the disturbances of the base line and the appearance of the non-defined 'ghost' peaks on chromatograms when working with the non-specific detectors. Simultaneously, the peak shape and the retention volume of the sample can also be influenced.

In gel chromatography of biological systems, water-based eluents are dominant. The main reason for using the aqueous milieu is both the low solubility and susceptibility to the irreversible denaturation of many biologically active substances in organic solvents. However, pure water is applied only exceptionally, e.g., when separating saccharides. In the majority of applications, one works with the aqueous solutions of various substances, which means with the mixed eluents, and therefore the above mentioned problems are often encountered.

Many samples of biological origin require an optimal pH range, beyond which they lose their biological activity and may also change some of their physical properties. That is why the aqueous eluents are usually buffered. When choosing buffers, it is necessary to consider not only the nature of the sample but also the nature of the gel: e.g., the borate buffers should not be used with the polysaccharide gels. For preparative purposes, volatile buffering compounds are preferred since they can easily be removed from the fractions by lyophilization. The shielding effect of the buffers towards charged groups of the gel and the sample is also very important, i.e., the suppression of the ionic effects mentioned in Section 4.6.2.2 [48], as well as the association and aggregation of the sample molecules and their sorption on the gel. Neutral salts are often added to the mobile phase in order to adjust its ionic strength.

Further additives to the mobile phase involve the structure-breaking, solubilizing and dissociation promoting agents that prevent the excessive solute-solute interactions or ensure the appropriate and conformationally stable structure of biological macromolecules, e.g., random coil conformation. These additives are most often urea

TABLE 4.6.2

Some important polymer reference materials

A	Proteins and viruses		$M_M \times 10^3$, $g \times mol^{-1}$
1.	Insulin B-chain (bovine)		3.4
2.	Glucagon		3.5
3.	Insulin		5.7
4.	Kallikrein		6.5
5.	Cytochrome *C*		12.3
6.	Ribonuclease A (from bovine pancreas)		13.7
7.	Methemoglobin		17.0
8.	Apomyoglobin		17.2
9.	Myoglobin (equine)		17.8
10.	Trypsin inhibitor (from soybean)		22.0
11.	α-Chymotrypsin (from bovine pancreas)		25.0
12.	Chymotrypsinogen A (from bovine pancreas)		25.0
13.	Tropomyosin (from rabbit muscle)		33.5
14.	Pepsin		35.5
15.	β-Lactoglobulin B		38.0
16.	Ovalbumin		45.0
17.	Bovine serum albumin	(monomer)	66.0
		(dimer)	132.0
		(trimer... etc.)	
18.	Hemocyanin (cross-linked, monomer)		70.0
		(dimer)	140.0
		(trimer... etc)	
19.	Transferrin (human)		80.0
20.	Aldolase (rabbit muscle)		158.0
21.	Alcohol dehydrogenase		150.0
22.	γ-Globulin (bovine)		170.0
23.	Catalase (from bovine liver)		232.0
24.	Fibrinogen (human)		340.0
25.	Ferritin (from horse spleen)		560.0
26.	Turntip yellow mosaic virus		3.6×10^3
27.	Tobacco mosaic virus		39×10^3

The above reference materials are available from the following companies: Boehringer, Mannheim, FRG; BDH Chemicals Ltd., Poole, UK; Calbiochem, Lucerne, Switzerland; Fluka AG, Buchs, Switzerland; Intermed-Export-Import, Berlin GDR; Koch-Light, Laboratories, Colnbrook, UK; E. Merck, Darmstadt, FRG; Pharmacia Fine Chemicals, AB, Uppsala, Sweden; Reanal Fine Chemicals, Budapest, Hungary; Serva Feinbiochemica, Heidelberg, FRG; Sigma London Chemical Company, Ltd., Poole, UK.

B Narrow fractions of natural and synthetic polymers

Name	M_M range, g × mole^{-1}	Suppliers
1. Polystyrenes	6×10^2–2×10^7	1, 2, 3, 4, 5, 6
2. Polymethylmethacrylates	10^3–2.5×10^5	5, 6
3. Polyvinylacetates	9×10^3–10^6	7
4. Polyvinylalcohols	2×10^4–4×10^5	7
5. Dextrans		
also coloured high molecular fractions available – Blue Dextran and Orange Dextran	10^4–2×10^6	8
6. Polyethyleneoxides	2.4×10^2–1.2×10^6	2, 6

Suppliers: 1, Pressure Company, Pittsburgh, PA, USA; 2, Toyo Soda Manufacturing Co., Ltd., Tokyo, Japan; 3, Waters Assoc. Inc., Milford, MA, USA; 4, National Bureau of Standards, Washington DC, USA; 5, Chrompack Nederland B.V., Middelburg, The Netherlands; 6, Polysciences Inc., Warrington, PA, USA; 7, National Physical Laboratory, Teddington, UK; 8, Pharmacia Fine Chemicals AB, Uppsala, Sweden.

Note: Polystyrenes, polymethylmethacrylates, polyvinylacetates and polyethyleneoxides are soluble in many organic solvents; polyvinylalcohols, dextrans and polyethyleneoxides are soluble in water and water solutions of salts

Only products and companies known to the authors are given here and the list is necessarily incomplete and should in no case be considered as implying a recommendation of any company or product.

and guanidinium chloride, surface active substances like sodium dodecyl sulphate, etc. Polar organic solvents are sometimes added to aqueous eluents in order to suppress the hydrophobic interactions of the sample molecules with the gel.

Finally, the stabilizing agents such as antioxidants, bacteriostatics and viscosity enhancing substances are added to the aqueous mobile phases when necessary.

In some special applications, organic solvents are also used as eluents. The highly polar solvents like formamide, dimethylformamide or dimethylsulphoxide dissolve many biopolymers, while for chemically modified biological macromolecules even less polar solvents can be used. Formamide and dimethylformamide also dissolve numerous inorganic salts like LiBr, LiCl, KSCN, etc., and the corresponding solutions can be used — similar to the water solutions of salt — as eluents suppressing the ion effects in gel chromatography of the charged samples.

4.6.4.3 Reference materials — standards

The reference materials are widely used for the calibration of gel chromatographic instruments, namely for determination of the dependences: logarithm of sample molecular dimensions versus retention volume; detector response versus sample concentration; zone broadening versus operation variables or versus molecular dimensions of sample.

The low molecular reference materials are pure substances prepared by the conventional or chromatographic procedures. The higher molecular reference materials possess narrow or at least known distribution of the molar masses and the mean values of their molar masses are precisely determined by independent methods. They are isolated from the raw materials by various methods of purification and fractionation, often also by gel chromatography. Synthetic macromolecular reference materials (mostly polystyrenes) are prepared by special, controlled polyreactions, e.g., by anionic polymerization.

When using molar mass as the calibration parameter, one has to consider the shape of macromolecules in solution during the data evaluation: e.g., it is not possible to calculate directly the molar mass of the randomly coiled proteins from the calibration curve obtained by means of the globular proteins.

The basic data on commercially available macromolecular reference materials with an emphasis on the water soluble polymers are collected in Table 4.6.2.

4.6.5 Areas of applications

Biological materials usually represent rather complex heterogeneous systems. Their behaviour in vitro is, therefore, frequently hard to predict, and a flexibility in the selection of conditions for gel chromatography is required if the biological activity of the materials is to be preserved. Since biological materials remain in a native or reversibly denatured state only under certain suitable conditions, selection of the latter frequently is of decisive importance for the results of chromatography. Of the conditions which have to be taken into account, the following should be mentioned: pH, ionic strength, presence of certain substances in the environment, temperature and osmotic pressure. Chromatography is frequently carried out in the presence of high concentrations of urea, of guanidinium chloride, detergents or some inorganic salts, or at increased or lowered temperature, etc.

Because in gel chromatography steric exclusion is the dominating mechanism, the most common application of this type is the purification of a mixture of substances differing from one another by particle size. We think of purification in a broad sense depending on the degree which can be reached by the separation process itself or which is required. Various modes of GPC application will be discussed from this point of view.

Group separation. This designates separation of substances into two groups of low and high molar mass, respectively. In practice this means that a mixture of macromolecules is separated from various low molar mass substances, in particular from salts. The most simple and at the same time the most efficient way of applying this procedure is in the case of mixtures of one high molar mass substance with one or several low molar mass substances.

Isolation means separation of a substance, usually a macromolecule, from a mixture of other substances of high and low molecular mass. This procedure is methodically more pretentious than group separation.

Fractionation is more complicated than either of the two preceding procedures because it means separation of two or more substances or fractions from a mixture in the course of a single operation.

Several other applications can be derived from these three basic applications, for example:

Determination of the size of macromolecules. The uses follow from the principle of the method itself. Determination is relative, because it is derived with reference to substances of known molecular size. A prerequisite is a uniform conformation of all substances tested, including the reference substances; moreover, analysis has to be carried out in the same environment. Particle size is often expressed by Stokes radius (cf., section 4.6.2.3).

Determination of the molar mass (M_M) has been applied more frequently than the preceding procedure. As in the latter, the conformation of the test and reference substances has to be the same and the same environment is required for the analysis. The analysis is based on the evaluation of the dependence of log M_M on the retention volume. If analysing a mixture of substances which has to be fractionated, the distribution of molar masses can be determined in a single run.

In addition to these most important uses of gel chromatography some further applications may be mentioned. If we are interested in the degree of purity of a substance in a crude extract or in the course of purification, i.e., if we wish to determine the extent to which it is contaminated by substances of other M_M values, GPC can be used to determine the purity of the product. In substances possessing the same biological importance or effect, their diversity (or identity) may be determined by GPC. This approach has been employed, e.g., for the characterization of calcium-binding proteins in different organs [49], in the study of isoenzymes [50,51] and toxins [52], etc. The hit effect on the macromolecule studied of certain substances, temperature, pH, irradiation, oxygen, etc. can also be followed by GPC.

Of especial importance, in particular in medicine, is the screening examination, mainly of body fluids (like serum, urine, cerebrospinal fluid, saliva). In such analyses the chromatographic pattern in a healthy person is compared with that in a patient. A deviation is indicative of a pathological condition and may be specific. GPC is very suitable for such purposes because it is highly reproducible and a single column may be repeatedly used.

Besides the basic or conventional applications mentioned above, gel chromatography has also been used in numerous less common, non-conventional measurements:

By means of gel chromatography, structure of macromolecules can be elucidated (e.g., their branching [53]), and in some cases also their tacticity [54]. GPC was applied fairly often to the study of the interactions of the macromolecules such as association and aggregation [54–59], the association with ligands [60,61], or the preferential solvation of macromolecules in mixed solvents [62], and the interactions in the polyelectrolyte-salt-solvent [63] systems. GPC can provide information concerning the polymer-solvent interactions [64] as well as about the changes in the macromolecular dimensions with concentration in solution [65,66].

Porath [67] and Halász et al. [68] suggested the use of gel chromatography for the characterization of the porous structure of gels (inverse gel chromatography). GPC data can be utilized also for calculation of limiting viscosity numbers $[\eta]$ of macromolecules [69] and for estimation of the corresponding constants $\underline{K}$ and $\underline{a}$ in Mark-Houwink equation [70]. The application of gel chromatography is interesting because of the size separation and characterization of the colloidal dispersions [71].

A detailed discussion of the non-conventional applications of gel chromatography can be found in the monograph by Berek et al. [6].

All applications mentioned above can be done on an analytical or preparative scale depending on the size of the column used. In preparative applications, the great advantage of GPC over many other analytical methods should be stressed, namely the fact that the conditions of analysis in no way affect the structure or any other property of the macromolecules. By contrast, GPC may be carried out in whatever environment is most suitable for the test substance. As a consequence the method not only allows an analysis of the biological material but also preserves its biological activity, so that the fractions can be subjected to further study of their biological properties.

Recently, high speed GPC has become used more and more frequently. It is carried out under increased pressure of the eluent and, therefore, special equipment has to be employed. Analytical uses thus prevail. The time required for analysis is several times shorter and more accurate results are obtained than with the conventional procedure. For details see section 4.6.3.9.

Reports dealing with the various applications of GPC are too numerous to allow a survey offering a picture of the whole spectrum of all the possibilities used so far. We offer at least a selection of a few topics giving some hints in this respect. The examples will be presented for the individual types of biological material. For comprehensive reviews see also Refs. 1, 3, 5, 6, 36, 72 (see also Note following References, p. 320).

4.6.5.1 Proteins and peptides

In studies on polypeptide chains the proteins are dissociated by 6–12 M urea, 4–6.5 M guanidinium chloride in a buffer solution, usually in moderately alkaline region, in the presence of 2–5% 2-mercaptoethanol causing reduction of disulphide bonds. Among other reagents, one that is frequently used is surface-active sodium dodecyl sulphate (SDS) in a 0.1–5% concentration. To prevent association or even aggregation of polypeptides, the reagents mentioned above cannot be excluded in the course of chromatography. This results in a situation mentioned in the introduction to this section that the whole process of chromatographic separation has to be carried out in an environment equal or similar to that used for isolation of the individual structural subunits; one of the important steps to prevent artifacts (cf., Sections 4.6.2.2 and 4.6.4.2).

On the other hand, denaturation by such reagents results in loosening of the original protein structure and the polypeptide chain configuration becomes a random coil which leads to changes in retention volume. This fact has to be taken into account especially in molar mass determinations. SDS is bound to protein proportionally to the size of the protein molecule so that its use results in an increase in the apparent molar mass. Nevertheless, acceptable results may be obtained by standardizing the column with suitable reference materials of the same steric configuration, and in the same environment as the test substance. In this way differences due to an eventual effect of the mobile phase on the gel may be eliminated. Different kinds of gel also may affect the behaviour of the solute. For example, a comparison of the patterns of protein mixtures on three different gels, namely Sephadex G-75 superfine, Ultrogel AcA 54 and Bio-Gel P-100 minus 400 mesh, showed that the latter gave the best resolution when 0.05 M Tris-HCl buffer (pH 8.5) containing 1 M KCl was used. With Bio-Gel P-100, the flow rate was 3 ml/h. This slow flow rate was considered by the authors [73] sufficient to a certain extent, e.g., from the point of view of minimal protein denaturation. The efficiency of separation on three kinds of type TSK gels was investigated on an artificial polypeptide mixture in the native state [74], and in the presence of 6 M guanidinium chloride [75]. Imamura et al. [76] studied the conditions of analysis of some polypeptides by high-speed gel chromatography on TSK GEL G 3000SW, using Na-phosphate buffers of different concentrations. Basic proteins were adsorbed on the gel surface at low buffer concentration in native state, in the presence of 8 M urea as well as in the presence of SDS which forms polypeptide-SDS complexes with a high negative charge. But the K_D values of polypeptides in 6 M guanidinium chloride were independent of buffer concentration. According to the authors guanidinium chloride eliminates electrostatic interactions between the polypeptide and gel surface. Lin and Castel [77] demonstrated the effect of pH and salt concentration on the structure, and consequently the properties of agarose gel and thus their effect on separation. Increasing concentration of buffer or salts (e.g., ammonium sulphate) in the protein sample leads either to broadening of the protein peak, which means an inferior separation efficiency, or to translocation of the peaks into positions which would correspond to much smaller hydrodynamic volumes of the macromolecule.

Blood serum, separated by gel chromatography on Sephadex G-200, yields three main fractions, one formed by a mixture of lipoproteins and α- and β-globulins, the second containing mainly γ-globulin and the third consisting of albumin and partially transferrin. Of course, human plasma contains a number of further components with specific functions, for example, kinin, which occurs in two forms with low and high M_M, respectively, was purified in 9 steps [78]. The individual purification steps consisted of ionex chromatography on DEAE-Sephadex and CM-Sephadex, and gel chromatography on Sephadex G-150 and G-100. The yield of purified preparation with respect to serum was 1.2%, but the specific activity increased from 0.033 in serum to 8.28. The M_M, determined by chromatography on a Sephadex G-100 column, was 78 000 g/mol. When assayed by polyacrylamide gel electrophoresis, kinin purified in this way gave a single homogeneous zone. It also proved to be homogeneous on immunoelectrophoresis and ultracentrifugation. The effect of temperature on the efficiency of chromatographic separation was demonstrated in separating fibrin and fibrinogen from plasma [79]. Purified ^{125}I-labelled fibrin monomer was added to rabbit plasma containing [^{131}I]fibrinogen. At 20°C fibrin was partially precipitated and about 50% of it was eluted; at 37°C the yield reached 90%. With fibrinogen, sorption on the gel occurs at 37°C, but at 20°C only 3% is eluted. Fibrin may be separated at an appropriate temperature regimen, but considerable errors in quantitative estimates of the proportion of the two components may occur with the use of separation by GPC.

In dogs with experimental coronary thrombosis samples of blood were taken from the aorta and coronary sinus at intervals of 2, 4 and 24 h and subjected to gel chromatography [80]. A deviation from the normal pattern was observed as early as after 2 h. Such an early change in the characteristics of the chromatographic elution curve is a prerequisite for GPC being used in early diagnosis with eventual application in man.

Gibb et al. [81] compared the properties of normal sera from various fish species by sedimentation analysis and gel chromatography. By the latter, the sera were separated into three main fractions. The serum from each species gave a specific elution curve.

deLigny et al. [82] investigated the effects of NaSCN and Na_2SO_4 in the eluent at low pH on the distribution coefficient of bovine serum albumin in GPC and found that the distribution coefficient changed depending on the nature of electrolyte present in the eluent.

In the characterization and especially purification of enzymes GPC has become indispensable but could not replace older procedures like ammonium sulphate precipitation and ionex chromatography; the latter two methods still remain very efficient purification procedures. Electrophoresis, especially in polyacrylamide gel, has gained a firm position, especially for characterization of substances. A combination of gel chromatography with other methods proved to be very efficient. A few examples follow: thiamin pyrophosphokinase was purified 3000-fold by two cycles of ionex chromatography and subsequent gel chromatography with a yield of 12% [83]; 1,4-β-D-xylan synthase was purified 28-fold by chromatography on CM-Sephadex and Bio-Gel P-10 [84]; and bacterial lipoprotein lipase was purified 34-fold with a yield of 40% by gel chromatography on Sephadex G-50 and G-200 [85]. Argininosuccinate-lyase was purified by ammonium sulphate precipitation, ionex chromatography on DEAE-cellulose and gel chromatography on Sephadex G-200. The final preparation was purified 46-fold with respect to crude extract [86]. A 940-fold purification of metmyoglobin was reached by gel chromatography on Sephadex G-25, ammonium sulphate precipitation and rechromatography on Sephadex G-75 [87]. A 15-fold purification of crude β-galactosidase from bacterial cells and commercial urease was obtained after one run on a TSK-gel G3000SWG column [88]. Recoveries of enzymic activities were almost 100%.

Gel chromatography may prove useful for the separation of biomacromolecules at different degrees of polymerization, as is the case with synthetic polymers; for example, Nakasaki et al. [89] purified alkaline phosphatase from rat small gut. One of the purification steps was gel chromatography on a porous glass column-filling material which resulted in separation of the enzymatic activity into three fractions. Further analysis of the fractions by polyacrylamide gel electrophoresis in the presence of SDS showed that all three peaks had the same molar mass of 110 000 g/mol. It was calculated that the individual fractions corresponded to the monomer, dimer and a higher aggregate, respectively. Gel chromatography and isopycnic density gradient ultracentrifugation revealed the presence of a monomer-octamer mixture of cytochrome b_5. At the same time, the Stokes radius of the monomer, 2.6 nm, was determined by gel chromatography [90].

Polyacrylamide gel electrophoresis has greatly contributed to the elucidation of enzyme structure in the sense that enzymatic activity may be displayed not by one but by several substances with distinct chemical structure and molar masses, the so-called isoenzymes. Data obtained by electrophoresis usually are of an analytical character. GPC offers the possibility of obtaining isoenzymes of different M_M on a preparative scale. Three isoenzymes of different M_M were identified and prepared by GPC in purified polyphenol oxidase from tobacco [50]; a multicomponent character was demonstrated by GPC in exo-xylanase and endoglucanase [51]; and chromatography on Sephadex G-200 made it possible to isolate isoenzymes also in sphingomyelinase [91] and α-D-fucosidase [92]. Kanaya and Uchida [93] checked the purity of a ribonuclease T_2 preparation specially purified by affinity chromatography on 5′-adenylate-aminohexyl-Sepharose. Gel chromatography on Sephadex G-75 revealed a heterogeneity in molar masses: 36 000 and 80 000 g/mol. Ribonuclease of the higher M_M was separated on Sephadex G-150 into a further five fractions.

GPC has become widely used for determining the molar masses of enzymes. Only a few examples can be mentioned here: of the synthetases oestrogen [94], 7-phosphate synthetase [95]; of the dehydrogenases aldehyde dehydrogenase from the yeast *Saccharomyces cerevisiae* [96,97], malate dehydrogenase [98], *N*-methyl glutamate dehydrogenase [99]; of the other enzymes β-D-galactosidase [100], cAMP-dependent protein kinase [101], endonuclease [102] and collagenase [103].

GPC proved of great value in the preparation of viral proteins, because the fractions obtained could be subjected to further analysis. A study of viral proteins is of interest from the point of view of not only the virion structure but also the immunological properties.

After disintegration and gel chromatography in the presence of 6 M guanidinium chloride, 10 polypeptides were obtained from visna virus. Serological examination showed that the polypeptide occurring in the fraction of an M_M of 25 000 g/ml is responsible for the immunological response [104]. An analysis of the subunits by GPC in combination with electrophoresis proved to be even more efficient. When the polypeptides were prepared by treatment with the nonionic detergent Nonidet P-40, and the fractions obtained by gel chromatography further analysed by polyacrylamide gel electrophoresis, the results suggested that visna virus contains 25 structural polypeptides [105].

Michailova et al. [106] used GPC on Sepharose 4B for a comparison and characterization of adenovirus type 1 and 2 hexons.

Interesting results were obtained with the use of GPC in toxin analysis. Some toxins (e.g., snake venom, fungal and bacterial toxins) are known to consist of two components of different M_M and their effect is based on a joint action of these components. One of the possible mechanisms of action may be derived from some plant toxins (e.g., abrin and ricine) in which it was found that one polypeptide possesses the toxic activity and the other has the function of binding to the cell. Both peptides are connected together by the disulphide bonds. The joint action of two toxin components of different origin, namely crototoxin B from snake venom, and volvatoxin A_2 from an edible mushroom was studied by Jeng and Fraenkel-Conrat [52] with the use of GPC. The results showed that the two components studied exert a joint neurotoxic action and that the two proteins form a physical complex. Complex formation was assayed by GPC so that the individual components were identified based on the retention volume. With haemolytic toxin from *Bacillus alvei*, GPC was employed for its purification [107]. By combining gel with ion exchange chromatography a 3000-fold purification of cellular retinol-binding protein with a yield of over 40% was possible. The M_M, determined by gel chromatography, was 14 800 g/mol [108].

A very good separation of thiamine-binding protein from various rat organs was achieved on a TSK-gel G 3000SW column [109]. The method proved to be rapid and sensitive.

Analysis of histones is frequently carried out in the presence of detergents which may bind to them. Binding of the nonionic detergent Triton X-100 to histones from calf thymus can be analysed by gel chromatography. The binding of Triton to histones was abolished if 8 M urea was present in 0.01 M HCl or 0.9 M acetic acid [110]. Subsequently [111], the effects of the cationic detergent cetyltrimethyl ammonium bromide (CTAB), and the anionic detergent sodium dodecyl sulphate (SDS), in GPC and electrophoresis were compared. The results showed that, in the presence of CTAB, the values of the retardation and distribution coefficients were linear in respect to the molecular mass of histones, while in the presence of SDS, deviations from this type were observed.

A search was made for optimal conditions of insulin separation by comparing three different silica-based chromatographic supports in GPC under denaturing conditions [112]. The separation patterns were identical with those obtained with soft gels.

α-Fetopterin from rat allantoic fluid has an M_M very similar to that of albumin and its purification is, therefore, very difficult. A break in this situation followed after it was found that in GPC on a column of Ultragel AcA-44, measuring 2.5×150 cm and with 0.01 M phosphate buffer (pH 7.4) used as eluent, α-fetopterin is eluted in a distinct fraction [113]. GPC has been employed for the purification of a number of other substances, e.g. of skeletin [114], subfragments of thymus leukaemia antigen [115] and chloroplasts [116].

4.6.5.2 Nucleic acids and nucleotides

Nucleic acids are linear polymers and as such differ considerably by their structure from proteins. The steric configuration of nucleic acids is very strongly affected by the environment, e.g., ionic strength of the solution. It was shown experimentally that the secondary structure of nucleic acid affects the retention volume more than the molar mass. GPC analysis of nucleic acids becomes more complicated with increasing size of their molecules. To separate nucleic acids with molar masses of 1 and 2 million g/mol, gel with a pore diameter from 500 to 1000 nm is necessary [117]. Therefore, reports in the literature usually concern an analysis of nucleic acids with a smaller molar mass.

In spite of this GPC of nucleic acids is frequently affected by various working conditions to such extent that it does not follow the expected course, the method has been used for solving various problems. Several tRNA species were successfully separated and their distribution constant in the presence of CTAB was determined [118]. tRNAs were separated from plant tissues [119], as were mRNAs from a mixture of various RNA types [120]; the degree of DNA degradation by endonuclease and exonuclease, etc., was followed [121]. An elegant example is the investigation of the effect of carcinogen (9-^{14}C)-labelled *N*-acetoxy-*N*-acetyl-2-aminofluorene on DNA [122]. Hydrolysate of the modified DNA was separated by chromatography on Sephadex LH-20 into two fractions: 85% DNA occurred with the preparation bound in the form of *N*-(deoxyguanosine-8-yl)-*N*-acetyl-2-aminofluorene, which caused conformational changes in the DNA double helix and denaturation of certain segments of DNA. The remaining 15% of the DNA was modified as 3-(deoxyguanosine-N^2-yl)-*N*-acetyl-2-aminofluorene which caused no changes in the native structure of DNA.

Recently, Japanese silica-based aqueous gels (TSK-GEL) of the SW series possessing minimal sorption properties, have become widely employed. They proved to be very efficient in GPC of proteins. Since these gels are not compressible, they

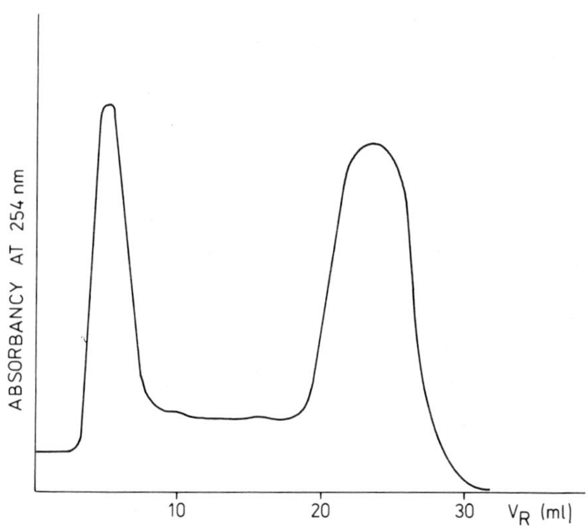

Fig. 4.6.10. Separation on a Sephadex G-25 column of RNA from phenol dissolved in the aqueous phase (after Ref. 117).

can be used for high-speed GPC. Uchiyama et al. [123] employed TSK-GEL 3000 SW for the separation of low M_M RNAs. They obtained good separation of three RNA species (5.8 S RNA, 5 S rRNA and tRNA) from each other and from RNA with greater M_M. With 200 mM Na-phosphate buffer (pH 7.0) containing 0.1% SDS used as eluent, 5.8 S rRNA from rat liver cells was eluted at another position than the same RNA from the yeast *Saccharomyces cerevisiae*; these two RNAs have the same chain length but another sequence. Molka et al. [124] used, for the separation of synthetic deprotected oligonucleotides, a I-125 Protein Analysis Column (Waters Assoc.). With 0.1 M triethylammonium acetate (pH 7.0) and a flow rate of 1.04 ml/min they obtained a linear plot of log M_M versus V_R for nucleotides with a chain length of from 2 to 20 nucleotides. The increasing buffer molarity causes lowering of the retention volume of higher oligomers. The removal of phenol from the aqueous phase by GPC on Sephadex G-25 after phenol extraction has become a routine procedure (Fig. 4.6.10).

4.6.5.3 Nucleoproteins

Nucleoproteins are comparatively large particles so that their separation on gel columns has become possible after gels with large pore diameters were introduced into laboratory practice. Soft compressible agarose gels were followed first by hard gels such as controlled pore glass (CPG), and later by other silica-based gels, etc. Already the first report on the use of CPG by Haller [125] dealt with the separation of plant viruses. A subsequent work concerned purification of some bacteriophages [126]. Marcinka [127,128] used CPG for purification of some plant viruses.

In spite of unfavourable sorption properties exerted by silica-based carriers due to a negative surface charge, these materials are still in use. But from the very beginning there has been a tendency to eliminate this charge. At first, columns were washed with polyethelene glycol solution; a minimal concentration of polyethylene glycol was maintained in the mobile phase throughout the whole procedure. In this way it was possible to elute rabies virus [129] which previously had remained completely adsorbed on to the column. Polyethylene glycols have been, for a considerable time, the most frequently used eliminators of the glass charge; other reagents like albumin and succinic acid anhydride were also employed [130]. The binding of these substances to the glass, however, was very weak and they were comparatively quickly eluted from the column. It was Regnier and Noel [131] who described the preparation of gels with a permanently bounded glycophase to the glass surface (cf. Section 4.6.4.1).

Božko et al. [132] studied the effect of divalent cations on the stability of thymus deoxyribonucleoprotein. The degree of dissociation after gradient dialysis and reconstitution was assayed by gel chromatography.

McCole et al. [133] elaborated a technique based on gel chromatography by which microsomal membranes can be rapidly and gently separated from free polyribosomes and soluble proteins. Lui and Ecobichon [134] used GPC for the isolation of microsomes from guinea pig liver.

4.6.5.4 Saccharides

Saccharides play an important role in life processes and their analysis at various steps of metabolic processes is, therefore, of great interest.

Constantopulos et al. [135] found that acid mucopolysaccharides are eluted from Sephadex G-200 column in the order of their molar masses and that — as in proteins — there is a correlation between log M_M and V_R. But the retention of saccharides differs from that of other kinds of biomacromolecules. For example chondroitin-4-sulphate with an M_M of 13000 g/mol is eluted in approximately the same volume as albumin, the M_M of which is 69000 g/mol [136].

Monosaccharides can easily be separated using common microporous gels and aqueous eluents. However, due to solute-gel interactions the secondary processes are rather pronounced and the calibration dependences log r versus V_R are not well defined. On the other hand, the *oligosaccharides* behave fairly regularly so that well-defined calibration dependences can be constructed for particular series of polymer homologues, and thus the results of the separation can be evaluated quantitatively. However, even in this case, the course of the calibration dependences differ for each chemically different series of oligosaccharides. Schmidt and Enevoldsen [137] reported that oligosaccharides are eluted from the column in an approximately reversed order of their molar masses, and that their position in the elution profile is structure-dependent.

Malto-oligosaccharides can be studied by gel chromatography on Bio-Gel P-2 and a multi-column system Bio-Gel P-2 and P-6 at increased temperature [138]. Under these conditions oligosaccharides have different retention volumes and the method is suitable also for preparative separation of linear and branched oligosaccharides.

Recently, both mono- and oligosaccharides have often been separated by methods of retention chromatography [142]. On the other hand, gel chromatography is still the most important method for separation of *polysaccharides.* Dextrans, starches, pectins, natural resins and gums, agaroses, hemicelluloses, celluloses, etc., were successfully analysed or preparatively separated in various gel-eluent systems. The most important problem encountered with separation of cellulose is its poor solubility in common solvents, so that either special multicomponent solvents must be applied (e.g., cadoxen) or cellulose must be derivatized by nitration, acylation, carbanilation, carboxymethylation, hydroxyethylation, etc. The resulting derivatives are, depending on their polarity, easily soluble either in organic solvents like tetrahydrofuran, chloroform or toluene, or in water and may be separated in the usual way.

Other applications which have been reported include purification of lipopolysaccharides from bacterial cell walls [139] and analysis of glycopeptides from mouse brains following pronase treatment [140].

An overview of gel chromatography of saccharides up to 1970 was presented by Churms [141]. Several applications of gel chromatography for separation of saccharides were collected in the reviews by Rinaudo [169] and Segal [143].

55 Ackers, G.K. (1967) J. Biol. Chem. 242, 3026 (Part 1); Halvorson, H.R. and Ackers, G.K. (1974) J. Biol. Chem. 246, 967 (Part 2).
56 Zimmerman, J.K. (1975) Biophys. Chem. 3, 339.
57 Melenevskij, A.T., El'kin, G.E. and Samsonov, G.V. (1978) J. Chromatogr. 148, 299.
58 Booth, C., Naylor, T.D., Price, C., Rajab, N. and Stubberfield, R.B. (1978) J. Chem. Soc. Faraday Trans. I. 74, 2352.
59 Ackers, G.K., Brumbaugh, E.E., Ip, S.H.C. and Halvorson, H.R. (1976) Biophys. Chem. 4, 171.
60 Hummel, J.P. and Dreyer, W.J. (1962) Biochim. Biophys. Acta 63, 530.
61 Sébille, B., Thuaud, N. and Tillement, J.P. (1978) J. Chromatogr. 167, 159.
62 Berek, D., Bleha, T. and Pevná, Z. (1976) J. Polym. Sci. Polym. Lett. Ed. 12, 323.
63 Lindström, T., de Ruvo, A. and Söremark, C. (1977) J. Polym. Sci. Polym. Chem. Ed. 15, 2029.
64 Bleha, T., Bakoš, D. and Berek, D. (1977) Polymer 18, 897.
65 Van der Linden, C. and Van Leemput, R. (1978) Macromolecules 11, 1273.
66 Bleha, T., Mlýnek, J. and Berek, D. (1980) Polymer 21, 798.
67 Porath, J. (1963) Pure Appl. Chem. 6, 233.
68 Halász, I. and Martin, K. (1975) Ber. Bunsenges. Phys. Chem. 79, 731; (1978) Angew. Chem. Int. Ed. 17, 901.
69 Cases, J. and Dobbins, R.J. (1970) J. Polym. Sci. B 8, 785.
70 Morris, M.C. (1971) J. Chromatogr. 55, 203.
71 Kirkland, J.J. (1979) J. Chromatogr. 185, 273.
72 Deyl, Z., Macek, K. and Janák, J. (1975) Liquid Column Chromatography, Elsevier, Amsterdam.
73 Rutschmann, M., Kuehn, L., Dahlmann, B. and Reinauer, H. (1982) Anal. Biochem. 124, 134.
74 Kato, Y., Komiya, K., Sasaki, H. and Hashimoto, T. (1980) J. Chromatogr. 190, 297.
75 Kato, Y., Komiya, K., Sasaki, H. and Hashimoto, T. (1980) J. Chromatogr. 193, 458.
76 Imamura, T., Konishi, K., Yokoyama, M. and Konishi, K. (1981) J. Liq. Chromatogr. 4, 613.
77 Lin, A.W.S.M. and Castel, D.O. (1978) Anal. Biochem. 86, 345.
78 Sakamoto, W. and Nishikaze, O. (1979) J. Biochem. 86, 1549.
79 Krell, W., Mahn, I. and Mueller-Berghaus, G. (1979) Thromb. Res. 14, 299.
80 Chowdhury, O. and Moschos, C.B. (1975) Thromb. Diath. Haemorrh. 34, 504.
81 Gibb, B., Becker, I. and Kraemer, M. (1974) Biol. Zentralbl. 93, 537.
82 deLigny, C.L., Gelsema, W.J. and Roozen, A.M.P. (1981) J. Chromatogr. Sci. 19, 477.
83 Arenkevič, I.M., Voskobojev, A.I. and Ostrovskij, J.M. (1977) Vopr. Med. Khim. 23, 203.
84 Baker, C.J., Whalen, C.H. and Bateman, D.F. (1977) Phytopathology 67, 1250.
85 Vovk, V.A., Levčuk, T.P., Baljasova, I.G. and Jakovlev, V.A. (1975) Bioorg. Khim. 1, 512.
86 Matagne, R.F. and Schloesser, J.P. (1977) Biochem. J. 167, 71.
87 Al-Shaibani, K.A., Price, R.J. and Brown, W.D. (1977) J. Food Sci. 42, 1013.
88 Kato, Y., Komiya, K., Sawada, Y., Sasaki, H. and Hashimoto, T. (1980) J. Chromatogr. 190, 305.
89 Nakasaki, H., Matsushina, T., Sato, S. and Kawachi, T. (1979) J. Biochem. (Tokyo) 86, 1225.
90 Calabro, M.A., Katz, J.T. and Holloway, P.W. (1976) J. Biol. Chem. 251, 2113.
91 Yamaguchi, S. and Suzuki, K. (1977) Biochem. Biophys. Res. Commun. 77, 999.
92 Wiederschain, G.Ya. and Beyer, E.M. (1977) Arch. Biochem. Biophys. 182, 335.
93 Kanaya, S. and Uchida, T. (1981) J. Biochem. 90, 473.
94 Bellino, F.L. and Osawa, Y. (1978) J. Steroid. Biochem. 9, 219.
95 McCandliss, R.J., Poling, M.D. and Herrman, K.M. (1978) J. Biol. Chem. 253, 4259.
96 Bostian, K.A. and Betts, G.F. (1978) Biochem. J. 173, 773.
97 Tamaki, N., Nakamura, M., Kimura, K. and Hama, T. (1977) J. Biochem. 82, 73.
98 Tyagi, A.K., Siddiqui, F.A. and Venkitasubramanian, T.A. (1977) Biochim. Biophys. Acta 485, 255.
99 Bamforth, C.W. and Large, P.J. (1977) Biochem. J. 167, 509.
100 Kuo, C.H. and Wells, W.W. (1978) J. Biol. Chem. 253, 3350.
101 Podesta, E.J., Dufau, M.L. and Catt, K.J. (1978) Biochemistry 17, 1566.
102 Bibor, V. and Verly, W.G. (1978) J. Biol. Chem. 253, 850.
103 Stricklin, G.P., Eisen, A.Z., Bauer, E.A. and Jeffrey, J.J. (1978) Biochemistry 17, 2331.

104 Mehta, P.D., Lin, F.H. and Thormar, H. (1976) Infect. Immun. 13, 1728.
105 Lin, F.H. (1978) J. Virol. 25, 207.
106 Michailova, E.H., Dyašenko, N.S., Vančak, N.P., Popenko, N.N. and Žovnovala, V.L. (1977) Mikrobiol. Zh. 39, 368.
107 Kiredjian, M. (1977) Ann. Biol. Clin. 35, 325.
108 Ong, D.E. (1982) Cancer Res. 42, 1033.
109 Kimura, M. and Itokawa, Y. (1981) J. Chromatogr. 211, 290.
110 Hamana, K. and Iwai, K. (1976) J. Biochem. 79, 125.
111 Hamana, K. (1981) J. Biochem. 90, 1591.
112 Welinder, B.S. (1980) J. Liq. Chromatogr. 3, 1399.
113 Lamar, C. Jr. and Cantley, J.R. (1977) J. Chromatogr. 137, 471.
114 Stigbrand, T., Eriksson, A. and Thornell, L.E. (1979) Biochim. Biophys. Acta 577, 52.
115 Stanton, T.H., Bennett, J.C. and Wolcott, M. (1975) J. Immunol. 115, 1013.
116 Cobb, A.H. (1977) Protoplasma 92, 137.
117 Marcinka, K., unpublished results.
118 Armstrong, D.W. and Fendler, J.H. (1977) Biochim. Biophys. Acta 478, 75.
119 Cornelis, P. (1978) Plant Sci. Lett. 11, 3.
120 Woo, S.L.C., Harris, S.E., Rosen, J.M., Chan, L., Means, A.R. and O'Malley, B.W. (1974) Prep. Biochem. 4, 555.
121 Miyagawa, T., Anai, M. and Urabe, H. (1975) Arch. Dermatol. Res. 254, 79.
122 Yamasaki, H., Pulkrabek, P., Grunberger, D. and Weinstein, I.B. (1977) Cancer Res. 37, 3756.
123 Uchiyama, S., Imamura, T., Nagai, S.-I. and Konishi, K. (1981) J. Biochem. 90, 643.
124 Molko, D., Derbyshire, R., Guy, A., Roget, A., Teoule, R. and Boucherle, A. (1981) J. Chromatogr. 206, 493.
125 Haller, W. (1965) Nature 206, 693.
126 Gschwender, H.H., Haller, W. and Hofschneider, P.H. (1969) Biochim. Biophys. Acta 190, 460.
127 Marcinka, K. (1971) Acta Virol. 15, 316.
128 Marcinka, K. (1972) Acta Virol. 16, 53.
129 Hiatt, C.W., Shelokov, A., Rosenthal, E.J. and Galimore, J.M. (1971) J. Chromatogr. 56, 362.
130 Krasilnikov, I.V., Elbert, L.B., Mamonenko, L.L., Pogodina, V.V., Krut'janskaja, G.L., Bresler, S.E., Kolikov, V.M., Mčedlišvili, B.V., Borisova, V.N., Golovina, N.C. and Nachapet'jan, L.A. (1977) Vopr. Virusol. 6, 685.
131 Regnier, F.E. and Noel, R. (1976) J. Chromatogr. Sci. 14, 316.
132 Božko, H.Ch., Domanskij, M.M., Puida, N.H. and Karenina, T.I. (1977) Dokl. Akad. Nauk. Ukr. SSR Ser. B. 6, 532.
133 McCole, N., Palmer, D.N. and Williams, D.J. (1979) J. Biochem. 180, 437.
134 Lui, E.M.K. and Ecobichon, D.J. (1977) Res. Commun. Chem. Pathol. Pharmacol. 17, 345.
135 Constantopulos, G., Dehuban, A.S. and Caroll, W.R. (1969) Anal. Biochem. 31, 59.
136 Wasteson, A. (1971) J. Chromatogr. 59, 87.
137 Schmidt, F. and Enevoldsen, B.S. (1976) Carlsberg Res. Commun. 41, 91.
138 Kainuma, K., Nogami, A. and Mercier, C. (1976) J. Chromatogr. 121, 361.
139 Romanovska, R. (1970) Anal. Biochem. 33, 383.
140 Quarles, R.H. (1976) Biochem. J. 156, 143.
141 Churms, S.C. (1970) Adv. Carbohydr. Chem. Biochem. 25, 13.
142 Heyraud, A. and Rinaudo, M. (1980) J. Liq. Chromatogr. 3, 721.
143 Segal, L. (1975) Adv. Chromatogr. 12, 31.
144 Mains, R.E. and Eipper, B.A. (1975) Proc. Nat. Acad. Sci. USA 72, 3565.
145 Estivariz, F.E. and Iturriza, F.C. (1976) Acta Physiol. Lat. Am. 26, 154.
146 Berthou, F., Picart, D., Bardou, L. and Floch, H.H. (1976) J. Chromatogr. 118, 135.
147 Bulbena, O., Rodriguez Farre, E. and Valdecasas, F.G. (1977) Arch. Pharmacol. Toxicol. 3, 70.
148 Croizer, A., Zaerr, J.B. and Morris, R.O. (1980) J. Chromatogr. 198, 57.
149 Assandri, A. and Semenza, G. (1977) J. Chromatogr. 135, 25.

chymotrypsin and glycin attached to 1 ml of hydroxyalkyl methacrylate copolymers of various pore sizes on the exclusion molecular weights and the different specific surface areas is shown [25] in Table 4.7.1. It is obvious that the amount of bound chymotrypsin depends directly on the specific surface area, which is largest with Spheron 300 and 500. The amount of bound glycine indicates that there are relatively small differences in the number of reactive groups. The specific surface area is the main factor determining the capacity of the biospecific adsorbent. Therefore we must choose the solid support with the largest surface area, but with sufficient porosity for the formation of the individual biospecific complexes.

Rigidity and suitable shape of the particles is important for good flow properties, i.e., the eluent should penetrate the support column at a sufficient rate even if the affinant is bound on it. The particle size of the matrix should not exceed 200 μm and should not be less than 5 μm.

A sufficient number of chemical groups that can be activated or modified in such a manner that they may be able to couple the affinant is a necessary condition for the preparation of specific adsorbents. The activation or modification should take place under conditions that do not change the structure of the support. No less important are the chemical and mechanical stabilities of the carrier under the conditions of attachment of the affinant, and also at various pH values, temperatures and ionic strengths, in the possible presence of denaturating agents, etc., which may be necessary for a good sorption and elution of the isolated substance. The specific adsorbent should not be attacked by microorganisms and enzymes. These stabilities are important primarily for repeated use of specific adsorbents.

The hydrophilic character of a solid support is desirable not only because of the necessity of minimizing non-specific sorption but also because the hydrophobic character of a support can decrease the stability of some bound proteins on the basis of denaturation analogous to that produced by organic solvents.

4.7.2.2 Choice of affinity ligands for attachment

All compounds are suitable affinity ligands to isolate biologically active products which bind these products specifically and reversibly. Hence, depending on the different nature of biologically active products, affinants represent very different types of chemical compounds. Affinants used for the isolation of enzymes, inhibitors, cofactors, antibodies, antigens, agglutinins, glycoproteins and glycopolysaccharides, nucleic acids, nucleotides, binding, transport and repressor proteins, hormones and their receptors, lipids, cells, viruses and other substances are discussed in greater detail in section 4.7.5.

Two main criteria determine the selection of an affinity ligand.

(1) The affinant to be immobilized must possess a functional group that can be modified for attachment to the solid support without impairing or abolishing its recognition by the complementary macromolecule.

(2) The affinity ligand should have an adequate affinity for the macromolecule to be purified. Most satisfactory results were obtained with specific complexes of equi-

librium constant K_L in the region 10^{-4}–10^{-8} M. However, it is important to mention here that the K_L values are considered for the complex of the substance to be isolated with the immobilized affinant, and that therefore they need not agree with the values of the equilibrium constant determined in solution. For the determination of K_L, the most commonly used method is so-called quantitative affinity chromatography, which is based on the elution of biological macromolecular substances from an affinity matrix with soluble affinant solutions of various concentrations [26,27].

Many macromolecules have several affinity ligands which satisfy the above requirements and in these instances the reaction sequences should be taken into account. According to the molecular weight of affinants we can distinguish the affinity systems with small ligands and with macroligands. Low molecular weight synthetic affinants are advantageous mainly owing to their stability and better accessibility. The specific adsorbents prepared from them are usually better characterized, because they are attached via a pre-defined functional group. In order to increase their steric accessibility, usually a spacer is inserted between them and the surface of the solid support. High molecular weight affinity ligands are predominantly proteins or nucleic acids. They can undergo denaturation leading to an irreversible loss of their activity. As a rule, the way of their attachment cannot be unambiguously defined.

According to the specificity of bioaffinity adsorbents we can distinguish affinants with narrrow specificity and so-called 'group-specific' or 'general ligands', which interact specifically and reversibly with a range of complementary proteins. Group-specific affinity matrices, though lacking in the selectivity, have numerous advantages: the chemistry of the immobilization of the affinant is restricted to one compound, as opposed to a whole range of substances for each individual biomolecule. The bioaffinity matrix is applicable to the purification of more than one complementary biomolecule. At present, many of these 'group-specific' matrices are available commercially. The range of dissociation constants with respect to the immobilized affinant is broad enough for us to achieve a separation of several like biomolecules through elution, using gradients of ionic strength, pH, temperature, organic solvents, competing ligands, allosteric effectors, coenzymes and chaotropic agents [28]. In this category we can include immobilized pyridine and adenine nucleotides [29], and recently adsorbents with nucleotide or coenzyme analogues — immobilized triazine dyes have become increasingly popular [30–33]. Group-specific affinity ligands for proteolytic enzymes are haemoglobin [34], casein [35] or bacitracin and gramicidin [36]. For the purification of thiol-protein and peptides, organomercurial-agarose derivatives can be used [37,38]. Carbohydrates, glycoproteins, cell surface receptors or cells can be purified by the use of immobilized lectins [39,40]. Immobilized phenylboronic acid associates with *cis*-diols containing acidic groupings, namely tRNA and adenosine polyphosphates [41], serine proteinases [42], α-glucosidase [43], glycosylated haemoglobins [44], etc. Bouriotis et al. [45] described applications of immobilized phenylboronic acids as supports for group-specific ligands (nucleotides) in the affinity chromatography of enzymes ($NADP^+$-dependent enzymes).

4.7.2.3 Affinant — solid support bonding

The bonding between the monomeric affinant, for example the inhibitor I, and the enzyme E is expressed by the equilibrium constant of the reaction, K_I, on the supposition that the enzyme exists in a single tertiary conformation:

$$E + I \underset{k_2}{\overset{k_1}{\rightleftharpoons}} EI; \quad K_I = \frac{[E][I]}{[EI]}; \quad \Delta G_I = -RT \ln K_I$$

When the affinity ligand is bound to the solid support, the equilibrium constant, K_I, is affected to a certain extent. A modification of the affinant by the attachment to the support results in an increase in K_I as a consequence of the limitation of the steric accessibility of the affinant. On the other hand, non-specific adsorption of the enzyme to the solid support and to the molecules of the already adsorbed enzyme causes a decrease in K_I. Assuming that a single enzyme of the crude proteins has an affinity for the specific adsorbent, the equilibrium between the attached affinant L and the isolated enzyme E is given by the equation

$$E + L \underset{k_2}{\overset{k_1}{\rightleftharpoons}} EL; \quad K_L = \frac{[E][L]}{[EL]}; \quad \Delta G_L = -RT \ln K_L$$

The difference between ΔG_I and ΔG_L is given by the change in the steric accessibility of the affinity ligand after its immobilization, by its modification due to its attachment on the solid support, by the nature of the solid support, etc.

The success of bioaffinity chromatography depends primarily on how much the created conditions are similar to the native, biological interaction of the affinity ligand and the macromolecule in free solution. Therefore the question of sufficient freedom for the formation of the complex is a factor of great importance. The primary condition for the formation of the specific complex is sufficient porosity of the solid support (see section 4.7.2.1). The second limitation imposed by the matrix relates to the local steric interference generated by the solid support backbone on the microenvironment of the immobilized affinant. For the purification by bioaffinity chromatography to be successful, the affinant must be placed at a sufficient distance from the solid support surface to minimize the steric interference. The importance of the increased distance of the low-molecular inhibitor from the solid matrix surface was already illustrated by Cuatrecasas et al. [46] in one of the first successful applications of bioaffinity chromatography to the isolation of enzymes. Fig. 4.7.2 represents the bioaffinity chromatography of α-chymotrypsin, both on Sepharose coupled with ε-aminocaproyl-D-tryptophan methyl ester (A), and on Sepharose coupled with D-tryptophan methyl ester (B), in comparison with the chromatography on unsubstituted Sepharose (C). In the first instance (A), the coupled inhibitor has a high affinity for α-chymotrypsin and the enzyme can be released from the complex only by decreasing the pH of the eluting buffer. By using 0.1 M acetic acid of pH 3.0, the chymotrypsin fraction is eluted as a sharp peak and the volume of the eluted

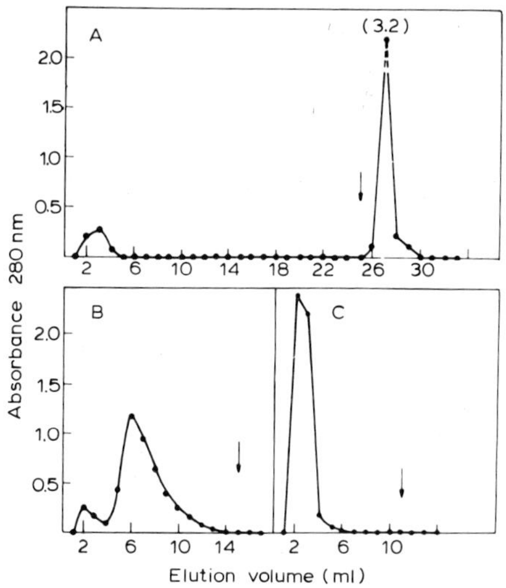

Fig. 4.7.2. Affinity chromatography of α-chymotrypsin on inhibitor Sepharose columns. The columns (50×5 mm) were equilibrated and run with 0.05 M Tris-hydrochloric acid buffer (pH 8.0). Each sample (2.5 mg) was applied in 0.5 ml of the same buffer. The columns were run at room temperature with a flow-rate about 40 ml/h and fractions containing 1 ml were collected. The arrows indicate a change of elution buffer (0.1 M acetic acid, pH 3.0). (A) Sepharose coupled with ε-aminocaproyl-D-tryptophan methyl ester; (B) Sepharose coupled with D-tryptophan methyl ester; (C) unsubstituted Sepharose. The first peaks in A and B were devoid of enzyme activity. Reproduced with permission from Ref. 46.

chymotrypsin does not depend on the volume of the sample applied to the column. In the second instance (B), the inhibitor coupled directly on Sepharose has a much lower affinity for the isolated α-chymotrypsin, owing to the steric hindrance. In this instance a change of the buffer is not necessary for enzyme elution and, as can be seen from Fig. 4.7.2, the enzyme is eluted in a much larger volume close after the inactive material. In order to verify that non-specific adsorption on the solid support did not take place under the given experimental conditions, the chromatography of α-chymotrypsin on an unsubstituted carrier was carried out (C).

However, as was observed, the described chromatography of chymotrypsin on unsubstituted Sepharose does not provide sufficient evidence of non-specific sorption. On the contrary, Hofstee [47] demonstrated for Sepharose with coupled ε-aminocaproyl-D-tryptophan methyl ester that it sorbed, for example, serum albumin or γ-globulin quite non-specifically. Thus, it was found that a series of substances contained hydrophobic regions on the surface of their molecules, by which they were capable of being bound to hydrophobic spacers, such as hexamethylenediamine or ε-aminocaproic acid. The utilization of this phenomenon for the separation of a number of biological macromolecules gave rise to a new technique, the hydrophobic chromatography [14,15].

In view of the different structures of the substances isolated, no general rule exists on the minimum distance between the affinity ligand and the surface of the solid matrix. Lowe et al. [48] considered that the use of an extension arm at least 0.5 nm in length enabled the nucleotide to traverse the barrier imposed by the micro-environment associated with the hydrophilic polymer. This can be caused by the ordered layer of water molecules surrounding the matrix backbone and limiting diffusion in this region, or by the vibrational movement of the lattice. In any event, the region of the solvent in close proximity to the surface of the solid matrix represents a real barrier for the interaction of the macromolecule with the complementary affinant, especially if an interaction with weak affinity is concerned.

The interfering effect of non-specific hydrophobic interactions with the spacer molecule is avoided if hydrophilic spacers are used [28,49]. The use of hydrophilic spacers also prevents another undesirable situation that may occur when a hydrophobic affinant is coupled to a long, flexible hydrophobic chain. When Holguin and Cardinaud [50] used Sepharose-N^6-*p*-amino-*n*-hexyladenine, they assumed that under the effect of the flexible chain the active part of the affinity ligand came into close proximity of the solid support and that therefore an effective interaction with the transfer enzyme could not take place.

On the contrary, Inoue et al. [51] found that the hydrophobic character of the spacer moiety was important for the specific interaction of some enzymes. Among various spacers tested, the lysyl spacer was found to be the most specific for glutathione *S*-transferase. Klyashchitsky and Mitine [52] recommended the use of polysaccharide spacers.

In order to minimalize non-specific sorption of inert compounds to affinity adsorbents and to exploit as much as possible the affinity ligand attached, its lowest possible density in the specific adsorbent should be used. The importance of the low concentration of the immobilized affinant and the effect of the uneven surface of the solid support are illustrated in Fig. 4.7.3. In the upper part of this figure the affinity ligand attached to a solid matrix does not form a specific complex with the enzyme to be isolated (there is no complementarity of the binding site of the enzyme to the immobilized affinant). However, the molecules of this enzyme are sorbed onto the modified support by non-specific multi-point bonding. In the central part of the figure the specific complementary 'one-to-one' bonding of the enzyme isolated to the immobilized affinant is represented. At a low density of the attached affinant multiple non-specific bonds cannot operate, therefore only biospecific bonding of the molecules via complementary active sites takes place, unless — of course — this bond is sterically hindered. The figure shows a macroreticular hydroxyalkyl methacrylate copolymer represented by aggregated globules. Due to the effect of the uneven surface of the matrix, well accessible, less accessible and even sterically hindered affinity ligands can be distinguished after the attachment of the affinant through a spacer. These steric hindrances account not only for the low saturation of the immobilized affinant molecules with the isolated enzyme, but also for their heterogeneity in bioaffinity. In the lowest part of the figure multiple bonding of the enzyme is again shown. Although the enzyme contains the complementary site for the immobilized affinant, the orientation of the bonding need not necessarily be correct. The non-specific bonds can be electrostatic or hydrophobic interactions, or a

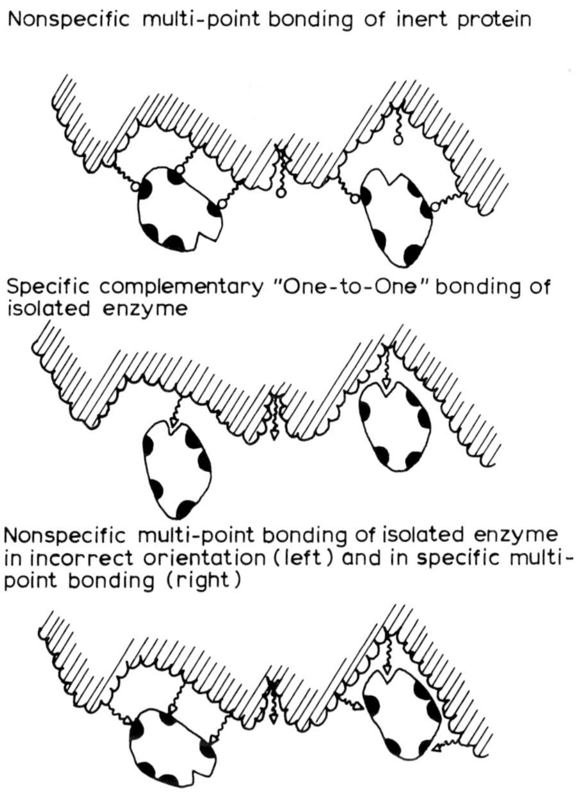

Fig. 4.7.3. Schematic drawing of effects of concentration of immobilized affinity ligands and of uneven support surface on nonspecific and specific sorption.

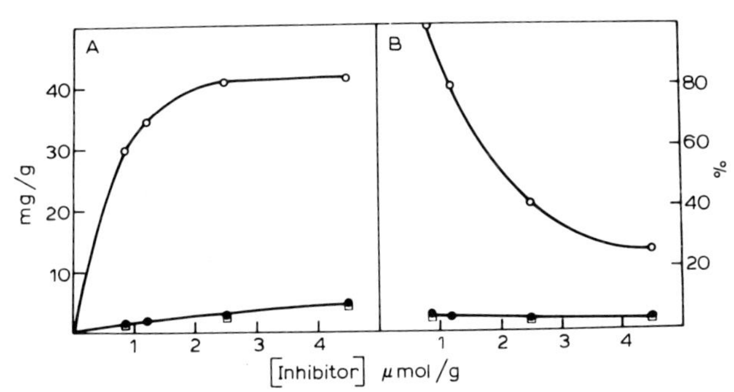

Fig. 4.7.4. (A) Capacity of immobilized inhibitor sorbent (ε-aminocaproyl-L-Phe-D-Phe-OMe-Separon) in mg of pepsin per g of dry sorbent, and (B) portion of immobilized inhibitor molecules involved in specific complex formation (in %) with respect to immobilized inhibitor concentration (in μmol of inhibitor per g of dry sorbent). -○-, porcine pepsin; -●-, chicken pepsin, and -□-, human pepsin.

combination of both. The enzyme can be bound non-specifically to groups of their own affinant or to the spacer, or even to the solid support. These non-specific multiple bonds may be stronger than one complementary bond between the isolated enzyme and the immobilized affinant. In addition to this specific complementary bond, these non-specific multiple bonds increase the strength of the bonding in a specific complex; this results in the elution of the same enzyme in several fractions [15,53] or in the difficulties with the elution of the enzyme from the specific sorbent [54]. Turková et al. [53] determined the maximum capacity of ε-aminocaproyl-L-Phe-D-Phe-OMe-Separon for porcine, chicken and human pepsin (Fig. 4.7.4A) and the

TABLE 4.7.2

Comparison of the binding of various enzymes to N^6-(5-aminohexyl)-5′-AMP-Sepharose

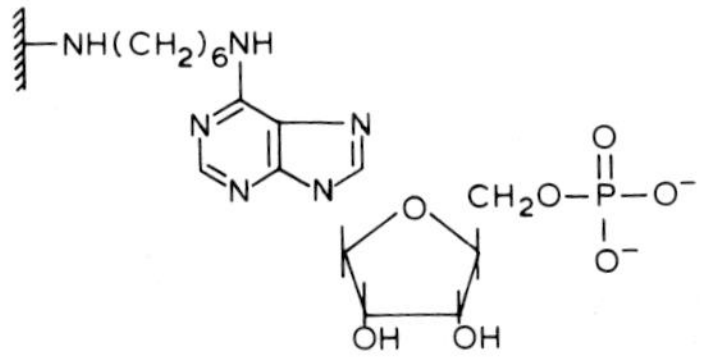

and P^1(6-aminohexyl)-P^2-(5′-adenosine)-pyrophosphate-Sepharose

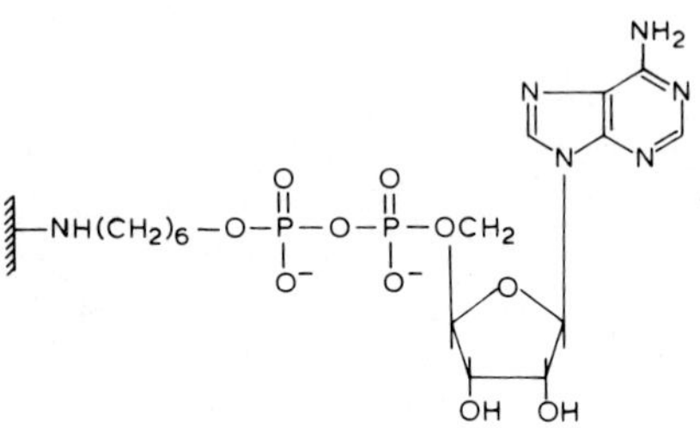

Enzymes		Binding (β) *	
Code number	Name	I	II
EC 1.1.1.27	Lactate dehydrogenase	>1000 **	>1000 **
EC 1.1.1.49	Glucose 6-phosphate dehydrogenase	0	170
EC 1.1.1.37	Malate dehydrogenase	65	490
EC 1.1.1.1	Alcohol dehydrogenase	400	0
EC 1.2.1.12	p-Glyceraldehyde 3-phosphate dehydrogenase	0	>1000 **
EC 2.7.2.3	3-Phosphoglycerate kinase	70	260
EC 2.7.1.40	Pyruvate kinase	100	110
EC 2.7.1.1	Hexokinase	0	0
EC 2.7.4.3	Myokinase	0	380
EC 2.7.1.30	Glycerokinase	122	0

* Binding (β) is the KCl concentration (mM) at the centre of the enzyme peak when the enzyme is eluted with a linear gradient of KCl.

** Elution was effected by a 200 μl pulse of 5 mM NADH.

portion (%) of immobilized inhibitor molecules involved in a specific complex formation (Fig. 4.7.4B) with respect to the immobilized inhibitor concentration (solutions of pepsins were applied to the bioaffinity columns continuously until the effluent showed the same activity as the solution applied). From a comparison of the curves for the individual pepsins it is evident that ε-aminocaproyl-L-Phe-D-Phe-OMe-Separon is a very good sorbent for porcine pepsin only. All the molecules of the inhibitor attached to the solid support take part in the formation of the specific complex, with the pepsin to be isolated only in the specific adsorbent with the lowest density of the immobilized inhibitor. With the increasing content of the attached affinant the portion of the inhibitor molecules involved in the specific complex with pepsin sharply decreased. Analogous results were obtained by Twining and Atassi [55], when they studied the amount of antibody bound in dependence on the concentration of immobilized antigens (peptides of myoglobin antigenic sites or myoglobin). Different sorption of the acetylcholinesterases from bovine erythrocytes and from electric eel to the *N*-methylacridinium-Sepharose columns on different affinant concentrations were described by Sekar et al. [56].

In addition, the capacity of the bioaffinity columns also depends on the purity of the sample applied. The purer the sample, the higher the capacity. Thus, the more impure the fraction to be chromatographed, the larger the column required [57].

The main principle on which specific interactions of biological macromolecules are based is the complementarity of the binding sites. Harvey et al. [58] demonstrated the basic importance of the way of attaching nucleotides onto the solid support for the efficiency of the bioaffinity chromatography of kinases and dehydrogenases depending on pyridine nucleotide. Adsorbent, N^6-(6-aminohexyl)-AMP-Sepharose, contains AMP coupled to Sepharose by means of the N^6 adenine part, in the adsorbent. P^1-(6-aminohexyl)-P^2-(5′-adenosine)-pyrophosphate-Sepharose, AMP is coupled by means of 5′-phosphate. The binding of various dehydrogenases and kinases onto these two adsorbents is shown in Table 4.7.2. From these results it may be inferred that while the free 5′-phosphate group is essential for the binding of alcohol dehydrogenase or glycerokinase, for example, it has a completely different role in the interaction of glyceraldehyde 3-phosphate dehydrogenase. In the latter case the adenosine part of the affinity plays a decisive role. The much stricter binding requirements of hexokinase and creatine kinase results in the fact that these enzymes are not bound to either of the adsorbents.

4.7.2.4 Sorption and elution conditions

In the binding of the complementary sites of the affinity ligand and of the isolated substances, ionic bonds or hydrogen bonds, hydrophobic interactions and Van der Waals-London forces may participate to various extents. Therefore, the optimum conditions for sorption and desorption will vary in particular instances. In general, the starting conditions for sorption should be selected so as to cause maximum sorption of the substance to be isolated. The choice of the starting buffer is completely dependent on the optimum conditions of a specific complex formation and, in addition to pH, ionic strength and temperature, it also depends on the content of metal ions [59,60] and of other specific factors [61].

The practical utilization of pH during the sorption of neutral proteinase from *Bacillus subtilis* on Sepharose with attached glycyl-D-phenylalanine (through a spacer 23 atoms long) is shown in Fig. 4.7.5. Walsh et al. [62] demonstrated that between pH 5 and 6.5 neutral proteinase was effectively adsorbed and thus separated both from subtilisin and other proteins present in the culture filtrate. At higher pH values, an effective separation of neutral proteinase from subtilisin does not take place.

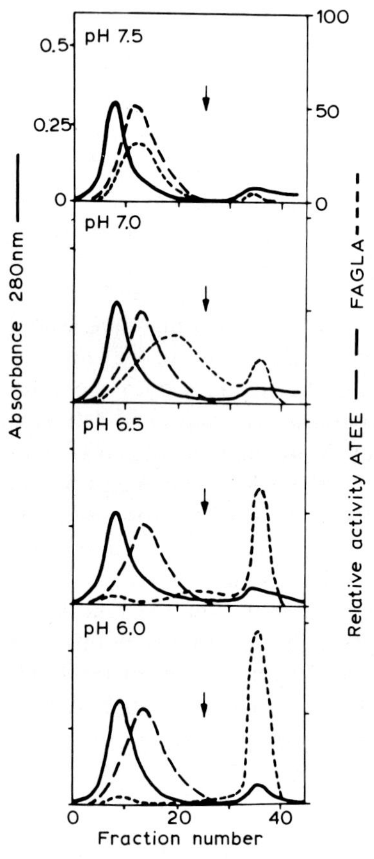

Fig. 4.7.5. Effect of pH on the adsorption of neutral protease on Sepharose 4B substituted with triethylenetetramine, succinic anhydride, triethylenetetramine and chloroacetyl-D-phenylalanine. Crude enzyme (10 mg) was dissolved in 0.1 ml of equilibrating buffer and applied to an affinity column (220 × 6 mm) equilibrated with 100 mM sodium chloride, containing 5 mM Tris (pH 7.5 or 7.0) or 5 mM 2-(*N*-morpholino)ethanesulfonic acid (pH 6.5 or 6.0). After elution for 1 h at 25 ml/h, each column was washed with 100 mM sodium chloride, 10 mM calcium chloride, 50 mM Tris (pH 9.0), as indicated by the arrows. Neutral protease and subtilisin are identified by catalytic activities toward 3-(2-furylacryloyl)-glycyl-L-leucinamide (FAGLA) and acetyl-*N*-tyrosine ethyl ester (ATEE). Reproduced with permission from Ref. 62.

TABLE 4.7.3
Effect of ionic strength of buffer on β-galactosidase binding on to a column of agarose substituted with p-aminophenyl-β-D-thiogalactopyranoside (bed dimension, 114×17 mm; flow-rate, 2.0 ml·min^{-1}·cm^{-2}).

Ionic strength	Protein bound (mg)	Activity bound (units)	Specific activity (units/mg)
0.010	75	6000	80
0.020	32	5500	190
0.035	14	3600	320
0.050	1.2	350	340

Neutral proteinase is set free from the complex with the immobilized affinant by increasing the pH to values at which the binding of the substrate is already weak, but at which the enzyme is still not denatured. The optimum pH for the elution of neutral proteinase depends not only on its affinity towards the affinant and on the pH dependence of this affinity, but also on the concentration of the immobilized inhibitor. Neutral proteinase is a metalloenzyme and is therefore inhibited by 1,10-phenanthroline and EDTA. The presence of these chelating agents in a concentration of 1–5 mM prevents the adsorption of the enzyme onto the specific sorbent mentioned.

If the high ionic strength of the starting buffer does not impair the formation of the bioaffinity complex, it is advantageous to use it because it decreases the non-specific electrostatic interactions. The effect of ionic strength on the binding of β-galactosidase on Sepharose with attached p-aminophenyl-β-D-thiogalactopyranoside is shown in Table 4.7.3. With increasing ionic strength, the amount of the sorbed enzyme decreases, but its specific activity, on the contrary, increases [64]. Smith et al. [65] described the elimination of non-specific adsorption of serum proteins by Sepharose-bound antigens by pre-elution washing procedures with borate-saline buffer (pH 8.5) containing a nonionic detergent, Tween 20 (0.1%) and a high salt concentration (1 M NaCl).

While the substances that have no affinity for the attached affinant are usually eluted with the hold-up volume, the specifically sorbed material mostly requires displacement with a specific elution agent or by a change in the solvent composition or temperature. Bioaffinity chromatography of pancreatic extract on agarose with attached soybean trypsin inhibitor, described by Porath [66], can serve as an example of biospecific elution with competitive inhibitors. Chymotrypsin was displaced with a solution of tryptamine and trypsin with a solution of benzamidine. From a theoretical point of view, a more advantageous elution with a competitive affinant has, however, numerous practical limitations because of availability and price. For example, lactate dehydrogenases from immobilized N^6-(6-aminohexyl)-5′-AMP-agarose can be eluted with a solution of a competitive ligand of NADH of relatively high concentration (1.5 mM), or with KCl solution of medium concentration (0.13 mM). Therefore, the elution of adsorbed substances by a change in temperature, pH or ionic strength is still the most commonly used procedure.

The temperature effects yield valuable information on the forces acting between solutes and the specific adsorbent and can greatly contribute not only to better understanding of the mechanism of the biospecific complex formation and dissociation, but also to practical chromatography [67]. Harvey et al. [68] used a linear temperature gradient for the differentiation of a mixture of yeast alcohol dehydrogenase, glycerokinase, hexokinase and lactate dehydrogenase in bioaffinity chromatography on N^6-(6-aminohexyl)-5′-AMP-Sepharose. The advantage of this elution is that the eluted enzymes are not contaminated with elution reagents. Analogous enzymes, which can also be utilized directly for kinetic studies, were also obtained by Lowe et al. [69] using an elution with a pH gradient. A linear gradient of salt was used by Nüsslein and Heyden [70] for the isolation of RNA polymerase on DNA-agarose. On the contrary, Bureau and Daussant [71] used distilled water for desorption in immunoaffinity chromatography. However, desorption with water became effective if the elution procedure involved an interruption of a few hours. Conditions for desorbing DNP-substituted proteins from their immunoadsorbent were studied in various hydro-organic media at different pH values and temperatures, including the sub-zero range by Andersson et al. [72]. Chaotropic ions (KSCN) as a suitable agent for the dissociation of antigen-antibody complexes have been discussed by Shimizu et al. [73]. Chaotropic ions, urea or guanidine used for the decomposition of a specific complex by their steric modification can cause irreversible changes in the tertiary structure. Therefore, before eluting proteins with these agents, the stability of the proteins involved should always be tested with the selected elution buffer.

A widely applicable, non-chaotropic elution technique in bioaffinity chromatography is electrophoretic desorption [60,74,75]. If the affinant is coupled to the solid support by an azo bond or by thiol- or alcohol-ester bonds, the complex of the affinant with the isolated substance can be detached from the solid matrix and then the affinity ligand separated by dialysis or gel filtration.

4.7.3 Solid matrix support and the most common methods of coupling

4.7.3.1 Survey of the most common solid supports

Virtually any water-insoluble substance that satisfies the required characteristics discussed in Section 4.7.2.1 can serve as a solid support. However, no ideal matrix has been introduced yet. But an important factor is the rapid development of solid supports. If, at present, some carrier is neglected owing to some undesirable quality, it may be suitable after appropriate modification. It is almost impossible to find a matrix material which does not interact with biological active molecules by hydrogen bonding or ionic or hydrophobic interaction. Occasionally, these interactions fortuitously improve adsorption selectivity, but according to the opinion of Egly and

Porath [76] they should be minimized to allow better control of the biospecific interaction. In the choice of the solid support and the method of coupling not only should the properties following from the nature of the bioaffinity chromatography be taken into consideration, but also their field of use. For example, there remains an important question of whether the practical application of a biospecific adsorbent on a larger scale is feasible from the economic point of view. According to Angal and Dean [77] the choice of a matrix for bioaffinity chromatography remains an empirical one. The effects of changing the matrix can be unexpected and dramatic.

Agarose and its derivatives

At present the most commonly used carrier in bioaffinity chromatography is agarose. It is a linear polysaccharide which consists of monomer units of D-galactose and 3,6-anhydro-L-galactose. The characteristics of gel have been explored by Porath and co-workers [78,79]. Agarose meets almost all of the requirements of an ideal carrier. As regards its stability, it is stable in the pH range of 4–9; and temperatures below 0°C or above 40°C are not recommended. The stability of an agarose matrix can be considerably increased by crosslinking with epichlorohydrin [78], divinyl sulphone [79] or bis-oxiranes [80]. Through crosslinking, the gels acquire improved stability towards extremes of pH, temperature, high concentrations of organic solvents and chaotropic agents. The main producers* of agarose are Pharmacia (Uppsala, Sweden, under the trademark Sepharose), and Bio-Rad Labs. (Richmond, CA, U.S.A., under the trademark Bio-Gel A).

Polyacrylamide

Polyacrylamide gels are synthesized by copolymerization of acrylamide and *N,N'*-methylene bisacrylamide. The acrylamide thus formed consists of a hydrocarbon skeleton onto which carboxamide groups are bound:

$$\begin{array}{l}-CH_2-\underset{\displaystyle CO\text{-}NH_2}{\underset{|}{CH}}-CH_2-\underset{\displaystyle CO\text{-}NH_2}{\underset{|}{CH}}-CH_2-\underset{\displaystyle CO\text{-}NH_2}{\underset{|}{CH}}-CH_2-\end{array}$$

Polyacrylamide is chemically stable in the pH range of 3–10, and at temperature below 35°C, but cannot withstand some organic solvents [76]. The advantage of this support is the high density of the chemically reactive carboxamide groups. Unfortunately, however, after chemical modification the porosity of the beads is reduced to the detriment of the effectiveness of the gel in the purification of macromolecules. The main producer of polyacrylamide gels is Bio-Rad Labs. (Richmond, CA, U.S.A., under the trademark Bio-Gel P).

---

* Only firms known to the author are mentioned in the text, therefore the list is necessarily incomplete and it should in no case be considered as implying a recommendation of any firm or product.

Copolymers of agarose and polyacrylamide

In order to minimize the reduction of porosity of polyacrylamide during activation the copolymers of agarose and polyacrylamide are produced. This support combines the advantages of each individual polymer, whilst extending the potential range of derivatisation procedures by virtue of the availability of both amide and hydroxyl groups of activation [81]. Copolymers are produced by LKB-Produkter AB (Bromma, Sweden). A support of similar properties, dextran crosslinked with *N,N′*-methylene-bisacrylamide [82], is produced by Pharmacia (Uppsala, Sweden, under the trademark Sephacryl).

Hydroxyalkyl methacrylate copolymers

The copolymerization of hydroxyalkyl methacrylate with alkylene dimethylacrylate gives rise to heavily crosslinked microparticles of xerogel, which subsequently aggregate and yield a macroporous structure of spheroids [83]. Because of this structure, the matrix has some chemical properties in common with agarose [25], and at the same time, because of its macroreticular structure, with inorganic supports. Hydroxyalkyl methacrylate copolymers do not change in volume with changes in pH or after the addition of organic solvents. They are not attached by microorganisms. The advantage of these supports lies in their chemical and mechanical stability and possibility of using high pressures. Poly(hydroxyethylmethacrylate) is produced by Lachema (Brno, Czechoslovakia, under the trademark Spheron), or by Lab. Instr. Works (Prague, Czechoslovakia, under the trademark Separon H). The glycosylation of Spheron was described by Filka et al. [84].

Silica-based supports

The advantages of porous glass and silica are their resistance to acids, organic solvents and microbial attack, their rigidity even under pressure, and they are also readily sterilized. The shortcoming in the use of porous glass and silica is the non-specific adsorption of some proteins, carbohydrates or nucleotides on the derivatized surface [28,85]. To diminish the undesirable non-specific sorption, the glass and silica are covered with a thin neutral organic layer that excludes sensitive compounds from contacting the silica surface [28,86]. Specific adsorbents prepared from these materials have a high capacity, are extremely stable and have excellent flow properties. Due to their inorganic nature, they can be easily regenerated for re-use by pyrolysis [28].

The silica-based supports are supplied by: Bio-Rad Labs (Richmond, CA, U.S.A., under the trademark Bio-Glass); by Electro-Nucleonics, Inc. (Fairfield, NJ, U.S.A.) and Pierce Chem. Co. (Rockford, IL, U.S.A. under the name CPG (controlled-Pore Glass)); by Waters Associates (Milford, MA, U.S.A. under the name Porasil); by E. Merck (Darmstadt, G.F.R., under the names Fractosil or LiChrospher); by DuPont Instrument Products Div. (Wilmington, DE, U.S.A., under the name SE Series or Zorbax), or by Rhone Poulenc (France, under the name Spherosil). The modified silica-based supports are supplied by: Electro-Nucleonics, Inc. (Fairfield, NJ, U.S.A.) or Pierce Chem. Comp. (Rockford, IL, U.S.A., under the trademark Glycophase-

CPG); by E. Merck (Darmstadt, G.F.R., under the names LiChrospher or LiChrosorb DIOL); by Sun Chrom., Inc. (Linden, IN, U.S.A., under the name SynChropak); by Waters Ass. (Milford, MA, U.S.A., under the name Bondagel or Protein 1125 or 1250); by Toya Soda Man. Comp. Ltd. (Tokyo, Japan) as derivatives of gels produced under the trademark TSK.

Other matrices
Cellulose, linear polymers of β-1,4-linked D-glucose units with an occasional 1,6-bond, was the first solid support used in bioaffinity chromatography [87]. The advantage of cellulose is the low price, the disadvantages are its physical structure, which may be the reason for the sterical hindrance of the permeation of proteins and nucleic acids [76] and its vulnerability to an attack, for example, by microbial cellulases. Flow and some other properties of cellulose were improved by preparing cellulose in the bead form [88]. Dextran, a branched-chain glucose polysaccharide, crosslinked with 1-chloro-2,3-epoxypropane in alkaline solution, is the well known solid support Sephadex, produced by Pharmacia (Uppsala, Sweden). The gels are very stable to chemical, but not to microbial attack. Sephadex is a suitable material for preparing specific adsorbents with small affinants for bioaffinity systems in which both the affinant and the solute are large molecules [76]. Collagen was used as an inexpressive solid support by Vieth et al. [89]. New promising solid supports were introduced by Japanese firms Toya Soda Manufacturing Comp. Ltd. under the trademark TSK-PW (hydroxylated gel of unknown structure) [90,91], and by Showa Denko K.K. under the name Shodex (probably methacrylate glycerol copolymer) [92,93].

4.7.3.2 *Survey of the most common coupling procedures*

For the selection of the method of attachment, the primary consideration is which groups of the affinity ligand can be used for the linkage to a solid support without affecting its binding site. The attachment should not introduce non-specifically sorbing groups into the specific sorbent. From this point of view, it is better to couple the spacer to an affinant and, only after it has been modified in this manner, to attach it to a solid support. The linkage between the surface of a solid support and an affinant should be stable during adsorption, desorption and regeneration. In the choice of this method, the dependence of the stability of the affinant on pH should also be borne in mind, as should the fact that the coupling procedure may be greatly influenced by both the nature of the solid matrix and the substance to be attached [94]. When bifunctional compounds are used for the coupling, complications arising from the crosslinking of both the carrier and the proteins with one another can be expected, and therefore suitable reaction conditions, such as pH, temperature and time, should be maintained carefully during the coupling.

Coupling of affinants on cyanogen bromide-activated support
The most widely used activation procedure of agarose, dextran, but less common of cellulose and hydroxyalkyl methacrylate gel, employs cyanogen bromide [95]. The

mechanism of activation by CNBr and the subsequent coupling of the affinant was elucidated by Kohn and Wilcheck [96]:

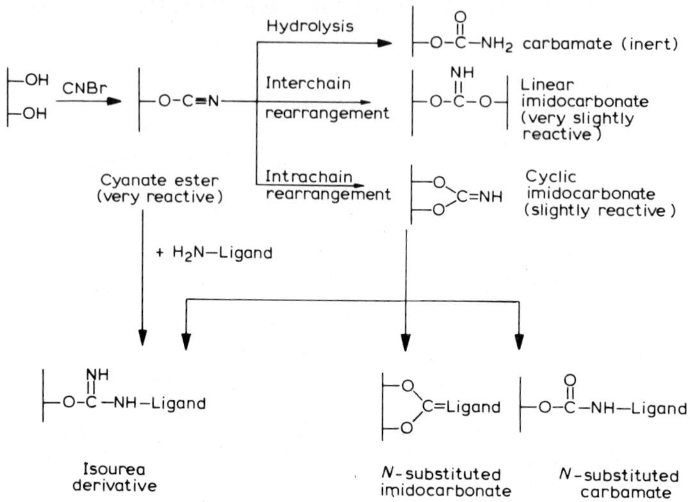

They developed analytical methods for the determination of cyanate esters and imidocarbonates [96]. Cyanate esters are most stable in acid, decreasing in stability as the pH is raised. Imidocarbonates behave in an opposite fashion, they are most stable in 0.1 M NaOH. During a short acid hydrolysis of an activated matrix all imidocarbonates will be removed quantitatively, while over 95% of the cyanate esters will remain intact. On the other hand, a short hydrolysis in 0.1 M NaOH leads to a rapid hydrolysis of all cyanate esters, leaving a substantial amount of imidocarbonates. Based on the assumption that the coupling of an affinant to cyanate esters and to imidocarbonates occurs simultaneously and independently, the total coupling capacity of an activated support can be predictable according to the equation:

$$\text{Total coupling capacity} = 0.80\ (\text{cyanate esters}) + 0.15\ (\text{imidocarbonates})$$

The initially formed cyanate esters are stable on Sepharose, but rearrange to cyclic imidocarbonates on Sephadex. The coupling of an affinant occurs predominantly via the free amino groups. The amine reacts in the unprotonated state. The following alkalinities are appropriate for coupling: aliphatic amines, pH ~ 10; amino acids, pH ~ 9; aromatic amines, pH 7–8. At pH 6 ribonucleotides are fixed, probably by an intramolecular rearrangement [97] to give

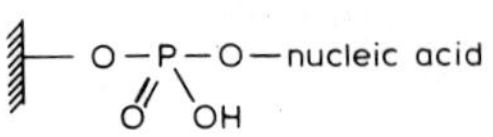

The attachment of amines on cyanogen bromide-activated Sepharose produces

N-substituted isourea, which is capable of protonation in the neutral and alkaline region. In the presence of primary amines and ammonia this linkage is unstable [98]. This is the primary flaw of this method when it is used for a single attachment.

Epoxide-containing supports

Epoxide-containing supports can be obtained directly by copolymerization (glycidyl methacrylate copolymer [99]) or by the use of epichlorohydrin [100,101] or bisoxiranes (1,4-butanediol-diglycylether) [80,102]. The epoxide group of the carrier may react with amino, carboxy, hydroxy and sulphydryl groups, with some aromatic nuclei, such as indole, imidazole, etc. [103], saccharides [104], polynucleotides [105], adipic acid dihydrazide [106]. The S-C, N-C and O-C bonds formed are extremely stable [76,101]. By selecting an appropriate long-chained bisoxirane, a long hydrophilic spacer can be introduced. Matsumoto et al. [107] converted epoxy-activated agarose into aminoderivatives using ammonia solution, which after succinylation were activated with N-hydoxysuccinimide.

Sulfonyl chlorides-containing supports

A simple method for the activation of hydroxyl group carrying supports such as agarose, cellulose, diol-silica, glycophase-glass or hydroxyalkyl methacrylate copolymer with 2,2,2-trifluoroethanesulfonyl chloride (tresyl chloride) was introduced by Nilsson and Mosbach [108]:

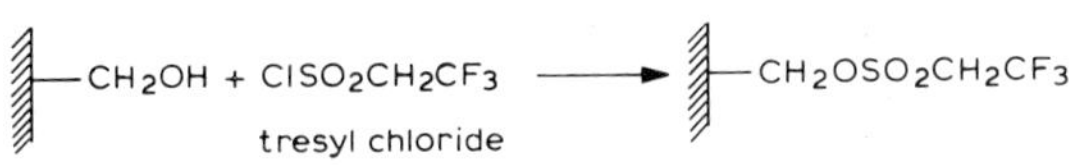

coupling:

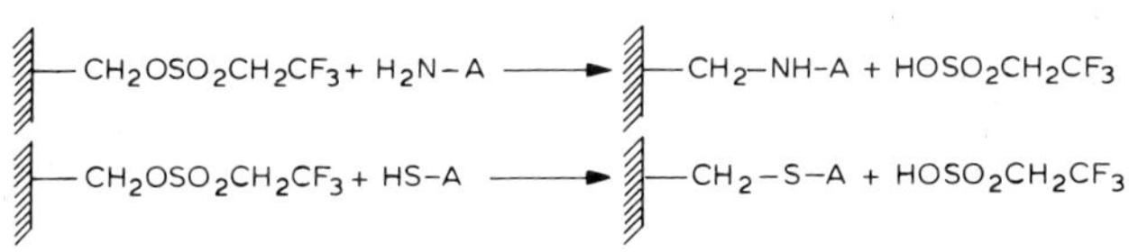

The advantage of tresyl chloride is that it is more reactive than the previously described tosyl chloride [109], and therefore it allows efficient coupling under very mild conditions, and to a number of different supports including those used for HPLC.

Immobilization via reductive alkylation

After oxidation of the saccharide molecules of polysaccharide [110], glycophase-CPG [111], glycosylated Spheron [112], etc., with $NaIO_4$, aldehyde groups are formed which bind the amino groups of proteins as Schiff's bases. The bond is further stabilized by addition of $NaBH_4$, which also eliminates the unreacted aldehyde groups.

Condensation reaction with the carbodiimide-promoted method

The affinants carrying primary aliphatic or aromatic amines can be attached to the support containing carboxyl groups, and on the contrary, the affinant carrying carboxyl groups can be coupled to the support containing an amino group with the carbodiimide-promoted method [113,114]. For the synthesis of adsorbents water-soluble carbodiimides are predominantly used, such as 1-ethyl-3-(3-dimethyl-aminopropyl)carbodiimide hydrochloride or 1-cyclohexyl-3-(2-morpholino-ethyl)carbodiimide metho-*p*-toluenesulphonate. Their main advantage is that their corresponding urea derivatives are soluble in water and thus they can be easily eliminated from the affinity adsorbent by washing with water. Boschetti et al. [115] presented the advantage of *N*-ethoxycarbonyl-2-ethoxy-1,2-dihydroquinoline.

Covalent attachment by using bifunctional reagents
Weston and Avrameas [116] developed a method for the direct attachment of affinants on to polyacrylamide gels using glutaraldehyde, which, if present in excess, reacts via one of its two aldehyde groups with the free amide group present in the polyacrylamide gel. The remaining free active group then reacts with the amino group of the affinant added during the subsequent binding reaction. Neurath and Strick [117] stabilized this bond by the reduction with sodium dithionite. The formation of disulfide and thioether linkages by heterobifunctional reagents was described by Inman [118]. In immunoaffinity chromatography glutaraldehyde-insolubilized antibody preparations were also used [119].

Other methods
Of the many other methods we can mention the preparation of biospecific adsorbents utilizing trichloro-*S*-triazine [120], coupling via thiol-disulfide interchange with activated thiolated support [121] or via azide formation [122]. The attachment of affinants containing the aromatic residues can be carried out by means of diazonium groups [87,123]. The coupling of the affinant onto the silinized glass after the reactions with *p*-nitrobenzoylchloride followed by reduction with sodium dithionite was described by Weetall and Filbert [124].

4.7.3.3 Blocking of unreacted groups

When the attachment of the affinity ligand on to the solid support is terminated, it is necessary that the remaining active groups that are capable of coupling should be eliminated. Some of the complications encountered in the application of bioaffinity methods could be due to the neglect of deactivation [76]. The coupling yield is never quantitative. When large affinants are being attached to the solid support, the concentration of excess reactive groups is particularly large. Further coupling can occur during the chromatographic procedure, which may result in the irreversible adsorption of the substance to be isolated [125]. The impurities coupled to the unreacted groups of a biospecific adsorbent can act as non-specific adsorption centers in the subsequent use of the adsorbent.

The problem of excess reactive groups is encountered in the use of all coupling methods. This problem can be solved by one of two methods.
(1) By coupling a highly penetrable substance with no effect on the adsorption-desorption procedure. In the methods that make use of the amino groups for the

coupling of affinants, ethanolamine is most frequently used as an inactivating reagent.
(2) By solvolysis. This method may be used when the affinants are stable in alkaline solution at elevated temperature; these conditions affect the hydrolytic removal of the activated groups. The excess epoxide groups may be hydrolysed to diols by treatment with 0.1 M perchloric acid [101] or with 10 mM HCl [126].

4.7.4 Experimental techniques

4.7.4.1 Classic bioaffinity chromatography

Bioaffinity chromatography (considered as one type of adsorption chromatography) is a relatively simple procedure, in which the active species are specifically adsorbed by the immobile reagent and subsequently desorbed from the specific adsorbent by an eluent. If a small amount of a substance is isolated from a crude mixture by means of an immobilized affinant of high affinity, the batchwise arrangement can be employed with advantage, sometimes combined with elution after a transfer of the material into a column. The isolation of thyroxine-binding globulins by Sepharose with attached thyroxine may serve as an example. In order to obtain a higher yield, Pensky and Marshall [127] carried out the first part of the sorption in a batchwise arrangement by stirring normal human serum with the immobilized affinant overnight. After washing it with 0.1 M sodium hydrogen carbonate solution, they transferred the insoluble carrier with the adsorbed isolated substances into a column from which two thyroxine-binding globulin fractions were than eluted with 0.002 M potassium hydroxide solution. By this combination the time necessary not only for the contact of substances to be isolated with the immobilized affinant is thus prolonged, but also the time necessary for a correct orientation of binding sites. In the columnar arrangement, the prolongation of the time of contact of the substance isolated with the immobilized affinant is achieved by stopping the flow through the column after sample application for a certain time. Manen and Russell [128] stated that if the flow of the solvent through the column during the chromatography of *S*-adenosyl-L-methionine decarboxylase on Sepharose with attached *p*-chloromercuribenzoate was not stopped for 2 h, the enzyme was eluted after several fractions. A sufficiently low flow-rate through the column of a specific sorbent is of prime importance in the isolation of high molecular weight substances, especially if they occur in high concentrations.

The rate of equilibrium attainment mainly during the batchwise arrangement is strongly affected by the concentration of a substance to be isolated in the sorption solution. Fig. 4.7.6 shows the rate of adsorption of trypsin on to soybean trypsin inhibitor-agarose particles in suspensions in solutions containing equal amounts of trypsin but of a different concentration [129]. The rate of sorption is determined on the basis of the residual activity of trypsin measured at various time intervals.

Although simple batch processing can be used in applying a biospecific ad-

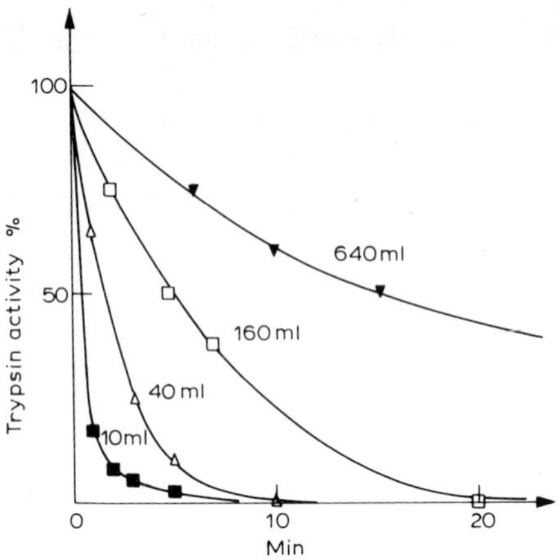

Fig. 4.7.6. Rate of adsorption of trypsin on to soybean trypsin inhibitor-agarose particles in suspensions of different concentrations. Reproduced with permission from Ref. 129.

sorbent, the column chromatographic method offers many advantages [125]. It is inherently more efficient and adaptable to a variety of related separations and more readily automated. A fact of importance for analytical determination of biological active substances in animal tissues or fluids by bioaffinity chromatography, or for their large scale isolation, is that the sorption to columns of a biospecific adsorbent is independent of the concentration of the biologically active substance in the sample applied [101]. In our laboratory we applied 50 mg of crude porcine pepsin dissolved in 10, 100 or 1000 ml to the column of 6-aminocaproyl-L-Phe-D-Phe-OMe-Separon. With the increasing dilution of the solution of the enzyme, the peak of the first fraction of inert proteins undergoes spreading (and completely disappears at the lowest concentration). On the contrary, the peak of the pepsin fraction (eluted by the desorbing buffer of a high ionic strength) is sharp in all three cases and is in no way affected by the dilution [101].

It is conventional practice to operate the column at a constant flow rate, which causes the sample, buffer washes, and eluent volumes to be directly related to application time intervals. Since the flow rate is obviously a function of the column dimensions, and is limited by the column resistance and the tendency of the softer porous gels to compact under hydrodynamic stress, the use of a more rigid support is more advantageous. The flow rate and exposure time influence the volume of the buffer required to lower concentration of ballast proteins to acceptable levels. For most practical applications the flushing of the column can be monitored with a UV flow cell, return to baseline volumes being indicative of the washing efficiency. The equilibrium constant of the biospecific complex influenced the amount lost by the dissociation of the specifically adsorbed material during the exposure to the washing buffer and the heterogeneity of the substance isolated. Using two or three column volumes of washing buffer, Eleveigh and Levy [125] determined the heterogeneity of the antibody preparation to be less than 2% in the majority of applications. These

TABLE 4.7.4
Effect of cycling a 5-ml column with excess antigen and eluting with 15 ml of 2.5 M NH_4SCN

Cycle	Amount (mg)			Eluted/bound (%)	Amount retained (mg)
	Applied	Bound	Eluted		
1	5.87	4.192	3.204	76.4	0.988
2	5.97	3.316	3.019	91.0	0.297
3	5.89	3.239	2.988	92.2	0.251
4	5.88	3.179	2.964	93.2	0.215
5	5.86	3.135	2.995	95.5	0.140
6	5.91	3.089	2.965	96.0	0.124

authors [125] also studied the effect of cycling the immunosorbent, which is shown in Table 4.7.4. The poor recovery of bound antigen from the column in the first cycle can be partly explained by an insufficient volume of eluent to completely dissociate the complex, but a comparison with the recovery in the subsequent cycles indicates that the lower yield represents an apparently irreversible reaction of the antigen. This 'first-cycle effect' has been observed with many supports and with a variety of immunological systems, and must be considered in defining the immunosorbent capacities [125].

4.7.4.2 High performance liquid bioaffinity chromatography (HPLAC) of proteins

High performance liquid affinity chromatography (HPLAC) combines the inherent speed and resolving power of HPLC with the biological specificity of bioaffinity chromatography [130]. Lowe et al. [126] demonstrated the ability of Cibacron Blue F3G-A bonded silica to resolve selectively and rapidly complex mixtures of complementary proteins. The chromatography of a crude yeast extract on Cibacron Blue F3G-A silica with on-line monitoring of protein (280 nm), and of hexokinase (HK) and 3-phosphoglycerate kinase (PGK) activity (340 nm) shown in Fig. 4.7.7 may serve as an example. The experimental set-up for on-line monitoring of the elution profiles from the HPLAC adsorbent is shown in Fig. 4.7.8. Enzymes eluted from the HPLAC column were detected with an on-line detector system comprising a reagent pump, post-column reactor equilibrated to 40 ± 0.1°C in a water bath and a UV/visible monitor. The post-column detection of HK and PGK was accomplished with a 'universal' assay medium comprising the reactions:

$$\text{D-glucose} + \text{ATP} \xrightarrow{\text{HK}} \text{D-glucose-6-phosphate} + \text{ADP}$$

$$\text{D-glucose-6-phosphate} + \text{NADP}^+ \xrightarrow{\text{G6PDH}} \text{6-phosphogluconate} + \text{NADPH} + \text{H}^+$$

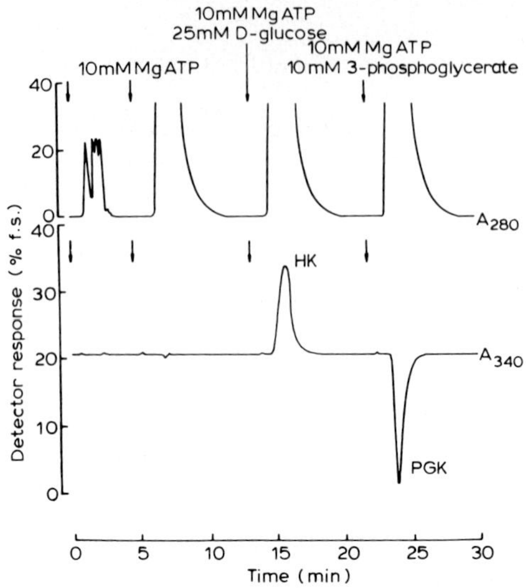

Fig. 4.7.7. Chromatography of a crude yeast extract on Cibacron Blue F3G-A silica with on-line monitoring of hexokinase (HK) and 3-phosphoglycerate kinase (PGK) activity. Sample applied at first arrow: crude yeast extract (2 μl); column irrigant, 0.1 M Tris-HCl buffer (pH 7.3) containing 0.5 mM EDTA, 5 mM $MgCl_2$ and 0.5 mM 2-mercaptoethanol; flow rates, 1 ml/min, both pumps; post-column reactor temperature, 40 ± 0.1°C; assay mix temperature, 0°C; temperature of analytical column, 22°C; eluents (400 μl), as indicated by the arrows; UV-detector (290 nm), 0.32 AUFS; enzyme activity detector, 0.5 AUFS. Reproduced with permission from Ref. 126.

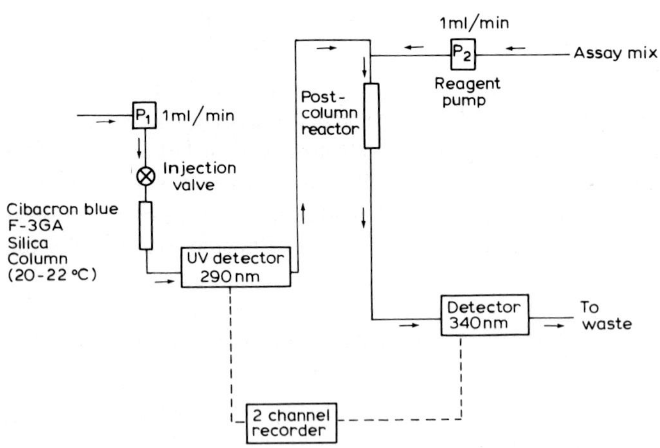

Fig. 4.7.8. The experimental set-up for on-line monitoring of elution profiles from HPLAC adsorbents. Reproduced with permission from Ref. 126.

for hexokinase activity and the reactions:

$$\text{3-phosphoglycerate} + \text{ATP} \xrightarrow{\text{PGK}} \text{glycerate-1,3-bisphosphate} + \text{ADP}$$

$$\text{glycerate-1,3-bisphosphate} + \text{NADH} + \text{H}^+ \xrightarrow{\text{G3PDH}} \text{glyceraldehyde-3-phosphate} + \text{NAD}^+ + \text{phosphate}$$

for 3-phosphoglycerate kinase activity.

The presence of HK in the column effluent produces an increase in absorbance at 340 nm, the presence of PGK decreases the absorbance at 340 nm.

It can be expected that the HPLAC will find wide application in the resolution of enzymes and other proteins of clinical or biotechnological significance especially when coupled to the high-speed fully automated chromatographic profiling systems reported here.

4.7.4.3 Automatic time-based instrument for preparative application

The immunoaffinity method of separation is basically very simple, being a direct application of adsorption-desorption chromatography. However, the practical success of bioaffinity chromatography in large scale preparation of biologically active molecules depends on many critical factors. The factors involved in the preparation of specific antibodies were successfully solved by Eveleigh [131]. According to him, some form of automated system becomes essential for routine immunoaffinity separation. The major components of an automated immunoaffinity system are schematically presented in Fig. 4.7.9. For immunoaffinity separation, conventional chromatography columns with a high cross-sectional area to lengthen the ratio, and preferably adaptors to reduce dead space, are quite suitable. All fluid lines, connectors and valve bodies should be of an inert plastic, as thiocyanate and acid are corrosive. A normally closed two-way sample valve is important, in order to prevent

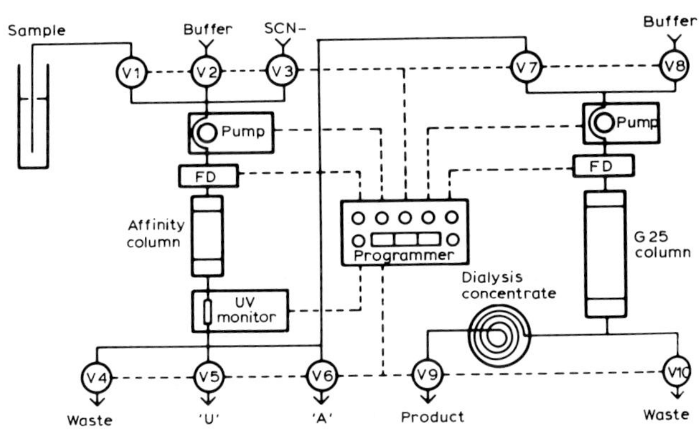

Fig. 4.7.9. Schematic diagram of immunoaffinity system. With permission from Ref. 131.

a disastrous loss of the sample by syphoning under these circumstances. A mechanical, programmable, multi-way device designed for the purpose forms the basis of an automated instrument built by Eveleigh [132]. The peristaltic pump is most suitable, as chaotropes of corrosive nature are used. A conventional chromatography UV monitor with its associated recorder is important in optimising the operating conditions and, in retrospect, monitoring of unattended repetitive separations. For the same reason it is useful to incorporate active fluid sensors at critical points within the fluid circuitry to detect line breakages and unanticipated loss of buffers, etc. Other protective devices desirable for long-term unattended operation are pressure sensors to indicate impending column blockage or valve malfunctions, and power failure interruptions to avoid a potential loss of synchronization. For a

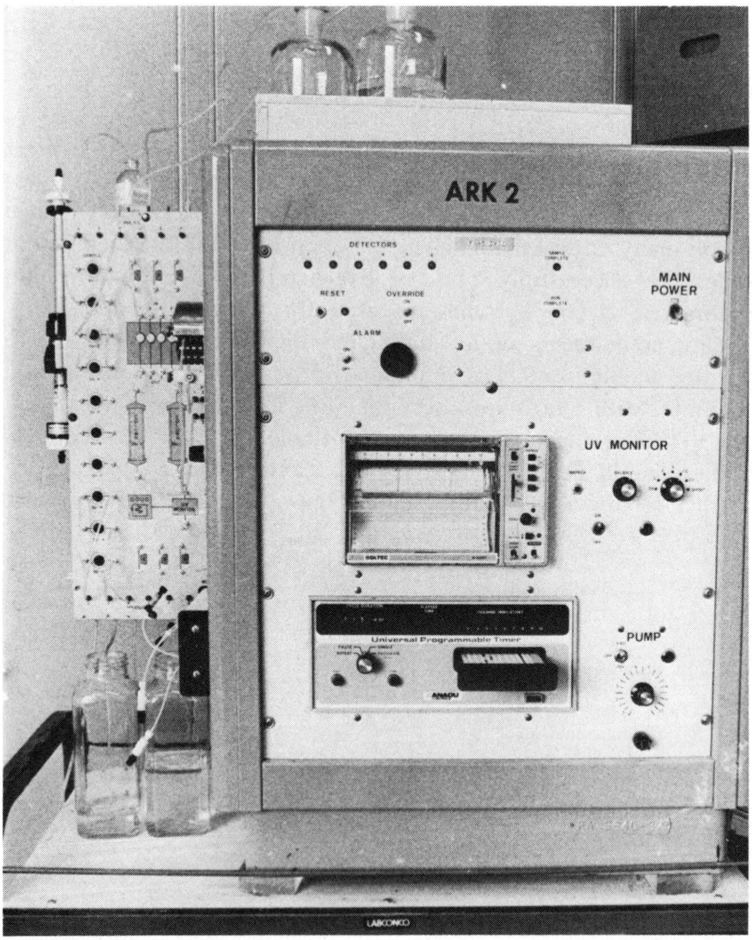

Fig. 4.7.10. An automatic time-based instrument based on the schematic representation of immunoaffinity system. Reproduced with permission from Ref. 131.

versatile and reliable control system the microprocessor based programmer has been introduced. An instrument based on the schematic representation of the immunoaffinity system previously discussed is illustrated in Fig. 4.7.10. An automatic immunospecific affinity chromatographic system for a continuous operation has also been described by Folkersen et al. [133], and repetitive semi-automatic immunoaffinity chromatography by Römer and Rauterberg [134].

By adopting the systematic practical approaches, separations and purifications of biologically active molecules can become routine procedures rather than an art.

4.7.4.4 Extracorporeal removal of substances in vivo

The specific removal of circulating immune reactants from the host could be utilized as a therapy for some immunologically mediated diseases. Terman et al. [135] described the specific removal of circulating bovine serum albumin (BSA) antibodies in a canine host with a BSA collodion charcoal extracorporeal immunoadsorbent. A schematic representation of an extracorporeal immunoadsorption system is illustrated in Fig. 4.7.11. The arterial catheter was placed in the arterialized segment of the jugular vein of dogs anaesthetized with sodium pentobarbital. Heparin was injected intravenously and the catheters were then connected to a continuous flow plasma cell separator. Arterial blood entering the plasma cell separator was separated into plasma and formed elements by centrifugation. The plasma was then

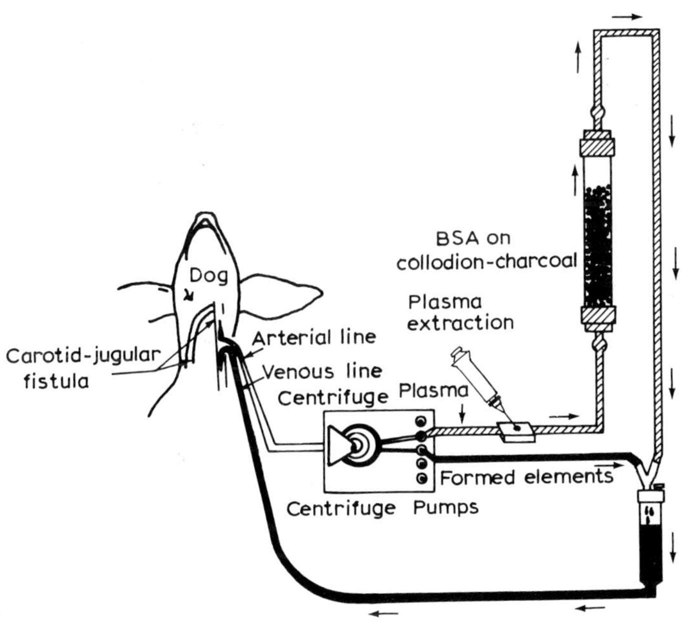

Fig. 4.7.11. Schematic representation of extracorporeal immunoadsorption system. Reproduced with permission from Ref. 135.

pumped through chambers containing antigen-coated charcoal. The formed elements of the blood which were separated from plasma in the plasma cell separator were pumped to a site where they rejoined the plasma coming from the immunoadsorbent column. All the recombined blood passed through a drip chamber and a bubble trap, and was then returned to the host. The extracorporeal volume in this system was approximately 170 ml. Due to the creation of arteriovenous fistulae between common carotid and internal jugular blood vessels this sequential extracorporeal immunoadsorption was possible without morbidity to the experimental animals. Lupien et al. [136] used the removal of cholesterol from blood by bioaffinity binding to heparin-agarose to lower plasma cholesterol levels in two patients suffering from homozygous familial hypercholesterolemia.

4.7.5 Areas of applications (see also Note following References)

Bioaffinity chromatography is primarily an efficient method for the isolation of biologically active substances. In my book [5] I presented a bibliographic review of the use of affinity chromatography for the isolation of biologically active products up to the end of 1976. Out of 1541 references included in this review 52.4% dealt with enzymes and their subunits, inhibitors and cofactors, 12.2% with antibodies and antigens, 8.8% with lectins, glycoproteins and saccharides, 6.9% with receptors, binding and transport proteins, 6.4% with nucleic acids and nucleotides, 3.2% with viruses, cells and their components, 2.4% with specific peptides and 7.6% with other components such as factors from human plasma, interferons, toxins, repressors, etc. The ratio of the individual groups of substances isolated by bioaffinity chromatography in recent years is practically the same. However, the use of bioaffinity chromatography is not limited only to the isolation of biologically active molecules. By its nature, bioaffinity chromatography is a suitable method for the study of interactions in biochemical processes. For example, interactions of proteins with nucleic acid [137] or peptides [138], or interactions of nucleotides with amino acids and peptides [139] have been studied.

Bioaffinity chromatography is also a useful tool for the solution of the mechanism of enzymatic processes [140]. Akanuma et al. [141] employed this method for the study of the binding site of bovine carboxypeptidase B on the basis of a complex formation with immobilized substrate analogues of basic and aromatic amino acids. The problem of determining peptides of the active sites of enzymes or antibodies can be solved by the isolation of labelled specific peptides as is shown in section 4.7.5.7. Bioaffinity chromatography was also applied to the study of the molecular structures of human fibroblast and leucocyte interpherones [142].

On the basis of the elution volumes of the enzyme eluted from the column with an immobilized inhibitor, using the solutions of various concentrations of a soluble inhibitor, we can determine the inhibition constants both with the immobilized inhibitor and with the soluble inhibitor employed [26,27]. Using the same inhibitor for the immobilization and elution, we can draw direct conclusions about the effects

TABLE 4.7.5

Use of bioaffinity chromatography for the isolation of biologically active products

Substances isolated	Affinity ligands	Solid supports or immobilized affinity ligands	Refs.
1. *ENZYMES*			
Acetylcholinesterase	*N*-methylacridine	Sepharose 2B	56
Adenylosuccinate synthetase	Hadacidin	Sepharose 4B with 1,6-diamino hexane	145
Alkaline phosphatase	Cibacron blue 3GA	Sepharose 6B	33
Amine oxidase	1,6-diamino hexane	Sepharose 4B	146
Aminoacyl-tRNA synthetase	AMP	Agarose-hexyl-adenosine-5′-phosphate	60
Aminopeptidase	Amastatin	AH-Sepharose 4B	147
Aryl sulfatase	Aryl sulfatase A monomer subunit	CNBr-activated Sepharose 4B	148
Aryl sulfatase (variant form)	Concanavalin A	ConA-Sepharose	149
Asparaginase	*N*-(ω-aminohexyl)-L-aspartic acid	Sepharose 4B (modified by spiropyran)	150
Aspartate aminotransferase	Citric acid	Aminoethyl-Sepharose 4B	151
Aspergillopepsin A	Gramicidin C	Sepharose 4B	152
Calmodulin-dependent NAD kinase	Calmodulin	CNBr-activated Sepharose 4B	153
Chloroplast α-1,4 glucan phosphorylase	Starch	Sepharose	154
Cholesterol ester hydrolase	Concanavalin A	ConA-Sepharose	155
Chymotrypsin	*N*-benzyloxycarbonylglycyl-D-phenylalanine	Spheron with 1,6-diamino hexane	156
Cobalt-activated acylase	ε-Aminopentyl-α-hydroxy-isocaproyl-DL-tyrosine ethyl ester	Sepharose 4B	157
Collagenase	L-arginine	CH-Sepharose	158
2′ : 3′-Cyclic nucleotide 3′-phosphodiesterase	Lectins	Agarose beaded plant lectins	159
Dihydropteridine reductase	AMP	5′-AMP-Sepharose 4B	160
DNA gyrase	Novobiocin	Epoxy-activated Sepharose 6B	161
DNA polymerase	Polynucleotide	Sepharose 6B activated with bisoxirane	105
Glucose-6-phosphate dehydrogenase	$NADP^+$	Sepharose 6B with 3-amino-phenylboronic acid	45
α-Glucosidase	*m*-Aminophenylboronic acid	Matrex-gel PBA (PBA-agarose)	43
Glutathione *S*-transferase	Glutathione	Sepharose CL-4B with aliphatic diamines	51
Hexokinase	*N*-Aminoacylglucosamine derivatives	CNBr-activated Sepharose 4B	57
Heparinase	Heparin	AH-Sepharose 4B	162
IMP dehydrogenase	Procion dyes	Sepharose 4B	163
β-Mannanase	Mannan	AH-Sepharose	164
Methylchymotrypsin	*Streptomyces subtilisin* inhibitor	Sepharose 4B	165

TABLE 4.7.5 (continued)

Substances isolated	Affinity ligands	Solid supports or immobilized affinity ligands	Refs.
Pectic enzymes	Cross-linked polyuronides		166
Pepsin	ε-Aminocaproyl-L-Phe-D-Phe-OMe	Separon with epichlorohydrin	53
Peroxidase	Vanillic aldehyde	γ-aminopropyltriethoxysilane glass	167
Phenylalanine ammonia lyase	L-Phenylalanine	CNBr-activated Sepharose 4B	168
Phospholipase A_2	Antiphospholipase antibodies	CNBr-activated Sepharose	169
Proteolytic enzymes	Bacitracin	Aminosilochrom	36
Proteolytic enzymes	Hemoglobin	Sepharose 4B	34
Pyruvate dehydrogenase	Thiochrome	Sepharose 2B with 1,4-butanediol-diglycidylether	59
Serine proteinases	*p*-(ω-Aminomethyl)phenyl boronic acid	CH-Sepharose	42
Trypsin	*p*-aminobenzamidine	Spheron with 1,6-diamino hexane	156
Trypsin	Ovomucoid	Spheron	156
Uroporphyrinogen I synthetase	Porphobilinogen	AH-Sepharose 4B	170
Enzyme subunits			
Regulatory subunit of histone kinase	8-(γ-Carboxypropylthio)adenosine-3′,5′-cyclic phosphate	Sepharose 4B with polylysine	171
Subunits of acid synthetase	Pantetheine	Sepharose with ε-aminocaproic acid	172
Enzyme inhibitors			
α-Antitrypsin	Heparin	CNBr-activated Sepharose 4B	173
Catalase inhibitor	Catalase	CNBr-activated Sepharose 4B	174
Chymotrypsin inhibitor	Chymotrypsin	Spheron	25
Trypsin inhibitor	Subunits of inhibitor	CNBr-activated Sepharose	175
2. *ANTIBODIES*			
Antibodies to human serum albumin (HSA)	HSA	CNBr-activated Sepharose 4B	73
Antibodies to antigenic sites of myoglobin	Peptides of five myoglobin antigenic sites	Sepharose CL-4B	55
Antibodies to Staphylococcal enterotoxin A and E	Enterotoxin A or E	CNBr-activated Sepharose 4B	176
Antimorphine antibodies	Morphine	AH-Sepharose	177
Antispermine antibodies	Spermine	Activated CH-Sepharose 4B	178
Antithyroglobulin antibody	Thyroglobulin	Sepharose 4B	179
Rat IgG subclasses	Staphylococcal protein A	Protein A-Sepharose CL-4B	180
Antigens			
Barley β-amylase	Anti-barley antibody	CNBr-activated Sepharose 4B	71
Antigens of *Fasciola hepatica*	Concanavalin A	ConA-Sepharose	181

TABLE 4.7.5 (continued)

Substances isolated	Affinity ligands	Solid supports or immobilized affinity ligands	Refs.
DNP-bovine serum albumin	Anti-DNP antibodies	Sepharose 4B	72
Escherichia coli heat-labile enterotoxin	Anti-cholera toxin immunoglobulin	CNBr-activated Sepharose 4B	182
Estradiol-17β (removing from plasma)	Anti-estradiol antibodies	CNBr-activated Sepharose 4B	183
Hepatitis B surface antigen	Polymerized human serum albumin	CNBr-activated Sepharose 4B with ethylenediamine	117
Immunoreactive radio-iodinated antigen	Glutaraldehyde-insolubilised antibody preparations		119
Platelet plasma membrane proteins labelled with DNP	Anti-DNP antibodies	Sepharose 4B	184
Polyclonal IgA or IgM	Anti-IgA or -IgM antibodies	Tri-ethoxysilyl-propylamine-CPG	134
Vaccines (poliovirus)	Cross-linked antibodies	Sepharose CL-4B	185
3. *LECTINS*			
Lectin from *Dolichos biflorus*	*N*-Acetyl-D-galactosamine	Spheron	84
Lectin from *Glycine sojae*	D-Galactose	Spheron	84
Lectin from *Phaseolus vulgaris*	Concanavalin A	ConA-Sepharose	186
Lectin from *Ricinus communis*	Water insoluble *S*-β-galactopranosyl polyacrylamide derivative		187
Limulus haemagglutinin	Fetuin	Sepharose 4B	188
Non-specific phytohaemagglutinin from *Vicia crocca*	Dextran	Sephadex G-100	189
Glycoproteins and saccharides			
Axon plasma membrane proteins	Concanavalin A	Con A-Sepharose	190
Galactomannan from *Cassia alata*	α-D-Galactopyranosyl-binding lectin from *Banderaea simplicifolia*	Sepharose 4B	191
Glycogen	Concanavalin A	Concanavalin A polymerized with glutaraldehyde	192
Human fibroblast interferon	Concanavalin A	ConA-Sepharose	193
Orosomucoid (breast tumor-associated glycoprotein)	Concanavalin A	ConA-Sepharose	194
Sciatin (glycoprotein from chicken sciatic nerves)	Concanavalin A	Glycosylex-A (ConA-agarose)	195
4. *RECEPTORS, BINDING and TRANSFER PROTEINS*			
Activated glucocorticoid receptor	DNA	Cellulose	196
Auxin receptor proteins	2-Hydroxy-3,5-diiodobenzoic acid	Epoxy-activated Sepharose 6B	197
DNA-binding proteins	DNA	Sephadex G-10	198
Erythrocyte receptors for viruses	Wheat germ agglutinin	WGA-Sepharose	199

TABLE 4.7.5 (continued)

Substances isolated	Affinity ligands	Solid supports or immobilized affinity ligands	Refs.
Glucocorticoid receptor	Dexamethasone	Sepharose CL4B with diaminononane	200
Ligandin (protein binding organic anions)	Sulfobromophthalein	Sepharose 4B	201
Nuclear estrogen receptor	17β-Estradiol-17-hemisuccinate	Sepharose 4B with ovalbumin	202
5. *NUCLEIC ACIDS and NUCLEOTIDES*			
Cyclic GMP	Dihydroxyboronyl derivative	Affi-Gel 601 (polyacrylamide gel)	203
Nucleosides	Dihydroxyboronyl derivative	Sephadex A-25	41
Oligonucleotides	Acriflavine	Ultrogel A4R	204
Poly(A)	Poly(U)	Poly(U)-Sepharose 4B	205
Poly(A)-containing ribonucleoproteins	Oligothymidylic acid	Oligo(dT)-cellulose T_2	206
6. *VIRUSES, CELLS and their COMPONENTS*			
Bovine viral diarrhea virus	*Crotolaria juncea* lectin	Sepharose 2B	207
Influenza virus	α- and β-Ketosides of neuraminic acid derivatives	Sepharose 4B	208
Poliovirus	Antibody	Sepharose 4B	209
Sodium-dependent D-glucose transport system	Phlorizin	Phlorizin polymer	210
7. *SPECIFIC PEPTIDES*			
Actin-binding fragment of fibronectin	Actin	Sepharose 2B	211
Affinity-labelled peptide from the tryptic digest of staphylococcal nuclease	Nuclease	Sepharose 4B	212
Cysteine and methionine containing peptides (after reaction with N^{α}-bromoacetyl-N^{ε}-DNP-L-lysine)	Anti-DNP antibodies	Sepharose 4B	213
Murein precursors (lactyl-pentapeptides of the structure R-D-ala-D-ala)	Vancomycin	CH-Sepharose 4B	214
Ribonuclease-*S*-peptide synthesized by the solid-phase technique	Ribonuclease-*S*-protein	Sepharose 4B	215
Thiol containing peptides	Glutathione-2-pyridyl disulphide	Activated thiol-Sepharose 4B	38
8. *OTHERS*			
Amnion interferon	Bovine plasma albumin	Sepharose 4B	216
Clostridium botulinum type A neurotoxin	*p*-Aminophenyl-β-D-thiogalactopyranoside	CH-Sepharose 4B	217
Enantiomers of warfarin and tryptophan	Serum albumin	Sepharose 4B	218

TABLE 4.7.5 (continued)

Substances isolated	Affinity ligands	Solid supports or immobilized affinity ligands	Refs.
Fibrinogen	T-Proteins of *Streptococcus pyogenes*	AH-Sepharose 4B	219
Fractions of blood plasma	Lysine, ε-aminocaproic acid or heparin	Sepharose	220
Human factor VIII	Antibody to factor VIII	CNBr-activated Sepharose 4B	221
Human high density lipoproteins	Heparin	Sepharose 6B	222
Human low density lipoproteins	Concanavalin A	Con A-Sepharose 4B	223
Human pregnancy zone protein (PZP)	Anti-human PZP-immunoglobulin	CNBr-activated Sepharose 4B	224
Lipids (polyphosphoinositides)	Neomycin	Glycophase-CPG	111
Migration inhibitory factor	α-L-Fucose	Sepharose 6B	225
Thyrotropin	Anti-thyrotropin antibodies	CNBr-activated Sepharose 4B	226

of the bonds and the support on the interaction being studied from the agreement between, or the difference in, the dissociation constants determined [143,144].

This enumeration of the uses of bioaffinity chromatography does not exhaust its possibilities, which are many and varied. In the subsequent sections some examples of the possibilities will be briefly discussed. Various types of substances, isolated by bioaffinity chromatography are given in Table 4.7.5.

4.7.5.1 Enzymes, their subunits and inhibitors

Most papers dealing with bioaffinity chromatography describe the isolation of enzymes. From Table 4.7.5 it is evident that bioaffinity chromatography has been used for the isolation of enzymes of all classes. The affinity ligands for the isolation are inhibitors of low molecular weight [36,53,145–147,152,157] and high molecular weight [156], substrates [34,105,154,166], products [158], cofactors [45,60], coenzyme analogues as triazine dyes [33,163], effectors [151], regulator protein [153], subunit of enzyme [148], antibodies [169] or for enzymes, which are glycoproteins-lectins [149,155,159]. For the isolation of inhibitors enzymes are mostly used as the affinants.

4.7.5.2 Antibodies and antigens

The interaction of antibodies with their antigens is comparable, in its specificity, to the binding of substrates with enzymes. The dissociation constants are most commonly within the range of 10^{-5}–10^{-8} M. From the examples described in Table 4.7.5 it is evident that antigens and haptens are used as affinants for the isolation of antibodies. In contrast, immobilized antibodies are used for the isolation of antigens.

Antigens may belong to the most varied types of substances and therefore can be found in all the sections describing the isolation of the individual groups of biologically active compounds.

4.7.5.3 Lectins, glycoproteins and saccharides

Lectins are proteins or glycoproteins of vegetable or animal origin displaying a higher or lower selectivity for carbohydrates or a group of carbohydrates. These specific complexes resemble the specific complexes of antibodies. Lectins react with cell membrane components. They agglutinate erythrocytes, tumour and embryonal cells. Agglutinins with a defined specificity for sugars may thus be useful means for the study of the surface structures of the cells transformed by malignant tumours or by viruses.

The differing affinities of lectins for terminal carbohydrate residues characteristic of single glycoproteins enable glycoproteins to be isolated by means of immobilized lectins. Identification of the terminal sugar or sugars in the carbohydrate part of the glycoproteins is necessary for the selection of a lectin with a corresponding specificity. The isolation of a series of lectins and glycoproteins by bioaffinity chromatography is mentioned in Table 4.7.5. A review of the use of bioaffinity chromatography for the purification of lectins has been recently published by Lis and Sharon [227].

4.7.5.4 Receptors, binding and transfer proteins

During the last ten years significant progress has been made in the understanding of the nature and function of the cell surface receptors, binding and transport proteins of hormones and vitamins mainly due to the successful application of bioaffinity chromatography. As these molecules are usually present in minute quantities in biological tissues, conventional purification techniques may be inadequate for their isolation. Due to the extremely small quantities of these proteins present in a biological system, special strategies are necessary for each individual system. Many recent references about the hormone-receptors interactions are included in the review of Parikh and Cuatrecasas [228]. Some examples of the use of bioaffinity chromatography for the isolation of receptors, binding and transport proteins are given in Table 4.7.5.

4.7.5.5 Nucleic acids and nucleotides

Immobilized nucleic acid bases, nucleosides or oligonucleotides may be used for separation, fractionation and structure determination of various nucleic acids and enzymes participating in their synthesis and degradation. Schott et al. [139,229] made use of immobilized defined oligonucleotides for the selective separation of free nucleotides on the basis of a base-pairing mechanism. Complementary oligonucleotides in the mobile phase are selectively adsorbed on the immobilized template if chromatography takes place under the conditions necessary for base pairing. Desorption is then carried out with a temperature gradient. This so-called template chromatography permits the study of the specificity of oligonucleotides for the formation of bases, and the interaction of oligonucleotides with peptides. For the chromatography of mononucleotides, oligonucleotides and transfer RNA molecules sorbents containing a dihydroxyboryl derivative [41,200] or aromatic ligands [201] can also be used. Some examples from the field of nucleotide research are given in Table 4.7.5.

4.7.5.6 Viruses, cells and their components

Application of bioaffinity chromatography to the isolation of cells, cell organelles, cell membranes, phages and viruses was the subject of an excellent review article by Sharma and Mahendroo [230]. Some examples not included in this review are in Table 4.7.5.

4.7.5.7 Specific peptides

Peptides from the active sites of biologically active proteins can be isolated after the reaction of the amino acids of the active sites, for example of enzymes with analogues of substrates, of antibodies with haptens, etc. In these instances, the specific and reversible binding of analogues at the binding site of the protein must be followed by the formation of a covalent bond at or near to the site of the labelling. After the digestion of an affinity labelled protein only the affinity-labelled peptides are specifically adsorbed on the column of an immobilized native protein (enzyme or antibody). After washing the column, the specifically adsorbed labelled peptides can be eluted under conditions such that dissociation takes place. The isolation of affinity labelled peptides from tryptic hydrolysate of staphylococcal nuclease after reaction with the bromoacetyl derivative of deoxythymidine 3′-*p*-aminophenylphosphate 5′-phosphate on nuclease-Sepharose may serve as an example [212]. For the localization of some residues on the protein surface, bioaffinity chromatography of the enzymatic digest of a modified protein on immobilized respective antibodies can be successively used. Some examples of the isolation of the specific peptides are included in Table 4.7.5.

4.7.5.8 Others

Different substances, such as lipids, lipoproteins, hormones, fractions of plasma, toxins, etc., are included in the last part of Table 4.7.5. From the point of view of practical application of bioaffinity chromatography an important example is industrial blood plasma fractionation [220].

References

1 O'Carra, P., Barry, S. and Griffin, T. (1974) Methods Enzymol. 34, 108–126.

2 Jacoby, W.B. and Wilchek, M. (eds.) (1974) Methods in Enzymology, Vol. 22, Enzyme Purification - Part B, Affinity Techniques, Academic Press, New York.

3 Lowe, C.R. and Dean, P.D.G. (1974) Affinity Chromatography, John Wiley and Sons, London.

4 Ruoslahti, E. (ed.) (1976) Immunoadsorbents in Protein Purification, University Park Press, Baltimore.

5 Turková, J. (1978) Affinity Chromatography, Elsevier, Amsterdam.

6 Epton, R. (ed.) (1978) Chromatography of Synthetic and Biological Polymers, Vols. 1 and 2, Ellis Horwood, Chicester.

7 Lowe, C.R. (1979) An Introduction to Affinity Chromatography, North Holland, Amsterdam.

8 Dunlap, R.B. (ed.) (1974) Immobilized Biochemicals and Affinity Chromatography, Adv. Exp. Med. Biol., Vol. 42, Plenum Press, New York.

9 Hoffmann-Ostenhof, O., Breitenbach, M., Koller, F., Kroft, D. and Scheiner, O. (eds.) (1978) Affinity Chromatography, Biospecific Sorption — The First Extensive Compendium on Affinity Chromatography as Applied to Biochemistry and Immunochemistry, Proc. 1st Int. Symp., Vienna, 1977, Pergamon Press, Oxford.

10 Sundaram, P.V. and Eckstein, F. (eds.) (1978) Theory and Practice in Affinity Techniques, Proc. Int. Symp., Göttingen, 1978, Academic Press, London.

11 Egly, J.M. (ed.) (1979) Affinity Chromatography and Molecular Interactions, Proc. 3rd Int. Symp., Strasbourg, 1979, INSERM, Vol. 86, Paris Cedex: INSERM-Publications.

12 Gribnau, T.C.J., Visser, J. and Nivard, R.J.F. (1982) Affinity Chromatography and Related Techniques. Proc. 4th Inter. Symp., Veldhoven 1981, Elsevier, Amsterdam.

13 Shaltiel, S. (1974) Methods Enzymol. 34, 126–140.

14 Jennissen, H.P. and Heilmeyer, L.M.G. (1975) Biochemistry 14, 754–759.

15 Hofstee, B.H.J. (1976) in Methods of Protein Separation (Catsimpoolas, N., ed.) Vol. 2, pp. 245–278, Plenum Press, New York.

16 Ochoa, J.L. (1978) Biochimie 60, 1–15.

17 Porath, J., Carlsson, J., Olsson, I. and Belfrage, G. (1975) Nature (London) 258, 598–599.

18 Von der Haar, F. (1973) Methods Enzymol. 34, 163–171.

19 Bøg-Hansen, T.C. (1973) Anal. Biochem. 56, 480–488.

20 Hořejší, V. and Kocourek, J. (1974) Methods Enzymol. 34, 178–181.

21 Kula, M.R. (1979) in Applied Biochemistry and Bioengineering (Wingard, L.B., Katchalski-Katzir, E. and Goldstein, L., eds.) Vol. 2, pp. 71–95, Academic Press, New York.

22 Egly, J.M. and Porath, J. (1979) J. Chromatogr. 168, 35–47.

23 Porath, J. (1974) Methods Enzymol. 34, 13–30.

24 Lowe, C.R. and Dean, P.D.G. (1971) FEBS Lett. 18, 31–34.

25 Turková, J., Hubálková, O., Křiváková, M. and Čoupek, J. (1973) Biochim. Biophys. Acta 322, 1–19.

26 Chaiken, I.M. (1979) Anal. Biochem. 97, 1–10.

27 Oda, Y., Kasai, K. and Ishii, S. (1981) J. Biochem. (Tokyo) 89, 285–296.

28 Lowe, C.R. (1977) Int. J. Biochem. 8, 177–181.

29 Dean, P.D.G. and Harvey, J. (1977) in Regulation of Enzyme Synthesis and Activity in Higher Plants (Smith, H., ed.) pp. 271–288, Academic Press, New York.

30 Land, M. and Byfield, P.G.H. (1979) Int. J. Biol. Macromol. 1, 223–227.

31 Dean, P.D.G. and Watson, D.H. (1979) J. Chromatogr. 165, 301–319.

32 Clonis, Y.D. and Lowe, C.R. (1981) Biochem. Biophys. Acta 659, 86–98.

33 Bouriotis, V. and Dean, D.G. (1981) J. Chromatogr. 206, 521–530.

34 Frith, G.J.T., Swinden, L.B. and Dalling, M.J. (1978) Plant Cell Physiol. 19, 1029–1041.

35 Inagami, T., Yokosawa, H. and Hirose (1978) Clin. Sci. Mol. Med. 55, 121s–123s.

36 Stepanov, V.M., Rudenskaya, G.N., Gaida, A.V. and Osterman, A.L. (1981) J. Biochem. Biophys. Methods 5, 177–186.

37 Brocklehurst, K., Carlsson, J., Kierstan, M.P.J. and Crook, E.M. (1973) Biochem. J. 133, 573–584.
38 Egorov, T.A., Svenson, A., Ryden, L. and Carlsson, J. (1975) Proc. Natl. Acad. Sci. U.S.A. 72, 3029–3033.
39 Kristiansen, T. (1974) Methods Enzymol. 34, 331–340.
40 Lis, H. and Sharon, N. (1973) Annu. Rev. Biochem. 42, 541–544.
41 Kolodkina, I.I., Evstigneeva, N.A., Mamontova, T.A., Gridneva, V.S. and Yurkevich, A.M. (1979) Bioorg. Khim. 5, 526–535.
42 Akparov, V.K. and Stepanov, V.M. (1978) J. Chromatogr. 155, 329–336.
43 Myöhänen, T.A., Bouriotis, V. and Dean, P.D.G. (1981) Biochem. J. 197, 683–688.
44 Mallia, A.K., Hermanson, G.T., Krohn, R.I., Fujimoto, E.K. and Smith, P.K. (1981) Anal. Lett. 14, 649–661.
45 Bouriotis, V., Galpin, I.J. and Dean, P.D.G. (1981) J. Chromatogr. 210, 267–278.
46 Cuatrecasas, P., Wilchek, M. and Anfinsen, C.B. (1968) Proc. Natl. Acad. Sci. U.S.A. 61, 636–643.
47 Hofstee, B.H.J. (1973) Biochem. Biophys. Res. Commun. 50, 751–757.
48 Lowe, C.R., Harvey, M.J., Craven, D.B. and Dean, P.D.G. (1973) Biochem. J. 133, 499–506.
49 Lowe, C.R. (1977) Eur. J. Biochem. 73, 265–274.
50 Holguin, J. and Cardinaud, R. (1975) Eur. J. Biochem. 54, 505–514.
51 Inoue, M., Hara, M., Nagashima, F., Matsui, S., Mitsuyasu, N. and Morino, Y. (1981) Biochim. Biophys. Acta 659, 362–369.
52 Klyashchitsky, B.A. and Mitina, V.Kh. (1981) J. Chromatogr. 210, 55–65.
53 Turková, J., Bláha, K. and Adamová, K. (1982) J. Chromatogr. 236, 375–383.
54 Holroyde, M.J., Chester, J.M.E., Trayer, I.P. and Walker, O.G. (1976) Biochem. J. 153, 351–361.
55 Twining, S.S. and Atassi, M.Z. (1979) J. Immunol. Methods 30, 139–151.
56 Sekar, M.C., Webb, G. and Roufogalis, B.D. (1980) Biochim. Biophys. Acta 613, 420–428.
57 Wright, C.L., Warsy, A.S., Holroyde, M.J. and Trayer, I.P. (1978) Biochem. J. 175, 125–135.
58 Harvey, M.J., Lowe, C.R., Craven, D.B. and Dean, P.D.G. (1973) Eur. J. Biochem. 41, 335–340.
59 Visser, J., Strating, M. and Van Dongen, W. (1978) Biochim. Biophys. Acta 524, 37–41.
60 Fromant, M., Fayat, G., Laufer, P. and Blanquet, S. (1981) Biochimie 63, 541–553.
61 Metcalf, E.C., Crow, B. and Dean, P.D.G. (1981) Biochem. J. 199, 465–472.
62 Walsh, K.A., Burstein, Y. and Pangburn, M.K. (1974) Methods Enzymol. 34, 435–440.
63 Kasai, K. and Ishii, S. (1975) J. Biochem. (Tokyo) 77, 261–264.
64 Robinson, P.J., Dunnill, P. and Lilly, M.D. (1972) Biochim. Biophys. Acta 285, 28–35.
65 Smith, J.A., Hurrell, J.G.R. and Leach, S.J. (1978) Anal. Biochem. 87, 299–305.
66 Porath, J. (1972) Biotechnol. Bioeng. Symp., No. 3, 145–166.
67 Korpela, T.K. and Mäkinen, E. (1979) J. Chromatogr. 174, 361–368.
68 Harvey, M.J., Lowe, C.R. and Dean, P.D.G. (1974) Eur. J. Biochem. 41, 353–357.
69 Lowe, C.R., Harvey, M.J. and Dean, P.D.G. (1974) Eur. J. Biochem. 41, 347–351.
70 Nüsslein, C. and Heyden, B. (1972) Biochem. Biophys. Res. Commun. 47, 282–289.
71 Bureau, D. and Daussant, J. (1981) J. Immunol. Methods 41, 387–392.
72 Andersson, K., Benyamin, Y., Douzou, P. and Balny, C. (1978) J. Immunol. Methods 23, 17–21.
73 Shimizu, F., Kist, M. and Vogt, A. (1978) Res. Exp. Med. 172, 231–238.
74 Igbal, M.J., Ford, P. and Johnson, M.W. (1978) FEBS Lett. 87, 235–238.
75 Morgan, M.R.A., George, E. and Dean, P.D.G. (1980) Anal. Biochem. 105, 1–5.
76 Egly, J.M. and Porath, J. (1975) in Affinity Chromatography, Biospecific Sorption (Hoffman-Ostenhof, O., et al., eds.) pp. 5–22, Pergamon Press, Oxford.
77 Angal, S. and Dean, P.D.G. (1977) Biochem. J. 167, 301–303.
78 Porath, J., Janson, J.C. and Låås, T. (1971) J. Chromatogr. 60, 167–177.
79 Porath, J., Låås, T. and Janson, J.C. (1975) J. Chromatogr. 103, 49–62.
80 Sundberg, L. and Porath, J. (1974) J. Chromatogr. 90, 87–90.
81 Doley, S.G., Harvey, M.J. and Dean, P.D.G. (1976) FEBS Lett. 65, 87–93.
82 Belew, M., Porath, J. and Fohlman, J. (1978) J. Chromatogr. 147, 205–212.
83 Čoupek, J., Křiváková, M. and Pokorný, S. (1973) J. Polym. Sci., Polym. Symp. 42, 182–190.

84 Filka, K., Čoupek, J. and Kocourek, J. (1978) Biochim. Biophys. Acta 539, 518–528.
85 Mizutani, T. and Mizutani, A. (1977) Anal. Biochem. 83, 216–222.
86 Regnier, F.E. and Noel, R. (1976) J. Chromatogr. Sci. 14, 316–320.
87 Campbell, D.H., Luescher, E.L. and Lerman, L.S. (1951) Proc. Natl. Acad. Sci. U.S. 37, 575–578.
88 Peška, J., Štamberg, J., Hradil, J. and Illavský, M. (1976) J. Chromatogr. 125, 455–469.
89 Vieth, W.R. and Venkatasubramanian, K. (1976) Methods Enzymol. 44, 243–283.
90 Hashimoto, T., Sasaki, H., Aiura, M. and Kato, Y. (1978) J. Polym. Sci., Polym. Phys. Ed. 16, 1789–1800.
91 Kato, Y., Komiya, H., Sasaki, H. and Aiura, M. (1980) J. Chromatogr. 193, 458–463.
92 Regnier, F.E. and Gooding, K.M. (1980) Anal. Biochem. 103, 1–25.
93 Barth, H.G. (1980) J. Chromatogr. Sci. 18, 409–429.
94 Zemanová, I., Turková, J. Čapka, M., Nakhapetyan, L.A., Švec, F. and Kálal, J. (1981) Enzyme Microb. Technol. 3, 229–232.
95 Axén, R., Porath, J. and Ernback, S. (1967) Nature (London) 214, 1302–1304.
96 Kohn, J. and Wilchek, M. (1981) Anal. Biochem. 115, 375–382.
97 Kempf, J., Pfleger, N. and Egly, J.M. (1978) J. Chromatogr. 147, 195–204.
98 Wilchek, M., Oka, T. and Topper, Y.J. (1975) Proc. Natl. Acad. Sci. U.S.A. 72, 1055–1058.
99 Švec, F., Hradil, J., Čoupek, J. and Kálal, J. (1975) Angew. Makromol. Chem. 48, 135–143.
100 Matsumoto, I., Mizuno, Y. and Seno, N. (1979) J. Biochem. (Tokyo) 85, 1091–1098.
101 Turková, J., Bláha, K., Horáček, J., Vajčner, J., Frydrychová, A. and Čoupek, J. (1981) J. Chromatogr. 215, 165–179.
102 Murphy, R.F., Conlon, J.M., Imam, A. and Kelly, G.J.C. (1977) J. Chromatogr. 135, 427–433.
103 Turková, J. (1980) in Industrial and Clinical Enzymology (Vitale, L. and Simeon, V., eds.) pp. 65–76, Pergamon Press, Oxford.
104 Uy, R. and Wold, F. (1977) Anal. Biochem. 81, 98–107.
105 Potuzak, H. and Dean, P.D.G. (1978) Nucleic Acids Res. 5, 297–303.
106 Ustav, M.B., Remme, J.L., Lind, A.J. and Villems, R.L.E. (1979) Bioorg. Khim. 5, 365–369.
107 Matsumoto, I., Seno, N., Golovtchenko-Matsumoto, A.M. and Osawa, T. (1980) J. Biochem. (Tokyo) 87, 535–540.
108 Nilsson, K. and Mosbach, K. (1981) Biochem. Biophys. Res. Commun. 102, 449–457.
109 Nilsson, K., Norrlöw, O. and Mosbach, K. (1981) Acta. Chem. Scand. B35, 19–27.
110 Sanderson, C.J. and Wilson, D.V. (1971) Immunology 20, 1061–1065.
111 Schacht, J. (1978) J. Lipid. Res. 19, 1063–1067.
112 Vančurová, D., Turková, J., Čoupek, J. and Frydrychová, A. (1979) Collect. Czech. Chem. Commun. 44, 3405–3410.
113 Salter, D.N., Ford, J.E. and Scott, K.J. (1970) FEBS Lett. 20, 302–306.
114 Cuatrecasas, P. (1970) J. Biol. Chem. 245, 3059–3065.
115 Boschetti, E., Corgier, M. and Garelle, R. (1978) Biochimie 60, 425–427.
116 Weston, P.D. and Avrameas, S. (1971) Biochem. Biophys. Res. Commun. 45, 1574–1580.
117 Neurath, A.R. and Strick, N. (1979) Arch. Virol. 60, 79–81.
118 Inman, J.K. (1982) in Affinity Chromatography and Related Techniques (Gribnau, T.C.D., Visser, J. and Nivard, R.J.F., eds.) pp. 217–233, Elsevier, Amsterdam.
119 Anderson, B., Donakowski, C., Entwistle, R. and Davis, L. (1980) J. Immunol. Methods 36, 309–314.
120 Finlay, T.H., Troll, V., Levy, M., Johnson, A.J. and Hodgins, L.T. (1978) Anal. Biochem. 87, 77–90.
121 Brocklehurst, K., Carlsson, J., Kierstan, M.P.J. and Crook, E.M. (1973) Biochem. J. 133, 573–584.
122 Hornby, W.E., Lilly, M.D. and Crook, E.M. (1966) Biochem. J. 98, 420–425.
123 Gundlach, G., Kohne, C. and Turba, F. (1962) Biochem. Z. 336, 215–228.
124 Weetall, H.H. and Filbert, A.M. (1974) Methods Enzymol. 34, 59–72.
125 Eveleigh, J.W. and Levy, D.E. (1977) J. Solid-Phase Biochem. 2, 45–78.
126 Lowe, C.R., Glad, M., Larsson, P.-O., Ohlson, S., Small, D.A.P., Atkinson, T. and Mosbach, K. (1981) J. Chromatogr. 215, 303–316.
127 Pensky, J. and Marshall, J.S. (1969) Arch. Biochem. Biophys. 135, 304–310.

128 Manen, C.A. and Russell, D.H. (1974) Life Sci. 14, 1907–1915.
129 Porath, J. and Kristiansen, T. (1975) In The Proteins (Neurath, H. and Hill, R.L., eds.) 3rd Edn., pp. 95–178, Academic Press, New York.
130 Ohlson, S., Hansson, L., Larsson, P.O. and Mosbach, K. (1973) FEBS Lett. 93, 5–9.
131 Eveleigh, J.W. (1982) In Affinity Chromatography and Related Techniques (Gribnau, T.C.J., Visser, J. and Nivard, R.J.F., eds.) pp. 293–303, Elsevier, Amsterdam.
132 Eveleigh, J.W. (1978) J. Chromatogr. 159, 129–145.
133 Folkersen, J., Teisner, B., Westergaard, J. and Svehag, S.-E. (1978) J. Immunol. Methods 23, 137–147.
134 Römer, W. and Rauterberg, E.W. (1980) J. Immunol. Methods 38, 239–250.
135 Terman, D.S., Garcia-Rinaldi, R., Dannemann, B., Moore, D., Crumb, C., Tavel, A. and Poser, R. (1978) Clin. Exp. Immunol. 34, 32–41.
136 Lupien, P.J., Moorjani, S., Lou, M., Brun, D. and Gagné, C. (1980) Pediat. Res. 14, 113–117.
137 Denburg, J. and DeLuca, M. (1970) Proc. Natl. Acad. Sci. U.S.A. 67, 1057–1062.
138 Gawronski, T.H. and Wold, F. (1972) Biochemistry 11, 442–448.
139 Schott, H., Eckstein, H., Gatfield, I. and Bayer, E. (1975) Biochemistry 14, 5541–5543.
140 Lowe, C.R. (1978) In Affinity Techniques (Sundaram, P.V. and Eckstein, F., eds.) pp. 55–75, Academic Press, London.
141 Akanuma, H., Kasuga, A., Akanuma, T. and Yamasaki, M. (1971) Biochem. Biphys. Res. Commun. 45, 27–33.
142 Jankowski, W.J., Davey, M.W., O'Malley, J.A., Sulkowski, E. and Carter, W.A. (1975) J. Virol. 16, 1124–1130.
143 Malaníková, M. and Turková, J. (1977) J. Solid-Phase Biochem. 2, 237–249.
144 Turková, J. (1978) Enzyme Engin. 4, 451–452.
145 Matsuda, Y., Shimura, K., Shiraki, H. and Nakagawa, H. (1980) Biochim. Biophys. Acta 616, 340–350.
146 Svenson, A. and Hynning, P.-A. (1981) Prep. Biochem. 11, 99–108.
147 Tobe, H., Kojima, F., Aoyagi, T. and Umezawa, H. (1980) Biochim. Biophys. Acta 613, 459–468.
148 Van Etten, R.L. and Waheed, A. (1980) Arch. Biochem. Biophys. 202, 366–373.
149 Ishibashi, T., Maru, A., Imai, Y., Makita, A. and Tsuji, I. (1980) Biochim. Biophys. Acta 616, 218–227.
150 Karube, I., Ishimori, Y. and Suzuki, S. (1978) Biotechnol. Bioeng. 20, 1775–1783.
151 Korpela, T.K., Kukko, E.I. and Hinkkanen, A.E. (1978) J. Solid-Phase Biochem. 3, 215–221.
152 Kovaleva, G.G., Yusupova, M.P., Balandina, G.N., Lysogorskaya, E.N. and Stepanov, V.M. (1977) Biokhimiya 42, 534–539.
153 Dieter, P. and Marmé, D. (1980) Cell Calcium 1, 279–286.
154 Steup, M. (1981) Biochim. Biophys. Acta 659, 123–131.
155 Dousset, J.C., Dousset, N., Foglietti, M.J. and Douste-Blazy, L. (1981) Biochim. Biophys. Acta 664, 273–277.
156 Turkovä, J. and Seifertová, A. (1978) J. Chromatogr. 148, 293–297.
157 Slowinska, R. and Szewczuk, A. (1979) Acta Biochim. Polon. 26, 229–238.
158 Emöd, I., Tong, N.-T. and Keil, B. (1981) Biochim. Biophys. Acta 659, 283–291.
159 Müller, H.W., Clapshaw, P.A. and Seifert, W. (1981) J. Neurochem. 36, 2004–2012.
160 Aksnes, A., Skotland, T., Flatmark, T. and Ljones, T. (1979) Neurochem. Res. 4, 385–398.
161 Staudenbauer, W.L. and Orr, E. (1981) Nucleic Acids Res. 9, 3589–3603.
162 Ototani, N., Kikuchi, M. and Yosizawa, Z. (1981) Carbohydr. Res. 88, 291–303.
163 Lowe, C.R., Hans, M., Spibey, N. and Drabble, W.T. (1980) Anal. Biochem. 104, 23–28.
164 McCleary, B.V. (1978) Phytochemistry 17, 651–653.
165 Fujiwara, K. and Tsuru, D. (1978) Biochim. Biophys. Acta 522, 195–204.
166 Rombouts, F.M., Geraeds, C.C.J.M., Visser, J. and Pilnik, W. (1982) In Affinity Chromatography and Related Techniques (Gribnau, T.C.J., Visser, J. and Nivard, R.J.F., eds.) pp. 255–260, Elsevier, Amsterdam.

167 Lobarzewski, J. (1981) Int. J. Biol. Macromol. 3, 47–81.
168 Ussuf, K.K. and Nair, P.M. (1980) Ind. J. Biochem. Biophys. 17, 335–337.
169 Gubenšek, F., Žunič, D. and Babnik, J. (1978) Period. Biol. 80, 97–100.
170 Shioi, Y., Nagamine, M., Kuroki, M. and Sasa, T. (1980) Biochim. Biophys. Acta 616, 300–309.
171 Severin, E.S., Kochetkov, S.N., Nesterova, M.V. and Gulyaev, N.N. (1974) FEBS Lett. 49, 61–64.
172 Lornitzo, F.A., Qureshi, A.A. and Porter, J.W. (1974) J. Biol. Chem. 249, 1654–1656.
173 D'Souza, S. and Ananthakrishnan, R. (1979) Austr. J. Exp. Biol. Med. Sci. 57, 245–250.
174 Sordenson, J.C. and Scandalios, J.G. (1975) Biochem. Biophys. Res. Commun. 63, 239–246.
175 Mosolov, V.V. and Zimačeva, A.V. (1980) Dokl. Akad. Nauk, SSSR 255, 1278–1280.
176 Lee, A.C.M., Robbins, R.N. and Bergdoll, M.S. (1978) Infect. Immun. 21, 387–392.
177 Wahid, F.A. and Isom, G.E. (1980) Res. Commun. Subst. Abuse 1, 451–458.
178 Bartos, F., Bartos, D., Campbell, R.A., Grettie, D.P., McDaniel, P., Davis, M. and Smejtek, P. (1979) Res. Commun. Chem. Pathol. Pharmacol. 3, 547–559.
179 Alvino, C.G., Tassi, V. and Ambrosio, G. (1978) Bull. Mol. Biol. Med. 3, 130–139.
180 Rousseaux, J., Picque, M.T., Bazin, H. and Biserte, G. (1981) Mol. Immunol. 18, 639–645.
181 Hillyer, G.V. and DeAteca, L.S. (1979) Infect. Immun. 26, 802–807.
182 Wolk, M., Svennerholm, A.-M. and Holmgren, J. (1980) Current Microbiology 3, 339–344.
183 Mc Bride, J.H. and Davis, R.J.H. (1981) Clin. Chem. 27, 612–614.
184 Rotman, A. and Linder, S. (1981) Biochim. Biophys. Acta 641, 114–121.
185 Van Wezel, A.L. and Van der Marel, P. (1982) In Affinity Chromatography and Related Techniques (Gribnau, T.C.J., Visser, J. and Nivard, R.J.F., eds.) pp. 283–292, Elsevier, Amsterdam.
186 Basu, P.S., Ray, M.K. and Datta, T.K. (1981) Ind. J. Biochem. Biophys. 18, 101–104.
187 Pipková, J., Hořejší, V. and Kocourek, J. (1978) Biochim. Biophys. Acta 541, 515–520.
188 Sela, B.A., Wang, J.V. and Edelman, G.M. (1975) J. Biol. Chem. 250, 7535–7538.
189 Aspberg, K., Holmén, H. and Porath, J. (1968) Biochim. Biophys. Acta 160, 116–117.
190 Marquis, J.K., Hilt, D.C. and Mautuer, H.G. (1980) J. Neurochem. 34, 1071–1076.
191 Ross, T.T., Hayes, C.E. and Goldstein, I.J. (1976) Carbohydr. Res. 47, 91–97.
192 Donnelly, E.H. and Goldstein, I.J. (1970) Biochem. J. 118, 679–680.
193 Friesen, H.-J., Steins, S., Evinger, M., Familleti, P.C., Moschera, J., Meienhofer, J., Shively, J. and Pestka, S. (1981) Arch. Biochem. Biophys. 206, 432–450.
194 Papsidero, L.D., Nemoto, T. and Chu, T.M. (1978) Cancer Biochem. Biophys. 3, 15–18.
195 Marhelonis, G.J. and Oh, T.H. (1981) J. Neurochem. 37, 95–98.
196 Eisen, H.J. and Glinsmann, W.H. (1978) Biochem. J. 171, 177–183.
197 Tappeser, B., Wellnitz, D. and Klämbt, D. (1981) Z. Pflanzenphysiol. 101, 295–302.
198 Bearden, J.C., Jr. (1980) J. Biochem. Biophys. Methods 2, 37–47.
199 Pardoe, I.U. and Burness, A.T.H. (1980) J. Virol. Methods 1, 285–298.
200 Lustenberger, P., Formstecher, P. and Dautrevaux, M. (1981) J. Steroid. Biochem. 14, 697–703.
201 Wolkoff, A.W., Bhargava, M.M., Chung, C. and Gatmaitan, Z. (1979) Proc. Soc. Exp. Biol. Med. 160, 150–153.
202 Schneider, W. and Gschwendt, M. (1980) Biochim. Biophys. Acta 633, 105–113.
203 Davis, C.W. and Daly, J.W. (1979) J. Cyclic Nucleotide Res. 5, 65–74.
204 Egly, J.M. and Boschetti, E. (1982) In Affinity Chromatography and Related Techniques (Gribnau, T.C.J., Visser, J. and Nivard, R.J.F., eds.) pp. 445–451, Elsevier, Amsterdam.
205 Ochoa, J.-L., Kempf, J. and Egly, J.M. (1980) Int. J. Biol. Macromol. 2, 33–38.
206 De Meyer, P., De Herdt, E., Kondo, M. and Slegers, H. (1980) J. Biochem. Biophys. Methods 2, 311–314.
207 Moreno-Lopez, J., Kristiansen, T. and Kårsvas, P. (1981) J. Virol. Methods 2, 293–300.
208 Holmquist, L. and Nilsson, G. (1979) Acta Path. Microbiol. Scand. Sect. B, 87, 129–135.
209 Brown, F., Underwood, B.O. and Fantes, K.H. (1979) J. Med. Virol. 4, 315–319.
210 Lin, J.T., Da Cruz, M.E.M., Riedel, S. and Kinne, R. (1981) Biochim. Biophys. Acta 640, 43–54.
211 Keski-Oja, J. and Yamada, M. (1981) Biochem. J. 193, 615–620.
212 Wilchek, M. (1970) FEBS Lett. 7, 161–163.

213 Wilchek, M. (1974) Methods Enzymol. 34, 182–195.
214 De Pedro, M.A. and Schwarz, U. (1980) FEMS Microbiol. Lett. 9, 215–217.
215 Anfinsen, C.B. (1968) Pure Appl. Chem. 17, 461–487.
216 Ferreira, P.C.P., Paucker, M., Golcher, R.R. and Paucker, K. (1981) Arch. Virol. 68, 27–33.
217 Moberg, L.J. and Sugiyama, H. (1978) Appl. Environ. Microbiol. 35, 878–880.
218 Lagercrantz, C., Larsson, T. and Denfors, I. (1981) Comp. Biochem. Physiol. 69C, 375–378.
219 Schmidt, K.H. and Köhler, W. (1981) Immunobiol. 158, 330–337.
220 Eketorp, R. (1982) In Affinity Chromatography and Related Techniques (Gribnau, T.C.J., Visser, J. and Nivard, R.J.F., eds.) pp. 263–273, Elsevier, Amsterdam.
221 Lane, J.L., Ekert, H. and Vafiadis, A. (1979) Trombos. Haemostas, 42, 1306–1315.
222 Weisgraber, K.H. and Mahley, R.W. (1980) J. Lipid Res. 21, 316–325.
223 Seganti, L., Mastromarino, P., De Stasio, A., Sinibaldi, L., Valenti, P. and Orsi, N. (1980) Acta Virol. 24, 311–316.
224 Folkerson, J., Teisner, B., Ahrons, S. and Svehag, S.-E. (1978) J. Immunol. Methods 23, 117–125.
225 Fahlbusch, B., Zschiesche, W. and Römer, W. (1981) Acta Biol. Med. Germ. 40, 147–152.
226 Pekonen, F., Williams, D.M. and Weintraub, B.D. (1980) Endocrinology 106, 1327–1332.
227 Lis, H. and Sharon, N. (1981) J. Chromatogr. 215, 361–372.
228 Parikh, I. and Cuatrecasas, P. (1982) In Affinity Chromatography and Related Techniques (Gribnau, T.C.J., Visser, J. and Nivard, R.J.F., eds.) pp. 79–92, Elsevier, Amsterdam.
229 Schott, H., Eckstein, H. and Bayer, E. (1974) J. Chromatogr. 99, 31–34.
230 Sharma, S.K. and Makendroo, P.P. (1980) J. Chromatogr. 184, 471–499.

Note

For additional information the reader is directed to the respective part in the Bibliography section of the *Journal of Chromatography*.

Deyl (ed.) Separation Methods

CHAPTER 5

Flat bed techniques

JOSEPH SHERMA [a] and BERNARD FRIED [b]

[a] *Department of Chemistry, and* [b] *Department of Biology, Lafayette College, Easton, PA 18042, USA*

5.1 General introduction

Thin layer chromatography (TLC) and paper chromatography (PC), which together comprise 'flat-bed' or 'planar' chromatography, are the easiest of all chromatographic methods to perform, and they require only simple and inexpensive equipment. Many of the techniques and principles of TLC and PC are the same. They both readily provide qualitative analytical information, and, with optimization of techniques and materials, can give quantitative data as well.

An initial zone of mixture is placed near one end of the stationary phase, a thin layer or paper sheet; the sample is dried; and the end of the stationary phase with the initial zone is placed into a mobile phase, usually a mixture of pure solvents, inside a closed chamber. The components of the mixture migrate at different rates during the movement of the mobile phase through the stationary phase, which is termed the development of the chromatogram. When the mobile phase has moved an appropriate distance, the stationary phase is removed, the mobile phase is rapidly dried, and the zones are detected by application of a suitable visualization reagent.

Differential migration is the result of differing degrees of affinity of the mixture components for the stationary and mobile phases, due to structural variations of the compounds. Different separation mechanisms are involved, the predominant forces depending upon the exact nature of the mobile and stationary phases and the analytes (the substances to be determined). Paper can act as an inert support for separations based on partition between the flowing solvent and water held within the cellulose fibers (the cellulose-water complex); as an adsorbent for polar compounds through hydrogen bonding and van der Waals forces; and as an ion exchanger because of the presence of carboxyl groups in the paper. On activated layers of silica gel with nonaqueous mobile phases, adsorption probably predominates, while partition may play a large role in separations on silica gel with aqueous mobile phases.

Sample collection, preservation and purification are common problems in both TLC and PC. For complex biochemical samples, the development step must resolve the analyte completely, or, more commonly, interferences must be removed in some manner prior to chromatography, for example, by a selective extraction of the compound(s) to be determined under conditions that result in their minimal loss. Frequently, substances are converted, prior to TLC or PC, to a derivative that is more suitable for separation and/or detection or quantitation than the parent compounds.

Detection is most simple when the compounds of interest are naturally colored or fluorescent or absorb ultraviolet (UV) light. However, application of a location or visualization reagent by spraying or dipping is usually required to produce color or fluorescence for most compounds. Absorption of UV light is common for many compounds, for example, aromatics and those with conjugated double bonds. This leads to a simple, rather universal detection method on layers or papers impregnated with a fluorescent indicator (fluorescence quenching detection). The use of chemical reagents is disadvantageous because they usually destroy the compounds being detected.

Compound identification in TLC and PC is based mostly on R_F values compared to authentic standards, where R_F is defined as:

$$R_F = \frac{\text{distance travelled by the zone from the origin}}{\text{distance traveled by the mobile phase from the origin}}$$

R_F values are not exactly reproducible from laboratory to laboratory or even on different runs in the same laboratory, so that they should be considered mainly as guides to relative migration distances and sequences. Factors causing R_F values to vary include: dimensions of the apparatus; nature and size of the layer or paper; direction of mobile phase flow; the volume and composition of the mobile phase; equilibration conditions; temperature; humidity; and sample preparation preceding chromatography.

Quantitation based on visual assessment of spot size compared to standards can be accurate within 10–30%, which may be adequate for many applications of TLC and PC. Quantitative results with accuracy and precision rivaling those in gas chromatography and high performance column liquid chromatography can be obtained by zone elution and microanalysis or by in situ densitometry.

When properly applied to analytical problems, TLC and PC are among the most versatile, rapid, and simple methods the biochemist can use. The following sections will outline the procedures and applications of these methods, with references to sources that can provide more details than space permits in this treatise.

Part A

5.2 Thin-layer chromatography

5.2.1 Introduction and history

5.2.1.1 Introduction

Thin-layer chromatography (TLC) is a rapid, simple, versatile, sensitive, inexpensive technique for the separation of substances. The mobile phase in TLC is a liquid containing a single solvent or a mixture of solvents, and the stationary phase is a solid, known as the sorbent. Numerous major sorbents such as silica, cellulose, alumina, polyamides and ion exchangers allow for considerable versatility in the types of substances that can be separated. TLC has advantages over paper chromatography (PC) since it is generally faster, provides better resolution, and is more sensitive than PC. TLC can be applied successfully to the separation of hydrophilic, lipophilic and inorganic compounds.

TLC is useful both as an analytical and a preparative technique, and substances tentatively identified by TLC may be further characterized by various analytical techniques such as nuclear magnetic resonance spectrometry, mass spectrometry, or gas liquid chromatography. Moreover, many specific chemical detection tests are available to help identify substances separated by TLC. TLC is a microanalytical procedure and provides for separations and at least tentative identification of substances in the milligram (mg), microgram (μg), and nanogram (ng) range. TLC can provide the biochemist with a method of eluting separated substances from plates for quantitative analyses. Recent studies indicate that elution techniques may not be the best alternative for quantitative analyses of many substances separated by TLC and that the preferred method may involve quantitative in situ densitometric analysis [1,2].

5.2.1.2 History

Stahl [3], Heftmann [4], Kirchner [5], and Pelick et al. [6] have reviewed the history of TLC. TLC is a relatively new discipline, and chromatographers usually date the advent of modern TLC from 1958. The review by Pelick et al. [6] tabulates significant early developments in TLC and provides translations of classical TLC studies by Izmailov and Shraiber and by Stahl. In 1938, Izmailov and Shraiber separated certain medicinal compounds on unbound alumina spread on glass plates. Since they applied drops of the solvent to a glass plate containing the sample and sorbent, their procedure was called drop chromatography. Meinhard and Hall in 1949 used a binder to adhere alumina to microscope slides, and these plates were used for the separation of certain inorganic ions using drop chromatography [5]. In the early 1950's, Kirchner and colleagues at the U.S. Department of Agriculture developed TLC mainly as we know it today. They used sorbents bound to glass plates with the aid of a binder, and plates were developed with conventional ascending procedures used in paper chromatography. Kirchner coined the term 'chromatostrips' for his findings. Stahl introduced the term 'thin-layer chromatography' in the late 1950's. His major contribution was the standardization of materials, procedures, and nomenclature and the description of selective solvent systems for the resolution of important compound classes. His first laboratory manual popularized TLC, and he obtained the aid of chemical manufacturers in offering standard materials for TLC [7]. Stahl recently provided a personal view of his early contributions to TLC [8].

Early TLC workers had to prepare their own plates, by procedures that were often messy, arduous, and non-reproducible. A major breakthrough in this field was the commercial availability of precoated plates in the early 1960's, the innovation of which has been discussed by Przybylowicz et al. [9]. Few chromatographers are currently making their own plates unless special layers are required.

The literature on TLC is extensive. Sherma [10] has examined the available literature based on books containing general TLC chapters, books specifically on TLC, books containing specialized materials plus bibliographies, TLC reviews, and major papers. Numerous analytical journals, including *Chromatographia, The Journal of Chromatography, Journal of Chromatographic Science,* and *Journal of High Resolution Chromatography and Chromatography Communications*, regularly publish articles on TLC. Important papers on TLC are reviewed biennially in the review issue of *Analytical Chemistry*, and the last review was by Sherma and Fried [11]. Abstracts of TLC articles appear in *Chemical Abstracts* and *Analytical Abstracts*. Some journals of interest to biochemists that contain articles on TLC are *Comparative Biochemistry and Physiology, Analytical Biochemistry, Lipids*, and *Journal of Lipid Research*. The review series edited by Florkin and Scheer, *Chemical Zoology* (Academic Press), has references to TLC procedures and applications.

5.2.2 Sorbents, layer preparation, and precoated plates

5.2.2.1 Sorbents

A. Silica gel Silica gel is the sorbent with the widest range of TLC applications. It is an amorphous, porous adsorbent often referred to as silica, silicic acid or porous glass. Particle sizes for TLC are typically in the 5 to 40 μm range. The preparation of silica gel is complex and involves the hydrolysis of silicates to polysilicic acid, and then further condensation and polymerization of polysilicic acid [12]. The OH groups attached to the silicon atoms are reactive and contribute to the adsorptive properties of silica gel. There are different types of active sites, which vary in strength of adsorption. Substances are adsorbed on silica by hydrogen bonding, and surface hydroxyls serve as hydrogen donors [13]. An 'ideal' condition for general TLC silica gel separations is 11–12% water (w/w) [14]. This level of water leaves available plentiful 'middle-activity' silica gel adsorption sites while the most active sites are attached to water. This optimal water level is achieved when silica gel is at equilibrium with air having a relative humidity of 50% at 20°C, conditions that often occur in modern analytical laboratories. Heat activation is usually not required prior to TLC. If precoated plates have been exposed to high humidity, it is good practice to heat the plates at 70–80°C for 30 min and then allow them to cool in an open, clean environment to attain the 11–12% water level. This procedure should be followed for home-made plates.

Physicochemical properties, that is the hardness and polarity of silica gel, are related to particle size (μm), surface area (m^2/g), pore size (Å), and presence or absence of additives. The surface area of silica gel adsorbent for TLC is typically 300–400 m^2/g, and pore sizes range from 20 to 150 Å. Adsorbent with larger surface area (smaller particle size) will generally give better resolution, but a slower development time. The usual binder for commercial TLC adsorbent powders is 5–20% gypsum (silica gel G), whereas most commercial, precoated 'hard' layers are bound to glass or plastic with an organic binder. Analytical layers are typically 250 μm in thickness, whereas high-performance precoated layers are often 200 μm.

During manufacture, inorganic and organic contaminants may be introduced into TLC silica gel. During slurry preparation or handling of a finished plate, additional contaminants may be added. Slurries or plates can be washed in chloroform-methanol (1 : 1 v/v) or the mobile phase to help reduce contaminants. Silica gel is usually prepared as an aqueous slurry of approximately 2 parts water to 1 part powder, prior to application to a glass plate. When properly purified, silica gel tends not to catalyze the reactivity of labile substances.

B. Alumina Several aluminum hydroxides serve as starting material for TLC alumina, and, by a series of dehydration processes, a variety of aluminas are obtained. The ones most suitable for TLC are the crystalline modifications of χ-Al_2O_3 and γ-Al_2O_3 [13]. Physicochemical properties and the precise details on the nature of adsorption sites of alumina are not well understood. It has been suggested that exposed Al atoms, strained Al-O bonds, and perhaps other cationic sites serve as adsorption sites, whereas, unlike silica, surface hydroxyl groups are probably not important [13]. Acids are probably retained by interaction with basic sites such as surface oxide ions. Physical properties of TLC alumina are somewhat similar to TLC silica gel in terms of particle size, surface area and average pore diameter. The selectivity of alumina in TLC adsorption chromatography is similar to that of silica gel, and this sorbent is useful for the separation of neutral and acidic lipophilic substances. A widely used alumina (Merck, alumina G, pH 7.5) is basic, and it can preferentially retain certain acids. Of the inorganic sorbents in TLC, alumina is second in frequency of use to silica. Aromatic hydrocarbons are more strongly retained on alumina than silica gel. Alumina has been used to separate compounds such as fat-soluble vitamins, alkaloids, certain antibiotics and chlorinated hydrocarbons. A drawback of alumina is that it may catalyze certain labile substances.

C. Cellulose Cellulose is a polymerized polysaccharide characterized by the cellobiose unit. The presence of free OH groups in cellulose permits hydrogen bonding with low-molecular-weight liquids such as alcohols or water. Cellulose is useful for the separation of hydrophilic substances primarily by the mechanism of normal-phase partition chromatography. For a discussion of separation mechanisms and the 'cellulose-water complex', see Ref. 176.

TLC cellulose powders consist of two basic types, native fibrous and microcrystalline. An example of the fibrous type is Machery-Nagel MN-300 and of the microcrystalline, 'Avicel'. The fibrous type is best for the transfer of procedures originally designed for paper chromatography. Because of the particulate nature of TLC microcrystalline cellulose powder, spots formed on plates of this cellulose are more compact than in paper chromatography, and separations are more rapid. Because of the fine particulate nature of microcrystalline cellulose, it may act partly by the mechanism of adsorption chromatography. Separation capabilities of both types have been reported [15]. Cellulose sorbent can be prepared as a 15–35%

aqueous slurry of the powder, followed by brief homogenization, and the coated plates should be air dried. A binder is not necessary because of the particulate nature of the cellulose powder. Impurities may be present in the cellulose powder and could affect separations. Corrosive visualization agents must be avoided in cellulose TLC. Precoated cellulose layers are available commercially.

D. Polyamide For a review of TLC on polyamide layers see Refs. 16, 17. Separations on polyamide layers are affected by hydrogen bonding between the solute and the functional amide groups of the sorbent [18]. A frequently used support is ϵ-polycaprolactam (= polyamide 6 = nylon 6). Layers are coated on glass plates or plastic films from a slurry consisting of 20 g of sorbent in 100 ml of 75% formic acid [16]. Precoated polyamide layers are available commercially. Corrosive visualization agents cannot be used in polyamide TLC, and visualization is made possible by the inherent fluorescent properties of the support. Iodine can also be used. Since minimal diffusion occurs on polyamide, spots are sharp and compact. Organic compounds that interact through a hydrogen bond with the polyamide support are suitable for polyamide chromatography [16]. Polyamide TLC has been used for the separation of phenols, carboxylic acids, amino acid derivatives, steroids, quinones, aromatic nitro compounds, *N*-terminal groups in polypeptide chains, acid dyes, alkaloids, heterocyclic compounds, nucleotides, bile pigments and pesticides.

E. Sephadex Sephadex layers are prepared from modified dextran gels for the separation of hydrophilic solutes such as nucleic acids and peptides. The mechanism of separation is partition chromatography governed by size exclusion in the swollen gel containing pores of controlled dimensions. The gels, a layer spreader and special equipment for developing layers are available from Pharmacia Fine Chemicals.

F. Ion exchangers Various ion-exchange materials such as resins, cellulose, and inorganics, are used in thin-layer chromatography [19,20]. Ion exchangers act through charge attraction forces, although attraction to the exchanger matrix by London (dispersion) forces and other physical interactions is also possible. The resins are high-molecular-weight cross-linked polymers with attached functional groups that can exchange their bound ions with ions of the same charge in the mobile phase. Commonly used resins for TLC are the strong acid and base Dowex exchanger. Most cellulose exchangers are made by chemically modifying cellulose via the addition of basic or acidic groups attached by ether or ester linkages. Cellulose and resin exchange layers can be prepared in the laboratory, or are available precoated. Inorganic exchangers (e.g., ammonium phosphomolybdate) layers are prepared in the laboratory. Cellulose ion-exchange chromatography has application in biochemistry for the separation of proteins, nucleic acids and their derivatives, steroid sulfates, glucuronosides and nucleotide coenzymes [19]. Resin layers (Fixion) are widely used for amino acid and inorganic separations.

Sephadex ion exchangers are suitable for separation of large biological molecules. These are prepared from the polysaccharide dextran cross-linked by reaction with epichlorohydrin, followed by the introduction of the desired functional groups on the fiber and pore surfaces [21].

G. Impregnated layers Acid, alkaline, or buffered layers can be prepared with suitable aqueous solutions of an acid, base or salt mixture instead of water to prepare the absorbent slurry used to cast the layer. Alternatively, various methods can be used to impregnate manually or commercially precoated layers of cellulose or silica gel. In one method, the dry plate is placed in a TLC tank containing the impregnation solution, for example, 5 ml of paraffin oil in 95 ml of petroleum ether, for preparation of a lipophilic reversed-phase layer. The solution is allowed to ascend for 15 cm, and the plate is air dried for 5–10 min before sample application. In another method, plates can be pretreated by immersing them in a solution of the impregnating agent in a volatile solvent contained in a tank or tray. After immersion for several seconds, the excess solution is allowed to drain, and the volatile solvent is evaporated prior to use. The range of applications on conventional and high-performance TLC is expanded through the use of different impregnation agents [22].

5.2.2.2 Layer preparation

Relatively few chromatographers now prepare their own plates (homemade plates) from commercially available sorbents. Reasons for the preparation of homemade plates include: (1) use of a large number of plates and the ability of personnel to prepare layers of adequate quality at low cost; (2) the need of a special sorbent or combination of sorbents not offered commercially; (3) the need for a backing material not offered commercially. The most uniform plates are obtained with commercial spreaders such as The Desaga Moving Hopper Spreader or The Camag Fixed Hopper Spreader. For details on the use of these spreaders, see Ref. 23. Although less satisfactory than using commercial spreaders, a slurry of sorbent can be poured on a glass plate and spread with a ruler or glass rod [24]. Laboratory spray devices can be used to coat glass plates with adsorbent slurry [25], or plates can be dipped into the slurry [26].

5.2.2.3 Precoated layers

A. Glass, plastic and aluminum backings Silica gel, alumina, cellulose, polyamide and ion-exchange precoated layers are available with backings of glass, plastic, or aluminum foil. Glass is the most popular support. For a listing of the companies that supply various sorbents on precoated layers see Ref. 27, and for a listing of the nomenclature and symbols associated with precoated plates see Ref. 28. Figure 5.1 is a photograph of precoated plates.

Most precoated layers on glass are 250 μm thick, whereas those prepared on plastic or aluminum foil are usually 100 μm thick. The usual sizes are 20×20 cm, 10×20 cm, and 5×20 cm. Charring procedures with corrosive sprays are best done on layers with glass supports, although charring at less than 120°C is possible on plastic-backed layers. For ease of handling and storing chromatograms, plastic and aluminum are preferred to glass. Fewer sorbents are available on aluminum than on glass or plastic, and aluminum-coated foils may curl irreversibly upon oven heating. Aluminum and plastic layers are unsuitable with certain solvent systems, and their thinner coating leads to lower sample capacity. Layers precoated on glass, plastic or aluminum backings usually contain proprietary organic binders that result in hard

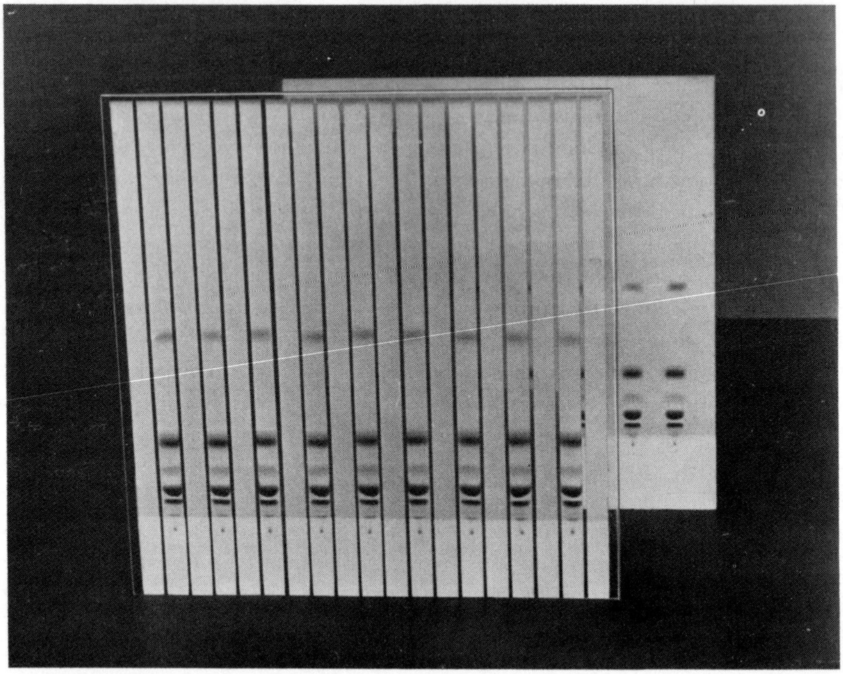

Fig. 5.1. Precoated plates (photo supplied courtesy of Analtech, Inc.).

layers stable to handling and shipping, rather than gypsum (silica gel G). However, different binders may affect a given separation, and trial and error is important in determining a suitable precoated plate for a given separation.

B. High-performance silica gel High-performance (HP) precoated silica gel layers are available as 10×10 cm or 10×20 cm precoated plates from several manufacturers. Plates from Whatman and Merck have an organic binder, whereas those from Analtech contain an inorganic binder. HP layers consist of a dense packing of particles of very uniform size (about 5 μm), with a smooth, homogeneous surface. HPTLC plates have advantages in faster analysis time, higher separation efficiency, and more reproducible chromatograms, with improved sensitivity of detection because of a smaller degree of zone diffusion. Development can be carried out in inexpensive, small rectangular chambers for a distance of 3–7 cm, or in a commercial radial U-chamber (Camag). The amount of developing solvent used is much smaller than that required for classical TLC. For a description of high performance radial chromatography see Ref. 29.

C. Reversed-phase chemically bonded plates Whatman KC_{18} plates consist of a layer of sorbent produced by reacting octadecyl monochlorosilane with the surface silanol groups of a special silica gel [30]. This presents a nonpolar hydrocarbon surface to the solutes for separations by reversed-phase TLC. The advantage of this chemically bonded medium compared to conventional impregnated nonpolar phases is that the 'liquid' phase cannot be stripped off during development, and no pains need be taken to saturate the mobile phase with the stationary liquid to attempt to prevent this loss. The practice and applications of TLC on Whatman KC_{18} reversed phase plates have recently been reviewed [31].

D. Combination layers Some precoated TLC and HPTLC plates (Whatman Linear K) contain silica gel with a bottom zone consisting of an inert preadsorbent [32]. The preadsorbent acts as a blotter or dispensing area that can accomodate relatively large volumes of standards and samples, including crude biological materials such as blood or urine. Some impurities may be retained in this zone, thereby achieving some sample cleanup, but materials soluble in the solvent system, such as salts, may still rise with the solutes. The major advantage of the preadsorbent is that variable, large-volume samples can be applied quickly as spots or streaks, and the solutes are concentrated at the preadsorbent-silica gel interface into uniform, tight bands. Whatman also supplies a combination silica gel-C_{18} reversed phase layer (Multi-K) for enhanced separations of complex mixtures by two independent mechanisms. For a further description of combination layers see Ref. 33.

5.2.3 Sample preparation, derivatization and solvent systems

5.2.3.1 Sample preparation

Samples for TLC should be dissolved in an appropriate solvent prior to application. Usually, volumes of 1–10 μl are applied to the origin. These volumes should contain sufficient sample so that the compound can be detected by the method of choice. Sensitivities of detection methods are usually in the 0.1–1 μg range. If a compound is present in low concentration in a complex sample such as biological materials, extraction, isolation and concentration procedures usually precede TLC. These steps are essential to the success of the separation because of the adverse effects of impurities on the chromatography. Sample preparation prior to TLC is similar to that required for gas chromatography and high-performance column liquid chromatography and may include extraction, solvent partitioning, column chromatography, desalting and deproteinization procedures.

If a compound is present in sufficient purity and concentration, it may be possible to spot the sample directly without extraction or clean-up procedures. The concentration should be such that the normal spotting volume of 1–10 μl for TLC will give a detectable zone. Impurities must not retain the compound of interest at the origin, distort the shape, or alter the R_F of the spot. For example, Sherma and Gray [34] determined caffeine in cola, tea and coffee by direct spotting of these beverages.

In place of typical TLC plates, those having layers with a lower spotting and clean-up strip to retain impurities, and an upper adsorbent area of silica gel for the analytical separation, can be used. Samples may be applied directly to the plate without clean-up procedures. For example, linear K (Whatman) preadsorbent silica gel plates incorporate a 3 cm wide diatomaceous earth prespotting area. These plates allow for fast, diffuse application of large volumes of crude sample material. As the solvent passes over the preadsorbent, material to be separated is moved to the solvent front and applied to the origin of the silica gel as a concentrated band. A similar effect can be attained after crude samples such as urine or serum are spotted on conventional layers and then predeveloped for a distance of 1–2 cm with a polar solvent. After drying, the plate is then developed a second time for a distance of 10 cm with a suitable but less polar solvent. Interfering materials such as proteins are retained at the original origin. An example of this in situ clean-up procedure used to determine lipid profiles in rat plasma and liver homogenate was reported recently [35].

When a compound is a major sample constituent, dissolving the sample in an appropriate amount of suitable solvent followed by direct spotting of a portion of the sample may be adequate for TLC. This approach was used to separate caffeine in APC tablets on C_{18} reversed-phase layers (Whatman) containing a fluorescent indicator [36].

Simple extractions can be used to separate a single tissue sample into major chemical fractions [37]. Starting with 1 g or less of fresh biological tissue sample, 10 ml of 70% ethanol is added and the sample allowed to sit overnight; the supernatant fluid is decanted and saved for analyses of free pool compounds (amino acids, organic acids and simple carbohydrates). The tissue is then homogenized in 10 ml of 70% ethanol and centrifuged. The precipitate will contain lipids, polysaccharides, proteins and nucleic acids. The supernatant is then decanted and combined with the original free pool material.

Extracts too impure for direct spotting may require partitioning with immiscible solvents as a clean-up step. The principle of differential partitioning is to leave impurities behind in one solvent, while extracting the compound of interest into another. This clean-up method may require some knowledge of wet chemistry, trial and error, and a good idea of the constituents of the sample. For a review of various liquid-liquid partitioning systems see Ref. 38.

Column chromatography can be advantageous for sample clean-up prior to TLC. A Bio-Rad 11A8 (Bio-Rad Laboratories) ion-retardation resin combines desalting and removal of peptides from urine samples before amino acid TLC. Urine is passed into the column and eluted with distilled water and interfering salts and peptides are retained on the column [39].

Two examples of modern clean-up methods that use columns that have been prepared commercially are the Extube (Analytichem International) and the Sep-Pak (Waters). The Extube was designed to isolate drugs from a complex biological sample by the principle of liquid-liquid extraction. Samples to be analyzed, such as urine, whole blood or bile, are introduced into a disposable Extube, which is packed with an inert fibrous matrix of large surface area. When a suitable solvent is poured into the Extube, it interphases film-on-film with the sample. The compound of interest is extracted by a solvent that passes freely through the matrix, whereas impurities such as water, pigment, particulate matter and other polar components are retained by the matrix. The Extube has been used to extract and purify cortisol from plasma by liquid-liquid partition with methylene chloride [40]. The Sep-Pak cartridge (Waters) was designed for clean-up prior to high-performance LC, but is also suitable for TLC. The Sep-Pak consists of a disposable tube or cartridge prepacked with either silica gel or C_{18} (reversed-phase) sorbent. The cartridge is fitted to a syringe with the plunger removed. Sample is poured into the barrel of the syringe and the plunger is returned to pump the sample through the cartridge. The process is repeated with an appropriate solvent or solvents to purify the sample.

5.2.3.2 Derivatization

For a review on the preparation of derivatives prior to TLC, see Ref. 41. Derivatization is frequently used in the TLC of amino acids where certain amino acids are easily converted into the so-called DANS forms (*d*imethyl *a*mino *n*aphthalene *s*ulfonyl) [42]. These authors have studied a variety of amino acid derivatives on silica gel. Derivatives are also used when a volatile compound can be changed into a nonvolatile compound, or when the detection sensitivity of the derivative is greater than that of the original compound. Derivatization is advantageous as a clean-up procedure when it is easier to clean up the derivative than the original compound.

5.2.3.3 Solvent systems

It is often possible to find a layer-solvent combination previously reported in the literature for compounds of interest, or at least similar compounds [5,43]. Since the mobile phase competes with the chromatographed substances for the sorbent sites, more polar substances will require a more polar solvent to cause migration on a silica gel or alumina layer. A stronger solvent is one that increases R_F values, often with an accompanying decrease in resolution.

Mobile phases should be as simple as possible and prepared from pure grades of solvent. Mixtures composed of more than four components should be avoided because of problems associated with reproducible preparation. Mobile-phase proportions are designated in parts by volume so that the sum is usually 100. Quantitative techniques should be used to prepare the mobile phase and the constituents should be mixed thoroughly before use. Multisolvent mobile phases should be discarded after use.

Solvents can be grouped for adsorption chromatography into a so-called elutropic series according to their elution strengths. The choice of solvent system is made in relation to the polarity of the adsorbent and the mixture to be separated, utilizing the elutropic solvent series. This series orders solvents according to their hydrophilic or hydrophobic character. A representative elutropic series beginning with the most hydrophobic solvent is as follows:

petroleum ether (most hydrophobic)—cyclohexane—toluene—benzene—chloroform—ether—ethyl acetate—*n*-butanol—phenol—*n*-propanol—acetone—ethanol—acetic acid—methanol—formamide—water (most hydrophilic).

Mixtures of solvents fit into this series according to the components and their proportions. For example, benzene-chloroform (1 : 1 v/v) is more hydrophilic (polar) than pure benzene, and the mixtures benzene-acetone (9 : 1 v/v) and benzene-ethyl acetate (8 : 2) increase further in hydrophilic nature in this order [44]. Hydrophilic solvents are used for the separation of hydrophilic substances, whereas hydrophobic solvents are most useful for hydrophobic compounds. In general, adsorption systems are especially suitable for lipophilic compounds.

5.2.4 Development modes and chambers

5.2.4.1 Development modes

A. Ascending Ascending chromatography is the most frequent development mode in TLC. Following sample application, a plate is placed in an appropriate chamber so that the solvent is below the point of sample application. The solvent is allowed to rise by capillary action usually from 10 to 18 cm above the origin on a 20 × 20 cm TLC plate and from 3 to 7 cm on a smaller, high-performance TLC plate.

B. Descending For descending development, the mobile phase is fed to the top of a vertical inclined plate through a wick arrangement. This arrangement has no advantage over ascending development relative to development time or resolution, and is rarely used except for thin-layer gel-permeation chromatography (TLG) of hydrophilic substances such as proteins, peptides, nucleic acids and nucleotides.

C. Two-dimensional Two-dimensional development is used to examine complex mixtures. Following application of the sample in one corner of a 20 × 20 cm plate, ascending development is carried out for the full length of the plate to achieve maximum resolution. The plate is then removed from the chamber, and air-dried to remove solvent vapors. The plate is then rotated through 90 degrees and redeveloped, usually with a different solvent. The line of partially resolved components from the first development becomes the origin for the second development. For a good description of two-dimensional TLC development, see Ref. 45.

D. Radial (circular) In radial or circular development the mobile phase is slowly applied to the center of a sample spot in the middle of a horizontal TLC plate. Sample components move outward in the form of concentric rings under the influence of a negative solvent gradient that causes the rear of each zone to move relatively faster than the front. This leads to compressed zones with better resolution, especially for compounds in low R_F regions. Radial development can be used to select a suitable mobile phase for linear or circular TLC by adding solvents from a pipet onto spots placed on microscope slides and observing the optimum pattern of rings [46]. A commercial device for testing 16 solvents on a single 20 × 20 cm plate by horizontal radial development is available from Schleicher and Schuell. This Selecta Sol apparatus accomplishes the same results as the pipet/microscope slide approach but in a more elegant manner.

E. Multiple (manual) In multiple (manual) development, following a single development in the ascending mode, the chromatogram is removed from the chamber, air-dried and then placed in the same solvent and redeveloped in the same direction. This process, which may be repeated numerous times, increases the resolution of components with R_F values below 0.5. The theory of unidimensional multiple development has been reviewed by Perry et al. [47].

F. Overrun (continuous) As in multiple development, continuous development is a method of lengthening the plate to improve the resolution of slowly moving solutes. Using a commercial tank with a slot in the lid (Shandon) the plate is allowed to protrude from the top of the tank so that the solvent continuously evaporates.

G. Programmed multiple development A sophisticated, automated device was designed to produce programmed multiple development (PMD). By means of an electronic programmer, a mobile phase can be advanced and removed from a TLC plate in a number of different modes. The advancing solvent is removed by evaporation with radiant heat or flow of inert gas. In a new cycle, solvent is returned to or below the origin, then is programmed to rise to a new height, and is subsequently caused to recede again. The advancement-removal solvent cycle can be repeated for different distances, as determined by the operator. The advantages are superior resolution and sensitivity because solutes are reconcentrated into elliptical zones during each solvent advance-removal cycle. For a discussion of the operation and limitations of PMD, see Refs. 47, 48. For an application of PMD in the determination of digitoxin and digoxin residues in biological fluids on high-performance plates at the 1–2 ng/ml level, see Ref. 49.

5.2.4.2 Chambers

A. Rectangular glass tanks (N-tanks) A rectangular glass tank is often used in TLC to develop 20 × 20 cm TLC plates (Fig. 5.2). Popular models have inner dimensions of 21 × 21 × 9 cm and can accommodate two plates using 50–100 ml of solvent to fill the tank to a depth of 5 mm. The tank is lined on three sides with thick filter paper that is thoroughly soaked with the solvent. The tank should be allowed to stand for 30–60 min to allow the inside atmosphere to saturate with solvent vapor

Fig. 5.2. Chambers for TLC (photo supplied courtesy of Analtech, Inc.).

and to attain equilibrium before the plate is inserted. Upon dipping plates into the mobile phase, the origins must be above the level of the solvent in the tank so that the sample does not diffuse away.

Some chromatographers use unsaturated tanks, that is, they pour in the mobile phase and then add the plate and begin development at once. Unsaturated tanks produce generally higher R_F values because of evaporation of mobile phase from the surface of the layer and subsequent higher mobile-phase flow through the sorbent. Superior resolution with multicomponent solvents has been reported in some situations. The improvements are caused by a concentration gradient and mostly depend on the differential rates of evaporation of the solvent components and their affinities for the sorbent layer [50].

B. Sandwich chambers (S-chambers) The sandwich chamber is very thin to accommodate a single 20 × 20 cm plate and minimal solvent for ascending chromatography. S-chambers with a plain glass cover plate are considered 'unsaturated,' whereas 'saturated' conditions are achieved by employing a counter plate coated with sorbent (often cellulose) and soaked with solvent. S-chambers offer advantages over rectangular tanks, that is, faster gas-phase equilibration and excellent reproducibility both due to the much smaller internal volume when compared with N-tanks [51].

Fig. 5.3. Camag U-Chamber (photo supplied courtesy of Applied Analytical Industries).

C. Pharmacia TLG chamber The TLG chamber is used for thin-layer gel-filtration chromatography. Sephadex gels used for molecular weight and size determinations, particularly for globular proteins, can be coated on 20×20 cm plates and are therefore of interest to the biochemist. TLC is carried out on a wet, swollen gel layer with continuous, descending flow. The Pharmacia TLG chamber permits simultaneous chromatography on two 20×20 Sephadex gel plates. The determination of molecular weight and molecular size of protein samples weighing less than 10 μg can be made on a plate. Thin-layer gel permeation can be used preliminary to column gel permeation. For further details consult the Pharmacia Corporation.

D. Camag U-chamber The Camag U-chamber is designed for high-performance radial chromatography on 5×5 cm plates (Fig. 5.3) [29]. Multiple spots of 100 or 200 nl volume are automatically applied in a concentric ring around the center of the plate using a platinum-iridium pipet in the Camag Nanomat spotter (Fig. 5.4). (The Nanomat also allows the precise application of these very small volumes to 10×10 or 20×20 high-performance TLC plates for conventional linear development.) The spotted plate is placed face down on the U-chamber body, and 100–250 μl of solvent is automatically fed underneath the apparatus by a motor-driven syringe through a capillary tube to the center of the layer. The constant, controlled rate of solvent delivery produces consistent, well-resolved arc or sector chromatograms of the several spotted samples in about 5 min. A special radial scanner for quantitation of the chromatograms is also available from Camag. Fotodyne, Inc. has designed a less expensive version of this radial high-performance TLC development apparatus.

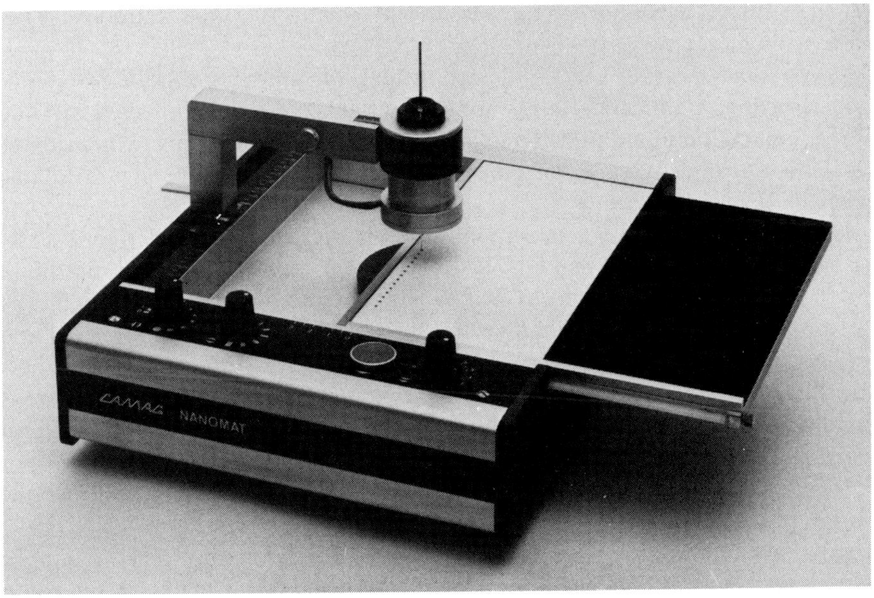

Fig. 5.4. Camag Nanomat Spotter (photo supplied courtesy of Applied Analytical Industries).

5.2.5 Detection

5.2.5.1 General

After development, chromatograms are removed from the chamber, dried to remove the mobile phase, and zones are then detected. Colored substances may be viewed in daylight without further treatment. Detection of colorless substances is simplest if compounds self-absorb in the short-wave ultraviolet region (254 nm) or if they can be excited to produce fluorescence by short- or long-wave (365 nm) ultraviolet radiation. Detection can also be achieved by means of chromogenic reagents (producing colored zones) or fluorogenic reagents (producing fluorescent zones), or by biological methods, such as enzyme inhibition.

5.2.5.2 Methods of detection

A. Ultraviolet lamps Plates can be viewed in a darkened room with suitable commercially available UV lamps. On sorbents not impregnated with phosphor, fluorescent substances absorb UV light and emit bright, longer-wave visible energy against the dark layer background and the emitted color may aid compound identification. Fluorescent solutes can be detected at levels of 10 ng or less. Inorganic phosphor incorporated in the adsorbent allows many organic compounds that absorb at or near 254 nm to be visualized as dark zones on a yellow-green background when the layer is irradiated with short-wave UV radiation. This process is termed fluorescence quenching or fluorescence diminution, but the detection limit is usually from 500 ng to several micrograms. Commercial plates or sorbents having this phosphor are generally designated with the symbols F or UV. For detailed discussion of fluorescence quenching, fluorescence, and other nondestructive TLC detection methods see Ref. 52.

B. Sprayers and spraying Usually detection reagents are applied to a chromatogram by spraying. Commercial glass sprayers attached to a compressed air line permit the application of a uniform mist to the plate. Glass atomizers with a rubber squeeze bulb attachment or spray guns using propellant cans are also available commercially (Fig. 5.5).

C. Dipping Various commercial dipping vessels are available and they usually consist of thin chambers that can be filled with a 200–500 ml volume of detection reagent and can accommodate one or two 20 × 20 cm plates. Glass trays can also be used as dipping vessels. In use, the detection reagent is poured into the vessel, and the developed and dried chromatogram is dipped in the reagent, usually for a few seconds. The plate is then removed to allow excess reagent to drain, and the plate is dried and heated to detect spots. The procedure of dipping in the detection reagent following development of the chromatogram is referred to as 'postdipping'.

D. Preimpregnation (predipping) Detection reagent may be impregnated into the layer prior to sample application and development. If this procedure is used, it is important that the reagent does not interfere with separation, and that the mobile phase does not remove the reagent during development. When plates are prepared manually, a detection reagent may be incorporated into the slurry. With the aid of a

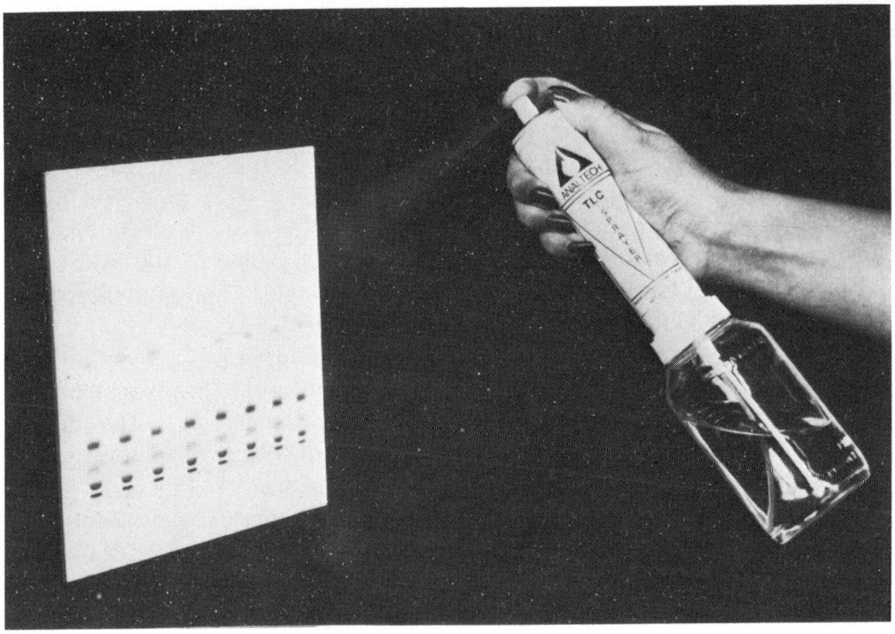

Fig. 5.5. Spraying a chromatogram (photo supplied courtesy of Analtech, Inc.).

dipping vessel, detection reagent may be applied to a plate prior to development and sample application.

Charring reagents have been incorporated by dipping layers to avoid spraying of corrosive chemicals. For example, silica plates have been impregnated with 4% sulfuric acid in methanol [53] or ammonium bisulfate [54].

5.2.5.3 Detection reagents

Detection reagents include: (1) general reagents that react with a wide variety of different compound types and can characterize an unknown sample; and (2) specific reagents that indicate the presence of a particular compound or functional group. There are nondestructive (reversible) and destructive reagents in each category. When a series of detection reagents is used on the same plate to enhance detection and solute identification, it is important to use nondestructive reagents prior to the destructive test.

A. Iodine The universal detection reagent iodine is used either as a 1% alcoholic spray, or the plate is placed in a closed jar or tank containing iodine crystals. The iodine vapor dissolves in, or forms weak charge-transfer complexes with organic compounds, which show up as brown spots on a pale yellow background within a few minutes. After marking zones for future reference, exposure of the plate to air causes the iodine to sublimate and the spots to fade, after which the plate can be sprayed with another reagent or the solute can be eluted from the plate for further analysis.

B. Charring reagents Many corrosive reagents are available for charring organic compounds upon spraying and then heating. These reagents are suitable for glass-backed layers with inorganic (e.g., gypsum) binders, but organic-bound layers should be tested for resistance to reaction with each individual reagent-temperature combination. Charring reagents will often produce colored or fluorescent zones when heating is carried out at relatively low temperatures for a short time. Heating at higher temperatures for longer periods will usually char the spots. A classic charring reagent is prepared by dissolving 5 g of potassium dichromate in 100 ml of 40% sulfuric acid. After spraying the solvent-free plate charring is achieved by heating at 200°C. Charring for TLC detection has been reviewed [55].

C. Phosphomolybdic acid Spraying of a chromatogram with a 5% solution of phosphomolybdic acid in ethanol followed by brief heating at 110°C gives blue-black spots against a yellow background with numerous organic compounds. The solution is sprayed until the layer is completely saturated, as indicated by a uniform yellow color visible on the back of the layer through the glass plate [56].

D. Water Spraying a plate with water until the layer appears translucent when held against light shows lipophilic substances such as steroids, hydrocarbons and bile acids as white opaque spots on a semi-transparent background. The spots are clearest if the plate is saturated by spraying and then allowed to dry slowly during viewing. After marking the spots and then drying the plate, the layer can be oversprayed with another reagent or the zones eluted for further analysis [57].

E. Selective reagents Stahl [58], Kirchner [5], a Merck booklet [59], and the CRC *Handbook of Chromatography* [43] each list about 300 spray reagents to which the reader is referred for details. These reagents will react more or less specifically with different functional groups to reveal natural products and organic or biochemicals as colored or fluorescent zones. Fluorescamine is an example of a very useful and widely used non-destructive, selective detection reagent [60–62]. When a developed chromatogram is sprayed with fluorescamine and viewed under long-wave UV light, compounds with primary amine groups will be detected as brightly fluorescent, greenish spots. Amino acids, pesticides and drugs have been detected with this reagent with sensitivities in the nanogram range.

F. Enzyme inhibition Enzymatic reactions can be surveyed on a TLC plate, and the end products detected. For example, cholinesterase-like enzymes in nervous tissue will catalyze the substrate indoxyl acetate to indoxol and acetic acid. In air, indoxol is converted to indigo blue, which can be visualized on a TLC plate. Some pesticides and insecticides inactivate enzymes associated with animal nervous tissue. The presence of minute amounts of such substances may interfere with the indoxyl acetate substrate reaction producing colorless spots on a blue background. For further details on enzyme inhibition TLC, see Ref. 63.

5.2.6 Identification

Positions of separated zones on thin-layer chromatograms are described by the R_F value of each substance as follows:

$$R_F = \frac{\text{distance traveled by the spot center of the solute}}{\text{distance traveled by solvent front}}$$

R_F values range from 1 for zones migrating at the solvent front to zero for a zone not leaving the origin. When R_F data are tabulated, it is more convenient to report $R \times 100$ or hR_F values. In conventional TLC using ascending development or in overrun (continuous) or multiple development where the solvent front is not measurable, R_X values can be recorded. R_X is defined as follows:

$$R_X = \frac{\text{distance traveled by the solute}}{\text{distance traveled by standard substance X}}$$

Neither R_F nor R_X values are true constants, but R_X values are more reproducible than absolute R_F values [64]. The reproducibility of R_F values depend on numerous factors such as sorbent, humidity, layer thickness, development distance and temperature [65].

The best method to obtain tentative identification of a substance is to spot the sample and reference compounds on the same chromatogram. Mobilities of all compounds are then compared under the same conditions, and a match in R_F values between a sample and standard is evidence for the identity of the sample. Ideally, experimental conditions should be chosen so that the compound to be identified moves to a point near the center of the layer (approximate R_F of 0.5). If R_F values are higher than desired, the polarity of the mobile phase can be reduced; for lower R_F values, the polar component of the mobile phase can be increased. If the spots of interest from the sample do not align with any standards, they are either not the same as any of these compounds or their mobilities have been affected by extraneous material from the sample. If the latter is the case, sample clean-up may be necessary. Identification of a compound based on mobility in one chromatographic system is only tentative and should be confirmed by other evidences. It is common to chromatograph the sample and standards in several thin-layer systems to compare their mobilities, but this provides useful corroborative evidence only if the systems are truly independent [66].

Additionally, instead of running the sample and standards side by side on the same plate, the sample can be mixed with the standard of the expected compound (co-chromatography). Non-separability of this mixed sample in several independent systems indicates that the compounds are identical.

TLC can provide other information critical in compound identification. Colors from selective chemical detection reactions, behavior in ultraviolet light, absorbance and/or fluorescence spectra obtained directly on the chromatogram by use of a spectrodensitometer or in solution after elution and R_F values of derivatives prepared by reaction before, during, or after development can be combined with the R_F values of the sample to increase the degree of probability of correct identification [67,68]. Infrared (IR) and mass spectrometry combined with chromatography can provide unequivocal identification if sufficient sample is available. This is often not the case in trace analysis where nanogram amounts of material may be separated and detected by TLC, whereas microgram quantities are needed for conventional IR spectrometric confirmation of the trace substances.

Identification procedures for large groups of compounds have been reported and termed 'systematic analysis.' These methods have been done mainly in paper chromatography, but the principles are also valid in TLC. For example, chromatographic spectra or profiles of compounds are plotted as R_F values in different solvents. Each profile is characteristic of one particular compound in the group, and, if solvents are carefully chosen, information about the presence of certain functional groups can be obtained [69]. For example, if the pH of the system is changed and R_F is plotted against pH, pH chromatograms characteristic of compounds can be obtained [70]. Instead of plotting data, R_F values in each of a series of solvent systems can be expressed by a letter, leading to a 'word' characterization and identification of each compound [71]. These types of systematic procedures lend themselves to computer analysis for data handling and retrieval. A comprehensive example of the systematic identification of 161 nitrogen-containing drugs has been published [72].

An advantage of substance identity by TLC is the ability to work with small amounts of sample. It is usually necessary to have available an authentic sample for comparison. If such a standard is not available, the structure of a substance can sometimes be deduced from chromatographic data using the principle of constant differences and R_M values [73].

5.2.7 In situ densitometry

In situ densitometry is rapidly becoming the method of choice for the quantitative analyses of substances separated by TLC [2]. This method of analysis involves the use of a commercially available densitometer for scanning TLC plates. Biochemists interested in using this procedure should contact commercial suppliers of densitometers such as Kontes or Schoeffel.

Densitometry is the instrumental measurement of absorbance, fluorescence or fluorescence quenching directly on the thin layer. Measurements are made either by transmission through the plate or by reflection from the plate, using either single- or double-beam operations of scanning instruments. Compounds that lack natural color or fluorescence can be derivatized prior to or after chromatography by treatment with a chromogenic or fluorogenic reagent.

The design of densitometers is similar in principle but differs in details. Most instruments have a light source, an optical system, a monochromator or filters or both, one or more photosensing detectors, a read-out system, and a means of moving the layer past the detector. Accessories may include a baseline corrector and integrator. The optical design can be single- or double-beam, and measurement modes are reflectance, transmittance or simultaneous reflectance-transmittance. Colored, quenched or fluorescent zones can be measured by most instruments, the latter being facilitated by use of excitation and emission monochromators or filters. For a review of the design and specifications of commercially available densitometers, see Ref. 2. Undoubtedly, the sum of the errors in producing a reproducible chromatographic spot far outweighs errors inherent in most densitometers.

Transmission measurements through glass plates are limited to wavelengths longer than 300–350 nm, and reflection measurement is superior for layers of uneven thickness [74]. Most experimental studies indicate the superiority of the transmission over the reflection mode, and double-beam operation over single-beam to obviate uneven reagent application or layer thickness [75]. Chromatographers differ on the theoretical superiority of the various measurement modes and the results of experimental studies have not been consistent [76]. In practice, it is best to optimize each determination with the scanner available by testing all possible measurement modes, different filters or monochromator settings, different slit sizes and positions [77] and different scan directions [78] to obtain the best signal-to-noise ratio for the peaks, the best linearity and slope of the calibration curve, and the highest degree of scanning reproducibility.

To do in situ densitometric analysis, several standards and samples are applied to the plate. Development and detection are then performed and the spots scanned. A calibration curve consisting of the scan areas of the standards (measured usually by height × width at half-height, or using an integrator) versus the amounts of analyte spotted is constructed, and the amount (weight) of analyte in the sample, represented by its scan area, is interpolated from the curve. Densitometers are designed to produce calibration curves that are typically straight lines passing through the origin or are slightly curved downward at higher concentrations. The concentration of

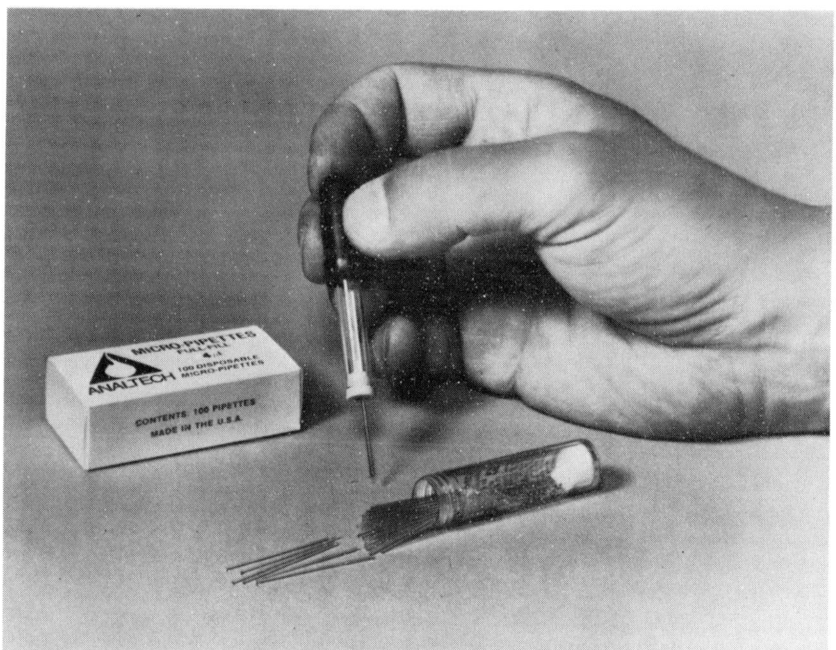

Fig. 5.6. Micropipet (photo supplied courtesy of Analtech, Inc.).

TABLE 5.1.

Selected examples of TLC separations of different compounds of interest to the biochemist

Compounds separated	Stationary phase [1]	Mobile phase [2]	Ref.
Aliphatic acids	C+0.1 M NaOH, pH9	B-80% HAc (100:17.5)	83
Alkaloids	SG	$CHCl_3$-AC-DEA (50:40:10)	84
	C	DMF-DEA-EtOH-EA (5:2:20:75)	85
	SG	B-AC-MeOH (7:2:1)	86
	SG	DMF-EA-EtOH (19:130:1)	87
	KC	MeOH-0.5M NaCl (65:35)	31
Amines and carboxylic acids	$AgNO_3$ *	EtOH-EA-NH_4OH (9:3:2)	88
Amino acids	Cation exchange resins	Li citrate buffers	89
	Ion exchange, Na^+-form	Na citrate buffer	90
	SG	96% EtOH-H_2O (63:37 w/w) or $CHCl_3$-MeOH-17% NH_4OH (2:2:1)	91
	C	BuOH-AC-DEA-H_2O (10:10:2:5) or P-H_2O (75:25)−3% NH_4OH for chamber equilibration	92
	SG	BuOH-HAc-H_2O-Pyr (15:3:12:10)	92
	KC	EtOH-H_2O (80:20)	31
Amino acids (dansyl)	KC	MeOH-2% HAc (75:25)	94
Amino acids (PTH)	KC	MeOH-0.1 M NaAc (pH 4.6) (1:1)	31
Anilines	CdS *	B-HAc (9:1) or B-CH (1:1)	95
Antibiotics	KG+phthalate buffer (pH 5.3)	CCl_4-IPA-H_2O (65:35:4)	96
Antihistamines	SG	MeOH-BuOH (3:2)	97
Antioxidants	SG+0.5N OA	B-EA (6:4)	98
	SG-KG (25:5 w/w)	H-HAc (4:1)	99
Aromatic acids	KC	AN-0.5 M NaCl (20:80)	175
Aromatic alcohol 3,5-dinitrobenzoates	SG	B-PE (1:1) or T-EA (9:1)	100
Aromatic aldehydes	SG	PE-EA (10:5) or $CHCl_3$-EA (98:2)	101
Aromatic phenols	SG	$CHCl_3$, PE-Pyr (95:5) or PE-HAc (95:5)	102
Artificial sweeteners	DEAE-C	0.5M NH_4 formate	103
Barbiturates	SG	$CHCl_3$-AC (9:1)	104
	KC	MeOH-0.5M NaCl (70:30)	31
Carbonyl DNPH's	$AgNO_3$ *	PE	105
Carboxylic acids	POEG	di-IPE-HCOOH-H_2O (90:7:3)	106
	EG or PEG	PE-AC (2:1) or E-98% HCOOH (50:1)	107
Carotene and Canthaxanthin	KC	MeOH-AC (1:1)	31
Carotenoid aldehydes	PO	MeOH	108
Ceramides	SG+2% Na arsenite	$CHCl_3$-MeOH (95:5)	109
Ceramide acetates	$AgNO_3$ *	$CHCl_3$-B-AC (80:20:10)	109

TABLE 5.1. (continued)

Compounds separated	Stati
Triglycerides (contd.)	TD
	AgN(
Ubiquinones	PO
Vitamin A isomers	SG
Vitamins D_2 and D_3	SG
Wax esters	SG
Xanthine derivatives	KC

* Impregnating agent used on silic

[1] *Stationary phases*
Al, aluminum oxide; BA, boric acic
dimethylformamide; DMSO, dimet
glycol; F, formamide; KC, Whatm
polyamide; PEG, polyethylene glyc
oil; POEG, polyoxyethylene glycol 1

[2] *Mobile phases* (all solvents given ii
AC, acetone; AN, acetonitrile; B, b(
dioxane; DBK, diisobutyl ketone; I
ethyl acetate; EtOH, ethanol; F, for
IP, isopropanol; IPA, isopropyl alco
chloride; MEK, methyl ethyl keto
propanol; Pyr, pyridine; T, toluene;

analyte in the sample is calcula
taken, and the portion of the f
Accurate and consistent ap
small, equally sized zones is c
accomplished manually using D
5.6), or by use of commercial s
increases the concentration of
applied sample to be reduced tc
maintaining detection intensity.
lower is now common, especiall
Many workers prefer to apply
etry and commercial plates with
for manual streak application. U
uniformly applied to unscored p
(Camag). A more uniform distri
usually result if it can be appli
polymer-bound layers are very su

TABLE 5.1. (continued)

Compounds separated	Stationary phase [1]	Mobile phase [2]	Ref.
Chlorinated insecticides	DMF	IO	110
Chlorophylls	KC	MeOH-THF-H_2O (75:20:5)	111
Chloroplast pigments	SG	IO-AC-E (3:1:1)	112
	Sucrose	PE-AC (95:5)	113
	$Ca(OH)_2$	PE-$MeCl_2$ (95:5)	114
	KC	MeOH-AC-H_2O (20:4:3)	30
	KC	BuOH-MeOH (8:7)	30
Cholesteryl esters	PO	MEK-AN (7:3)	115
	KC	AN-MEK-$CHCl_3$ (50:35:15)	31
Conjugated bile acids	KC	EtOH-0.3M $CaCl_2$-DMSO (25:25:2)	116
	KC	EtOH-KH_2PO_4(pH2.5) (1:1)	117
Corticosteroids	SG	$CHCl_3$-MeOH-H_2O (188:12:1)	118
	F	BuAc or BuAc-H_2O (300:15)	119
Coumarins	SG + pH 11 phosphate buffer	$CHCl_3$-B-HCOOH-acetylacetone (49:48:2:1)	120
Dicarboxylic acids	SG	B-D-HAc (90:25:4)	121
Diglyceride acetates	$AgNO_3$ *	$CHCl_3$-MeOH-H_2O (65:25:4)	122
2,4-Dinitrophenyl hydrazones	CW-400	PE	105
Fat soluble dyes	$AgNO_3$ *	B-E-HAc (90:10:0.1)	123
Fat soluble food colors	SG	$CHCl_3$ or H-EA (90:10)	123
Fatty acids	$AgNO_3$ *	PE-E (40:10)	124
Fatty acid esters	$AgNO_3$ * + SO *	MeOH-H_2O (95:5)	125
Fatty acid methyl esters	SO	AN-HAc-H_2O (70:10:25)	126
	KC	HAc-AN-THF (45:45:10)	127
	KC	MeOH-HAc-HCOOH (80:10:10)	127
	KC	AN-5% NH_4OH (95:5)	127
Flavonoids	SG + Na acetate (30:1.5 w/w)	B-EA-HCOOH (45:35:20)	128
	P	B-MEK-MeOH (30:13:7)	129
Flavonoid aglycones	KC	MeOH-H_2O-HCOOH (28:10:5 or 28:4:5)	130
Gangliosides	SG	$CHCl_3$-MeOH-2.5N NH_4OH (60:40:9)	131
Glycerides	BA *	$CHCl_3$-AC (96:4)	132
Glyceryl ethers	SG	H-E (95:5)	133
Glycosphingolipids	SG	$CHCl_3$-MeOH-H_2O (65:25:4)	134
Hallucinogenic drugs	Al	MeOH-$CHCl_3$ (1:1)	135
Higher alcohols and glycols	Al	H-AC (3 or 4:1), H_2O sat.	136
Isomers of hydroxy acids, esters, and alcohols	BA *	PE-E (6:4)	137
Lipids	SG	PE-E-HAc (70:40:2 to 80:20:1)	138
		or PE-E (95:5 to 80:20)	138

TABLE 5.1. (continued)

Compounds separated	S
Lipids (contd.)	S(
Lower aldehyde and ketone 2,4-DNPH derivatives	Zı
Nucleosides	DI
Nucleotides	PE
	PE
	FI:
	f.
Penicillins	KC
Phenols	Bio-
	oı
	D
Phenothiazine derivatives	SG
	Basi
Phospholipids	AgN
	SG
	SG
	SG
	SG
Phosphonolipids	SG
Serum lipids	SG
Simple lipids	SG
Steroids	SG
	F
Steroid glycosides	SG
Sterols	SG
	Al
	AgNO
	KC
	KC
Sterols and sterol acetates	Un
Sugars	BA *
	SG + N.
	KG + N
	KC
	Bisulfite
Terpenes	$AgNO_3$
Thiols	C
Tocopherols	SG
	Al
Triglycerides	SG

even some water-based reagents. A Thomas-Mitchell or Kontes dip tank is recommended for the dipping operation. The time-intensity (area) characteristics of the detection reaction should be determined, and all work should be done on a definite time schedule so that scanning is performed when the spot areas and intensities have become constant [81]. If heating is part of the detection procedure, the temperature and time of heating should be controlled. When quantitating by densitometry, it is best to apply the smallest amount of material that can be detected.

Densitometry of high-performance plates developed in the usual ascending mode is carried out similarly as on conventional layers, except that smaller volumes and spot sizes are applied, times and migration distances are less, and the slit length of the scanner light beam must be reduced. Densitometric baselines are usually flatter because of the more homogeneous HP layer thickness. Up to 18 tracks can be utilized for samples and standards on a 10 × 10 cm HPTLC plate [74]. Densitometry of circular HP chromatograms produced in the commercial U-chamber (Camag) requires a special scanning device [82].

5.2.8 *Applications* (see also Note following References, p. 413)

A voluminous literature exists on applications of TLC for the separation of numerous compounds [3,5,43]. A recent book that provides applications and laboratory experiments of interest for the biochemist is also available [174]. Table 5.1 provides TLC applications for selected compounds of interest to the biochemist.

References (Part A)

1 Touchstone, J.C. (ed.) (1973) Quantitative Thin Layer Chromatography. Wiley, New York.
2 Touchstone, J.C. and Sherma J. (eds.) (1979) Densitometry in Thin Layer Chromatography—Practice and Applications. Wiley, New York.
3 Stahl, E. (1969) In Thin Layer Chromatography (Stahl, E., ed.) 2nd Edn., p. 1, Springer-Verlag, Berlin.
4 Heftmann, E. (1975) In Chromatography—A Laboratory Handbook of Chromatographic and Electrophoretic Methods (Heftmann, E., ed.) 3rd Edn., p. 1, Van Nostrand Reinhold, New York.
5 Kirchner, J.C. (1978) Thin Layer Chromatography, 2nd Edn. Wiley, New York.
6 Pelick, N., Bolliger, H.R. and Mangold, H.K. (1966) In Advances in Chromatography (Giddings, J.C. and Keller, R.A., Eds.) Vol. 3, p. 85, Marcel Dekker, New York.
7 Stahl, E. (1962) Dünnschicht-Chromatographie, ein Laboratorium-Handbuch, Springer Verlag, Berlin.
8 Stahl, E. (1978) J. Chromatogr. 165, 59.
9 Przybylowicz, E.P., Staudenmayer, W.J., Perry, E.S., Baitsholts, A.D. and Tischer, T.N. (1965) J. Chromatogr. 20, 506.
10 Sherma, J. (ed.) (1975) Kontes Quant. Notes, 1,1.
11 Sherma, J. and Fried, B. (1982) Anal. Chem. 54(5), 45R.
12 Scott, R.P.W. (1978) Analyst 103, 37.
13 Snyder, L.R. (1975) In Chromatography (Heftmann, E., ed.) 3rd Edn., p. 46, Reinhold, New York.
14 Felton, H.R. (1979) Technical Report 7905, Analtech, Newark, DE.
15 Waring, P.E. and Ziporin, Z.Z. (1964) J. Chromatogr. 15, 168.

16 Wang, K.-T. and Weinstein, B. (1972) In Progress in Thin Layer Chromatography and Related Methods (Niederwieser, A. and Pataki, G., eds.), Vol. 3, p. 177, Ann Arbor Science Publishers, MI.
17 Wang, K.-T., Lin, Y.-T. and Wang, I.S.Y. (1974) Adv. Chromatogr. 11, 74.
18 Endres, H. (1969) In Thin Layer Chromatography. A Laboratory Handbook (Stahl, E., ed.) 2nd Edn., p. 41, Springer Verlag, New York.
19 Dorfner, K. (1969) In Thin Layer Chromatography. A Laboratory Handbook (Stahl, E., ed.), 2nd Edn., p. 44, Springer Verlag, New York.
20 Wollenweber, P. (1969) In Thin Layer Chromatography. A Laboratory Handbook (Stahl, E., ed.), 2nd Edn., p. 32, Springer Verlag, New York.
21 Steverle, H. (1966) Z. Anal. Chem. 220, 413.
22 Halpaap, H. and Ripphahn, J. (1977) Chromatographia 10, 613.
23 Stahl, E. (1969) In Thin Layer Chromatography. A Laboratory Handbook (Stahl, E., ed.), 2nd Edn., p. 86 Springer Verlag, New York.
24 Lees, T.M. and DeMuria, J. (1962) J. Chromatogr. 8, 108.
25 Morita, K. and Haruta, F. (1963) J. Chromatogr. 12, 412.
26 Stahl, E. and Mangold, H.K. (1975) In Chromatography (Heftmann, E., ed.), 3rd Edn., p. 164, Reinhold, New York.
27 Touchstone, J.C. and Dobbins, M.F. (1978) Practice of Thin Layer Chromatography, p. 49, Wiley-Interscience, New York.
28 Felton, H.R. (1979) Technical Report No. 7904, Analtech, Newark, DE.
29 Vitek, R.K. and Kent, D.M. (1978) Am. Lab., Jan. 1978, p. 71.
30 Sherma, J. and Latta, M. (1978) J. Chromatogr. 154, 73.
31 Sherma, J. (1981) Whatman TLC Technical Series, Vol. 1, p. 1.
32 Beesley, T.E. (1972) Am. Lab. Oct., p. 25.
33 Halpaap, H. and Krebs, K.F. (1977) J. Chromatogr. 142, 823.
34 Sherma, J. and Gray, D. (1979) Am. Lab. Nov., 21.
35 Kupke, I.R. and Zeugner, S. (1978) J. Chromatogr. 146, 261.
36 Sherma, J. and Beim, M. (1978) HRC-CC, J. High Resolut. Chromatogr. Chromatogr. Commun. 1, 309.
37 Graff, D.J. (1974) J. Parasitol. 51, 72.
38 Sherma, J. (1980) In Thin Layer Chromatography, Quantitative Clinical and Environmental Applications (Touchstone, J.C. and Rogers, D. eds.) p. 17, Wiley Interscience, New York.
39 Heathcote, J.G. (1979) In Densitometry in Thin Layer Chromatography—Practice and Applications (Touchstone, J.C. and Sherma, J., eds.) p. 153, Wiley, New York.
40 Touchstone, J.C. and Dobbins, M.F. (1979) In Densitometry in Thin Layer Chromatography—Practice and Applications (Touchstone, J.C. and Sherma, J., eds.) p. 633, Wiley, New York.
41 Gasparič, J. (1961) Chem. Listy 55, 1439.
42 Pataki, G. and Niederwieser, A. (1967) J. Chromatogr. 29, 133.
43 Zweig, G. and Sherma, J. (1972) Handbook of Chromatography, Vol. 1. CRC Press, Boca Raton, FL.
44 Randerath, K. (1963) Thin Layer Chromatography, p. 15, Academic Press, New York.
45 Randerath, K. (1963) Thin Layer Chromatography, p. 51, Academic Press, New York.
46 Stahl, E. (1958) Chem. Z. 82, 323.
47 Perry, J.A., Jupille, T.H. and Glunz, L.H. (1975) Anal. Chem. 47(1), 65A.
48 Jupille, T.H., and Perry, J.A. (1975) J. Chromatogr. Sci. 13, 163.
49 Faber, D.B. (1977) J. Chromatogr. 142, 422.
50 De Zeeuw, R.A. (1968) Anal. Chem. 40, 915.
51 Dallas, M.S.J. (1965) J. Chromatogr. 17, 267.
52 Barrett, G.C. (1974) Adv. Chromatogr. 11, 146.
53 Touchstone, J.C., Murawec, T., Kasparow, M. and Wortmann, W. (1972) J. Chromatogr. Sci. 10, 490.
54 Touchstone, J.C., Murawec, T., Kasparow, M. and Wortmann, W. (1972) J. Chromatogr. 66, 172.
55 Beesley, T.E. (1976) In Kontes Quant. Notes (Sherma, J., ed.) 1,3; 3,1, Kontes Inc., Vineland, NJ.
56 Tochstone, J.C., Balin, A.K., Murawec, T. and Kasparow, M. (1970) J. Chromatogr. Sci. 8, 443.

57 Chavez, M.N. (1979) J. Chromatogr. 162, 79.
58 Stahl, E. (1969) Thin Layer Chromatography. A Laboratory Handbook, 2nd Edn., p. 568, Springer Verlag, New York.
59 Merck, E. (1975) Dyeing Reagents for Thin Layer and Paper Chromatography. Available from E. Merck, Darmstadt, Germany.
60 Sherma, J. and Touchstone, J.C. (1974) Anal. Lett. 7, 279.
61 Sherma, J. and Marzoni, G. (1974) Am. Lab., Oct., p. 21.
62 Touchstone, J.C., Sherma, J., Dobbins, M.F. and Hansen, G.R. (1976) J. Chromatogr. 124, 111.
63 Mendoza, C.E. (1972) Residue Rev. 43, 105.
64 Mendoza, C.E., Wales, P.J. and Bray, D.F. (1968) Analyst 93, 688.
65 Lederer, M., Macek, K. and Hais, I.M. (eds.) (1968) Reproducibility in Paper and Thin Layer Chromatography, Elsevier, New York.
66 Connors, K.A. (1974) Anal. Chem. 46, 53.
67 Dallas, M.S.J. (1970) J. Chromatogr. 48, 193.
68 Hais, I.M. (1970) J. Chromatogr. 48, 200.
69 Reio, L. (1970) J. Chromatogr. 48, 11.
70 Betina, V. (1961) Chem. Zvesti 15, 661 and 750.
71 Goldbaum, L.R. and Kazyak, L. (1956) Anal. Chem. 28, 1289.
72 Macek, K. (1965) In Stationary Phase in Paper and Thin Layer Chromatography (Macek, K. and Hais, I.M., eds.) p. 309, Elsevier, New York.
73 Gasparič, J., Gemzova, I. and Snobl, D. (1966) Collection Czech. Chem. Commun. 31, 1712.
74 Hezel, U. (1978) Am. Lab., May, p. 91.
75 Touchstone, J.C., Levin, S.S. and Murawec, T. (1961) Anal. Chem. 43, 858.
76 Pollak, V. (1979) In Densitometry in Thin Layer Chromatography—Practice and Applications (Touchstone, J.C. and Sherma, J., eds.) pp. 11, 47, Wiley, New York.
77 Shellard, E.J., and Alam, M.Z. (1968) J. Chromatogr. 33, 347.
78 Turano, P. and Turner, W.J. (1972) J. Chromatogr. 64, 347.
79 Brian, K.R. and Turner, T.D. (1971) J. Chromatogr. 61, 157.
80 Emanuel, C.F. (1973) Anal. Chem. 45, 1568.
81 Pataki, G. (1968) Chromatographia 1, 492.
82 Janchen, D. and Schmutz, H. (1978) High Resolut. Chromatogr. Chromatogr. Commun. 1, 315.
83 Miyazaki, S., Suhara, Y. and Kobayashi, T. (1968) J. Chromatogr. 39, 88.
84 Waldi, D., Schnackerz, K. and Munter, F. (1961) J. Chromatogr. 6, 61.
85 Telechrom Topics (1974) Applied Science Laboratories, Inc., Vol. 1, No. 1, p. 1. State College, PA.
86 Pfeifer, S. (1966) J. Chromatogr. 24, 364.
87 Hohmann, T. and Rochelmeyer, H. (1964) Arch. Pharm. 297, 86.
88 Tabak, S., Mauro, A.E. and Del'Acqua, A. (1970) J. Chromatogr. 502, 500.
89 Varadi, A. (1975) J. Chromatogr. 110, 166.
90 Elodi, P. and Karsai, T. (1980) J. Liq. Chromatogr. 3, 809.
91 Brenner, M. and Niederwieser, A. (1960) Experentia 16, 378.
92 von Arx, E. and Neher, R. (1963) J. Chromatogr. 12, 329.
93 Ehrhardt, E. and Cramer, F. (1962) J. Chromatogr. 1, 405.
94 Macek, K., Deyl, Z. and Smrž, M. (1980) J. Chromatogr. 193, 421.
95 Yasuda, K. (1970) J. Chromatogr. 60, 144.
96 Hellberg, H. (1968) J. Assoc. Off. Anal. Chem. 51, 552.
97 Cochin, J. and Daly, J.W. (1962) Experentia 18, 294.
98 Seher, A. (1960) Nahrung 4, 446.
99 Meyer, H. (1961) Dtsche. Lebensm. Rundsch. 57, 1970.
100 Dhont, J.H. and deRooy, C. (1961) Analyst 86, 527.
101 Sundt, E. and Saccardi, A. (1962) Food Technol. 16, 89.
102 Klouwen, M.H. and terHeide, R. (1962) Parfum. Kosmet. 43, 195.
103 Nagasawa, K., Yoshidome, H. and Anryu, K. (1970) J. Chromatogr. 52, 173.

104 Uhlmann, H.J. (1964) Pharm. Ztg. 109, 1998.
105 Badings, H.T. and Wassink, J.G. (1963) Neth. Milk Dairy J. 17, 132.
106 Knappe, E. and Peteri, D. (1962) Z. Anal. Chem. 188, 184 and 352.
107 Lupton, C.J. (1975) J. Chromatogr. 104, 223.
108 Winterstein, A., Studer, A. and Ruegg, R. (1960) Chem. Ber. 93, 2951.
109 Karlsson, K-A. and Pascher, I. (1971) J. Lipid Res. 12, 466.
110 Kovacs, M.F. (1966) J. Assoc. Off. Anal. Chem. 49, 365.
111 Siouffi, A.M., Wawrzynowicz, T., Bressolle, F. and Guiochon, G. (1979) J. Chromatogr. 186, 563.
112 Strain, H.H. and Sherma, J. (1969) J. Chem. Educ. 46, 476.
113 Colman, B. and Vishniac, W. (1964) Biochim. Biophys. Acta 82, 616.
114 Stahl, E., Bolliger, H.R. and Lehnert, L. (1963) Wiss. Veroeff. Dtsch. Ges. Ernähr. 9, 129.
115 Kaufmann, H.P., Makus, Z. and Deicke, F. (1961) Fette, Seifen, Anstrichm. 63, 235.
116 Touchstone, J.C., Levitt, R.E., Levin, S.S. and Soloway, R.D. (1980) Lipids 15, 386.
117 Touchstone, J.C., Levitt, R.E., Soloway, R.D. and Levin, S.S. (1979) J. Chromatogr. 178, 566.
118 Bennett, R.D. and Heftmann, E. (1962) J. Chromatogr. 9, 348.
119 Yawata, M. and Gold, E.M. (1964) Steroids 3, 435.
120 Daenens, P. and vanBoven, M. (1971) J. Chromatogr. 57, 319.
121 Petrowitz, H.J. and Pastuska, G. (1962) J. Chromatogr. 7, 128.
122 Renkonen, O. (1968) Lipids 3, 191.
123 Copius Peereboom, J.W. and Beekes, H.W. (1965) J. Chromatogr. 20, 43.
124 Morris, L.J. and Wharry, D.M. (1965) J. Chromatogr. 30, 27.
125 Paulose, M.M. (1966) J. Chromatogr. 21, 141.
126 Malins, D.C. and Mangold, H.K. (1960) J. Am. Oil Chem. Soc. 37, 576.
127 Sherma, J., Krywicki, R. and Regan, T.E. (1981) Am. Lab. 13(1), p. 117.
128 Hansel, R., Langhammer, L., Frenzel, J. and Ranft, J. (1963) J. Chromatogr. 11, 369.
129 Vanhaelen, M. and Vanhaelen-Fostre, R. (1980) J. Chromatogr. 187, 225.
130 Hiermann, A.J. (1979) J. Chromatogr. 174, 478.
131 Ledeen, R. (1960), J. Am. Oil Chem. Soc. 43, 57.
132 Thomas, A.E., Scharoun, J.E. and Ralston, H. (1965) J. Am. Oil Chem. Soc. 42, 789.
133 Schmid, H.H.O., Jones, L.L. and Mangold, H.K. (1967) J. Lipid Res. 8, 692.
134 Svennerholm, E. and Svennerholm, L. (1963) Biochim. Biophys. Acta 70, 432.
135 Genest, L. and Hughes, D.W. (1968) Analyst 93, 485.
136 Kučera, J. (1963) Coll. Czech. Chem. Comm. 28, 1341.
137 Roomi, M.W., Subbaram, M.R. and Achaya, K.T. (1966) J. Chromatogr. 24, 93.
138 Mangold, H.K. (1961) J. Am. Oil Chem. Soc. 38, 708.
139 Pernes, J-F., Nurit, Y. and DeHeaulme, M. (1980) J. Chromatogr. 181, 254.
140 Urbach, G. (1963) J. Chromatogr. 12, 196.
141 Randerath, K. and Randerath, E. (1964) J. Chromatogr. 16, 111 and 126.
142 Beardsley, G.P. and Abelson, H.T. (1980) Anal. Biochem. 105, 311.
143 Randerath, K. (1963) Thin Layer Chromatography, p. 196, Academic Press, New York.
144 Tomasz, J. (1980) Chromatographia 13, 345.
145 Walker, K.C. and Berozoa, M. (1963) J. Assoc. Off. Anal. Chem. 46, 250.
146 Lepri, L., Desideri, P.G., Tanturli, G. and Landine, M. (1975) J. Chromatogr. 108, 169.
147 Awe, W. and Schulze, W. (1962) Pharm. Ztg. 107, 1333.
148 Eiden, F. and Stachel, H.D. (1963) Dtsch. Apoth. Ztg. 103, 121.
149 Kaufmann, H.P., Wessels, H. and Bondopadhyaya (1963) Fette, Seifen, Anstrichm. 65, 543.
150 Skipski, V.P., Peterson, R.F. and Barclay, M. (1964) Biochem. J. 90, 374.
151 Gardner, H.W. (1968) J. Lipid Res. 9, 139.
152 Nichols, B.W. (1963) Biochim. Biophys. Acta 70, 417.
153 Kapoulas, V.M. (1969) Biochim. Biophys. Acta 176, 324.
154 Zollner, N., Wolfram, G. and Amin, F. (1962) Klin. Woch. 40, 273.
155 Story, J.E. and Tuckley, B. (1967) Lipids 2, 501.

156 Barbier, M., Jager, H., Tobias, H. and Wyss, E. (1959) Helv. Chem. Acta 42, 2440.
157 Vaedtke, J. and Gajewoska, A. (1962) J. Chromatogr. 9, 345.
158 Stahl, E. and Kalfenbach, U. (1961) J. Chromatogr. 5, 458.
159 Copuis Peereboom, J.W. and Beekes, H.W. (1962) J. Chromatogr. 9, 316.
160 Ditullio, N.W., Jacobs, C.S., Jr. and Holmes, W.L. (1965) J. Chromatogr. 20, 354.
161 Copius Peereboom, J.W. and Beekes, H.W. (1965) J. Chromatogr. 17, 99.
162 Stahl, E. (1960) Arch. Pharm. 293, 531.
163 Lajunen, K., Purokoski, S. and Pitkanen, E. (1980) J. Chromatogr. 187, 455.
164 Stahl, E. and Kaltenbach, U. (1961) J. Chromatogr. 5, 351.
165 Adachi, S. (1965) J. Chromatogr. 17, 295.
166 von Schantz, M., Jovonen, S. and Hemming, R. (1965) J. Chromatogr. 20, 618.
167 Ito, S. and Fujita, K. (1980) J. Chromatogr. 187, 418.
168 Litchfield, C., Ackman, R.G., Sipos, J.C. and Eaton, C.A. (1971) Lipids 6, 674.
169 Kaufmann, H.P. and Das, B. (1962) Fette, Seifen Anstrichm. 64, 214.
170 Burns, D.T., Stretton, R.J., Shepherd, G.F. and Dallas, M.S.J. (1969) J. Chromatogr. 44, 399.
171 von Planta, C., Schwieter, U., Chopard-Dit-Jean, L., Ruegg, R., Kofler, M. and Isler, O. (1962) Helv. Chim. Acta 45, 548.
172 Janecke, H. and Maas-Joebels, I. (1960) Z. Anal. Chem. 178, 161.
173 Holloway, P.J. and Challen, S.B. (1966) J. Chromatogr. 25, 336.
174 Fried, B. and Sherma, J. (1982) Thin Layer Chromatography — Techniques and Applications, p. 1, Marcel Dekker, New York.
175 Brinkman, U.A.Th. and deVries, G. (1980) J. Chromatogr. 192, 331.
176 Sherma, J. and Zweig, G. (1971) Paper Chromatography. Academic Press, New York.

Part B

5.3 Paper chromatography

5.3.1 History and Introduction

Paper chromatography began before the year 1900 with the discovery of the technique of capillary analysis by Runge and the subsequent application and development of the principle by Schoenbein and Goppelsroeder. This early history of paper chromatography has been reviewed elsewhere [1,2]. Paper partition chromatography as it is practiced today evolved from the development of column partition chromatography in 1941 by Martin and Synge [3] for the separation of amino acid derivatives. The column consisted of silica gel with fixed water, and the mobile phase was chloroform or some other immiscible solvent. Consden et al. [4] extended partition chromatography to a micro-technique using filter paper (an 'open' column) as the support for the immobile water phase. The immiscible solvent moved by the capillarity of the paper and gravity, and acetylation of amino acids was no longer necessary since the compounds were directly detectable on the paper by treatment with ninhydrin. From this beginning in 1944, paper chromatography grew to be an essential analytical technique in all laboratories in the 1940's and 1950's and is still in use today despite the advent of TLC.

The impetus for development of TLC using inorganic adsorbents arose because the separation of lipophilic materials by PC was not completely successful, even with the preparation of adsorption media by impregnation of cellulose and later glass fiber paper, or the reversal of partition phases by impregnation and chemical modification of cellulose fiber. It was soon found that adsorbent TLC has greater separation speed, easier application to preparative scale work, and the possibility of using corrosive detection reagents. Therefore, despite increased costs and somewhat more complex procedures, the use of TLC grew rapidly in relation to cellulose paper chromatography, especially for separation of lipophilic compounds. However, paper chromatography is still often and successfully used, especially for the separation of hydrophilic organic and inorganic compounds. Separations of organic compounds of interest to the biochemist will be stressed in this section.

Many of the basic techniques in paper and thin-layer chromatography are essentially the same, for example sample spotting, ascending development, detection of spots, and their direct scanning. Some techniques are simpler in PC than in TLC, namely circular, horizontal, and descending development methods (due to the ability to fold the paper) and methods for recovery of spots from the chromatogram (the paper, as well as flexible thin layers, can be cut with scissors before elution, while layers must be scraped from glass plates). Migration behavior in both PC and TLC is described by the R_F value of each compound. The basic difference between PC and TLC is the character of the stationary phase. The existence of one method does not exclude the use of the other, and, in fact, the methods are complimentary in many respects. Those aspects of PC that differ from TLC will be stressed in the following subsections. Common techniques such as sample preparation, sample application, and spot detection and identification will not be described again in detail.

5.3.2 Chromatography papers

Papers used for chromatography are short-fiber [5] cellulose filter papers with high purity and other special properties, or glass fiber papers. The papers can be modified by chemical reaction or impregnation in a number of different ways.

A wide variety of commercial papers have been used for chromatography, and it is important that the degree of separation can differ considerably depending upon the exact paper used [6,7]. Whatman No. 1 is a standard, strong, medium-fast, pure cellulose paper that has probably been the most widely used. Comparable standard papers are Schleicher and Schüll 2043b and Machery Nagel (MN) 260. Other papers are thicker and have higher capacities for preparative work (e.g., Whatman 3MM, S and S 2071).

The speed of mobile phase development of paper depends upon the density and strength of the cellulose fiber network. Rapid flow rates due to a looser fiber network are achieved on Whatman 4 and 31ET and on S and S 2043A. Whatman

No. 4 has resolving power and purity similar to No. 1, but it is thicker, so that more sample can be applied to improve detectability. The following papers are denser and smoother and, therefore, less permeable and less transparent: Ederol 202 s, S and S 2045b Gl. Whatman No. 54 is a washed, fast paper. It is important to select the proper paper for the analytical problem at hand.

Commercial papers contain more or less impurites (e.g., inorganic salts, lipids, amino acids) that may or may not interfere with the analysis being performed. Where interference is possible, a prewashing step should be incorporated to minimize impurities. As much as 0.05% by weight of diethyl ether-soluble material has been washed from good grades of filter papers. Dilute hydrochloric acid, methanol, complexing agents (EDTA), methanol and acetone can also remove significant amounts of material. Touching the paper and exposure to vapors should be kept to a minimum, kinks should be avoided, and proper storage is a necessary precaution. Many papers are commercially available in a variety of rectangular or square sheet, circle, strip and roll sizes and other special shapes (e.g., slotted discs, wedge shaped strips, etc.), or papers can be cut to other sizes and shapes to suit individual applications. Paper strips should be cut with the grain for proper flow.

Besides pure cellulose papers, many types of modified chromatography papers can be prepared or obtained from commercial sources, but the use of these papers is decreasing. Special commercial papers with a high content of carboxyl groups serve for the separation of cations such as protonated amines and amino acids [8]. More common commercial ion exchange papers are prepared by incorporation of ion exchange resins (Amberlite papers), cellulose ion exchange groups (carboxymethyl, phosphate, DEAE and ECTEOLA cellulose), or inorganic exchangers [9–11].

Cellulose can be esterified with acetic anhydride to form acetylated paper with hydrophobic properties [12]. These papers, which are commercially available from Schleicher and Schüll with 5–40% of acetyl residues, serve for the reversed-phase chromatography of lipophilic substances such as steroids [13]. It has been suggested that these papers are suitable for separation of racemic mixtures due to the migration of optical antipodes at different rates on the chiral acetyl cellulose [14].

The more usual way of preparing reversed phase papers is by impregnation with silicones, rubber, or liquid ion exchangers [15]. These papers, only some of which are available commercially, can be used to separate hydrophobic substances such as amines, fatty acids, sterols, triglycerides, vitamins and pesticides.

Papers impregnated with kieselguhr, aluminum oxide, silica gel, and calcium carbonate are also commercially available from Schleicher and Schüll and Whatman. These adsorbent-loaded papers have been used to separate moderate and low polarity substances such as lipids, sterols, steroids, insecticides and pigments [16]. They function generally in a manner similar to thin layers of the same adsorbent, but, in some cases, the presence of cellulose along with the adsorbent can provide unique separations [17,18]. Applications of adsorbent paper chromatography have been reviewed [19].

Glass fiber papers are fast, have very low adsorptivity, and can be used with detection reagents too corrosive for cellulose paper. They are used either untreated

[20], or impregnated with aqueous buffers [20], silica gel [21], alumina [22] or nonvolatile liquids [23]. Glass fiber papers containing silica gel and silicic acid are available from the Gelman Corp. as ITLC (instant thin-layer chromatography) media [24]. A book by Haer reviews chromatography on glass fiber paper [25].

5.3.3 Sample preparation and application

The concentration of the sample solution spotted must not be below the detection limit of the analyte on the chromatogram after development. Neither should too much of the sample be applied or the paper will be overloaded. Concentration or dilution steps and the volumes of samples (and standards for comparison) spotted must take these considerations into account.

Before chromatography, unstable or volatile solutes must be converted to appropriate derivatives, e.g., alcohols to 3,5-dinitrobenzoates. Derivatives may be easier to detect, separate or identify than the parent compounds, e.g., fluorescent dansyl derivatives of amino acids.

Samples should also be as pure as possible, and desalting of biological samples is often required to produce satisfactory chromatograms. Other types of interfering large molecules may be present, e.g., proteins, lipids and plant pigments. These interferences can cause streaked zones, double spots, spots with abnormally low or high R_F values, or impaired detection, all of which render chromatograms difficult to interpret. Desalting is accomplished electrolytically, by ion exchange, by gel chromatography or by organic solvent extraction. Precipitation, extraction and ion exchange or adsorbent column chromatography can remove proteins or lipids.

Some chromatographic systems have been designed in which untreated biological samples can be applied directly to the paper. The impure sample on the paper is chromatographed with hexane, which elutes the lipids with the solvent front. After drying, the paper is developed with pure water (if the analyte is water insoluble) to elute salts, sugars, etc. at the front. Finally, development with the proper mobile phase for the analytical separation is carried out, to a point below the purification developments.

Some preliminary PC runs will usually be required to determine the optimum degree of sample concentration or dilution, any necessary cleanup steps, and the best solvent polarity for dissolving the sample and preparing the mobile phase. Sample preparation methods for planar chromatography were reviewed by Sherma [26].

The sample, in a small volume of volatile solvent, is applied to the paper on a starting line (origin) 1 to 4 inches from the end. A calibrated capillary pipet (e.g., Drummond Microcap) or a micro-syringe is usually used for application. For qualitative work with aqueous solutions, a platinum loop fixed in a glass rod can be used. A typical sample is a volume of 5–25 μl containing 0.1–100 μg of substance. The spot diameter should be in the 0.5–1 cm range. If the initial spots are too large, the concentration of analyte may be below the detection limit because of the large zone that will be formed. If the initial spot is too small to be concentrated, the mobile phase may be unable to completely and quickly dissolve the sample, leading to tailed spots. For application of larger volumes without increase in spot size, portions of the sample can be applied to the origin sequentially, with drying (warm air or IR light for thermo-stable substances) in between. For most biological samples, it is necessary to apply a number of drops of sample to each origin to achieve a sufficient concentration of substances.

Samples are usually applied as round spots. However, application in a thin line or band is best when especially high resolution is required. The starting spot or line must be as narrow as possible to obtain sharply separated zones. Wide initial zones can be concentrated by holding the paper in a trough of polar solvent such as acetone or methanol and allowing the front to travel up to the starting line. The paper then must be dried before development, which is carried out by placing the paper so that a few millimeters dip into the mobile phase but the dried sample does not.

5.3.4 Mobile phase (solvent) systems

There are many mobile phases that can be used for paper chromatography, depending upon the nature of the solutes and the stationary phase. In practice, mobile phases are chosen by consulting the literature for systems in which the analyte, or substances with closely related structures, have already been chromatographed. For example, the same solvents suitable for amino acid separations are also useful for indoles and imidazoles. An ideal mobile phase has the following properties: the individual solvents are easy to obtain, pure and inexpensive; stable in air when mixed with small quantities of acids or bases; easy to prepare for immediate use or in advance with storage until needed; components are relatively non-volatile or of similar volatility during chromatography; easily removed from the chromatogram before the detection test; remains homogeneous at the temperature of use; non-reactive toward the solutes; will spread out (resolve) the substances of interest along the paper.

To experimentally determine a PC mobile phase for an unknown, a number of solvents from the middle part of the elutropic series (see TLC Section 5.2.3.3) are tried, after equilibrating the paper with water vapor. If the R_F value is too high, the mobile phase is too polar relative to the solutes and a solvent closer to the top of the table should be used. If the R_F value is too low, the solvent is too hydrophobic compared to the solute, and a more polar solvent is needed. The R_F region between 0.1 and 0.7 is known to be best for optimum resolution [27].

When a solvent is found that gives R_F values for the solutes near the middle of the paper but separation is not adequate, other single solvents or solvent mixtures that maintain a similar elution strength (polarity) should be tried until one that provides better resolution is found. For example, if benzene provides R_F values that are slightly too low, the more-polar solvent chloroform can be tried, or a small amount of ethanol or methanol can be added to benzene to increase polarity. In either case, different interactions will be involved in the chromatographic system due to the new solvent, and resolutions may be improved. Solvents can conveniently be tested by making runs of 3–4 cm in a large test tube containing the solvent and closed with a cork from which a faster paper strip (e.g., Whatman No. 4) is suspended.

It sometimes happens that a mobile phase is found that gives suitable R_F values, but the spots are tailed or otherwise distorted. This can be due to poor solubility in the mobile phase (choose another solvent with similar polarity but in which the solute has better solubility); overloading at the origin (find a more sensitive detection reagent); irreversible adsorption on the cellulose (add a more-polar solvent to the mobile phase or impregnate the paper with a polar solvent, see below); or dissociation (add a volatile stronger acid or base to the mobile phase to render the solute completely ionized or unionized). When solutes can ionize, pH control is very important in choosing the solvent system (polar if ionized, nonpolar if unionized) and achieving good resolutions.

It is necessary to use very pure solvents for preparation of mobile phases, since the presence of impurities can drastically alter R_F values or interfere with color development. For example, the polarities of absolute chloroform and chloroform with 2% ethanol stabilizer are very different. Herein lies the cause of many of the problems in reproducibility of countless methods that are described in the literature.

A brief description of the six most popular types of solvent systems for paper chromatography follows.

1. Straight systems are used for separation of hydrophilic compounds such as amino acids and sugars. The stationary phase is the water in the cellulose (‘water-cellulose complex’), and the solvent is an aqueous and organic mixture such as phenol saturated with water, butanol/acetic acid/water (4 : 1 : 5 v/v), or ethyl acetate/pyridine/water (12 : 5 : 4 v/v).

2. Reversed-phase systems are for separation of lipophilic compounds such as fatty acids. The paper is rendered nonpolar by acetylation or impregnation with a 5–25% solution of paraffin oil or vaseline in benzene, ether, $CHCl_3$, etc. Development is with polar solvents such as various mixtures of acetic acid, methanol, acetone and water. Presumably, the paper acts as an inert support.

3. Nonaqueous partition systems are for compounds which are slightly water-soluble but too polar for reversed-phase systems. The Bush and Zaffaroni systems for separation of steroids are included here. Polar stationary phases such as methanol-water [28], DMF, formamide or propylene glycol [29,30] combined with a suitable less polar mobile phase such as benzene, cyclohexane or $CHCl_3$ are examples.

4. Buffered partition systems are used to separate weak acids and bases. The degree of ionization is kept constant by impregnation of the paper with a suitable buffer and/or by adding acid or base to the aqueous-organic solvent [31,32].

5. Adsorption systems are governed by adsorption rather than a partition mechanism. Compounds adsorbed by cellulose may be separated on unmodified cellulose developed with water or salt solutions [33]. Compounds adsorbed by adsorbents such as silica or alumina are separated on papers impregnated with these adsorbents and developed with organic solvents.

6. Chemically reactive systems are those in which a special effect, in addition to partition and/or adsorption, lends added selectivity to the system. Examples include use of chemically modified cellulose (carboxyl paper), ion-exchange papers [34], or papers impregnated with a selective reagent (e.g., boric acid [35] for separation of various hydroxyl compounds), or the mobile phase may contain a reactive or complexing reagent in combination with normal or modified cellulose [21].

Additionally, Macek [36] has listed a diverse series of standard solvents for analysis of compounds with a wide range of structures.

The classical two-phase systems for paper chromatography are prepared by shaking the organic solvent with the stationary phase (water) in a separatory funnel, separating the phases, and using the organic layer as the stationary phase and the aqueous phase for chamber saturation. If the developer contains a high percentage of water, presaturation of the paper is probably unnecessary because it will take up the necessary amount of water during development. If the mobile phase contains a small amount of water (ca. $< 10\%$), the paper should be presaturated by standing in a closed, humid chamber, by steaming, or by spraying with a fine mist of the stationary phase.

For impregnation with hydrophilic or hydrophobic compounds, the paper is usually drawn through a trough containing a solution of the compound in a volatile solvent, after which the solvent is thoroughly removed by air drying and blotting between filter papers [37]. Samples to be applied to impregnated papers should be contained in a solvent that will not dissolve the impregnating agent, if possible.

The need for elaborate precautions for saturation of the paper with water (or other stationary phase), the mobile phase with stationary phase, and the chamber with the vapors of both phases should be evaluated in each case to be sure they lead to improve reproducibility. For example, Cassidy [38] has shown that even after taking great pains to achieve maximum equilibration, the mobile phase composition varies along the paper, sometimes leading to a double solvent front. Especially with low volatility solvents and ascending development, prior saturation or equilibration is usually not now used for normal analytical work.

5.3.5 Development methods

The different modes and equipment for paper chromatography will be described in this section. Because of its simplicity, PC can be carried out in almost any available covered jar, tank or tube. The different geometric modifications, namely ascending, descending and horizontal, give generally the same results as the originally described development arrangement of Consden et al. [4], which is probably still the most frequently used technique.

5.3.5.1. Descending development

The end of the paper nearest the origins is placed in a trough supported in the upper part of a well-sealed tank. The paper is anchored by a glass rod and is folded over an antisiphon rod supported above the trough (Figs. 5.7 and 8). The solvent used as the

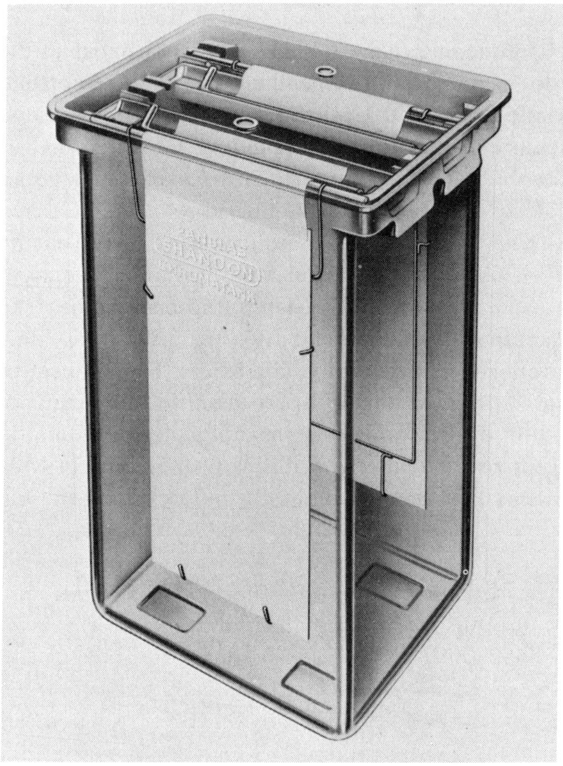

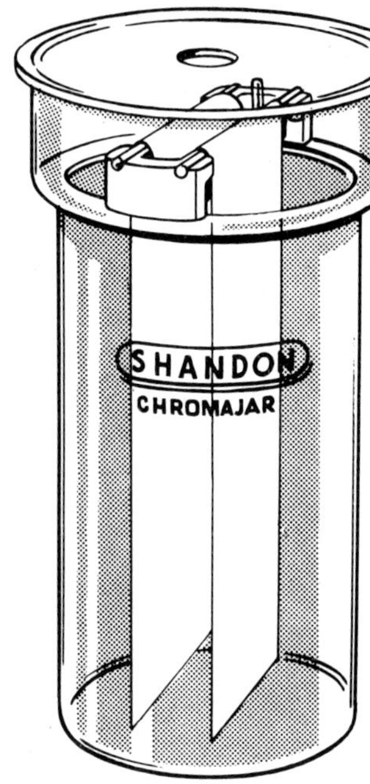

Fig. 5.7. 'Panglas' all-glass Chromatank Model 300 (Shandon Southern, Sewickley, PA) with troughs and antisiphon rods supported at the top for simultaneous descending development of two paper sheets.

Fig. 5.8. Shandon Chromajar tank for descending runs with two papers 10 × 30 cm (shown) or for ascending runs on a paper sheet 25 cm square that is clipped into a cylinder.

stationary phase can be placed in the bottom of the tank to equilibrate the spotted paper prior to adding the mobile solvent to the trough, or the development can be started at once. The mobile solvent flows down the paper under the influence of gravity, and development can be stopped after a certain distance or carried out continuously (overrun development). Many PC solvents develop at a rate of 25 cm in 2–4 h, while others require 10 h for a 20 cm run. In the latter case with compounds having R_F values below 0.8, overrun development overnight is usually convenient. The paper is cut in a saw-tooth design for this overrun development, and a colored standard should be chromatographed with the sample to indicate the movement of the solutes and against which to refer R_F values.

Various types of chambers, troughs, stands, holders, and lids have been devised for descending PC. Commercial metal cabinets for descending development of up to eight large (18 × 22 inch) sheets are available, or descending development can also be carried out in small tanks or large square or cylindrical glass units (Figs. 5.7, 8).

5.3.5.2 Ascending development

This procedure can be as simple as attaching (e.g., stapling) a strip of paper to the underside of a cork, applying a dot of sample a few centimeters from the bottom, and dipping the strip into the mobile phase in the bottom of a test tube, graduated cylinder or Erlenmeyer flask which is closed with the stopper.

Tanks similar to those for descending development can be used, with paper sheets, suspended from the lid, a glass rod, or a frame, dipping into solvent held in a tray on the floor of the tank. Alternatively, the paper can be stood in solvent by rolling it into a cylinder held together by staples or plastic clips [7]. This allows wider sheets to be developed in low volume cylinders or tanks. Special frames can accomodate several sheets for simultaneous development [39] (Fig. 5.9).

Spring-loaded hanger rods, available commercially, e.g., from Fisher, can be placed inside and at the top of standard TLC tanks. Papers are clipped to the rods and dipped into solvent contained on the bottom of the tank.

In ascending development solvent rises by capillarity but is restricted by gravity. The rate of ascent, therefore, slows as the distance increases, and development will

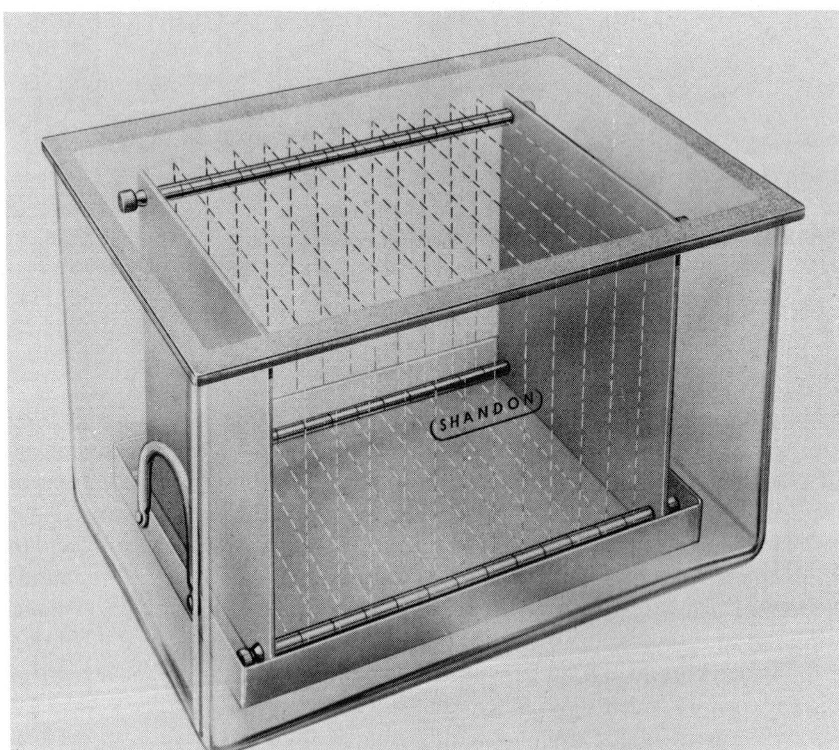

Fig. 5.9. Shandon Universal Apparatus assembled for ascending chromatography. The frame, holding up to 12 papers, sits in the solvent in the bottom of the tray.

stop completely in an equilibrated, sealed tank when the solvent reaches the top of the paper. If equilibrium has not been obtained in the chamber, development will continue as the mobile phase evaporates off the top of the paper. In this case, a standard should be included for calculation of relative R_F values. Squares of paper are conveniently developed by the ascending technique for two-dimensional chromatography. R_F values can vary significantly, especially with more-volatile mobile phases, when ascending or descending development is used. Usually either approach can be made to solve a given separation problem, and the choice usually depends upon personal preference and the equipment available.

5.3.5.3 Horizontal and radial development

In this method, a square or rectangle of paper is laid horizontally on glass rods in a shallow, covered tray-type tank ('Speedi-Cell', Schaar and Co., Chicago), and the end of the paper nearest the origins is dipped into a solvent trough [40]. The compact size of the apparatus allows development in an oven or refrigerator at various temperatures.

Development can also be between glass or aluminum plates. This method is especially advantageous for volatile compounds, and generally increased reproducibility of R_F values has been claimed [37]. A small chamber for fast, horizontal PC has been described [41].

Radial or circular development is a special adaptation of horizontal chromatography. Zones are especially narrow and well resolved, but development times are long because the solvent is fed through a wick. Samples are placed at the center of the paper at the junction of the wick, and development is carried out by stretching the paper between two petri dishes, the top one serving as a cover and the bottom one as a solvent reservoir into which the folded-down wick is dipped [42]. A desiccator can be used as a larger tank in the same manner. Spots migrate from the center toward the edge in the shape of circular zones [43].

If multiple samples are applied in a circle (ca. 1–2 inch diameter) about the center of the paper where the wick joins and solvents flow in, the individual samples form segments instead of concentric circles. This allows comparison of standards and unknowns on the same sheet [44,45].

A method combining the principles of linear and radial chromatography which produces semi-circular zones is the wedge-strip method of Matthias [46]. The wedge-shaped strip is not in a horizontal position but is mounted in a frame by means of two glass rods. The spot is placed on the narrow end of the strip, and solvent flows in ascending development through this spot and resolves difficult mixtures into narrow bands which are better separated than with usual linear development. A multiple or overrun development is often used, leading to lengthy analysis times. Specially shaped strips for this method are commercially available. Circular methods are today a rarity, despite their often improved resolution, because of slow speed and difficulty in comparing R_F values.

5.3.5.4 Multiple development

Unidimensional multiple development enhances the separation of compounds with low (< 0.2), very similar R_F values. The same solvent is used to repeatedly develop the paper in the same direction, with complete drying of the solvent between runs [47]. When the paper is impregnated with an organic liquid, the drying period is shortened to remove only the volatile mobile phase. The theory of unidimensional multiple development has been described [48], and the method has been used to analyze steroids [49] and amino acids [50] among other classes of compounds. Fractional chromatography is a special variation of multiple development in which a solvent of increasing polarity is employed for each subsequent development [51].

5.3.5.5 Two-dimensional development

The usual two-dimensional chromatography is a multiple development method in which the sample is spotted in the corner of a square of paper and ascending or descending development is made successively with two different solvents in two directions perpendicular to each other. Between developments, the paper is thoroughly dried to remove all traces of the first solvent. The best separations of complex mixtures are obtained if different types of solvents are employed.

Two-dimensional paper chromatography of standard mixtures produces chromatography 'maps' that may be kept as permanent records. Normal biological samples, e.g., urine, are run, and the maps are retained so that patterns of certain important compounds, such as amino acids, are available for reference. In this way, abnormal spots can be identified easily when they occur.

5.3.5.6 Miscellaneous techniques

Other methods only occasionaly used are centrifugally accelerated circular chromatography [52], low [53] and high [54] temperature development, development in gaseous atmospheres other than air [55], solvent [56] or stationary phase [57] gradient PC, and rear phase chromatography [58]. In the latter, the spot is applied after the solvent has moved some distance past the origin so that the effects of solvent demixing are obviated.

5.3.6 Drying of the chromatogram

Paper chromatograms developed with volatile solvents can be completely dried in 10–30 min if they are hung in a fume hood with an efficient draft. Papers impregnated with reagents such as formamide or developed with poorly volatile solvents such as phenol are better dried in a forced-air heated drying oven, with heated air flow from a hair dryer, or over a hot plate, if the solutes are not heat labile. The final traces of some solvents, such as acetic acid and hydrochloric acid, are particularly difficult to remove. In some cases, traces of impurities may inevitably remain that will interfere with some detection reagents, e.g., ammonium formate on formamide-impregnated paper, which interferes with the use of $SbCl_3$.

5.3.7 Detection of zones

The paper chromatographic zones must be detected or visualized if they are colorless, preferably without being destroyed, to allow the maximum number of experimental procedures. Many substances are visualized by absorption of or fluorescence in 254 or 365 nm ultraviolet light; this also provides some characterization. Other compounds can be detected only after a reaction in which they form a colored or fluorescent compound. The specific and nonspecific reactions are numerous [59,60]. The reagent can be applied by dipping or spraying. Dipping, which can be carried out in a shallow tray, gives uniform coloration but may spread the spots, especially if the papers are not thoroughly dry. Spraying, which should be carried out under a hood, can eliminate spreading but may not give the uniformity of spot density or background required for in situ quantitation.

Color reagents may be general or react specifically. The produce colors that are visible in daylight and, in some instances, under ultraviolet light. The maximum amount of information is obtained by applying, if possible, three or four compatible reagents in sequence to a single chromatogram (multiple dipping or spraying technique) (see below). Enzymatic, biologic, and radioisotopic detection methods may also be used.

Important detection reagents for paper chromatography include iodine vapor (general reagent giving yellow to brown spots); acidified or alkaline potassium permanganate (general oxidizing agent producing yellow to white spots on a violet background); aqueous or alkaline solutions of acid-base indicators (for general detection of acids or bases); diphenylhydrazine (reacts with aldehydes and ketones); ninhydrin (reacts with amines and amino acids); ammoniacal silver nitrate (gives brown to black spots for many reducing compounds); Dragendorff reagent (for alkaloids and basic substances); and Carr-Price reagent ($SbCl_3$ in chloroform) (for unsaturated lipophilic compounds). See Refs. 59 and 60 for a complete listing on detection reagents.

The solvent in which the detection reagent is dissolved can affect results significantly. For example, acetone appears to be the best solvent, in terms of rate of color formation, intensity and exact color produced, for ninhydrin reagent. Pyridine acts as a ninhydrin color stabilizer and so is often included when detecting amino acids. In general, acetone is a superior solvent for many PC detection chemicals with which it is compatible.

Multiple detection techniques [61] can be used to obtain the greatest possible amount of information on diverse types of chemical compounds present on a chromatogram. An example for nitrogen-containing solutes is the use of ninhydrin to locate amino acids, Ehrlich reagent to detect indoles, and anisidine for imidazoles and hydroxyanisoles. Spots should be marked with a pencil after each reagent is applied, and exact colors, color changes and times for appearance and fading of colors should be carefully noted for maximum reliability. Viewing under short- and longwave UV light at each stage adds to the information obtained by use of sequential detection reagents.

Chromatograms that disintegrate (i.e., those to which acid reagents have been applied) or whose spots fade with time may be documented by photography or xerography, or by copying by hand on thin transparent paper using colored pencils to faithfully reproduce colors. The copy of the chromatogram should record the size and shape of zones, their intensity, and their position relative to the solvent front or a reference compound.

5.3.8 Qualitative identification of zones

After visualization of zones, R_F values are measured, and tentative identification of analytes is made by comparison of these values with those of standards processed on the same chromatogram or under the same conditions. The use of selective detection reagents, alone or in sequence, aids in confirmation of identification, as does elution followed by UV, IR, NMR or mass spectrometry or gas chromatography, if sufficient sample can be recovered.

R_X values are often used instead of R_F values to describe the migration of a substance, where R_X is the ratio of the distance traveled by the substance divided by the distance traveled by a standard substance X spotted on the same chromatogram. R_X values must necessarily be used in overrun chromatography, where there is no solvent front that can be measured, and they are used for better reproducibility in all modes of PC. This is because factors that influence the reproducibility of R_F values, such as temperature, grade of paper, direction of development, and length of solvent flow, will affect both the analyte and standard more or less uniformly, thereby allowing the R_X value to be less variable than the absolute R_F.

5.3.9 Quantitative PC

Paper chromatography is mostly employed as a qualitative analytical tool. Semi-quantitation can be achieved by visually comparing the size and/or intensity of sample and standard spots chromatographed on the same paper. Accuracy and precision range from ca. 10–30% for the visual approach. More exact results are obtained by measuring spot areas, eluting the spot and determining by a microanalytical technique, or scanning the spot in situ with a densitometer. Accuracy and precision within $\pm 5\%$ are possible with the latter two approaches using optimized techniques and equipment. The major source or error in quantitation by densitometry on paper will be the application of the sample. Application with a Microcap pipet or a commerical automatic spotter or streaker can overcome this source of poor accuracy and precision [62].

With many scanners, the paper chromatogram is sandwiched between glass plates and then handled in a manner identical to a thin-layer chromatogram. Measurement of visible light, UV-excited fluorescence, or fluorescence quenching have been made. In general, reflected light leads to more regular baselines when scanning paper chromatograms. For transmission measurements, it may be advantageous to make the paper transparent by drawing it through paraffin oil in order to compensate for uneven background [63]. Calibration curves are constructed for each analyte, and the quantity in the sample is interpolated from the plot.

A unique feature of quantitative PC is the method for elution of substances from the paper. Unless the spot is colored, fluorescent or absorbs UV light, it must be located on pilot strips from the sides of the chromatogram on which the substance of interest was spotted. A detection reagent is applied to these strips to locate the spot, and the corresponding location on the chromatogram is then eluted. This is accomplished by swirling the cut-out spot with solvent in a small beaker, cutting the spot area to a point and eluting between microscope slides into a beaker [64], by descending development of the paper strip in a special elution cabinet [65], or by Soxhlet extraction of the area of paper. The eluted compound can be determined by a wide variety of microanalytical methods, e.g., electroanalytical, spectrometric, gas chromatography, biological or radiometric.

If the detection reagent is applied to the entire chromatogram directly, the spot can be immediately and accurately located by its absorbance (color) or fluorescence and the product determined by spectrophotometry after elution. The reagent must be uniformly applied, as by dipping the paper, and any heating carefully controlled.

Impurities which may be eluted from the paper and interfere with the determinative step should be removed by prewashing the paper with appropriate acid solutions and/or organic solvents.

Radioactivity can also be measured for quantitation of paper chromatograms. Paper chromatograms may be assayed for radioactivity by cutting into sections for separate measurement (termed zonal scanning). Sectioning can be at regular intervals over the whole chromatogram or as dictated by previous detection methods. Each section can be eluted and assayed, but this method is tedious and often subject to unreliable recovery. The paper can be combusted and the products collected and counted [66]. Or, the sections are counted directly with a Geiger-Mueller device or, more often, by liquid scintillation methods. Some detection reagents (H_2SO_4 for charring, silver nitrate at high concentration, iodine vapor) cause quenching and cannot be applied to the paper prior to assay [67]. A solvent-scintillator mixture which elutes the radioactive material from the paper completely should be used for highest efficiency. The paper section should be small and lie flat on the bottom of the counting vial for best precision in counting [68]. A 0.5 ml volume of scintillation liquid has been found sufficient for counting glass fiber paper sections [69]. Other aspects of scintillation counting for PC are similar to those for TLC, e.g., correction for self absorption and quenching [70] and selection of the scintillator mixture [67,70]. Scintillators can also be incorporated directly in the paper [71].

5.3.10 Applications (Tables 5.1 and 5.2) (See also Note following References, p. 413)

Space limitations preclude an exhaustive review of all applications of paper chromatography and the presentation of detailed procedures. The literature includes many thousands of references describing separations and determinations of all types of organic and inorganic mixtures, so that scientists interested in a particular

TABLE 5.2

Applications of paper chromatography to various classes of compounds

Compound	Derivative	Stationary phase [1]	Mobile phase [2]	Ref.
Aglycones of flavones and flavonols	–	–	Benzene-acetic acid-H_2O (125 : 75 : 3)	77
Alcohols	3,5-Dinitrobenzoates	Formamide or DMF	Hexane or cyclohexane	78
	Xanthates	10% K_2CO_3	*n*-Butanol-30% K_2CO_3	79
	4-Dimethylaminobenzene--4′-azobenzoates	DMF	Petroleum ether	80
	2,4-Dinitrobenzyl ethers	50% DMF	Cyclohexane	81
Aldehydes and ketones	2,4-Dinitrophenyl-hydrazones	25% DMF	Cyclohexane	82
	–	–	Isobutanol-benzene-H_2O (1 : 9 : 10)	83
Aliphatic amines	3,5-Dinitrobenzoates	Formamide	Hexane-benzene (2 : 3 or 3 : 2)	84
Aliphatic mono- and diamines	–	–	Ethanol-HCl (1 : 1); *n*-propanol-HCl (2 : 1); *n*-butanol-HCl (4 : 1); ethyl acetate-DMF (2 : 1)	85
	–	0.1 M HCl-0.1 M KCl--H_2O (2 : 4 : 3)	Impregnation solution-phenol (1 : 4)	86
Alkaloids	–	–	*n*-Butanol-acetic acid-H_2O (12 : 3 : 5)	87
		5% Sodium dihydrogen citrate	*n*-Butanol-5% citric acid (9 : 1)	87
Alkylbenzoquinones and hydroxybenzoquinones	–	40% DMF or formamide	Heptane or benzene	88
N-Alkyl and *N*,*N*′-dialkylthioureas	–	–	*n*-Butanol-ethanol-H_2O (4 : 1 : 1) or $CHCl_3$-methanol-H_2O (7 : 5 : 1)	89
Alkyl phosphates	–	–	*n*-Propanol-2 M NH_3 (7 : 3)	90
Amino acids	–	–	Phenol-H_2O (160 g: 40 ml); *n*-butanol-pyridine-H_2O (1 : 1 : 1); *n*-butanol-acetic acid-H_2O (12 : 3 : 5)	91
	Dansyl	–	$CHCl_3$-Ethanol-acetic acid (38 : 4 : 3)	92
Aminosulfonic acids	–	–	*n*-Propanol-NH_3 (2 : 1); *n*-butanol saturated with 2.5 M HCl	93

Anthocyanins	–	–	Butanol-acetic acid-H_2O (6 : 1 : 2)	94
Anthraquinones	–	Paraffin oil	60% Acetic acid	95
		1-Bromonaphthalene	Pyridine-H_2O (1 : 1 or 2 : 1)	96
Aromatic nitrohydro-carbons	–	Kerosene	Ethanol-acetic acid-H_2O (20 : 1 : 14)	97
Azaheterocyclic hydrocarbons	–	–	1% Acetic acid; 6 M HCl; conc. NH_3; pyridine-H_2O (1 : 1 or 2 : 1)	98
	–	–	Formamide; DMF-H_2O (35 : 65)	99
Azo derivatives of bilirubin	–	–	Isobutyl acetate-acetic acid-H_2O (10 : 8 : 5)	100
Carboxylic acids	–	–	*n*-Butanol saturated with 1.5 M NH_3	101
	2,4-Dinitrobenzyl esters	25% Formamide	Cyclohexane-benzene (20 : 1)	102
Carotenoids	–	Silica gel or alumina impregnated paper	Hexane; hexane-acetone (99 : 1-80 : 20)	103
Cholesterol and its esters	–	Undecane	Acetic acid-acetonitrile (1 : 3)	104
Corticosteroids	–	Ethylene glycol	Toluene; carbon tetrachloride	105
Coumarins	–	–	*n*-Butanol saturated with 25% NH_3; *n*-butanol-acetic acid-H_2O (4 : 1 : 5)	106
Digitalis glycosides	–	Formamide	Benzene-$CHCl_3$ (7 : 5)	107
Estrogens	–	Formamide	Benzene-$CHCl_3$ (4 : 1)	105
Fat soluble dyes	–	Paraffin oil	Methanol or ethanol-H_2O (4 : 1); acetone-H_2O (7 : 3)	108-110
Glycerides	–	Undecane	Acetone-acetonitrile (7 : 3 or 8 : 2)	111
Hydroxy and dicarboxylic acids	–	–	Ethanol-NH_3-H_2O (80 : 4 : 16)	112
Indole derivatives	–	–	*n*-Propanol-NH_3-H_2O (10 : 1 : 1)	113
Keto-acids	–	–	H_2O saturated *n*-butanol-formic acid (95 : 5)	114
	2,4-Dinitrophenylhydra-zones	–	*n*-Butanol-ethanol-0.5 M NH_3	115
Mono-, di- and triglycerides	–	Silica gel impregnated paper	Petroleum ether-diethyl ether (95 : 5)	116
Napthylaminesulfonic acids	–	–	*n*-Butanol-acetic acid-H_2O (4 : 1 : 5); *n*-butanol-pyridine-H_2O (3 : 1 : 1)	117
Nitrobenzoic acids	–	–	Butanol-pyridine-H_2O (5 : 3 : 3)	118
Nitro esters	–	Formamide or dioctyl phthalate	Hexane-ethyl acetate (10 : 1); ethanol-H_2O (7 : 3)	119

TABLE 5.2 (continued)

Compound	Derivative	Stationary phase [1]	Mobile phase [2]	Ref.
Nucleic acid derivatives	–	–	Butanol-acetone-5% NH_3-acetic acid-H_2O (9 : 3 : 2 : 2 : 4)	120
Phenols	–	–	Acetonitrile-0.1 M ammonium acetate (7 : 3 or 6 : 4), pH 7 or 4.5	121
	–	25–50% DMF	Hexane	122
	–	Paraffin oil	Methanol-H_2O (3 : 2 or 4 : 1)	122
	–	Formamide	Xylene	123
	–	Silica gel	Toluene-propionic acid-formic acid (5 : 4 : 1)	124
	–	Stannic molybdate	1.0 M $NaNO_3$, pH 6	125
Phospho- and glycolipids	–	Silica gel impregnated paper	Diisobutyl ketone-acetic acid--H_2O (40 : 25 : 5)	126
Polycyclic aromatic hydrocarbons	–	Acetylated paper	propionic acid-methanol-H_2O (1 : 1 : 1)	127
	–	DMF	Hexane	128
	–	1-Bromonaphthalene	90% acetic acid	129
Polyphenols	–	Formamide	$CHCl_3$-Ethyl acetate (1 : 1)	130
	–	–	*n*-Butanol-benzene-acetic acid-H_2O (2 : 10 : 2 : 1)	131
Primary aromatic amines	–	Formamide	heptane; benzene	132
Purines, pyrimidines, nucleosides	–	–	*n*-butanol-acetic acid-H_2O (50 : 25 : 25); ethyl acetate-formic acid-H_2O (7 : 2 : 1)	133
Pyridine and homologs	–	–	*n*-Propanol-HCl-H_2O (20 : 1 : 2)	134
	–	–	*n*-Butanol-HCl-H_2O (20 : 1 : 2)	135
Quinolines	–	–	1% acetic acid; 6 M HCl; conc. NH_3; 10% aqueous pyridine	136
Saponins (acid)	–	–	*n*-Butanol-ethanol-NH_3 (10 : 2 : 5)	137
Saponins (basic)	–	–	Ethyl acetate-pyridine-H_2O (3 : 1 : 3)	137
Saponins (neutral)	–	Formamide	$CHCl_3$-THF-Pyridine (10 : 10 : 2)	137
Secondary aromatic amines	–	1-Bromonaphthalene	5% Acetic acid	138
	–	Dimethyl phthalate; chlorinated paraffin	Heptane; isopropanol	139

Steroids	–	–	Dichloroethane-ethanol (92.5 : 7.5) and diisopropyl ether-ethanol (95 : 5) (2D PC)	140
	–	Triethylene glycol	Cyclohexane; tetrachloromethane; benzene	141
Steroid acids	–	–	*n*-Propanol-NH_3-H_2O (90 : 2 : 8); $CHCl_3$-methanol-NH_3 (1 : 1 : 1)	107
Sterols	–	Paraffin oil	84% Acetic acid; 85% pyridine	105
Sugars	–	–	Phenol saturated with H_2O (NH_3 atmosphere)	142
	–	–	*n*-Butanol-pyridine-H_2O (6 : 4 : 3)	143
	–	–	Ethyl acetate-pyridine-H_2O (2 : 1 : 5)	144
	–	–	Butanol-acetic acid-H_2O (4 : 1 : 5, upper layer)	142
	Phenylosazones	–	Dioxane-heptane-H_2O (2 : 1 : 1)	145
Sulfides	*p*-Nitrophenylsulfonyl-sulfimides	Formamide	Benzene; benzene-cyclohexane (3 : 2)	146
	p-tolylsulfonylsul-fimides	Paraffin oil	$CHCl_3$-H_2O-paraffin oil (25 : 25 : 1)	147
Sulfonamides	–	–	Isopropanol-NH_3 (4 : 1); $CHCl_3$--acetone-NH_3(18 : 100 : 20)	148
s-Triazines	–	–	*n*-Propanol-NH_3 (2 : 1)	149
Vitamins A	–	Kerosene	Acetic acid-H_2O (91 : 9)	150
	–	Vaseline	Methanol-H_2O (90 : 10)	151
Vitamins B	–	–	*t*-Amyl alcohol-H_2O-*p*-toluene--sulfonic acid (60 : 30 : 2)	152
Vitamin C	–	–	Isopropanol-H_2O (7 : 3); *n*-butanol-acetic acid-H_2O (4 : 1 : 5)	153
Vitamin D	–	Olive oil	Acetone-H_2O (75 : 25)	154
Vitamin K	–	Paraffin oil	Methanol; ethanol	155

[1] Plain, unimpregnated cellulose paper unless an impregnating agent is indicated.
[2] (v/v) proportions unless otherwise indicated.

Deyl (ed.) Separation Methods

CHAPTER 6

Electromigration techniques

Z. DEYL [a] and J. HOŘEJŠÍ [b,*]

[a] *Physiological Institute, and* [b] *Institute of Molecular Genetics, Czechoslovak Academy of Sciences, Prague, Czechoslovakia*

6.1 Introduction

The world of electromigration separations is sharply divided into two areas. Zone electrophoresis on paper and related procedures have (in spite of their wide applicability to diverse organic compounds) already passed their period of favour. The other branch is represented by the more recent techniques some of which have already became widely accepted (such as isoelectric focusing or separations in polyacrylamide gel) and the others that are at the moment in the centre of a rapid development like displacement electrophoresis (isotachophoresis). This chapter is devoted mainly to analytical procedures such as these which are governing the area of electromigration separations at the moment with a single exception: flow deviation (curtain) electrophoresis which will be discussed in more detail because it offers several new dimensions in the separation field. The other preparative procedures are summarized only briefly.

The approach to any separation (including electromigratory) is governed by the possibilities of the laboratory; some of the electromigration techniques suffer from the fact that appropriate commercial equipment is not available. Therefore some rather simple but efficient procedures are included in this survey and examples of some home made equipment are presented.

6.2 Zone electrophoresis

6.2.1 Paper electrophoresis

Paper electrophoresis represents one of the most early developed electromigration techniques. The first references go back to 1950 [1–6]. The flourishing era of paper electrophoresis has already passed; the main reason being a shift to other more

* Dr. J. Hořejší contributed the section on affinity electrophoresis (6.8)

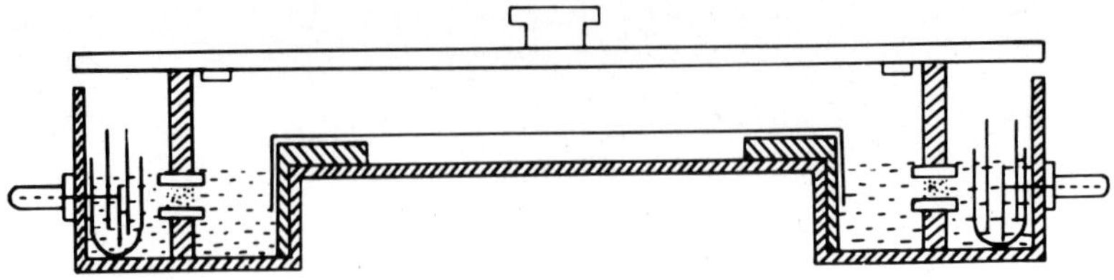

Fig. 6.1. Schematic representation of horizontal paper electrophoresis cell (according to Ref. 6).

convenient media for zone electrophoresis (paper was certainly not the most suitable medium for the analysis of large molecules like proteins, nucleic acids, etc.) On the other hand it may keep its dominating position in the separation of low molecular weight compounds were there not more less versatile and rapid procedures available (such as HPLC).

6.2.1.1 Equipment for low and high voltage paper electrophoresis

Apparatuses used for paper electrophoresis consist of two electrode jars and a system that holds the paper sheet between them and prevents, at least partly, the imbedded buffer from evaporation. The equipment used could be categorized into two groups with either horizontal or vertical arrangement of the paper strip (Figs. 6.1, 6.2 (Ref. 7, 8)). In the horizontal arrangement the paper strip is clamped between at least two glass rods or nylon threads and covered with a plexiglass lid. Its ends are equally immersed into the electrode jars, that are filled with an appropriate buffer to equal height (to prevent siphoning and distorting of the electrophoretic pattern). In the vertical version the filter paper strip is folded in two and held in this position by a suitable trough keeping thus a profile of inverted 'V'. The supporting trough is made of glass or a suitable plastic that withstands the action of electrode

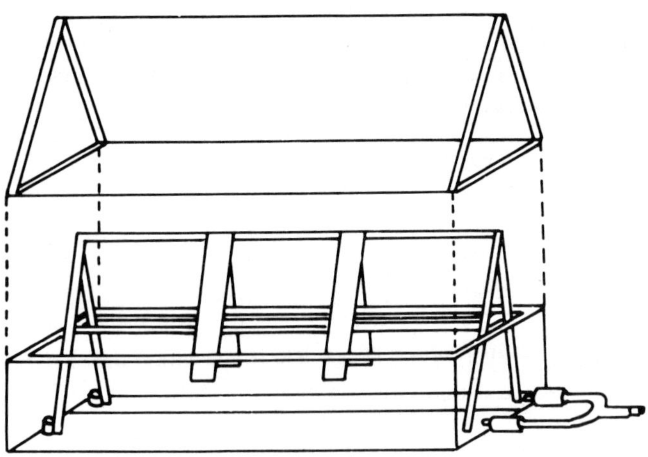

Fig. 6.2. Hanging strip type apparatus (vertical electrophoresis cell) (according to Ref. 8).

buffers. The whole assembly is covered by a plexiglass lid. In both versions evaporation is not hindered completely but it is kept at a reasonably low level.

Electrodes are usually made of platinum wire but other materials may also be used (graphite, silver/silver chloride, palladium). Certainly, the choice of the electrode material depends on the buffer system used and the popularity of platinum is caused by its practically universal application.

If the system works with a potential gradient less than 20 V/cm the procedure is called low voltage electrophoresis. There are no additional needs to control evaporation of liquid and warming up the filter paper strips. The filter paper strip is wetted with the electrophoresis buffer and positioned into the electrophoresis equipment. The sample is applied either in the form of drops or streaks. Commercially available applicators can be used for this purpose. Streaks are used to improve resolution of two closely moving zones while dots are preferred if repeated application is needed because of a large volume of diluted sample. After electrophoresis the paper strip is removed from the apparatus and dried in a ventilated oven at 100°C.

On the contrary if the electrophoretic separation is carried at voltages above 20 V/cm (up to 200 V/cm) the procedure is designed as high voltage separation. The advantage of this procedure is shortening of the separation time; therefore there is also no time for diffusion distortion of the separated zones. The heating effect of the current applied is, however, much higher. This technique frequently allows separations that cannot be achieved by the low voltage version. Therefore all apparatuses designed for high voltage electrophoresis are equipped with an intense cooling system. Cooling is effected on principle in two different ways. In the first system [9,10] the filter paper strip is immersed into a buffer immiscible non-conducting liquid that is cooled by a cooling coil (Fig. 6.3). The second system is represented by a strip of filter paper placed between two cooled metal plates. In the immersed sheet system carbon tetrachloride, chlorobenzene, heptane, Varsol (Esso white spirit 100)

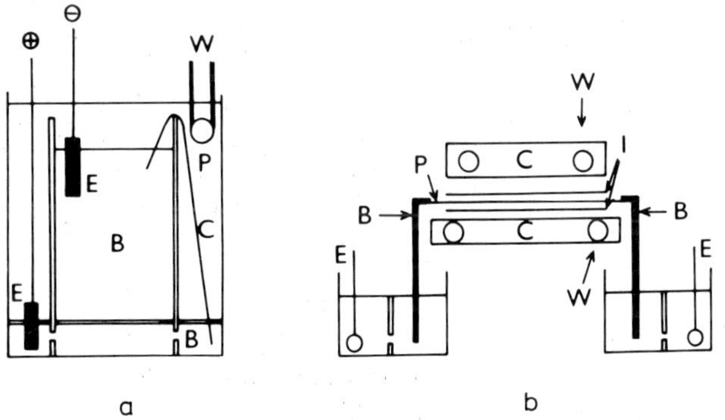

Fig. 6.3. Schematic representation of high-voltage electrophoresis apparatus. (a) Immersed paper strip (sheet) type. B, Buffer compartments; P, filter paper strip; C, cooling liquid; E, electrodes; W, cooling coil. (b) Enclosed paper strip (sheet) type. P, filter paper; B, filter paper wicks; I, polythene insulating foil; C, cooling metal plates; W, cooling coil systems.

or low viscosity silicon oils can be used as coolants. At lower voltage gradients the large volume of the coolant is sufficient to remove excess heat from the filter paper sheet. However, if the voltage is further increased, a special cooling coil, through which cold tap water is pumped, is needed.

The enclosed strip apparatus requires more skillful construction, because the cooling plate on which the filter paper sheet is positioned must be manufactured with a uniform flatness to achieve a precision within 0.002 mm [11,12]. The plate itself is usually made of aluminium and the filter paper sheet is uniformly pressed to it during separation by a pressure of about 1 atm/cm^2. Cooling is effected by tap water running through the labyrinth of channels of the cooling plate. Buffer vessels are placed at each end of the plate and are equipped with screened platinum electrodes. Contacts between the filter paper sheet on the plate and electrode vessels is materialized by means of thick filter paper wicks (Whatman 3 MM) moistened with the buffer solution. The filter paper sheet is not positioned directly on the cooling plate, but is isolated from both sides by a polyethylene insulation foil.

In the original version two cooling plates, one located below and another one above the filter paper sheet, were used. Later, however, a modification has been introduced that uses only a simple cooled plate [13–16]. The apparatus consists of a thin V shaped glass plate, which is located in the inverted position between the electrode compartments. On both sides two other plates are fitted tightly to this plate to form a cooling space (a box) into which the coolant is pumped. The whole system is covered with a plexiglass lid equipped with switching off relay as in the double plate cooled system. This apparatus can work at voltages up to 8000 V (Fig. 6.4).

6.2.1.2 Two-dimensional separations

The principle of two-dimensional separations was described by Durrum [17] and Strain [18]. There are several modifications of this technique which differ mainly in the driving force used in individual directions. Thus, one may distinguish between two dimensional electrophoresis in which electromigration occurs in two perpendicular directions in two subsequent steps, or electrophoresis in one direction may be combined with chromatography in the other direction or finally, separation can be carried out in a single run under the influence of two driving forces acting simultaneously at right angles.

The technique of two dimensional electrophoresis with two identical procedures applied in a right angle to each other, does not offer additional resolution and would be therefore useless unless there is some reaction that alters the charge of some members in the mixture of separated components performed between individual runs. Then the reacted members would deviate from the diagonal where all the other non-reacted components would be localized. An example of this procedure is that of Mikeš and Holyšovský [19] and Hartley et al. [20]. Using this procedure disulphide peptides from a tryptic hydrolysate of an enzyme were analyzed. The hydrolysate is separated in the first run on a filter paper strip. The separated peptides are then oxidized with performic acid vapour by means of which the S-S bridges are broken

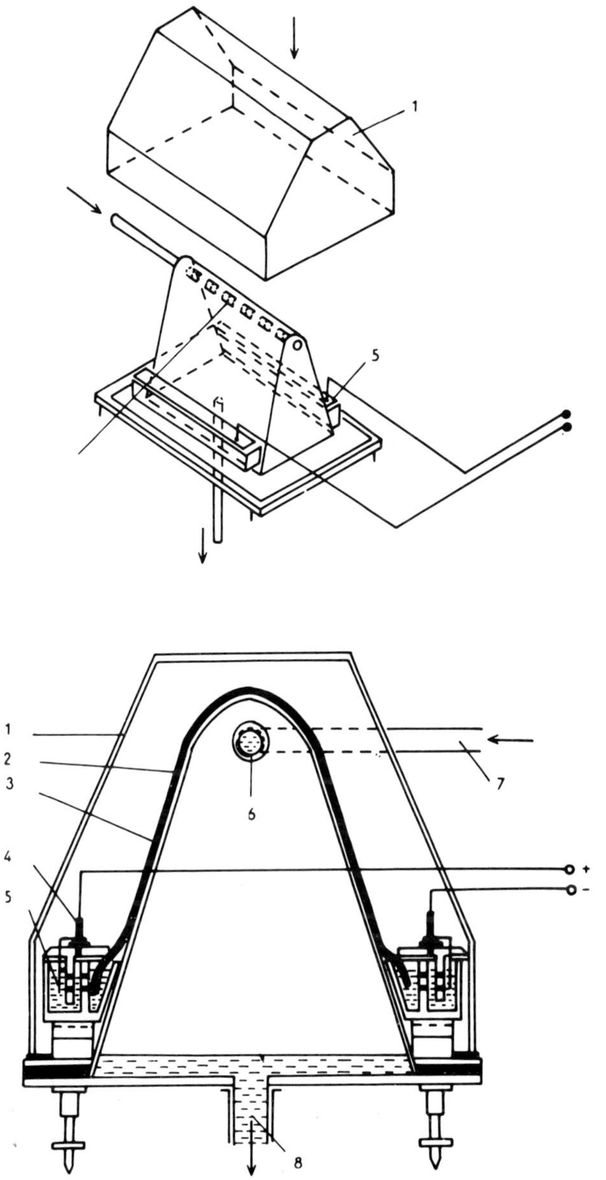

Fig. 6.4. High-voltage electrophoresis tank with one cooling plate. (a) Perspective view, (b) cross sectional view. 1, Lid from Plexiglas; 2, filter paper sheet; 3, thin glass plate bent as an inverted V; 4, platinum electrode; 5, buffer compartment; 6, 7, metal tube with small holes; 8, outlet of cooling liquid.

under the formation of cysteic acid residues. The strip is then dried to remove excess of reagent, stitched to another sheet of paper, and electrophoresed again. All the oxidized peptides will deviate from the diagonal due to the presence of an extra

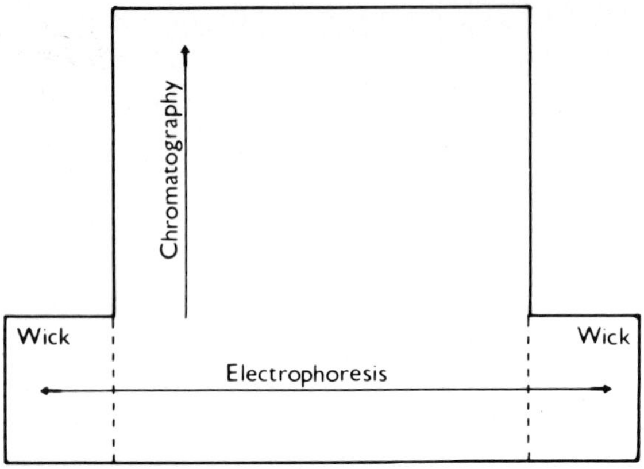

Fig. 6.5. Diagrammatic representation of cutting the paper sheet for two dimensional separation: electrophoresis and chromatography.

negative charge. This procedure is not only generally applicable to disulphide-containing peptides, but with an appropriate selective reaction it can be applied to a wide variety of compounds. Another example is treating the separated mixture of peptides after the first run with carboxypeptidase B to release the C-terminal arginine or lysine residues [21]. Again, the peptides that have undergone this alteration would deviate from the diagonal in the second dimension run.

Combination of electrophoresis and chromatography was perfected by Ingram [22], who used this technique for the separation of tryptic peptides of haemoglobin A and S, and was able to demonstrate the different peptides (and thus different sequences) in both proteins. The technique was called 'fingerprinting'; today however the term is used in a more general (perhaps incorrect) sense describing not only paper or gel two-dimensional separations but also the application of combinations of, e.g., two different chromatographic steps; the term 'peptide mapping' is also used. In the original version of fingerprinting the sample of peptides was placed on a sheet of paper cut as shown in Fig. 6.5, and separated by high voltage electrophoresis. This was followed by ascending paper chromatography, at a right angle, in *n*-butanol/acetic acid/water (200 : 30 : 75 v/v/v).

The two-dimensional techniques can be further expanded into a 'three-dimensional' procedure. This was introduced originally for the separation of peptides [23], and applied later for the separation of e.g., dinitrophenyl derivatives of amino acids. In principle the mixture of peptides is separated in the first electrophoretic run (pH 6.5) into three zones, of neutral, acidic and basic peptides. The paper sheet is then cut into three pieces containing always one category of peptides. The acidic and basic ones are separated by a second electrophoretic run at pH 2.1, while the neutral fraction is separated by peptide mapping (see above). The three (and more) dimensional separations are quite useful in specified situations, but they never reached a stage of wide popularity simply because it is frequently possible to find a less complicated system for the separation of a particular mixture.

The two-dimensional separations in which the different driving forces are employed simultaneously are reported here for completeness as, except for curtain electrophoresis, they never reached a wide range of applicability: there are (except the already mentioned curtain electrophoresis) electrophoresis with crossed currents employed simultaneously at right angles [24,25], centrifugally accelerated electrophoresis [26] and electromagnetophoresis [27].

Crossed electrophoresis with two currents acting perpendicularly was described by McDonald and Urbin [24]. The electrophoretic chamber has four electrodes, two of which are positive, two are negative. After the potential has been applied, the substances to be separated start to move along the diagonal. Therefore this technique is sometimes referred to as continuous diagonal electrophoresis. The technique was applied to the separation of DNP amino acids.

Centrifugally accelerated electrophoresis, as described by McDonald et al. [26], is based on the fact that centrifugal force drifts the electrolyte (mobile phase) from the centre of the circular paper sheet to the collecting cups, and the electric field that acts perpendicularly to the direction of the centrifugal force drifts the charged particles to the positive and negative poles.

The same author (McDonald et al. [25]) used a combination of electric and magnetic fields for the separation of neutral and charged particles in free solution. The technique was later perfected by Kolin [27]. Another possibility, namely the application of electric and magnetic fields in perpendicular fashion was used by Kowalczyk and Pompanski [28] for the separation of molecules and charged particles. The electric field used was 20 Vcm^{-1}, and the magnetic field had an intensity of 10 000 Gauss.

Cross paper electrophoresis is a technique that is suitable to discover possible interactions between two components [28–34]. The interacting components are forced by the electric force to migrate to cathode and anode. When the paths of these components cross there is a deviation if they interact, if there is no interaction the components move along unchanged paths. There are two variations of this

A

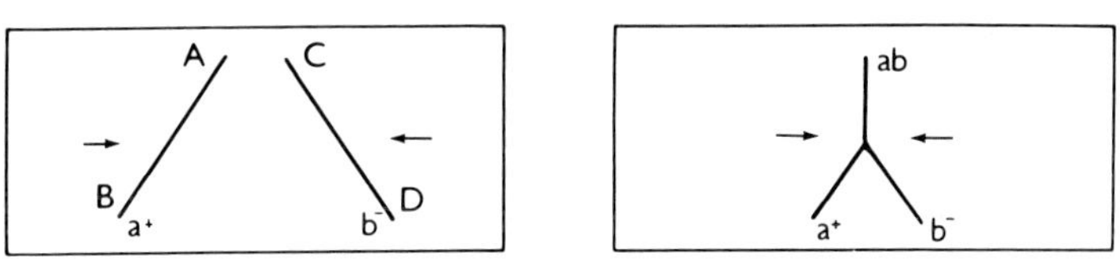

B

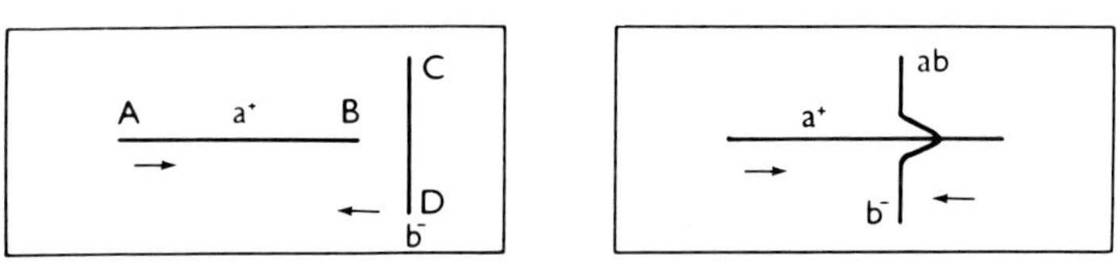

Fig. 6.6. Diagrammatic representation of the principle of cross electrophoresis. A, crossing at a moving point; B, crossing at a fixed point.

technique: one- and two-dimensional cross paper electrophoresis (Fig. 6.6). The supporting media are not limited to paper only. In the one-dimensional variation the reactants are applied to the paper sheet in the form of two streaks inclined to the electric field with an angle of about 90° between them. In the other variation the two reactants are applied also in the streak form, but one of them is parallel, the other perpendicular with respect to the electric field.

The two-dimensional version of cross paper electrophoresis resembles two dimensional separations with a buffer change before the second run. This version is applicable when one or both reactants are not pure compounds. Crossed electrophoresis has found its wide applicability in immunochemical studies, in studies about enzyme-substrate interactions, and in interactions of proteins with a number of low molecular weight compounds.

6.2.1.3 Cellulose and cellulose acetate membranes

Because of unfavorable sorption effects on paper that cause tailing, materials with lower adsorptivity were sought. Thus, cellulose acetate [35] and nitrocellulose [36,37] membranes were introduced. Cellulose acetate can be either prepared in the laboratory by treating cellulose with acetic anhydride, or it may be purchased from commercial sources. Cellulose acetate membranes are readily soluble in phenol, glacial acetic acid, dichloromethane and acetone. In part they can be solubilized in several solvent mixtures: e.g., chloroform/ethanol (9 : 1 v/v). For detection (optical scanning) the foil can be made translucent by immersion in cottonseed oil, decalin, liquid paraffin or Whitemore oil 120.

Nitrocellulose membranes must, prior to electrophoresis, be treated for 5 min with 2% Tween 60 in a veronal/citrate/oxalate buffer. Nitrocellulose membranes exhibit a selective absorptivity for proteins and nucleic acids which can be deliberately made use of. Detergent-treated nitrocellulose membranes do not absorb proteins at all and can be used at high potential gradients (up to 20 V/cm), for rather long periods of time (10–15 min). The best detection is effected by nigrosine staining.

The equipment used for both cellulose acetate and nitrocellulose separations is virtually the same as that for paper electrophoresis, allowing separations at both high and low voltages. In order to minimize evaporation, glycerol is sometimes added to the electrophoretic buffer. The applicability of high voltages requires cooling of the electrophoresis apparatus. The foils are usually quite small 5 × 1 cm, require small amounts of samples (e.g., 0.25 μl of blood for the separation of blood proteins), and the runs rarely exceed one hour running time.

6.2.1.4 Ion exchange papers

The possibility to use ion exchange papers like CM-, DEAE- and ECTEOLA-cellulose for electrophoretic separations was demonstrated by Yaron and Sober [38]. Huens and Jakubovic [39] used CM cellulose for the separation of haemoglobin variants. Several studies using paper impregnated with inorganic ion exchangers were reported.

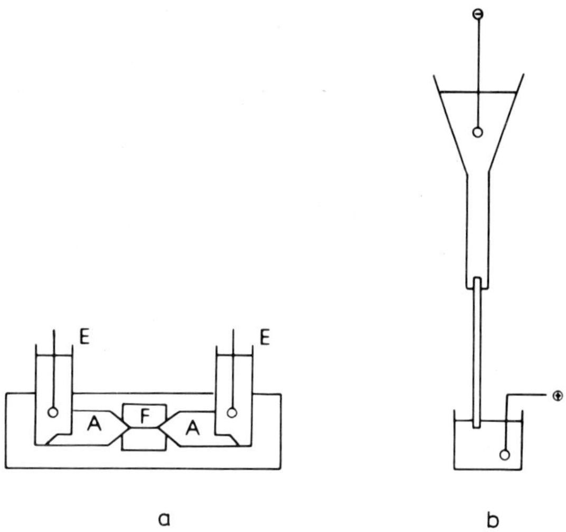

Fig. 6.7. Schematic representation of two types of apparatus for ultramicro-electrophoresis. (a) Ultramicro-electrophoresis on a single fibre of polyacrylamide gel. A, agarose bridges; E, electrode vessels; F, polyacrylamide gel fibre on the microscope cover glass. (b) Ultramicro-electrophoresis in polyacrylamide gel in a vertical capillary.

In general, most of the electrophoretic separations on ion exchange papers has been used for the separation of oligonucleotides, although in the original version of Sanger et al. [45] the electrophoretic separation was done on cellulose acetate strip, and the final resolution was achieved by a second electrophoretic step on DEAE cellulose, in further papers DEAE-cellulose was used for the separation in both directions. Attempts were also reported to separate peptides on the same ion exchanger, and other ion exchange materials, such as polyethyleneimine cellulose or aminoethylcellulose, were introduced [46,47]. The advantage of these materials is much lower fragility and higher wet strength than with DEAE-cellulose. Because of this reason paper-thin layers of this ion exchanger are used instead of DEAE cellulose. The main field in which ion exchange materials are still in use is the separation of oligonucleotides and sequencing of nucleic acids and their fragments [48–55].

6.2.1.5 Ultramicroelectrophoretic methods (Fig. 6.7)

These procedures require such a low amount of material that proteins or nucleic acids present in a single cell can be separated [56–57]. For the separation of RNA and DNA cellulose fibres (20 nm × 10–20 μm) are placed on a quartz glass, washed with alkali and treated with a viscous buffer, and imbedded in a paraffin oil chamber; the sample is applied by means of a micromanipulator and separation is carried out within 5–10 min at 1800 V. Detection is done by photographing the thread in UV light. Cellulose acetate microstrips can also be used for microelectrophoretic procedures: Marchalonis and Nossal [58] used this technique for the separation of immunoglobulins from a single myeloma cell, Tufts [59] and van Dijk et al. [60] separated inorganic cations. Detection was done either by radioactivity measurement (immunoglobulins) or by spraying the microstrips with rubeanic acid (inorganic cations). Also, threads of polyacrylamide gel were used in this category of separations [61]. In this case the gel thread is placed between two microscope slides and connected to electrode jars through agarose bridges. Separation is run at 1000 V/cm for 4–5 min. Detection is done by observing the gel in monochromatic light. Hyden et al. [62] used polyacrylamide gels packed in capillaries (50 mm long, 200 μm in diameter, vol 1–5 μl). Voltage gradient used was 80 V/cm and the separation time was 15 min. The amount of separated proteins is $3 \cdot 10^{-11}$ g and may be further increased if tritium-labelled proteins are used.

Aside from the above described home-assembled equipment, there is one, called Microzone equipment (Beckman) that is available commercially. It is constructed for microelectrophoresis on cellulose acetate, and miniaturization is achieved simply by shortening the strip and using relatively high voltages for separation (30–40 V/cm). The procedure takes usually 20–40 min and is derived from the constructions of Grunbaum and Kirk [63] and Grunbaum et al. [64]. Similar equipment is available under the name Pheno-slide system (Milipore Co.). However the strips used here are 2.5×7.5 cm in size which is considerably larger (and requires more sample) than the above described ultramicro equipment.

6.2.1.6 Electrophoresis in non-aqueous buffers

Electrophoresis in non-aqueous buffers introduces several new aspects into the separation process [65].

(1) It can decrease dissociation of compounds to be separated and may result in better differentiation of individual species in the electromigratory separation.
(2) Low dielectric constant of the buffer system may promote ion pairing between charged groups of the solute and counterions of the buffer system. These complexes may help better separation of individual compounds.
(3) Non-polar solvents help the resolution of relatively hydrophobic compounds that are not susceptible to separations in pure aqueous media (constituents of cell membranes).
(4) Because of the low conductivity high voltages can be applied without special requirements for cooling.

In principle two types of electrode buffers are in use: dry, non-aqueous solvents, or mixtures of organic solvents with aqueous buffers. Paper serves as supporting medium for electrophoresis.

The first type of electrophoretic buffer (nitromethane-acetic acid, absolute ethanol-dimethylformamide, sucrose in anhydrous formamide) has been used for the separation of sterols, fatty acids, dyes, nucleotide phosphates and oligonucleotides. High voltage electrophoresis in liquid ammonia has been applied to the separation of complexes of inorganic cations [66–69].

Organic buffer systems with decreased polarity are mainly used for the separation of amino acids, peptides and nucleotides. Pyridine, ammonia, acetic acid, formic acid, formamide, phenol, glycerol and polyethylene glycol are the most frequent components of these buffers [70–74].

6.2.2 *Thin-layer electrophoresis*

The advantages of the thin layer over the paper electrophoretic technique are to a considerable extent similar to those seen when comparing paper and thin-layer chromatography, e.g., better resolution, smaller amount of the sample, and the possibility to use aggressive detection reagents. In addition there are some features typical for electrophoresis: better resolution results in better fingerprints and separations can be carried out at relatively high potentials at relatively high voltages.

Equipment used for thin-layer electrophoresis is practically the same as that used for separations on paper, cellulose acetate and ion exchange papers. Usually cooling is introduced into the equipment. Thus, e.g., a water-cooled flat aluminium (dural, brass) block is insulated by means of a replaceable glass plate or plastic film. The thin-layer plate is located on the insulated block and paper wicks ensure the contact with electrode vessels. A polyethylene sheet is placed between the top of the layer and the lid. This protects the surface of the layer from drops of condensed moisture that condense on the lid. A sheet of plate glass covers the whole chamber. If precoated plastic thin-layer sheets are used, the apparatus of the immersed type can be used [78]. Elution of spots, and the general practice in thin-layer electrophoresis is the same as in thin-layer chromatography.

Practically all types of commercially available thin-layer plates and precoated sheets can be used: silica gel, cellulose, kieselguhr, alumina and glass fibres. It is, perhaps, not necessary to emphasize that both one- and two-dimensional arrangements can be materialized as well as different combinations with chromatography and multidimensional procedures [65].

There is, however, one technique, that is especially suited for the separation of various polymers. These compounds can be separated by thin-layer gel chromatography (Sephadex G-25–G-200 superfine) prepared by the usual spreading technique (see page 368); then the plate is placed into a suitable apparatus and the compounds are separated in the first run by chromatography. After the chromatographic separation has ceased, the plate, which is inclined during the chromatographic run, is leveled off and electrophoresis is effected in the perpendicular direction; detection is carried out by the printing technique, e.g., the plate is carefully covered with Whatman No 1 paper, which is removed after about 5 min, dried and stained for the particular type of separated compounds [79].

6.2.3 *Electrophoresis in fused salts*

This category of separations is done on glass fibre asbestos or quartz papers, on thin layers or in columns [80–84]. In any case the support and equipment used must withstand high temperature. All materials must be purified before electrophoresis in order to remove impurities and to ensure better wettability by the melted salt:

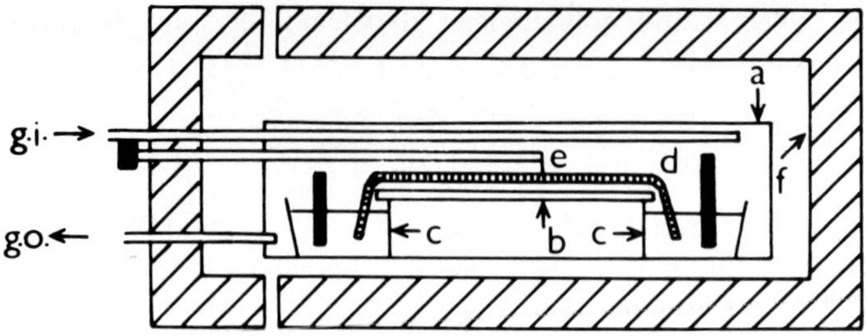

Fig. 6.8. Cross-sectional view of apparatus for electrophoresis in fused salts. (a) Electrophoresis chamber; (b) Pyrex glass plate; (c) Pyrex vessels; (d) glass fibre paper; (e) capillary with a screw to move it in all directions; (f), furnace with heating and temperature regulating system; g.i., gas inlet tube; g.o., gas outlet tube.

washing with mineral acid followed by distilled water and high concentrated salt solution (4–6 M $KMnO_4$) are usually sufficient.

The procedure itself is materialized in a special tank made of Pyrex glass (Fig. 6.8). The whole apparatus is placed into a thermostatted furnace which is being washed with neutral gas in order to remove products of electrolysis. The electrodes are made of metals which can withstand the drastic conditions of the procedure, platinum for the anode and tungsten, nickel, copper or graphite for the cathode. The impregnated support (fibreglass paper) is equilibrated for about 1 h with the molten salt before the actual electrophoresis. The sample is applied from a capillary that can be moved from the outside of the furnace.

The electrolytes used here are mainly nitrates and chlorates of alkaline metals or their eutectics in order to achieve low melting points. The procedure is applied mostly to the estimation of electrophoretic mobilities of inorganic anions and cations and for the separation of isotopes. Of theoretical interest is the fact that electro-osmotic flow in molten salts is practically negligible because of the small electric double layer at the temperatures used.

6.3 Moving boundary electrophoresis

The principle of this technique can be explained on the following example. Let us assume that there are three anions (A, B, C) sandwiched between anode and cathode. In the front, three zones will be formed after introducing the current, namely A, (A, B) and (A, B, C). Also at the other end there will be three zones, namely C, (B, C) and (A, B, C) (Fig. 6.9). The boundaries between the individual zones are determined by the concentration and effective mobilities of the sample constituents, the choice of the anodic and cathodic buffer and to a considerably lesser extent by viscosity, the nature of the solvent and temperature. The boundaries seen in both sides of the separation can be either zone or isotachophoretic boundaries. Thus, e.g., if the anion buffer has a lower electrophoretic mobility than the most

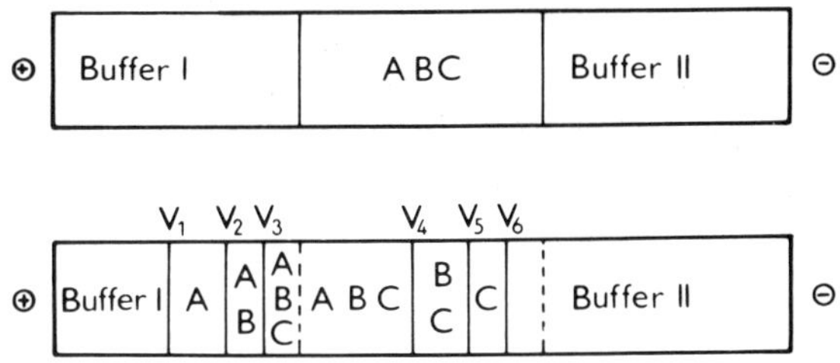

Fig. 6.9. Principle of moving boundary electrophoresis. Separation of three anionic species A, B and C.

anodic constituent of the mixture (e.g., component C), the last boundary is isotachophoretic in nature. On the contrary if the anion buffer has a mobility equal to or higher than that of the most anionic component then the boundary is that of zone electrophoresis [85–89]. More detailed information has been presented [90,91].

In comparison to other electrophoretic techniques moving boundary electrophoresis has only limited applications and can be used for testing, among other things, purity of substances. Its advantage is based on the fact that rather simple equipment can be used (in comparison with isotachophoretic apparatus); zone boundaries are detected with a conductivity and a photometric detector and the whole separation is carried in narrow bore tubes.

6.4 Electrophoresis in gel media

6.4.1 Starch gel electrophoresis

The procedure was developed by Smithies [12] and represents the first application of an electrophoretic procedure in combination with a sieving effect. The apparatus for this purpose is very simple. It consists of two buffer vessels with platinum electrodes and a gel tray placed between them. The contact of the gel layer with electrode buffers is ensured by wicks of thick chromatographic paper. The gel tray is composed of a plastic base (about 10 × 15 cm) and two identical frames placed on top of each other. The whole system is held together by rubber bands. There are several commercially available instruments of this type (e.g., Desaga Disaphor, Savant Model HGE 312) all provided with a cooling system.

The applicability of starch gel electrophoresis is today rather limited (for review see Ref. 93) as it has been superseded by a more efficient method: polyacrylamide gel electrophoresis. However, there are areas of application such as typing of haemoglobin variants, where starch gel electrophoresis still plays its irreplaceable role.

Gel layers are prepared from hydrolysed starch for electrophoresis (Sigma St. Louis, MO, Serva Feinbiochemica, Heidelberg, FRG) by suspending in the gel buffer, degassing the mixture and heating it to the moment when viscosity drops down. Then, the starch solution is poured between the frames. The gel is permitted to cool, solidify and finally it is covered with a cellophane foil. The separation is run at 15 V/cm for about 4 h. After the electrophoretic run is terminated, the gel frame is removed and the gel is cut in two; the cutting string is inserted in the slot between the two tray frames. Detection is done by staining and destaining in a bath. Heating of the stained gel in glycerol for 20–30 s at 70–80°C provides transparency of the gel.

6.4.2 Polyacrylamide gel electrophoresis

In polyacrylamide gel electrophoresis the gel density and degree of cross linking are of decisive importance. It should be noted that polyacrylamide gel is created by a polymerization reaction of acrylamide with *N,N'*-methylenebisacrylamide (Bis). The gel density is determined by the total concentration of both components, gel cross linking is given by the percentage of Bis. The degree of cross linking determines the mechanical properties, i.e., its elasticity and fragility. The most frequently used concentrations are 5–10% gels with 3–5% cross linking. The polymerization reaction is promoted by a suitable redox system: ammonium persulphate or riboflavin in combination with *N,N,N',N'*-tetramethylene diamine (TEMED) or dimethylaminopropionitrile (DMPN) are most common.

The original technique developed by Weber and Osbron [94] exists today in two variations with regard to the spatial arrangement: in the form of disc (discontinuous) electrophoresis on gel rods, or in the form of slab gel separation.

6.4.2.1 Disc electrophoresis — general considerations and solutions

The principle of this method goes back to the premise of Ornstein [95] and Davis [96] that resolution is better the more narrow the bands formed; therefore it would be of advantage to apply the sample in as narrow band as possible. This is materialized in the original version by using three gels of different concentrations on top of each other. The separation gel is overlayed by the spacer gel (large pore gels) on top of which is placed the sample gel (large pore gel). After starting the electrophoresis, sharpening and stacking of the zone occurs as the result of the Kohrausch principle. The separation itself is run in Tris-HCl-glycine buffer (pH 8.9), while the electrode vessels are filled with the same buffer at pH 9.3. Sample and separating gels are prepared in buffer at pH 8.3. If electric current is applied to the system, anions start to move to the anode. Chloride ion moves first, followed by macromolecules in order of their mobility; the last anion to move in the system is glycinate. Each macroionic zone concentrates ad maximum. On entering the separation gel (pH 8.9) the mobility of the last anion (glycinate) increases so that it almost reaches the speed of the chloride anion. Thus, the glycinate-chloride boundary is formed and starts to traverse through the column leaving a uniform electric field behind, in which the macromolecular species are separated according to their molecular size and charge. Later, however, it was proved that the sample buffer is not necessary to obtain the sharpening effect and was therefore omitted. In order to avoid diffusion, the sample is now applied in the form of concentrated sucrose solution. Because of the wide popularity of this method, the composition of stock solutions and buffers are listed below.

Stock solutions: alkaline gel (pH 8.9), 7.5% gel (for the separation of biopolymers p*I* below 7.5).

(A) HCl 1 M	24 ml
Tris	18.2 g
TEMED	100 ml
H_2O	to 100 ml
(B) HCl 1 M	48 ml
Tris	6 g
TEMED	0.46 ml
H_2	to 100 ml
(C) Acrylamide	29.2 ml
N,*N*′-methylenebisacrylamide	0.8 g
H_2O	to 100 ml
(D) Acrylamide	10 g
N,*N*′-methylenebisacrylamide	2.5 g
(E) Riboflavin	0.004 g
H_2O	to 100 ml
(F) Ammonium persulphate	0.15 g
H_2O	to 100 ml
(G) Sucrose	40 g
H_2O	to 100 ml
Electrode buffer (pH 8.3)	
Tris	6 g
Glycine	28.8 g
H_2O	to 1 l

Before use the electrode buffer is diluted 1 : 10. For the preparation of the separation gel (7.5%) the above solutions are mixed in following proportions:
$A:C:F:H_2O = 1:1:1:1$ (for the preparation of the lower gel) or
$B:D:E:G = 1:2:1:4$ (for the preparation of the upper gel).
The anode is joined to the bottom, the cathode to the upper part of the device.

For some special separations (for the separation of biopolymers with p*I* above 7.5) acidic gels can be used; in this case the following composition of solutions is recommended.

(A) KOH 1 M	48 ml
Glacial acetic acid	17.2 ml
TEMED	4 ml
H_2O	to 100 ml

(B) KOH 1 M	48 ml
Glacial acetic acid	2.9 ml
TEMED	0.45 ml
H_2O	to 100 ml
(C) Acrylamide	29.2 g
Bis	0.8 g
H_2O	to 100 ml
(D) Acrylamide	10 g
Bis	2.5 g
H_2O	to 100 ml
(E) Riboflavin	0.004 g
H_2O	to 100 ml
(F) Ammonium persulphate	0.28 g
H_2O	to 100 ml
Electrode buffer (pH 4.5)	
β-alanine	31.2 g
Glacial acetic acid	8.0 g
H_2O	to 1 l

Lower gel, 7.5% (pH 4.3)/$A:C:H_2O:F = 1:1:1:4$
Upper gel, $B:D:E:H_2O = 1:2:1:4$

Polarity of the system is reversed in comparison with the alkaline system, e.g., the anode is mounted at the top, the cathode to the bottom of the apparatus.

For the purpose of studying, e.g., protein subunits, the so-called urea-polyacrylamide gels are in use. The separation in the presence of urea can be done in basic and acidic media, and the solutions for preparation of both types of gels are naturally different. The alkaline gel system with urea-alkaline (for denatured protein subunits) is prepared in the following way (7.5% gel, pH 8.3).

(A) HCl 1 M	10 ml
Tris	18.1 g
TEMED	0.25 ml
Urea (10 M)	to 100 ml
(C) Acrylamide	29.2 g
Bis	0.8 g
Urea (10 M)	to 100 ml
(F) Ammonium persulphate	0.3 g
Urea (10 M)	to 100 ml

Solutions B, D and E do not differ from the alkaline system described above.
The buffer for the electrode vessels is prepared from:

Tris	3.0 g
Glycine	14.4 g
H_2O	to 1 l

Lower gel (7.5%, pH 8.3) is prepared by mixing solutions $A:C:F = 2:4:10$

Upper gel is prepared by mixing solutions B : D : E : 10 M urea = 1 : 2 : 1 : 4
Polarity of the system is conventional, e.g., cathode on top, anode at the bottom of the gel.
The acidic version of the urea gel is prepared in the following way (7.5% gel, pH 4.3).

(A) KOH 1 M 12.5 ml	12.5 ml
Glacial acetic acid	53.2 ml
TEMED	1.2 ml
Urea (10 M)	to 100 ml
(C) Acrylamide	29.2 g
Bis	0.8 g
Urea (10 M)	to 100 ml
(F) Ammonium persulphate	2.8 g
Urea (10 M)	to 100 ml

Solutions B,D and E are identical with those described with the acidic gel system.
Electrode buffer (pH 4.08)

Glycine	28.1 g
Glacial acetic acid	3.06 ml
H_2O	to 1 l

Lower gel (7.5% gel, pH 4.3) is prepared in the following way: A : C : F = 2 : 1 : 1
The composition of the upper gel is B : D : E : 10 M urea = 1 : 1 : 1 : 4
Polarity is reversed in comparison with the preceding system, e.g., the anode is joined to the top, the cathode to the bottom of the gel.

6.4.2.2 Rod shaped gel system (Fig. 6.10)

The apparatus consists of two electrode jars that fit on top of each other. The upper jar is equipped with several (usually six or eight) holes fitted with rubber grommets for fixing the gel tubes. In the centre is a plastic rod, which passes through the bottom to the lower reservoir. It has a drilled groove and fixes the upper and lower electrode. The electrodes can also be mounted directly through the wall of each reservoir, however the danger of leaking leads to the preference of the first arrangement.

Commercial apparatuses of this type are available from: e.g., Desaga, Heidelberg, FRG; Savant Instruments, Hicksville NY, USA; or BioRad Laboratories, Richmond, CA, USA. Many types of the apparatus have been described in the literature [96–99].

Before electrophoresis the gel tubes (usual dimensions are 50–100 × 5 mm) are placed into a rack (vertically) with the lower ends blinded with rubber caps or by plastic rods. The solutions needed are equilibrated at room temperature, the solutions for the lower gel are mixed and pipetted into the tube (to give a gel height of about 50 mm). The surface is carefully overlayed with water (the surface should be smooth and equal, not disturbed by addition of the overlaying water) and permitted to polymerize for half an hour. If the polymerization is too fast, the concentration of persulphate should be lowered, as otherwise the gel would be inhomogenous and

device and gels with graded density can be prepared. The top of the separation gel is flattened by overlaying with water and the system permitted to polymerize. Then the water layer is removed, the stacking gel is poured in a similar manner and a PTFE well-forming comb is inserted.

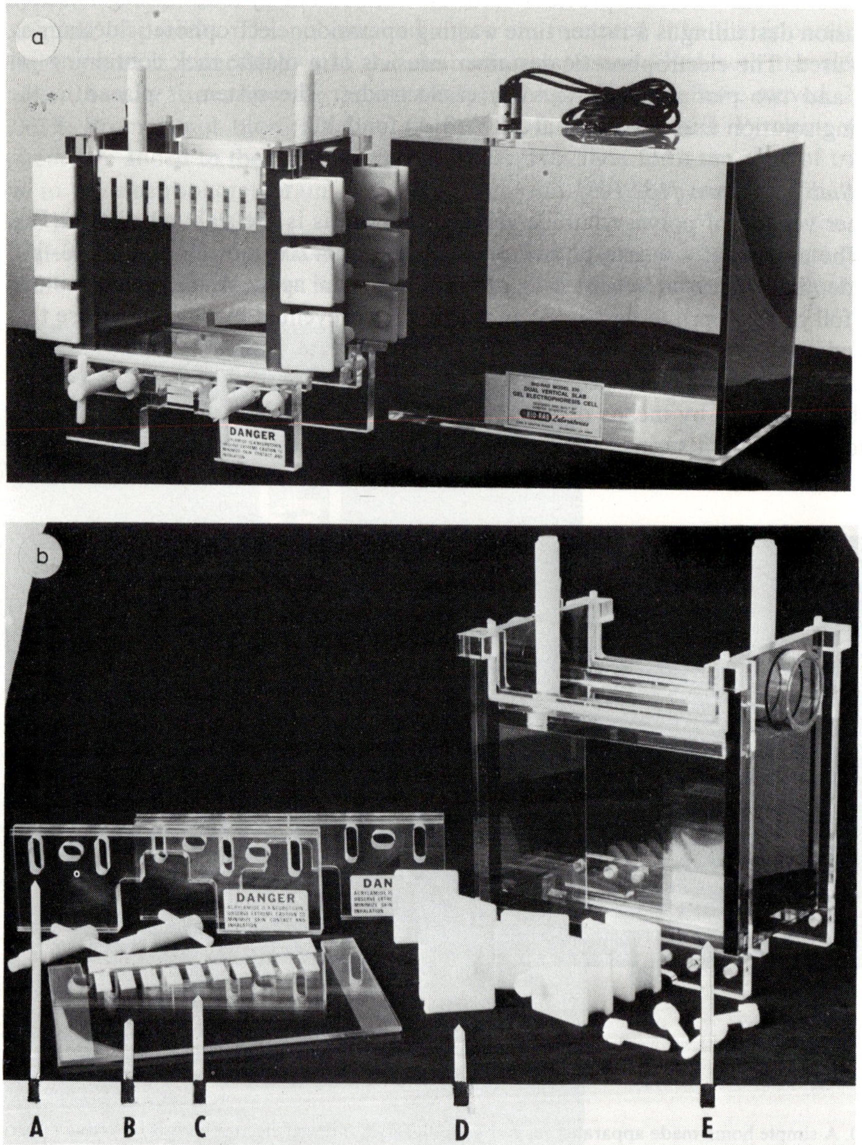

Fig. 6.11. Apparatus for slab gel electrophoresis (Bio Rad). (a) Assembled; (b) disassembled. (A) Lower part of the covering lid (B). (C) Sample comb. (D) Clamps. (E) Upper electrode jar and holder.

The photopolymerization of the spacer gel is effected, the comb is carefully removed (in order not to destroy the surface of the wells), the plates are mounted into the electrophoretic apparatus, and samples are placed in the wells. After having filled the electrode vessels with buffer the electrophoretic separation can be started. Diffusion staining and destaining of the gel is done after disassembling the plates.

On the contrary to rod shaped gels, slab gels can be dried on a paper sheet after staining and destaining and stored in this rather conventional form. Commercial instruments for slab gel electrophoresis and the destaining accessories can be obtained from: e.g., Pharmacia Fine Chemicals, Uppsala, Sweden; Desaga, Heidelberg, FRG; or Bio-Rad Laboratories, Richmond, CA, USA.

6.4.2.4 Gradient gel electrophoresis

The main problem of discontinuous electrophoresis is caused by diffusion; the initially sharp zones loose their sharpness with increasing time. Therefore, gradient gels are prepared so that the density of the gel increases in the sense of electrophoretic migration. The density gradients may be linear or non-linear, and the separation itself occurs on the basis of molecular sieving effects. The electric current provides here only the means of promoting the particles through the gel layer. The interaction of biopolymers with the gel matrix increases with increasing gel density, and therefore the migrating zones are focused on the opposite side from the starting point. The separation is left to proceed until the migration of zones has seemingly stopped, and during an additional period the trailing molecules migrate to the level of the zone front (focus), and logically the separation is more clear in comparison with a matrix of uniform gel concentration.

The most frequently used version is that with a linear gradient gel that can be prepared by mixing two differently concentrated gel solutions in two connected jars [104]. Thus, for a 4–26% gel gradient the following solutions are recommended (ammonium persulphate is added to the mixture immediately before mixing is started).

Solution (A) (20% gel)

Acrylamide	76.6 g
Bis	4.0 g
Sucrose	12.4 g

Tris-EDTA-borate (pH 8.28) to 302 ml
Dimethylaminopropionitrile (DMPN) 0.03 ml
Ammonium persulphate (10% solution) 7.75 ml

Solution (B) (4% gel)

Acrylamide	11.8 g
Bis	0.6 g
Sucrose	3.1 g

Tris-EDTA-borate (pH 8.28) to 302 ml
Dimethylaminopropionitrile (DMPN) 0.09 ml
Ammonium persulphate (10% solution) 7.75 ml

The recommended electrode and gel buffer (pH 8.28) is prepared in the following way.

Tris	10.75 g
EDTA (Na_2 salt)	0.93 g
Boric acid	5.04 g
H_2O	add 1 litre

A slab gel of gradient density is prepared by starting with feeding the gel of lower concentration from the bottom of the plate (through the sealing bar), and supplying the solution of higher concentration from the gradient forming device. As no stacking gel is used in this case, pre-electrophoresis at 125 V for 15 min is applied. The rest of the procedure and detection do not differ from those used with gels of uniform density.

6.4.2.5 SDS-Gel polyacrylamide gel electrophoresis

The finding of the ability of SDS (sodium dodecyl sulfate, sodium lauryl sulphate) to equilibrate charge differences in a wide variety of proteins resulting in nearly constant charge-to-mass ratio and thus identical mobilities in free solution is widely exploited [105]. In gel media the mobilities of proteins are then governed solely by molecular size. The amount of SDS (sodium dodecyl sulphate, sodium lauryl sulphate) bound per 1 g of protein is nearly constant (1.4 g SDS per 1 g protein). This binding results in conformation changes and dissociation of subunits (except those bound by S-S bonds). Shapiro et al. [106] have demonstrated that in SDS-polyacrylamide gels the electrophoretic mobility of reduced and carboxymethylated proteins is inversely proportional to the logarithm of molecular mass [107]. Because of the necessity to compare the mobility of an unknown protein with a standard series of known relative molecular mass, the slab gel procedure is preferred to rod shaped gels as the comparison in the former procedure is more easy and more

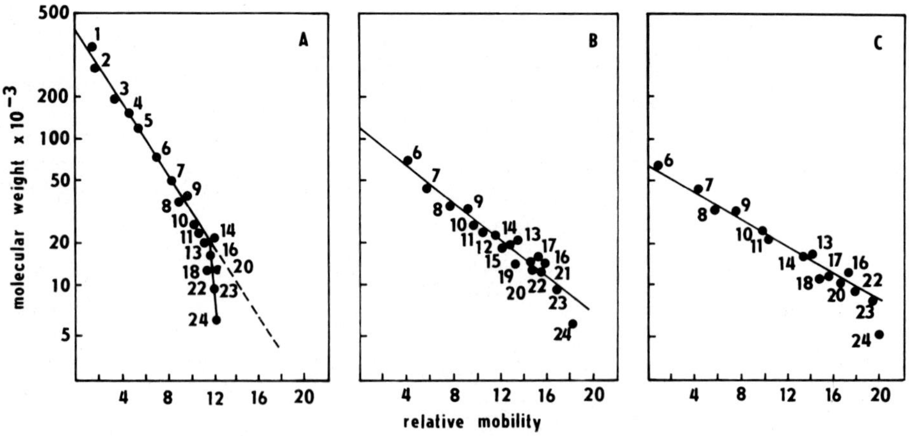

Fig. 6.12. The dependence of rel. mol. mass upon electrophoretic migration. The numbers refer to proteins listed in Table 6.1. A, B and C refers to 5%, 10% and 15% polyacrylamide gels containing 1% SDS

precise (Fig. 6.12). The following solutions are employed:

(A) Acrylamide 29.2 g
Bis 0.8 g
H_2O to 100 ml

(B) Sodium phosphate buffer (pH 7.2), 1 M

(C) Sodium dodecyl sulphate 10% w/v

(D) Ammonium persulphate 0.1 g/m (freshly prepared)

These solutions serve as stock solutions and are used in the actual gel preparation according to the following scheme:

Acrylamide solution	16.7 ml
Sodium phosphate buffer (pH 7.2), 1 M,	10 ml
SDS, 10% solution	1 ml
H_2O	70 ml
TEMED	0.05 ml
Ammonium persulphate	1 ml
H_2O	to 100 ml

As electrode buffer solution 0.1 M sodium phosphate buffer (pH 7.2) containing 1% SDS (0.6 ml per 100 ml) is used. Samples are prepared by dissolving 1 mg of protein in 0.1 ml of denaturation solution (0.1 M sodium phosphate buffer (pH 7.2) containing 1% SDS and 1% 2-mercaptoethanol *), incubated for 3 h at 37°C in small stoppered test tubes, and then 0.9 ml of water are added to each test tube. 50–100 μl of this solution are loaded on the gel.

If an appropriate calibration is carried out it is possible to extend molecular weight determinations on 5% gels down to the 20000–5000 range [108]. However, the steepness of the slope requires a rather precise determination of mobility. A survey of results obtained from gels of different concentration is presented in Table 6.1. This table can also serve as a guide for choosing appropriate standards. It is also possible to select a series of proteins that can simultaneously act as molecular weight and isoelectric point markers [108a]. In general, the precision of this procedure is about 2% of the molecular weight (rel. mol. mass) value determined, but clearly with several proteins larger discrepancies occur, which can be attributed to small but significant variations in the mobility vs. weight relationship in the different proteins. It is likely that such variations result from differences in conformation or intrinsic charge, and from differences in the proportion of detergent anions bound to the structure. With, e.g., lysozyme and pepsin, the anomalies are attributed to differences in the detergent bound, but no such explanation is feasible for ribonuclease A. It has been also shown that the state of folding can also have a considerable effect on mobility in non-SDS gels. In the presence of the detergent such effects are small. As noted by Williams and Gratzer [109], the SDS method is likely to fail for strongly acidic proteins. Here, cationic rather than anionic detergents should be used, e.g.,

* 2-Mercaptoethanol is added only when cleavage of S–S bonds in the sample protein is required.

TABLE 6.1

Molecular weight estimates of standard series of proteins on 5, 10 and 15% gel

Protein	Molecular weight	Data from 5% gels		Data from 10% gels		Data from 15% gels	
		Apparent molecular weight	Deviation (%)	Apparent molecular weight	Deviation (%)	Apparent molecular weight	Deviation (%)
1. Bovine serum albumin (pentamer)	330000	305000	7.6				
2. Bovine serum albumin (tetramer)	264000	260000	1.5				
3. Bovine serum albumin (trimer)	198000	210000	6.0				
4. γ-Globulin	160000	164000	2.5				
5. Bovine serum albumin (dimer)	132000	134000	1.5				
6. Bovine serum albumin	66000	62000	6.5	62000	6.5	61000	8.2
7. Ovalbumin	46000	46500	1.1	46400	1.0	46800	1.7
8. Pepsin	35500	37500	5.6	37000	4.2	38000	7.0
9. Carboxypeptidase A	34400	31000	9.9	31300	9.0	32000	7.4
10. Chymotrypsinogen A	25741	26300	2.0	26300	2.1	26200	1.9
11. Trypsin	23800	24500	2.9	24100	1.3	24200	1.7
12. Bromegras mosaic virus	20300	N.D.		20800	2.4	N.D.	
13. β-Lactoglobulin	18000	17000	5.5	16000	11.0	16200	10.0
14. Myoglobulin	17600	17500	0.5	17500	0.5	17500	0.5
15. Tobacco mosaic virus	17400	N.D.		18600	6.0	N.D.	
16. Lysozyme	14400	13800	4.2	13900	3.5	13600	5.5
17. β-Hydroxyethyl lysozyme	15000	N.D.		13700	8.7	N.D.	
18. Ribonuclease A	13680	18500	36.0	16500	21.0	15800	15.0
19. Acetylcystaminyl ribonuclease A	14620	N.D.		15500	6.0	15000	2.6
20. Chymotrypsin B chain	13927	14000	1.4	14000	0.7	14100	1.44
21. R 17 virus	13729	N.D.		13500	1.6	N.D.	
22. Cytochrome *c*	12400	12000	3.6	13600	9.7	13300	7.3
23. Chymotrypsin C chain	10157	9500	6.0	10500	3.9	10600	4.8
24. Insulin	5700	6000	5.3				

trimethylammonium bromide. In these detergents the plot of the logarithm of molecular weight (rel. mol. mass) versus relative mobility (relative to the tracking dye or some component of mixture taken for internal standard) is similar to that obtained with anionic detergents except for some curvature at higher molecular weights. Equally, at a molecular weight of 10^4 the relationship becomes parallel to the molecular weight axis, i.e., all particles move with the same mobility regardless of their molecular weight and no estimates of the latter can be made. Highly charged proteins (e.g., ferredoxins) with too many negative charges have a lower mobility than those of other proteins of similar size, indicating a low binding of the detergent, while those with too many positive charges bind too much SDS, resulting in a higher relative mobility. Reducing the number of charges by, e.g., esterification makes the mobilities in question fit the linear plots of logarithm of molecular weight versus mobility. For more detailed description of the limitations of molecular weight estimation by electrophoresis the reader is directed to recent reviews on this topic [110–112].

Besides the very simple comparison of an unknown protein with a set of standards more sophisticated methods were elaborated with the aim to increase the precision of the method. A survey of these methods was recently published and discussed by Rothe and Purkhanbaba [112a]. In addition to gels of uniform concentration (Table 6.2) gradient polyacrylamide gels were used in order to improve resolution (Table 6.3).

6.4.2.6 Two-dimensional polyacrylamide gel electrophoresis and the Isodalt system

Two-dimensional polyacrylamide gel electrophoresis is essentially a combination of electrophoresis in rod like gels and in slab gel [113]. The sample is electrophoresed first in a thin gel rod (usually shortened Pasteur pipettes are used for this purpose); the gel is released from the tube by applying hydrostatic pressure and by adding 1% agarose to the top of a slab gel. Obviously the diameter of the rod gel and the thickness of the slab gel must be the same (usually 3 mm). The gel separation is run in the second dimension in a regular slab gel electrophoresis apparatus. Alternatively it is also possible to use thicker gels in the first dimension provided that they are appropriately sliced (Fig. 6.13) [115]. During two-dimensional electrophoresis the advantages of different polyacrylamide gel electrophoretic procedures are made use of in two consequent steps: the most common procedure is electrophoresis in a plain gel (without detergents rod gel) and in an SDS-containing gel (slab gel). Another combination is the separation in a urea-containing gel followed by SDS electrophoresis. Of extreme importance is the isoelectric focusing in rod gel followed by SDS electrophoresis in the second direction. This method is known under the name Isodalt system and offers the possibility to resolve up to 1000 proteins in a single slab gel.

Recently Altland et al. [116] published a procedure in which polyacrylamide gel electrophoresis is used first followed by isoelectric focusing. Moreover isoelectric focusing in the second dimension is carried out in a gel containing a gradient of from 0 to 8 M urea perpendicular to the pH axis.

TABLE 6.2

Methods for the determination of molecular weights of native proteins by use of homogeneous systems [112a]
The value of T is that presented by Ferguson [96a].

1st Derivation	2nd Derivation	Range $\times 10^3$	Buffer systems
log R_f vs. $\% T$ and determination of the slopes of the regression lines, K_R	K_R vs M_W	50–500 (degree of accuracy = $\pm 5\%$)	Discontinuous gel and electrode buffer systems. Electrode buffer: asparagine, Tris-HCl, pH 7.3. Small pore gel: Tris-HCl, pH 7.9. Large pore gel: imidazole-HCl, pH 5.7
log R_f vs. $\% T$ and determination of the slopes of the regression lines, K_R	K_R vs. log M_W	10–50 (degree of accuracy = $\pm 5\%$)	Buffer and gel system (according to Ref. 96, but omitting a small pore gel)
log R_f vs. $\% T$ and determination of the slopes of the regression lines, K_R	VK_R vs. R (R, radius of a weight-equivalent sphere)	40–50	Several different discontinuous buffer systems at pH 10.2, 8.88, 7.8, 5.5
log $\% T$ vs. R_f and determination of the points of intersection of the regression lines with the log T-axis, T_{LIM}	R_S vs. $1/T_{LIM}$ (R_S, Stokes radius)	36–2300	Discontinuous buffer system (according to Ref. 96)

TABLE 6.3

Two-step methods used in the determination of molecular weights of native proteins in gradient gel electrophoresis [112a]

For the construction of the regression line in the log D vs. log M_W correlation, the linear parts of the first derivation, i.e., the constant D- per t-values are used; γ-globulin was found to be beyond the calibration curve. $\%T$ range = 5–30 (cyanogum 41). Using linear gel gradients in the range of 5–15 $\%T$, the VK_R vs. $\bar{R}$-relation was found to be consistent with the Ogston model developed for fibrous gels. The surface of molecular weight-equivalent spheres — instead of Stokes' radii — is related to migration distances of proteins. A linear correlation was found between Vt and D provided that the duration of electrophoresis exceeded 8 h. The carbohydrate content of proteins did not influence the second derivation, but the buffer pH value did. The value of T is according to Ferguson [96a].

1st Derivation	2nd Derivation	Range × 10^3	Buffer systems
D vs. t (D, migration distance t, Bolthours)	log D vs. log M_W	36–460	Tris-glycine (pH 8.3)
log v vs. $\%T$ or log R_f vs. $\%T$ and determination of the slopes of the regression lines, K_R (v = distance migrated per time; cm/h)	VK_R vs. $\bar{R}$ ($\bar{R}$ = radius of a weight-equivalent sphere)	40–450	Several different discontinuous buffer systems at pH 10.2, 8.88, 7.8
Vt vs. D and determination of the slopes of the regression lines, a electrophoretic duration, t: migration distance, D	log M_W vs. a	20–950	Tris-borate-EDTA (pH 8.2), or veronal-Tris (pH 9.8), or phosphate (pH 7.2)

Very recently a new principle has been introduced into two-dimensional polyacrylamide gel electrophoresis; after the separation in the rod like gel has been completed, the whole gel is subjected to a protein-modifying reaction, or to a specific protein clearing reaction (e.g., trypsinization, CNBr cleavage, etc.) and the resulting components are subjected to slab gel electrophoresis in the perpendicular direction [116]. This method allows to obtain data on the primary structure of a series of proteins without their isolation, and allows to solve such problems as whether or not

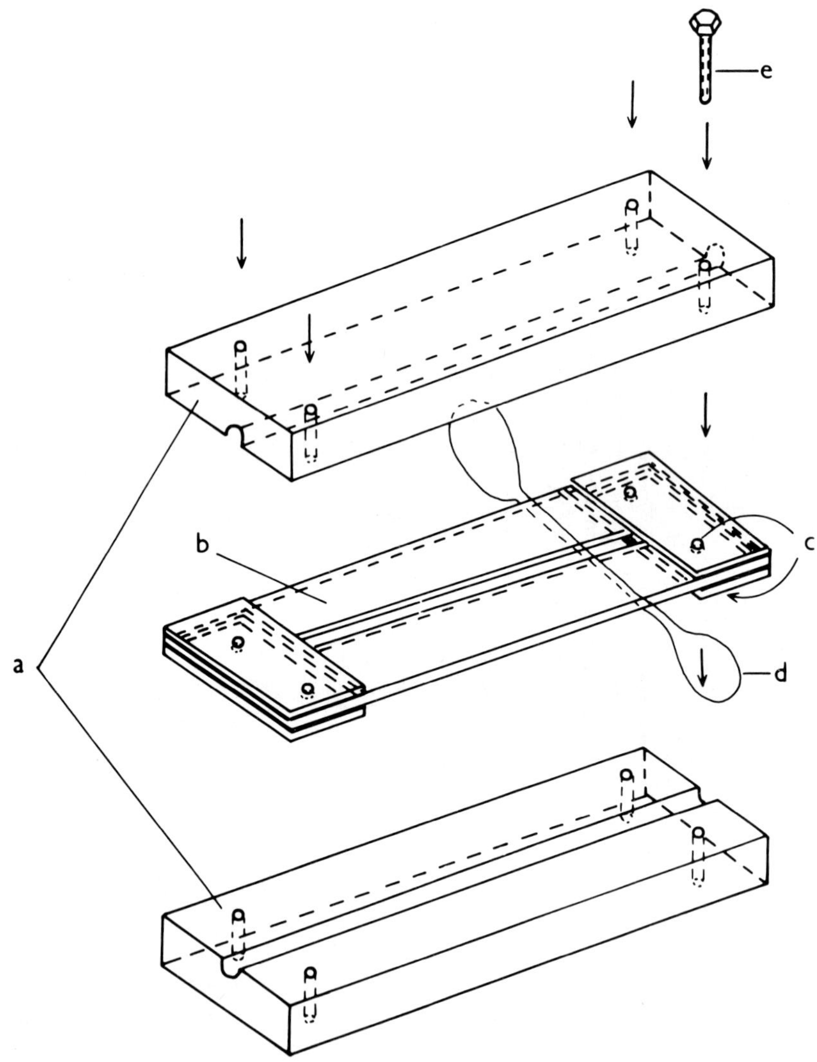

Fig. 6.13. Device for slicing rod gels. (a) Gel holder; (b) a mm spacer for gel strip; (c) 0.25 mm spacer for steel wire; (d) 0.2 mm slicing wire; (e) four connecting screws.

two proteins differing in relative molecular mass yield identical fragments (Fig. 6.14).

At present two-dimensional polyacrylamide gel electrophoresis in one of the most advanced electrophoretic procedures; the main obstacle with this method is a reliable interpretation of very complex electrophoretograms containing several hundreds of zones. The procedures for computer evaluation of such maps are available from more specialized literature [117]. Another problem is that the first-dimension run (rod gel) must be done with much less material than the amount required for optimum detection in the slab gel step. With Coomassie Blue staining (and related methods) this is bypassed by overloading the rod gel at the expense of poorer resolution. More sensitive detection methods seem to be the solution here (Fig. 5.15) [118].

6.4.3 *Agarose gel electrophoresis*

The advantages of agarose gel electrophoresis [119] over older methods like agar gel or paper electrophoresis are in negligible adsorption effects and lower content of charged groups resulting in lower endoosmosis. In comparison with polyacrylamide gel there is, however, lower resolution with no sieving effect involved. For routine examination, e.g., proteins of body fluids, this type of electrophoretic separation is

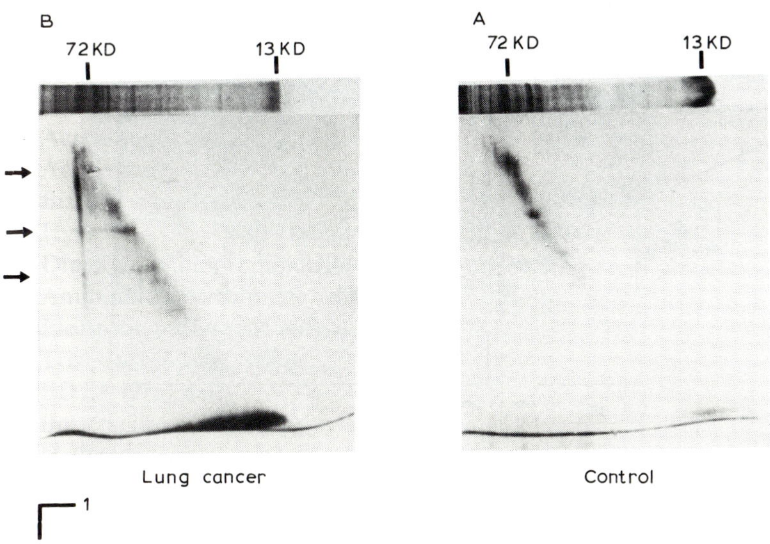

Fig. 6.14. Two dimensional electrophoresis of a mixture of cell surface antigens obtained from lung tissue. Pepsin treatment between the first and second run. A, Healthy tissue; B, neoplastic tissue. The arrows denote the presence of unusual pepsin fragments in the preparation from the neoplastic tissue (courtesy of Dr. Svojtková).

which is prepared in the following way.

5,5-Diethylbarbituric acid (Diemal)	224 g
Tris	443 g
Calcium lactate	5.33 g
Sodium azide	10 g
H_2O	add 10 l

This stock solution of ionic strength 0.1 is diluted 1 : 4 with water (to obtain ionic strength 0.02).

The gel is prepared in a 1% concentration in the above buffer: 2 g agarose are added to 200 ml of the diluted buffer and dissolved by gentle heating and magnetic stirring. Finally the solution is briefly boiled to ensure complete dissolution of agarose. The solution is kept at 56°C in a water bath and is ready for spreading after temperature equilibration. The solution is stable for several weeks at room temperature and can be re-used after being liquified by heating.

Each batch of agarose should be tested for electroendoosmosis (M_R value) which is obtained by subjecting a mixture of human albumin and polydextran to electrophoresis. The shift of the polydextran is measured (negative value for the cathodic shift, positive value for anodic one) and this distance is related to the total distance between the polydextran and the zone of human albumin. The values around 10% are satisfactory. 0.1 M sodium chloride is used for washing excess of protein out of the plate after immunoelectrophoresis. Staining and destaining are done with Coomassie Brilliant Blue and acidified ethanol respectively.

6.5.2 Crossed immunoelectrophoresis (Laurell's electrophoresis) [125,126]

1.5 mm thick agarose is prepared and five wells (2.5 mm) are punched into the plate. One of the wells is used to apply the marker, e.g., bromphenol blue stained albumin, other wells are loaded with samples. The plate is positioned onto the cooled surface of the electrophoresis apparatus, eight layers of Whatman 1 paper are used to prepare the connection-forming wicks and electrophoresis at 15 V/cm is started. The time for run in the first dimension should be about 55 min at 15°C. The electrophoresis is stopped after the marker has traversed 5.5 cm. The agarose gel is then cut into strips containing the separated entities and the well near the left hand edge (Fig. 6.17). This strip is then transferred (using a long blade) to another plate which has been coated with agarose and dried. The glass plate is then positioned onto the levelled table and 12 ml of the antibody-containing solution, consisting of 11 ml 1% agarose and 1 ml of anti-human serum, are poured onto the upper (dried) part of the plate. The plate is left to equilibrate and is placed into the electrophoresis apparatus (with another set of Whatman paper wicks). Second dimension electrophoresis is run at 3 V/cm for 18–20 h. After the separation has ended, excess proteins (that have not precipitated) must be removed from the plate: this is done by pressing the gel under filter paper, sample wells are filled with distilled water before pressing to avoid any damage caused by air bubbles. Usually six layers of paper are applied out of which the first one is applied wet. The remaining five sheets of paper

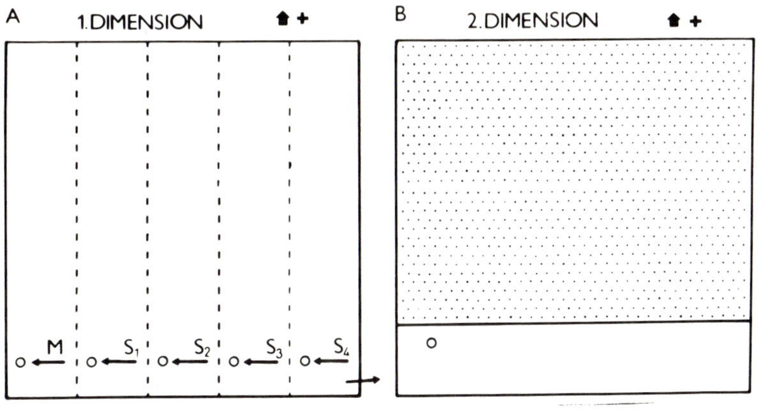

Fig. 6.17. Schematic representation of the arrangement of crossed immunoelectrophoresis. (A) Template for the first dimension gel. (B) Template for the second-dimension plate. The hatched area indicates the antibody-containing gel.

are removed after 5 min; equal pressure is obtained when the plate is positioned reversed on the filter paper sheets. To obtain low-staining background with poly-specific antibodies the plate should be rinsed for 15 min with 0.1 M sodium chloride, dried by means of filter paper sheets and this procedure should be repeated once. After rinsing, the last filter paper sheet is wetted with water and the liberated plate is dried in a stream of hot air and stained with Coomassie Brilliant Blue R 250 for 5 min and destained in the conventional manner by rinsing the plate at least three times with diluted ethanolic acetic acid. At the end of the procedure the plate is dried and stored. 2 μl of serum and 1 ml of antihuman serum are the amounts required for obtaining good electrophoretograms. The individual precipitated zones are identified either by means of monospecific antibodies or, alternatively, additional detections for e.g., lipids or specific enzymic activities, are used. If the plate is to be stained for enzymic activities, it is necessary to preserve them and dry the plate in a stream of cool air. Autoradiography as detection method for proteins binding radioactive labels can be used as well.

Quantitation is done by measuring the plate area below the peak in question. This procedure is generally applicable and can be computerized. The other method, quantitation on the basis of peak height, may not always be possible because of the asymmetry of the peak and inequality in the baseline. For electronic peak evaluation the electrophoretograms are frequently enlarged either directly or after being photographed. The data obtained are mostly relative ones taking a reference serum for the basis of calculations.

A variation of the above method is crossed immunoelectrophoresis with an intermediate gel. The first electrophoretic steps including plate preparations, running buffer and sample application are the same as with the conventional procedure of Laurell's. The modification with the intermediate gel has been worked out to find out which of the peaks appearing on the complex electrophoretic profile of serum paper clamps. The antibody containing gel is poured into the cuvette-like space by means of a pipette. After solidifying (about 10 min) the uncoated glass is removed by sliding it off. The U shaped frame is left on the gel covered plate until the end of the operation. Immunoelectrophoresis is started at 3 V/cm and run for the conventional 18–20 h. The plate is pressed and dried (with monospecific antibodies rinsing

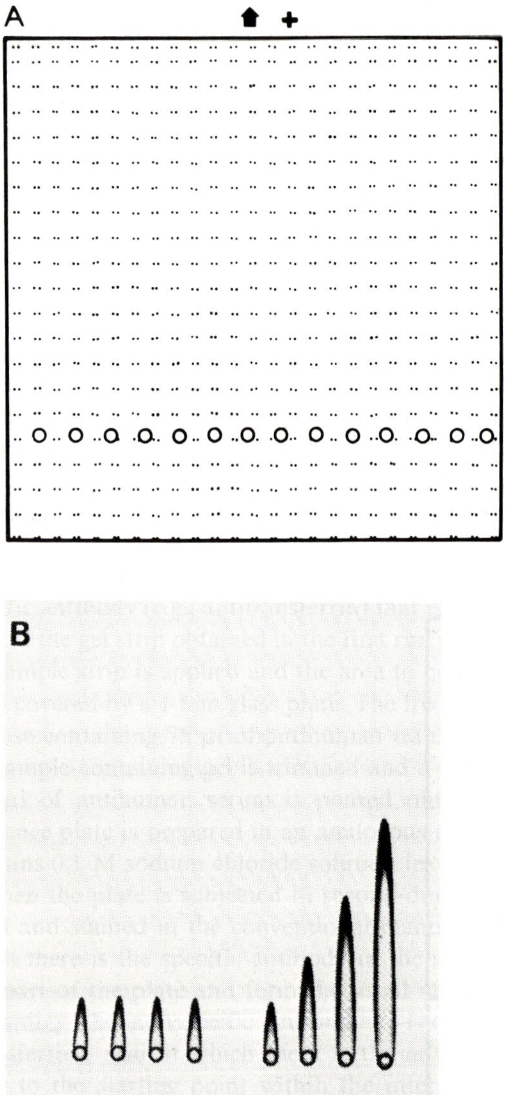

Fig. 6.20. Rocket immunoelectrophoresis. (A) template; (B) aliquots (5 μl) from a gel chromatography on Ultrogel AcA 44 (left four samples) were applied on the plate. Right (four samples) starting material of different dilution. Reaction with antihuman transferrin (according to Ref. 128).

is unnecessary) and finally dyed, destained and dried again. The calibration graph is obtained in a similar manner from the pooled sample from the preceding separation step (e.g., the sample that is loaded onto the chromatographic column) by diluting it to at least four different concentrations. In the final stage the rocket shaped precipitation lines between the standard series and the sample are compared: Laurell rocket immunoelectrophoresis offers an almost linear correlation between the antigen concentration and the area limited by the precipitation line, and hence also an almost linear correlation between the peak height and antigen concentration. In practice the latter is used for quantitation.

6.5.5 *Crossed line immunoelectrophoresis*

The template for the crossed line immunoelectrophoresis is prepared similarly to crossed immunoelectrophoresis with the intermediate gel. The sample gel (containing, e.g., human serum proteins) from the first run is transferred onto another coated plate, and a strip of about 1.2 cm above this gel is filled with 1.8 ml agarose containing 20 μl of the pooled fractions (sample from a preceding, e.g., gel chromatographic, separation) containing the pure compound in question (transferrin); the glass plate is removed, the gel is trimmed and 9.5 ml of agarose containing 900 μl of antihuman antiserum are poured onto the remaining part of the plate. After immunoelectrophoresis the plate is handled in the conventional way. The peak of the compound that has been added into the intermediate gel (sample from a chromatographic run) is elevated in comparison to a control plate where 20 μl of 0.1 M sodium chloride were added to the intermediate gel (Fig. 6.21). *Warning:* if the

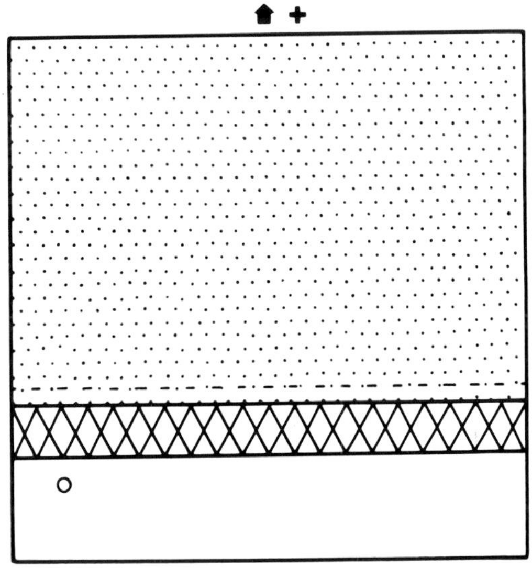

Fig. 6.21. Template for crossed line immunoelectrophoresis.

amount of the sample material added to the intermediate gel is too high, no precipitation line occurs as long as all antibodies are exhausted by being absorbed in the protein line (transferrin line) occurring close to the intermediate and non-specific antibody-loaded gel boundary.

6.5.6 Tandem crossed immunoelectrophoresis

In many respects this method resembles crossed immunoelectrophoresis. The difference is the fact that two wells 1 cm apart are punched into the sample gel (Fig. 6.22). The first one is filled with the whole of the mixture to be studied (e.g., serum), the other with the isolated material (pooled fractions) from a preceding separation step. The plate is left for diffusion to proceed. At this stage diffusion causes the contents of both wells to mix and after immunoelectrophoresis a double peak of the particular antigen appears.

6.6 Isoelectric focusing

The principle of this method is based on the fact that amphoteric macromolecules exhibit zero mobility in the electric field at their isoelectric points and tend to focus into narrow zones in places where the pH of the surrounding media equals p*I*. The process of isoelectric focusing involves two steps, namely the formation of a stable pH gradient that increases from the anode to cathode, and electrophoretic migration of the amphoteric molecules towards their p*I* positions with subsequent attainment

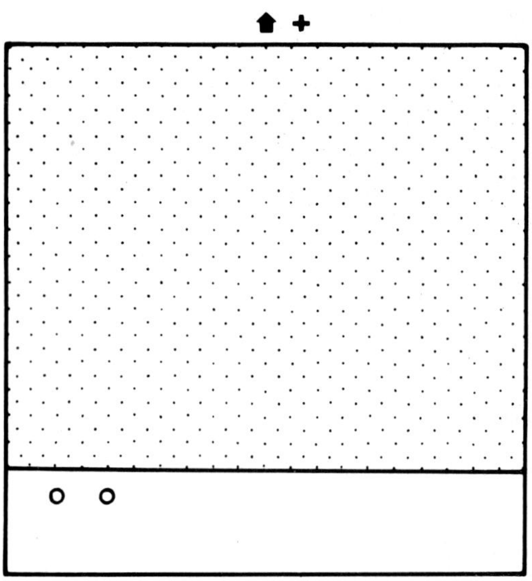

Fig. 6.22. Template for tandem cross immunoelectrophoresis.

of steady state. The concentration of a protein zone at its isoelectric point appears to be a steady-state process between electric mass transport and zone diffusion [130–132]. In practice it is necessary to ensure that the proteins will be negatively charged at the cathode and positively charged at the anode, so that if the protein was positioned anywhere between the electrodes, it would migrate to the place where the pH of the gradient formed between electrodes would be equal to the p*I* value of the protein. This situation is ensured by placing the cathode into a strong base and the anode into a strong acid [133,134]. The pH gradient is formed by applying the electric field to a mixture of amphoteric compounds called carrier ampholytes. This series of compounds must withstand the criterion that isoelectric points of all components of the carrier ampholyte mixture are in close proximity (and p*I* zones overlap), and so a continuous pH gradient can be formed. Naturally the carrier ampholytes obey the above described principle and focus themselves in the electric field. The ampholyte component with the lowest p*I* would migrate to the proximity of anode where it will condense in the p*I* position. This would be somewhat removed from the anode itself because of repulsive forces. The pH gradient is stabilized by using an anticonvection medium (density gradient, gel matrix). Proteins separate by this method according to their isoelectric points and therefore their p*I* values can be directly estimated. Detection is done by conventional means, e.g., by zone staining and UV absorbancy.

6.6.1 Carrier ampholytes

Ampholytes represent complex mixtures of homologous polyamino polycarboxylic or polysulfonic acids with close p*I* and p*K* values [135,136]. The differences in p*I* values represent inter-species differences, while the differences in p*K* values are related to a single species (intra-species differences). Usually the following ranges of pH are covered: 3.5–10.0, 2.0–4.0, 3.5–5.0, 4.0–6.0, 5.0–7.0, 5.0–8.0, 6.0–8.0, 7.0–9.0, 8.0–9.5 and 9.0–11.0. They exhibit good water solubility and low absorbancy at 280 nm. The buffering capacity and conductance differs with pH (which is not desired) and appears to be highest at extreme pH values and lowest between pH 5 and 8. The most disturbing effects (from which one can derive the requirements upon an ideal carrier ampholyte mixture) are as follows: (1) non-uniform conductance along the separation column; (2) variability in concentration of individual species of the carrier ampholyte mixture; (3) non-uniform viscosity; (4) non-uniform buffering capacity; (5) non-linearity of the pH gradient; and (6) instability of the gradient. The last property leads to flattening of the pH gradient and a 'drift' of the pH curve. Consequently this phenomenon causes electrophoretic migration of proteins towards the anode and irreproducibility of isoelectric focusing patterns [137–139].

When studied by gel permeation chromatography with respect to the distribution of their molecular weight, it was shown that most of the ampholyte components exhibit a relative molecular mass of between 300–1000, with few species found at a value of 5000 [140]. This is valuable information as long as it might be necessary to remove carrier ampholytes from an isolated protein. Based on the above facts this is done by simply passing the sample through a Sephadex G 50 column. This step should remove all ampholytes from proteins of rel. mol. mass higher than 10 000. Commercially available ampholytes are available under the names: Ampholine LKB; Biolyte, Bio-Rad Labs; Physiolyte, Brinkmann Instruments; Pharmalyte, Pharmacia, Uppsala; Servalyt Serva Heidelberg. These differ by little in quality of separation and focusing speed [141].

6.6.2 Isoelectric focusing in polyacrylamide gel

Isoelectric focusing in polyacrylamide gel today appears to be nearly the most popular version of this technique [142–145]. The essential components for the gel preparation are identical with those used with conventional polyacrylamide gel electrophoresis (see pg. 428). The only difference is that carrier ampholytes (2% w/w of the total gel volume) are incorporated into the gel solution before polymerization. The protein sample (free of salts) can be mixed with the sample gel solution or it can be loaded in the conventional way in a sucrose solution. In the latter case it is covered with a protective layer of ampholyte.

Both rod shaped and slab gels can be used for isoelectric focusing. Recently, miniature ultrathin-layer isoelectric focusing has been introduced [144]. The critical factor in this procedure is the gel concentration that should be selected in such a way as to exhibit minimum sieving effect upon proteins, as any restriction in the migration of proteins towards their p*I* results in serious disturbances of the isoelectric focusing patterns (the steady state is not achieved or even some protein species cannot be focused at all). From the nature of polyacrylamide gels it is impossible to prepare absolutely non-restrictive gel without sieving effects. However, for the isoelectric focusing of proteins up to rel. mol. mass 800 000, 3.75% gels with 3.33% cross linking offer satisfactory results. For proteins of lower rel. mol. mass, e.g., 15–60 000, 5% gels can be used. The tendency to use gels of higher concentrations is given by their higher mechanical strength and by their ability to form sharper bands because of diffusion restriction. The pH gradient at the end of the isoelectric focusing can be determined either directly by using a microelectrode, or the gel rod can be liberated from the gel tube, sliced, slices eluted with water, and the pH of the eluate can be measured in the conventional manner. Detection causes more difficulties in isoelectric focusing than in polyacrylamide gel electrophoresis, which is due to the property of dye used to bind to ampholines. Naturally specific staining techniques, like those based on enzymic reactions, radioactivity, fluorescence labeling, direct optical density measurement, etc., can be used.

Apparatuses for polyacrylamide gel isoelectric focusing are available from several leading manufacturers (LKB, ISCO, Hoeffer, Ortec, E.C.Med.Res.Apparatus, Buchler, Brinkman-Desaga). The advantages and disadvantages of slab gel and rod gel technique are practically the same as with conventional polyacrylamide gel electrophoresis.

6.6.3 Thin-layer isoelectric focusing

Despite the fact that polyacrylamide-covered plates can serve as stabilizing medium for thin-layer isoelectric focusing [145–147], a typical procedure of this category is that which uses Sephadex G 75 [145,146,148]. Isoelectric focusing is performed on 40×20 or 20×20 plates coated with Sephadex G 75 suspension in 1% carrier ampholyte solution. The plates are prepared in the conventional way by the spreading technique (0.75–1.0 mm thickness). Contacts with electrode vessels are materialized by Whatman paper (No. 3) wicks. The wicks are covered with a dialysis bag or similar foil to prevent liquid flow from the electrode vessels to the gel. Electrofocusing is run at 10–20 V/cm for a period of 6–8 h in the 20 cm plates, and at 5–10 V/cm for 18–24 h in the 40 cm plates. The imaging of the electric focusing process is done by making a print of the gel on Whatman 3MM paper. Handling the print and sample application are identical with procedures used in thin-layer gel permeation chromatography.

6.6.4 Density gradient isoelectric focusing

Stabilization of the medium against convection can be achieved by using a gradient of sucrose (0–50%), but other anticonvective materials like ethylene glycol or Ficoll [149,150] can be employed as well. The properties of the stabilizing medium were summarized by Catsimpoolas [151] in the following way: (1) the stabilizing agent must be non-ionic; (2) the stabilizing agent must exhibit a high solubility at low viscosity values in water; (3) the density differences produced should be beyond 0.12 g/cm^3; (4) it should not interact with the separated compounds and alter their biological activity; and (5) The compound should be easy to remove from the separated compounds.

The preparation of a density gradient is done either by means of a gradient-forming device or can be prepared manually by preparing a series of solutions of different density and stratifying them manually one on top of the other. Ampholines are added to the solutions at a concentration of 2 g per 100 ml of the solution. The protein sample can be either mixed with the whole gradient or added to one of the gradient solutions preferably to that which occurs in the neighbourhood of the focused main zone. This procedure cuts down the focusing time considerably. Other compounds used in protein chemistry like urea, Triton X 100, Bij 35, dithiothreitol can also be added to the gradient solutions. The arrangement of density gradient isoelectric focusing is mostly such that the anode is placed on the bottom and 5% orthophosphoric acid is used for electrolyte, while the cathode is located on top with 5% ethanolamine as electrode solution.

There are several drawbacks to density-gradient isoelectric focusing in comparison with other isoelectric fractionation methods.

(*1*) After the separation has ended, the column must be emptied and the fractions collected. Even if this is done with utmost care, the separation obtained in individual fractions is inferior to that seen in the column.

(*2*) Immediately after the electric current is switched off, diffusion starts to disturb the isoelectric pattern. Therefore, the quality of final separation depends on how fast the emptying of the column is done. On the other hand this step cannot proceed too fast as the high speed can lead to distortion of the gradient and hence to distortion of the zones as well.

(*3*) Considerable problems may be caused by proteins that precipitate at their p*I* since they might contaminate other proteins that are soluble at the same p*I*.

(*4*) It is difficult to determine the optimum focusing time. The main advantages are that large amounts of proteins can be prepared in a state of very high purity, and that protein subunits can be obtained; dissociating agents can also be added to the gradient-forming solutions.

6.6.5 Free solution isoelectric focusing

This technique, although described in the literature, has been given little attention. Isoelectric focusing without stabilizing media can be done in an apparatus similar to Tiselius free zone electrophoresis [152]. Separation itself occurs in a horizontal quartz tube that is rotated at 40 rpm to counteract convective forces [153]. The pH gradient is evolved without the addition of ampholytes, however this method leads to either too steep or too shallow gradients and is therefore not practically applicable. Another alternative method for free solution isoelectric focusing is the application of polyethylene coils; however this procedure makes use of ampholines [154].

6.6.6 Two-dimensional procedures involving isoelectric focusing

Obviously one of the most popular two dimensional electromigration procedures is the combination of isoelectric focusing (1st dimension) with SDS-polyacrylamide gel electrophoresis (2nd dimension) [113]. It should be noted that in the two-dimensional procedures isoelectric focusing represents nearly always the first step (see Isodalt technique). The success of this procedure is based on the fact that first dimension separation depends on the p*I* of the separated proteins, while the separation in the second run depends on the molecular size [144,155–157]. Other possibilities are offered by the combination of isoelectric focusing with immunodetection or immunoelectrophoresis [158–163]. Thus, the polyacrylamide gel rod can be embedded in agar gel and the separation is visualized by precipitation lines arising from the reaction with antiserum that is loaded into trenches cut parallel to the embedded gel. Further alterations depend on the nature of the experiment: if more antisera are necessary for detection, the gel rod can be sliced, longitudinally, into two semicylinders offering the possibility of using four different antisera. Alternatively, the gel rod can be sliced transversally (about 40 slices are required from a 6 cm long rod) and each slice can be subjected to immunochemical testing. It is also possible to alter the former method in such a way that the proteins separated by isoelectric focusing in the first dimension are electrophoresed perpendicularly after embedding into antibodies containing agarose beads (crossed electrophoresis).

The third variation that exploits unsliced gels is based on the following principle: the gel is, after isoelectric focusing, incubated with antiserum and the non-precipitated proteins are eluted. Further processing, e.g., staining and destaining is conventional. Methods that use slicing can be also modified in at least three different ways; the individual sections can be tested by double gel diffusion, radial diffusion or immunoelectrophoresis.

As a matter of fact any of the methods used in immunoelectrophoresis are applicable (see the corresponding section).

Moreover not only polyacrylamide, but also agarose gel, can be used as stabilizing medium for isoelectric focusing [164]. The procedure is done on microscope slides. The pH gradient developed during isoelectric focusing must be abolished before the plate is further processed. The main problem which prevents this technique becoming more popular is the various degrees of electroosmotic flow observed with various commercial agarose preparations.

6.6.7 Transient state isoelectric focusing

The method of transient state isoelectric focusing has been developed by Catsimpoolas and his co-workers (for review see Refs. 165–167). Until now isoelectric focusing results have been recorded and presented after the system has reached a steady state, and all the important parameters should remain constant with time. If, however, any of these parameters change, whatever the cause of instability, the system appears to be in a transient state and the laboratory worker is forced to follow the kinetics of isoelectric focusing for a particular mixture in several consequent time intervals. Naturally, any deviation from the steady state that may occur in the system is also recorded. The apparatus used for transient state isoelectric focusing is rather sophisticated and three variations have been described in the literature along with the necessary computerization. In principle the process of isoelectric focusing is performed in a quartz column, and the particles of partly separated proteins are scanned at a desired wavelength (280 or 230 nm); the current is applied throughout the scanning time and the quartz column may be cooled during focusing. The results are usually presented in the form of scanning isoelectric spectra. An important practical aspect (besides a considerable contribution to the theory of isoelectric focusing; for review see Ref. 151) is the possibility to establish minimal focusing time for a particular system. An approach that has been utilized in this respect is measuring of the peak area versus time at its p*I* position. The peak keeps increasing during the focusing process and reaches its maximum when the system has come to the steady state. For this experiment the protein in question is loaded uniformally throughout the isoelectric focusing system. It should be noted that in this arrangement two peaks start to migrate from both ends of the isoelectric focusing column that finally fuse into a single zone of the protein at its p*I* position when the steady state is attained (Fig. 6.23).

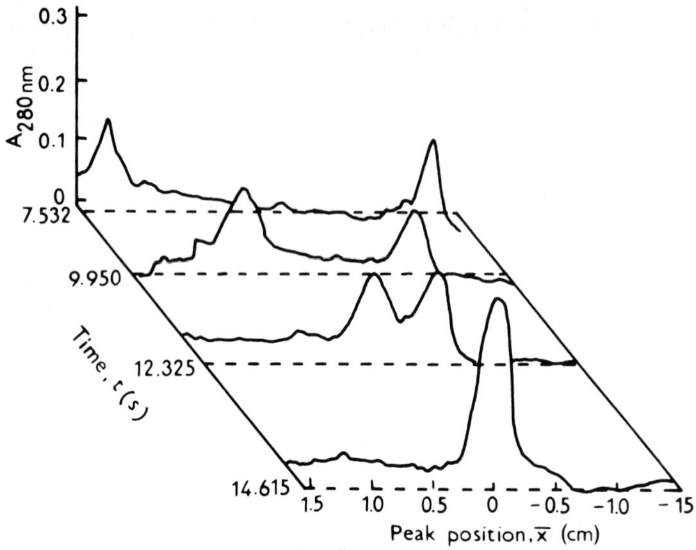

Fig. 6.23. Scanning electrophoretic patterns of histidyltyrosine peaks moving towards the isoelectric position. Peak 1 migrates away from cathode, peak 2 migrates away from anode. The isoelectric position is marked 'zero' on the x-axis.

6.7 Isotachophoresis

In isotachophoretic separations the sample ion is applied between the so-called leading and terminating electrolyte. It is absolutely necessary to make such a choice of the electrolytes, that the effective mobility of the leading electrolyte is the highest of the system, the terminating electrolyte should exhibit the lowest effective mobility (the effective mobility of the sample being intermediate). The leading electrolyte initially fills the separation compartment and the counter electrode compartment. Analytical isotachophoresis is generally performed in narrow bore tubes without stabilizing components [168–171]. The principle of isotachophoretic separation is described with an anionic species as an example because, due to the polarity of isotachophoresis, the choice of whether anionic or cationic compounds are to be separated must be done beforehand (Fig. 6.24). When the current is applied, each anionic species of the sample starts to move with a characteristic speed, differing from the velocity of other species present. The mixture will begin to differentiate in such a way that the species with the highest mobility will move first, and those with lower effective mobility will remain behind. Between the individual species mixed zones appear at the beginning of the separation, which is the result of the moving boundary procedure. In the series of mixed zones, individual sample species are arranged in order of decreasing effective mobility. The anionic species of the leading electrolyte cannot be overtaken by any of the sample species as it has been selected to be the fastest. Also, none of the sample components can be overtaken by the

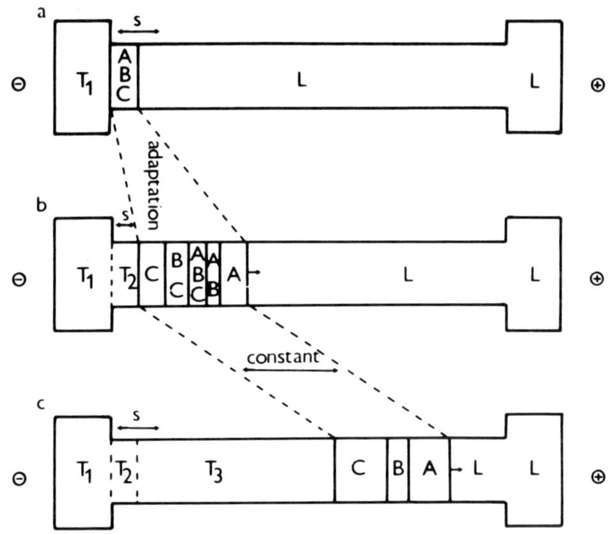

Fig. 6.24. Isotachophoretic separation of a mixture of anions. The sample (ABC) is introduced between the leading (L) and terminating (T) anionic species. A suitable cation is chosen as the buffering counter ion. (a) Original conditions; (b) after some period of time, mixed zones are obtained according to the moving boundary principle; (e) final stage, all zones are separated (ideal situation) (according to Ref. 171).

terminating electrolyte, as this anionic species has been chosen to be the slowest. Thus the steady state in isotachophoresis appears to be that one in which all ionic species (except the counter ion) with differences in effective mobility are separated according to these mobilities.

6.7.1 Apparatus for isotachophoresis

Analytical isotachophoresis is usually performed in a narrow bore glass or plastic tube (50–100 cm). The length of the tube needed for a particular separation is determined by the pair of ions that are most difficult to separate, and on their quantity, as this determines their zone length. In order to prevent extreme lengths of the isotachophoretic capillary, the counterflow technique was introduced. The counterflow system is based on the application of the leading electrolyte against the direction of migration of separated ion, thereby increasing the effective length of the tube. Moreover the whole tube is thermostatted and blinded at one end by a semipermeable membrane which decreases the endoosmotic flow to such an extent that no stabilizers are needed. Another apparatus consisting of a solid block with electrode compartments, sample inlet, capillary and detection cell has been described by Boček [172] (Fig. 6.25). The capillary used here was of rectangular cross-section (0.2 × 1 mm) and 25 cm total length. Commercially available apparatus is produced under the name Tachophor by LKB Bromma, Sweden.

Instead of capillary (narrow bore tube) isotachophoresis, this type of electromigration separation can be materialized on cellulose acetate strips [173]. Combinations with classical electrophoresis resulting in two dimensional separations have also been described [174,175].

Isotachophoresis of proteins can also be done in dilute polyacrylamide gel, in which sieving effects are oppressed. The results are comparable with isoelectric focusing or disc electrophoresis. The isotachophoretic profile is analogous to the isoelectric focusing pattern; this is because proteins are separated mainly according to their p*I* values. Chrambach [138] has shown that the advantage of isotachophoresis over isoelectric focusing is that, in isotachophoresis, no precipitation of proteins takes place. Carrier ampholytes were applied by Svendsen [128] to separate protein zone.

6.7.2 Detection in isotachophoretic separations

Detection in isotachophoresis is considerably different from all other electromigration techniques and therefore will be discussed here; moreover it requires special

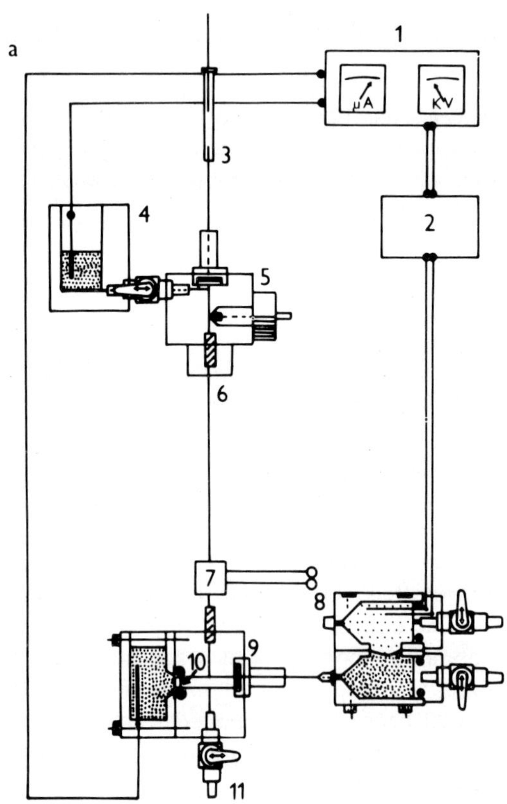

equipment. The principle of the detection technique is based on the fact that each zone has its own voltage gradient, and therefore the amount of Joules generated in various zones are different. This can be made use of for thermometric detection. Alternatively conductometric and potential gradient detectors can be applied. For compounds with distinct UV absorptivity spectrophotometric detection is the method of choice. This latter procedure is, on the contrary to the preceding ones, also used in other electromigration techniques (see page 468).

6.7.3 Buffer systems for isotachophoretic separation of serum proteins

To get some idea what composition of buffers can be used for protein separation by isotachophoresis the following examples are presented.

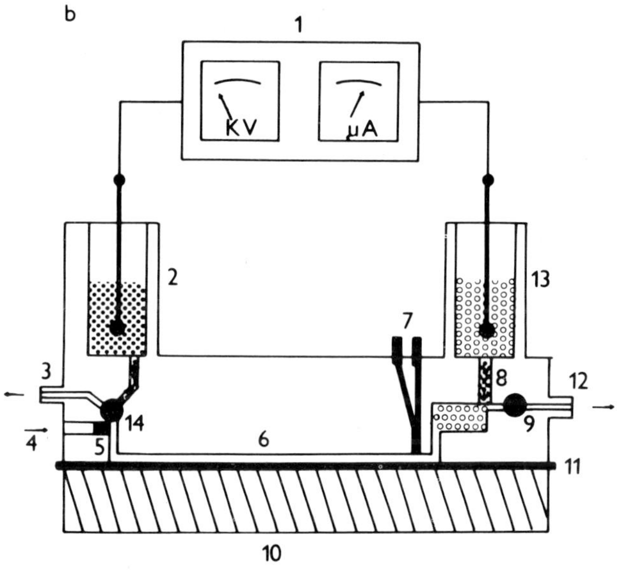

Fig. 6.25. Schematic arrangement of isotachophoretic equipment. (a) constructed of single components; (b) compact version (according to Ref. 171). (a) 1, Current stabilized power supply; 2, electronics for regulating the counter flow of electrolyte (if necessary), produced by a gas pump; 3, syringe for sampling via a septum; 4, reservoir filled with terminating electrolyte; 5, injection block; 6, separation compartment; 7, set of detectors (conductometric, potentiometric, UV absorption); 8, gas pump suitable for a regulated counter flow of electrolyte; 9, counter electrode compartment; 10, semi-permeable membrane; 11, Hamilton 1 mm tap for connection with reservoir filled with leading electrolyte. (b) 1, Current stabilized power supply; 2, reservoir filled with terminating electrolyte; 3, connection towards drain; 4, hole used for sampling with a syringe; 5, septum; 6, separation compartment, i.e., a groove in a block of organic glass; 7, measuring electrodes connectable to a conductometric or potentiometric detector; 8, porous glass filter; 9, PTFE tap; 10, aluminium block, which can be thermostatted; 11, PTFE sheet; 12, connection towards drain; 13, reservoir filled with leading electrolyte; 14, tap that can be used for rinsing and re-filling of the separation compartment and the reservoir with terminating electrolyte.

Buffer system according to Routs [176]. 0.03 M acetate, 0.011 M Tris-HCl (pH 4.5) as leading electrolyte, β-alanine buffer (pH 9) as terminating electrolyte. 3.5% gel with 5% cross linking prepared with 0.5 mg% riboflavin, 0.03 M acetate, 0.11 M Tris (pH 4.5), 10 μl of human serum albumin (sample) is applied with an equal amount of 40% carrier ampholyte pI 5–6 on top of the gel. Isotachophoresis is carried out for 2 h at 2 mA per gel tube.

Buffer system according to Catsimpoolas [177]. 0.24 M Tris-phosphate (pH 6.1) as leading electrolyte, 0.14 M β-alanine-Tris (pH 8.0) as terminating electrolyte. 3.7% gel with 3.33% cross linking, prepared with 0.5 mg% riboflavin, 0.03 M Tris-phosphate buffer (pH 6.1), 80 μl per 100 ml TEMED, 3.1% sucrose. 13 cm long gels are used. Proteins are dissolved in terminating buffer and carrier ampholytes (pI 5–8, 40%) are added (35 μl to 10 μl of serum). Isotachophoresis takes 2 h at 6 mA per gel.

Buffer system according to Davies and Rosenbaum [178]. This buffer system has been worked out for preparative purposes. 100 ml of acrylamide solution (prepared from 3.67% acrylamide, 5% bisacrylamide, in 0.012 M Tris, 0.013 M cacodylate (pH 7.0), 0.25 mg% riboflavin, 10 μl per 100 ml TEMED) are polymerized in a 30 cm column. Photopolymerization is ended after 3 h of irradiation. The lower electrode compartment and buffer reservoir are filled with the leading electrolyte (0.012 M Tris, 0.013 M cacodylate, pH 7.0). 0.032 M Tris, 0.188 M β-alanine (pH 8.95) serves as terminating electrolyte; this is placed on top of the gels and in the upper buffer vessel. 2 ml of 40% ampholine pI 5–7 are placed on the top of the column; separation is carried out at 10 mA with cathode above and anode at the bottom of the column. Ampholines are allowed to migrate into the gel, which takes 5 h. Then the current is switched off and the sample (2 ml of serum with 2 ml of terminating buffer) is loaded on top of the gel (by means of a capillary) and separation is carried on at 10 mA, and fraction eluted with the leading buffer are collected. Some additional information in this respect regarding amino acids and peptides can be gained from Table 6.4.

6.8 Affinity electrophoresis

Various modifications of affinity electrophoresis (for review see Ref. 179) are based on electrophoresis of proteins or other macromolecules in gel media containing immobilized or quasi-immobilized ligands capable of specific interaction with the electrophoresed substance.

Two major variations of the method are used at present. The first approach is closely related to some immunoelectrophoretic techniques, e.g., crossed immunoelectrophoresis or rocket electrophoresis [180,181], the only difference is that usually a lectin instead of an antibody is incorporated in the agarose gel; the conditions in the gel (electroendoosmosis and pH) are used such that the lectin has a very low electrophoretic mobility. The lectin reacts with specific glycoproteins and forms characteristic precipitation lines analogously to the reactions observed during immunoelectrophoresis between the antibody and the antigen. This method can be

TABLE 6.4

Some suggested electrolyte systems for the isotachophoretic separation of amino acids and peptides

Characterization of the buffer system	Leading ion	Counter ion	Terminating ion	Additions to the terminator	Notes
Neutral and acidic amino acids as anions	Cl^- (5–10 mM)	Ammediol (to pH 8.6–9.7)	β-Alanine	$Ba(OH)_2$ to pH 10	Sensitive to atmospheric CO_2
Neutral and acidic amino acids as anions	Formate (10 mM)	Collidine (2,4,6-trimethypyridine) (to pH 7.1)	Vanillin (10 mM)		Leading electrolyte contains 17.5% formaldehyde which by reacting with amino acids makes analysis at lower pH possible. Not applicable for UV detection
Special conditions	Cl^- (10 mM)	β-Alanine (pH 4) or histidine (pH 5)	Hexanoate (10 mM)		Neutral and basic amino acids migrate in the opposite direction than acidic amino acids; neutral and basic ones move to cathode
Special conditions for the basic amino acids as cations	Ba^{2+} (5 mM)	Valine (pH 9.9)	Tris (20 mM)	HCl to pH 8	Neutral and acidic amino acids migrate as anions. Specific for lysine and arginine
Cationic separations at low pH	K^+ (5 mM)	Acetate (to pH 5)	Alanine (5 mM)	Acetate to pH 5.1	
Cationic separations at intermediate pH	K^+ (5 mM)	Cacodylate (to pH 7)	Creatinine (10 mM)		
Anionic separations at low pH	Cl^- (5 mM)	Histidine methyl ester to pH 5.3	Cacodylate (10 mM)		
Anionic separations at intermediate pH	Cl^- (2.5–10mM)	Tris (to pH 48.3)	β-Alanine (20 mM)	$Ba(OH)_2$ to pH 10	
Anionic separations at high pH	Cl^- (5 mM) or MES * (5 mM)	Ammediol to pH 9.8	β-Alanine or phenol or 6-amino-hexanoic acid	$Ba(OH)_2$ to pH 11	

* 2-(N-morpholino)ethanesulphonic acid.

used for detection, partial characterization and quantitation of glycoproteins precipitating with the lectin.

The lectin can be incorporated also into the first-dimension agarose gel; under appropriate conditions (low concentration of the lectin and still much lower concentration of the glycoprotein in the sample) soluble complexes are formed with the mobility lower than that of the free glycoprotein. The decrease of mobility is simply related to the strength of interaction of the glycoprotein with the lectin (dissociation constant of the complex). The positions of the glycoprotein in the control gel (i.e., gel containing no lectin or the gel containing, in addition to the lectin, also the specific sugar blocking the carbohydrate-binding sites of the lectin) and in the affinity gel (containing incorporated lectin) are most conveniently determined by second-dimension electrophoresis into the gel containing precipitating antibodies against the glycoprotein. The observed differences between the mobilities of the glycoprotein in the control and affinity gels are used to demonstrate the presence of variant forms of glycoproteins with different affinity toward the lectin and for estimation of dissociation constants of the lectin-glycoprotein complexes.

The other major variant of affinity electrophoresis can be considered as an electrophoretic analog of affinity chromatography: the electrophoresis of a ligand-specific protein (monoclonal antibody, lectin, enzyme) is performed in a gel medium (usually polyacrylamide agarose gel or composite) containing an effectively immobilized ligand (hapten, sugar, inhibitor). The apparatuses, buffer systems and procedures are otherwise exactly like those used normally for polyacrylamide or agarose gel zone electrophoresis. Under appropriate conditions the interaction of the protein with the immobilized ligand results in a retardation of the protein zone as compared to the control (non-interacting) gel; mobilities of the proteins devoid of affinity toward the immobilized ligand remain practically unchanged. This specific retardation can be used for identification of the specific ligand-binding protein(s) in a complex mixture, for detection of inactive admixtures in the purified preparations, for monitoring the course and results of a chemical modification reaction affecting the ligand-binding site, and for optimization of conditions for preparative affinity chromatography. The degree of retardation is simply related to the concentration of the immobilized ligand (c_i) and dissociation constant of the protein-ligand complex (K_i), so that at known c_i the value of K_i can be determined using the simple equation

$$\frac{1}{d_o - d} = \frac{K_i}{c_i}$$

where d_o and d are the mobilities of the protein in the control and affinity gels, respectively. If the affinity gel contains, in addition to the immobilized ligand, also a free (mobile) ligand (in concentration c) capable of interaction with the protein, then the resulting mobility of the protein will depend on the ratio between c_i and c and between K_i and K (dissociation constant of the protein-mobile ligand complex). At constant c_i, the observable value of d depends on c; this concentration dependence can be used also for estimation of K using the following equation

$$\frac{d}{d_o - d} = \frac{K_i}{c_i}\left(1 + \frac{c}{K}\right)$$

Best results are achieved with uncharged mobile ligands (e.g., sugars) in this technique; the intrinsic charge of immobilized ligand molecules does not matter. Both above-mentioned simple equations applicable for estimation of K_i and K are valid exactly only under several conditions: (a) the protein molecule contains only a single ligand-binding site; (b) immobilization of the ligand is complete; (c) all molecules of immobilized ligand are accessible to interaction with the protein; (d) mobility of the protein-mobile ligand complex is identical with that of free protein; (e) molar concentration of the protein in the migrating zone is much lower than c_i and c; (f) complex formation and dissociation are very rapid reactions; and (g) microdistribution of the immobilized ligand is homogeneous. If some of these assumptions do not hold, more complicated equations are to be used for evaluation of K_i and K [182,183].

Several techniques were developed for immobilization of the ligands in affinity gels.

(*1*) Incorporation (by copolymerization) of a suitable derivative (allyl-, acryloyl-, etc.) of the ligand into the polyacrylamide gel network [184].

(*2*) Physical entrapment of a macromolecular derivative of the ligand into the polyacrylamide gel network; thus, ligand-substituted natural or synthetic soluble copolymers can be used as 'carriers' of the ligand [185–188].

(*3*) Incorporation of agarose (or other gel) beads substituted with the ligand into polyacrylamide gel [189].

(*4*) Preparation of composite polyacrylamide-agarose or purely agarose gels using suitable meltable derivatives of ligand-substituted beaded agarose [189].

A more detailed discussion of advantages and drawbacks of these various methods as well as affinity electrophoresis in general can be found elsewhere [179,189]. A technique closely related to affinity electrophoresis is affinity isoelectric focusing, which is based on isoelectric focusing in a gel containing immobilized ligand and can be used mainly for qualitative purposes [190]. In addition to the study of protein-ligand interaction nucleic acid-ligand interactions can also be followed by affinity electrophoresis [189,191].

6.9 General detection procedures

Detection methods for low molecular weight compounds in zone electrophoresis are identical with those used in flat bed chromatography. In isotachophoresis detection is done by special techniques that have been referred to in the respective chapter. Therefore in the following sections attention will be paid to detection of biopolymers (mainly to techniques used in gel electromigration techniques).

6.9.1 *Detection by ultraviolet absorbance*

In these detection procedures, which are certainly not the most popular, but most accurate, the absorbancy of the supporting medium plays an important role. Thus, monomeric acrylamide has a UV absorption peak at 280 nm which decreases upon polymerization. Polyacrylamide gel has a rather low extinction coefficient below 270 nm, but absorption rises rapidly, and at 270 nm is characterized by the extinction coefficient, 0.569. Therefore unpolymerized acrylamide must be absent in the gel if spectrophotometric evaluation of the gel is to be applied. To ensure this Watkin and Miller [193] recommended to use only 0.2% Bis in 7.5% gel.

Differences also exist with regard to the nature of separated compounds. Thus nucleic acids have a generally high extinction coefficient which does not depend upon base composition.Therefore they can be easily quantitated both in agar and polyacrylamide gel at 260 nm. Densitometers were constructed for scanning both agar gel (Tsanev and Staynov [194]) and polyacrylamide gel (Loening [195]). The detection limit in UV is 0.05 μg RNA in a 1 mm wide zone in 2.2–2.6% polyacrylamide gel. On the other hand the extinction coefficient for proteins at 280 nm is much lower than that of nucleic acids at 260 nm. Moreover it is dependent on the content of aromatic amino acids in the structure and therefore different proteins exhibit a considerable variability in extinction coefficients. Consequently ultraviolet densitometry of proteins does not allow direct quantitation and if used, special calibration graphs must be constructed for each protein band. According to Dravid et al. [192] the sensitivity limit of densitometric detection of proteins lies in the neighbourhood of 1 μg of protein. The sensitivity of UV detection could be increased by shifting the wavelength of the UV light to 230 mm (UV absorption band of the peptide bond), however practical application of this fact is grossly abolished by the high absorbancy of the polyacrylamide gel matrix in this region.

Scanning of polyacrylamide gel is done by using specifically constructed densitometers [196–200]. In many instances the marketed scanners are multi-purpose scanning devices suitable both for the evaluation of polyacrylamide gel electrophoretograms as well as flat bed chromatograms.

6.9.2 *Detection by fluorescence measurement*

Fluorescence can be used for detection in two different ways. Firstly, the separated compounds could be labeled with a fluorescent label (if not fluorescent as such), or UV absorbing compounds could be visualized as dark bands in matrices to which a fluorescent chromophore was added (fluorescence quenching). A variation of the second method was worked out by Eisinger [201] for polyacrylamide gel: the gel was placed between a quartz and a fluorescent glass plate and then illuminated with UV light. Ultraviolet absorbing compounds appear as dark bands on the glass plate. When electrophoresis is done on silica gel thin layer or cellulose, addition of a fluorescent indicator to the sorbent can be made use of in the conventional way.

In direct fluorescence of either labeled (dansylated) or unlabeled proteins one has to keep in mind that polyacrylamide gel itself is also a fluorescent material: its fluorescence appears when excited at 280 or 340 nm. These wavelengths are very close to the excitation wavelengths of native proteins and dansyl chloride or 8-anilino-1-naphthalenesulphonic acid-labeled proteins. On the other hand polyacrylamide exhibits maximum fluorescence emission at 340 and 460–520 nm, which is sufficiently far away from fluorescence emission of both unlabeled and labeled proteins. However, the broad fluorescence increases considerably the background. It is also possible to prepare non fluorescent polyacrylamide gels by, e.g., adding dithiothreitol or by replacing TEMED by sulphite [202].

Direct measurement of tryptophan fluorescence (280/340), as done by Easton et al. [202] or Isenberg et al. [204], has the sensitivity limit between 1–10 μg of protein. 8-anilino-1-naphthalenesulphonic acid exhibits bright fluorescence when adsorbed on proteins (or bound to them). In practice this can be exploited by carrying out polyacrylamide gel electrophoresis in the presence of this compound. The gels are suspended in ammonium sulphate (saturated solution diluted by 1/3 of its volume with water) and scanned. In labeling the proteins with dansyl chloride or fluorescamine, one has to keep in mind that considerable conformational changes may occur during the labeling reaction. Therefore SDS-polyacrylamide gel electrophoresis is the method of choice as long as it involves conformational alterations. The practical procedure for fluorescamine labeling runs as follows: protein (50–100 μg) is dissolved in 5% SDS, 5% sucrose, 0.011 M phosphate buffer (pH 8.5) and the solution is heated to 100°C for 5 min to ensure complexing with SDS. Then the solution is cooled off and 5 μl of a 1 mg/ml solution of fluorescamine in acetone is added. After electrophoresis the gels are evaluated with fluorescence excitation at 390 nm; fluorescence emission is measured at 475 nm. According to Ragland et al. [205] 0.5 μg of protein can be detected.

Fluorescence labeling of nucleic acids is limited to double-stranded structures. Both double-stranded DNA and RNA react with ethidium bromide under the formation of a fluorescent product (360/580) [206–208]. The sensitivity of this method is 50 μg for DNA and 100 μg for RNA. The differentiation between RNA and DNA can be done by using specific nucleases. According to Jovin [209] it is also possible to add ethidium bromide at a low concentration (0.1 relative to DNA phosphorus) before electrophoresis without any effect upon the subsequent separation.

6.9.3 Detection by staining

Staining of macromolecules with suitable dyes is no doubt the most popular method of detection in different types of gel electrophoresis. It can, however, be used also with other sorbents of definite structure like paper, cellulose acetate, etc. Because the efficiency of the method is based on non-specific sorption differences, the choice of the dye must be such that it will be strongly bound to the separated macromolecule but rather loosely bound to the stabilizing medium, e.g. polyacrylamide gel, agarose gel or cellulose supports. The dye can stain either specific classes of macromolecules (selective staining) or can stain several different classes of biopolymers simultaneously (nucleic acids, proteins, lipoproteins, lipids, glycoproteins, etc.) ("Stains-all", Dahlberg et al. [210]). The procedure itself is rather simple: the electrophoretogram (gel, paper strip) is immersed into the dye bath for a given period of time and then the excess of the dye not absorbed on the separated bands is washed out. Because diffusion starts immediately after the electrophoretic run has been finished, it is recommended to either precipitate the proteins within the gel or to dry the gel or other electrophoretic support. Specific devices are used for drying slab gels. The precipitation of proteins can also be done by adding a suitable fixative to the dye bath. There are many dyes commercially available, however even if the same type of dye is used for detection and quantitation, new calibration must be done with every new delivery as the individual batches of dye may differ quite considerably. The rate-limiting step of staining is the migration of the dye into the gel. This has a practical role namely during removing the excess of the dye from the unstained areas of the gel which might proceed slowly and be quite time consuming. Therefore, electrophoretic destaining was introduced by Schwabe [211]. In electrophoretic staining (according to the same author) the cathode compartment is filled with 2% Amido Black 10B in 5% acetic acid. 8 min electrophoresis is sufficient for staining and the removal of the excess of the dye is accomplished by exchanging the staining solution for dilute acetic acid and by further electrophoresis that is continued until the desired result is obtained. It is also possible to add the dye directly to the cathodic buffer, which is usually done a while after electrophoresis has been started [212]. Similar procedures were described in detail by Lushnikov and Balandin [213] and Holmberg and Huttunen [214]. Staining with countermigration is especially convenient where fixation of the bands is for some reasons impossible.

It has been mentioned already that destaining is positively the most time-consuming operation. Frequent exchanging of the destaining solution is necessary to ensure obtaining the results within a reasonable period of time. In a most primitive way electrophoretic destaining can be materialized in a beaker filled with destaining solution into which both electrodes and the gel to be destained are placed. The efficiency of this system is, however, low in comparison with the more sophisticated devices either commercially available or described in the literature [215–222].

The time necessary for protein staining can be considerably shortened especially in polyacrylamide gel isoelectric focusing [223]. The improvement is such that the staining time can be cut down to 3–5 min. This is achieved by staining dry gels on silanized supports instead of staining hydrated gels, by reducing the gel thickness, by using carrier ampholytes with good destaining properties (such as Servalyt), and using dyes with low affinity for carrier ampholytes (Serva violet 19 or 49). It is recommended to use 20% trichloroacetic acid for protein fixation as, e.g., 5–20% methyl acetate decreases the sensitivity of staining considerably. Adequate fixation is necessary particularly with basic, low molecular weight or hydrophobic proteins. The following procedure is recommended.

(*1*) Fixation of the gel at room temperature with 20% w/v trichloroacetic acid (30 s).

(*2*) Rinsing with the destaining solution (5 s).

(*3*) Drying (at 80°C for 90 s, silanized support).

(*4*) Staining with 0.5% Serva Violet 17 (or 49) in methanol-water-glacial acetic acid 25 : 65 : 10 (20 s) (freshly prepared).

(*5*) Destaining in methanol-water-acetic acid 25 : 65 : 10 v/v (1–2 min, the same solution as used in step 2).

(*6*) Drying as in step 3 or at room temperature (50 s).

We now present some widely used staining and destaining solutions.

(1) *Staining* of starch gel

Amido Black 10B	10 g
Methanol	450 ml
H_2O	450 ml
Glacial acetic acid	100 ml
Destaining solution	
Methanol	250 ml
Glacial acetic acid	100 ml
H_2O	750 ml

(2) *Staining* solution for polyacrylamide gels run in alkaline or acidic buffers and for urea-containing gels

1% Amido Black 10B in 7% acetic acid solution

Destaining and storage solution

7% acetic acid

(3) *Staining* solution for SDS-polyacrylamide gels

Coomassie Brilliant Blue R 250	1.25 g
Methanol	227 ml
Glacial acetic acid	46 ml
Water	to 500 ml

Staining is performed in test tubes (for rods) or troughs (for slabs) within 2–10 h.

Destaining solution	
Methanol	50 ml
Glacial acetic acid	75 ml
Water	to 1 l

(4) *Staining* solution for agarose gels

Amido Black 10B (or Coomassie Brilliant Blue R-250)	5 g
Ethanol (96%)	450 ml
Glacial acetic acid	100 ml
Water	450 ml

The dyes are added to the ethanol acetic acid mixture and left at room temperature overnight. The solution is then filtered and water is added.

TABLE 6.5

The scheme for the staining procedure for the silver based protein stain [226]

Step	Solution	Duration of agitation
1 Fixation	50% EtOH, 10% AcOH	2 h overnight
2 Washing	50% EtOH, 10% AcOH	2 h
3 Washing	25% EtOH, 10% AcOH	1 h, repeat once more
4 Washing	10% EtOH, 0.5% AcOH	1 h, repeat once more
5 Equilibrating	$AgNO_3$ (1.9 g/l)	2 h (maybe more)
6 Rinsing	H_2O	10–20 s
7 Reduction of silver	$NaBH_4$ (87.5 mg/l) NCHO (7.5 ml/l) in 0.75 M NaOH	10 min
8 Color enhancing	Na_2CO_3 (7.5 g/l)	1 h
9 Color enhancing	Na_2CO_3 (7.5 g/l)	1 h, suitable for storage

Destaining solution	
Ethanol	96%
Acetic acid	100 ml
Water	450 ml

6.9.3.1 Silver based staining of polypeptides

The individual steps required for this ultrasensitive stain are summarized in Table 6.5 [224]. The gels (as many as five at once) are placed in a pyrex tray and gently agitated. No special requirements with regard to temperature and light are required. During the procedure surface evaporation is prevented by covering the tray with a suitable lid or Saran wrap. After electrophoresis the polypeptides are fixed in the polyacrylamide matrix by means of 50% ethanol and 10% acetic acid (about fivefold excess over the volume of the gels). It is recommended to leave the gels in the fixative overnight and carry out the other steps next day. Care must be taken to remove SDS and carrier ampholyte. The solutions used in the staining step should first be degassed. Optimal color is obtained if a suitable proportion between gel thickness, protein concentration and silver nitrate concentration is ensured. Thus, for 1.5 m gels the recommended silver nitrate concentration is 1.9 g/l. The equilibration of the staining solution takes about 2 h and the optimum gel-to-solution ratio appears to be 1 : 3. The gels are then rinsed with H_2O to remove excess silver nitrate and reduced with 0.75 M NaOH containing 7.5 ml per litre formaldehyde and 87.5 mg/l $NaBH_4$. The solution volume should be about in fivefold excess in comparison to the volume of the gel. Formaldehyde should be added freshly before staining; the addition of $NaBH_4$ is not absolutely necessary. Protein spots appear within 5–6 min. After 10 min of staining the gels should be, for optimum results, transferred to a color enhancing solution.

The described procedure is the only one that gives a variety of colors which may help the identification of polypeptides separated. It has, however, to be noted that some colors are dependent on the protein concentration. Comparison of the results of different variations of the silver stain method [225,226] and other staining procedures [227,228] proved that the method of Oakley et al. [227] is simple and quickly performed. In the Sammons procedures [224] (see above) the washing steps are rather time consuming but offer superior results. The most complicated procedure is that of Switzer et al. [225]. It must also be noted that the methods of Oakley et al. [227] and Switzer et al. [225] require ammoniacal silver solution which is more difficult to handle than silver nitrate. Also the above described method is the only one that allows more than one gel to be stained at once in the same tray.

The silver stain that was developed for staining proteins on polyacrylamide gels can also be used for detecting DNA with a sensitivity of 1 μg ($< 0.03\ \mu g/mm^2$); this is superior to ethidium bromide [229].

6.9.4 *Scanning of electrophoretograms*

There are numerous commercially available densitometers suitable, or specifically designed, for this purpose. Their operating principle is very simple: The electrophoretogram is moved at constant speed and at a certain point is illuminated by a monochromatic narrow light beam of stabilized intensity. The transmitted or reflected light is measured and recorded. The non-translucent electrophoretograms are measured in reflected light only. In translucent gels both reflected and transmitted light can be used. In order to get a good electrophoretogram scan, several conditions have to be met.

(*1*) The cross section of the light beam must be small enough to ensure that the densitometer measures one band at one moment only.

(*2*) The detector must exhibit high sensitivity in measuring light intensity (and light intensity changes).

(*3*) The scanning speed must be such not to influence resolution by mechanical inertia of the transport system.

(*4*) The quality of the scanned electrophoretogram must be excellent.

(*5*) Good quantitation requires linear response of the detection system to the band intensity.

Recently a method for the rapid and highly sensitive evaluation of electrophoretograms was described [200] by using photographs of stained or fluorescence-labeled gels. A flying spot scanner was used; the negatives were enlarged, projected and the area chosen for evaluation was automatically scanned vertically and horizontally by using a computer system.

6.9.5 *Detection by radioactivity counting*

6.9.5.1 *Autoradiography and fluorography*

Detection of separated zones by radioactivity measurement gained considerable popularity mainly because it increases sensitivity of the estimation. The most simple variation is perhaps radioautography that can be performed either directly with wet gels [230] (^{35}S-labeled proteins) or with dried matrix. The latter procedure results in a further increase in sensitivity so that even 300 dpm of ^{3}H radioactivity can be detected by using 3–5 μm thick dried polyacrylamide gels. On the other hand ^{32}P-labeled compounds are almost exclusively detected on wet gels [231].

The procedure of autoradiography itself is very simple. Dried gel, cellulose acetate or filter paper sheet is covered with X-ray film, clamped between two metal plates and allowed to stand for an appropriate period of time. Then, the X-ray film is developed and zones of the radioactive material are visualized (see also Ref. 232). Also, polaroid land film can be used for this purpose [223].

The detection of ^{35}P and ^{14}C compounds at a sensitivity 100–10000 times higher than with X-ray film can be achieved with a position-sensitive counter, and the data are obtained in digital form as intensity vs distance migrated (no scanning is needed) [233]. Dried gels can be also evaluated by semiconductography [234]. The method has the advantage of determining several radionuclides simultaneously. The method is applicable for ^{32}P and ^{14}C, but cannot be used for tritiated compounds. Radioautography can be carried out also in the presence of 2,5-diphenyloxazole (PPO); the resulting scintillation on interaction with β-particles increases the sensitivity of the method about ten times for ^{14}C and ^{35}S. Moreover, it allows to detect ^{3}H radioactivity almost as sensitively as with other labels. Bands containing 500 dpm of ^{3}H are detectable over 1 week exposure [235]. Because this method has a general applicability, the actual procedure is presented here.

About 20 volumes of dimethylsulfoxide are used for soaking the gels, which are left in the solution for half an hour and the procedure is repeated once more. Then the gels are immersed in 4 volumes of 20% (w/w) PPO in dimethylsulfoxide (22.2% w/v) for 3 h. Finally, the gels are immersed in 20 volumes of water for 1 h and dried under vacuum for 1 h. Exposure with the X-ray film is done at −70°C. A variation of this procedure was worked out by Ramsden [236] who applied fluorography for detecting radioactive bands in immunoelectrophoretograms. The PPO solution is, for this purpose, prepared in ether and poured over agarose plates. After the ether has evaporated, the gel is dried and exposed on X-ray film.

Warning: as dimethylsulfoxide is highly toxic and penetrates the skin, all operations must be made with rubber gloves.

6.9.5.2 Spark chamber detection

This is a highly sensitive method applicable to γ- and β-nucleides. A commercially available Radiochromatogram Camera (LKB, Bromma, Sweden) is used for this purpose. The spark camera itself is, as a matter of fact, a triode; the cathode is represented by the radioactive electrophoretic zone, the tungsten grid and anode are mounted into the chamber. The spark chamber pictures are obtained with an instant camera (Polaroid); resolution of 0.5–0.7 mm is possible. Sensitivity reaches 100 pCi/mm^2 with tritiated compounds, or 10 pCi/mm^2 with ^{14}C-labeled compounds. The advantage of this equipment over autoradiography is the speed of evaluation which is gained at the expense of rather high costs and somewhat lower resolution.

6.9.5.3 Direct counting

The practice of slicing the gels or other supports and counting the individual fractions that have thus arisen is widespread. Gels, paper or cellulose strips are dried and, after cutting, immersed into the scintillation solution. The method of Boyd and Mitchell [237] is described: it is based on replacing water in the gel with apolar solvents. Gel slices are soaked in a solvent containing 2.5 g of phosphotungstic acid for 5 h (or more), then in acetic acid-ethylene glycol monoethyl ether (1 : 1) for at least another 5 h, and finally in the scintillation solution (3 g PPO, 0.16 g POPOP in 1 litre of toluene) for at least 3 h. The gels are sliced and counted in 5 ml of the last solution.

Also, direct counting of gel slices without any pretreatment of the gel is possible [238–240]. According to Sung et al. [241] starch gels can also be measured directly in Bray's solution containing 5% Cab-O-Sil (Cabot Corp. Boston, Mass.). The gel slices should be, however, homogenized before counting in 0.4 ml of trimethylammonium hydroxide. Efficiency is 23% for ^{3}H, and over 80% for ^{14}C.

6.9.5.4 Elution or solubilization of radioactive material

The elution method is rather laborious. The slices of the gel are eluted with, e.g., 0.1% SDS, preferably directly into the scintillation vials. Then the vials are shaken with 10 ml of Instagel emulsifier (scintillant Packard) for 3–5 h at 37°C and counted. Efficiency is 7% for ^{3}H and 33% for ^{14}C. Alternatively the radioactive materials can be solubilized and thus liberated from the electrophoretic support. This can be materialized by, e.g., Bio-Solv BBS-3 (Beckman Instruments, acidic surfactant). The scintillation solution used is composed of 4 g PPO, 0.2 g POPOP, 170 ml Bio-Solv-BBS 3, 60 ml of water, and 770 ml of toluene [242]. Another solubilizer that can be applied is NCS (Nuclear Chicago Solubilizer); with this solubilizer the polyacrylamide or polyacrylamide-agarose gel slices have to be incubated at 65°C for 2 h. Tritiated ribonucleic acid fractions can be solubilized with 10% piperidine at 60°C. Gel slices are incubated in the scintillating vials with the solubilizer for several hours, allowed to dry, dissolved in distilled water (the gel actually swells at this stage), covered with water miscible toluene scintillation fluid and counted.

Solubilization of proteins for radioactivity counting can be also effected by non-specific proteases or by chaotropic agents (guanidinium chloride). In the latter case gel slices are incubated with 1 ml of 6 M guanidinium hydrochloride at 90°C for 12 h. Then, 33% Triton X scintillation fluid is added and samples are counted. Efficiency is 12% for tritiated compounds.

6.9.5.5 Counting after combustion

Combustion methods are applied with very low radioactivities. $^{3}H_2O$, $^{14}CO_2$ and $^{35}SO_2$ can be determined either directly in the gas phase [243] or after trapping in liquid scintillation fluids. The enumeration of commercially available apparatuses for this purpose is beyond the scope of this review (see Refs. 244–250).

6.9.5.6 Disruption of the gel structure

Sonication or rough chemical treatment can be used for solubilization of conventional gels. Thus, starch gel sections are susceptible to disruption by sonication, while polyacrylamide gels, if crosslinked by Bis, are generally difficult to solubilize. Most authors use 30% H_2O_2 [251,252] or a mixture of 30% H_2O_2 and NH_4OH (99 : 1). Cleavage is done by incubation at 37°C for 4–8 h. After cooling 10 ml of a solution consisting of 1 part of Beckman BBS-3 solubilizer and 5 parts of toluene-based scintillation fluid are added and the samples counted. Efficiency is 13% for tritium and 50% for ^{14}C.

If, however, quantitation via gel disruption is foreseen, it is advisable to replace the Bis cross-linking reaction with another one that yields more easily solubilizable gels. Ethylenediacrylate is used for this purpose [253]. Dissolution can be done either by the action of concentrated NH_4OH [254] or by piperidine-alcoholic hyamine [253] (9 : 1). In the latter case gels are dissolved by incubation at 37°C for 3–4 h. Dissolution can be accelerated by addition of another portion of piperidine.

Bis can be also replaced by *N,N'*-dialkyltartardiamine and the resulting gel can be solubilized in 2% periodic acid [255]. This is done by incubating the gel at 37°C for 10 min. The optimum composition of the counting liquid for solubilized gels is [256]: 0.3% PPO, 25% (v/v) Triton X-114 in xylene. POPOP may be added to a final concentration of 0.2 g/l. Efficiency is 47% for tritium, 93% for ^{14}C.

6.10 Preparative procedures

6.10.1 Electrophoresis in columns

A number of apparatuses were constructed for this purpose [257–264], some of which are commercially available. Here we describe the apparatus marketed by LKB (Bromma, Sweden) (see also Ref. 264) (Fig. 6.26). The apparatus consists of a inverted U-shaped tube that is composed of two parts connected by a joint in the upper part. One limb accommodates the electrophoresis column that is surrounded by a covering jacket, the other limb filled with buffer serves to make connection with the electrode vessel. At the lower end of the column a collecting funnel is attached. Liquid connection or disconnection between the two limbs can be done through a ground female joint. The stopcock above this joint connects the system to a buffer reservoir when the column is washed or when the zones are eluted. In the male joint of the top piece there is a hole that allows liquid to pass to the other limb. By turning the liquid stream can be disconnected at this point. The column size is 55 × 2 cm; Sephadex or cellulose serve as the supporting material. Separation is run at 500–1000 V and 25–50 mA.

Electro-osmosis, which is a troublesome effect in this system, can be counteracted by carefully adjusting buffer flow in the direction of electrophoretic migration. The separation is usually started in such a way that the sample is placed in the column (together with some dye to visualize the movement) by making use of the reservoir buffer. Only then is the electricity switched on.

Electrophoresis in supports that themselves exhibit sieving effect improves the resolution. Diluted agarose (0.12–0.2%) was used as support by Hjertén [265];

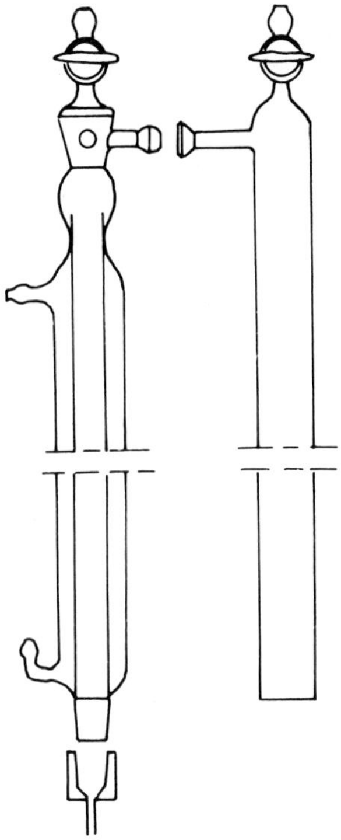

Fig. 6.26. Column for electrophoresis in granulated supports.

generally the agarose concentration is chosen just to prevent the sample from sedimentation. Agarose support is prepared by boiling the appropriate amount in water until all agarose has dissolved. Then the solution is left to cool and settle in a stoppered flask. Next day any film formed is removed, the solution is homogenized by gentle stirring. The lower end of the column is plugged by a piece of dialysis membrane, and the agarose homogenate is delivered to the column by means of a peristaltic pump. The sample should also contain agarose at the same concentration as the column support. The sample is applied to the column by means of a syringe equipped with a capillary tube to the optimal height of the column. After electrophoresis the contents of the column (the separated zones together with the support) are pumped out into a fraction collector. Agarose is then removed from individual fractions by centrifuging them at $10\,000 \times g$.

6.10.2 Preparative agar gel electrophoresis

Preparative agar gel electrophoresis can be done both in columns and thick layers. Most of the column equipment used for polyacrylamide gel electrophoresis can also be used for agar or agarose gel separations [226–268]. With the thin-layer technique the main problem is effective cooling; therefore Wieme [269] recommends to use long grooves instead of thick gels. The visualization of the separated components can be done either by UV photometry [270,194] or by the printing technique using Whatman paper. A sophisticated method for elution of the zones was elaborated by Gordon [111]. The stained paper print serves as a guide to localize the zones; then slits are made ahead of each zone, filled with electrophoresis buffer and electrophoresis is continued; at 2–4 h intervals the buffers are removed and tested for the presence of proteins. The elution is finished when no more protein material is eluted from the layer.

As long as it is carried out with dilute agarose gels, the elution and gel removal by, e.g., centrifugation is simple and offers satisfactory results.

6.10.3 Preparative electrophoresis in polyacrylamide gel

This category of separations is commonly carried in gel columns that are located between two electrode vessels. The elution chamber is positioned at the lower end of the column. The apparatuses are provided with either a cooling jacket or the column

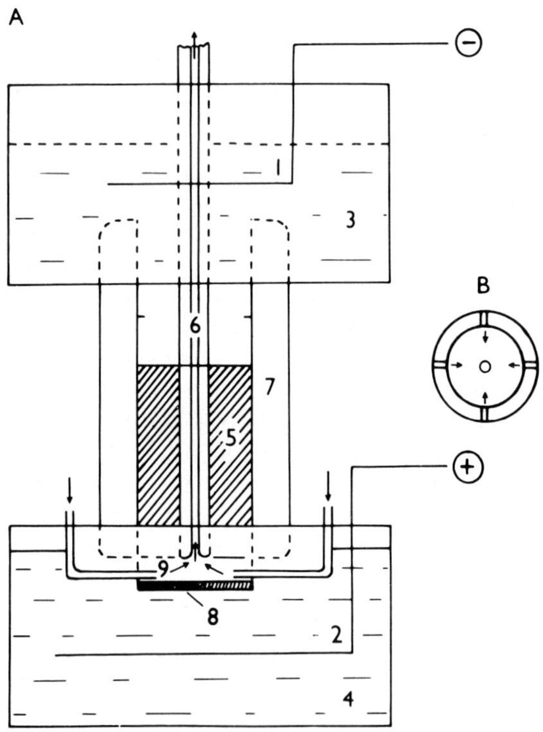

Fig. 6.27. A schematic representation of a preparative gel electrophoresis apparatus with continuous elution. 1, 2, Platinum electrodes; 3, upper buffer compartment; 4, lower buffer compartment; 5, separation gel; 6, cold finger with the central capillary; 7, cooling jacket; 8, glass membrane; 9, elution chamber; (B) birds eye cross-sectional view of the elution chamber.

has a central cooling system [271–274]. The latter authors described also the application of cylindrical gels, cast between two concentric tubes. Another method to ensure cooling is by using a cool elution buffer during the flow through the gel to fraction collector [275–277].

Besides cooling, the other important problem when using the continuous elution apparatus is the construction of the elution chamber so as to ensure rapid elution of the migrating components without excessive dilution (Fig. 6.27). For smaller devices (5–15 cm^2 cross-sectional area) allowing the separation of milligram amounts of protein mixtures, the apparatus constructed by Jovin et al. [276] is widely in use: the outer jacket and the inner finger provide cooling of the system during gel polymerization and electrophoresis. For the separation the elution chamber is attached to the lower end of the column replacing the originally present plastic bottom. The elution chamber consists of a porous glass plate covered with a dialysis membrane. Elution is effected from the periphery of the elution chamber of several inlets of the eluting buffer; the eluting buffer then traverses upwards and is led to a fraction collector. Both the upper and lower electrode vessels are provided with circular platinum electrodes. The following buffers and acrylamide solutions are used.

Separation gel (pH 8.9) (stock solutions)

(A) Acrylamide	30g
Bisacrylamide	0.8 g
H_2O	to 100 ml
(B) HCl (1 M)	24.0 ml
Tris	18.15 g
TEMED	0.23 ml
H_2O	add 100 ml
(C) Ammonium persulphate	0.14 g
H_2O	add 100 ml

Separation gel (7.5%) is prepared by mixing A : B : C = 1 : 1 : 2

Spacer gel (pH 6.9) (stock solutions).

(D) Acrylamide	5.0 g
Bisacrylamide	1.25 g
H_2O	to 100 ml
(E) H_3PO_4 (1 M)	12.8 ml
Tris	2.85 g
H_2O	to 100 ml
(F) Riboflavin	0.002 g
H_2O	to 100 ml

Spacer gel is prepared from the above solutions by mixing them in a ratio D : E : F = 2 : 1 : 1

Upper (cathodic) electrode buffer (pH 8.3)

Glycine	2.88 g
Tris	0.6 g
H_2O	to 1 l

Lower (anodic) electrode buffer (pH 8.1) (identical with the elution buffer)

HCl (1 M)	60.0 ml
Tris	to pH 8.1
H_2O	to 1 l

Electrode buffer (pH 8.3) according to Canalco

Tris	3.0 g
Glycine	14.4 g
H_2O	to 1 l

Elution buffer is identical with this but diluted with water 1 : 8

Continuous elution requires rather complex equipment, and dilution of zones, sometimes quite considerable, cannot be avoided. Therefore discontinuous elution systems are also in use. As an example of intermittent elution the apparatus of

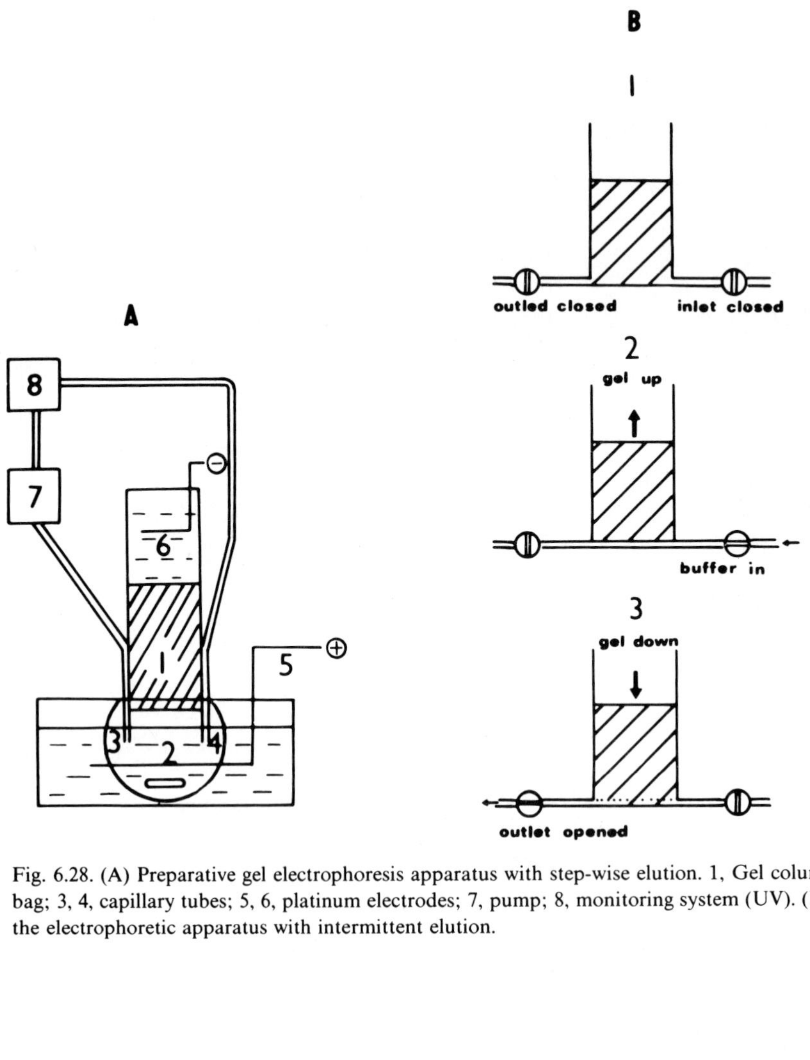

Fig. 6.28. (A) Preparative gel electrophoresis apparatus with step-wise elution. 1, Gel column; 2, dialysis bag; 3, 4, capillary tubes; 5, 6, platinum electrodes; 7, pump; 8, monitoring system (UV). (B) Principle of the electrophoretic apparatus with intermittent elution.

Marceau et al. [278] is described here (Fig. 6.28). Polyacrylamide gel is allowed to polymerize in a cylinder, the lower end of which is covered with a fritted disc covered with a dialysis membrane. The elution chamber is represented by the narrow space between the membrane and the lower surface of the gel. The buffer is closed by a valve. The Perspex material used for the construction of the cylinder allows displacement of the column along the longitudinal axis ensuring concomitantly a good contact with the wall. During elution the valve is opened and the buffer is expelled by hydrostatic pressure exerted on the top of the column.

Another arrangement of the discontinuous elution that allows the separation of up to 1 g of protein is materialized in a 5 × 14 cm gel cylinder [279]. The block is cooled at the upper and lower surfaces and the cylindrical part is provided with an insulating jacket to prevent heat losses. Intermittent elution and automated recycling of the eluted material lead to high resolution of the separated material.

There is a number of papers dealing with discontinuous elution in polyacrylamide gel electrophoresis and the reader looking for more detailed information is directed to Refs. 280–289.

In conclusion of this section it is necessary to mention the possibility to elute several polyacrylamide gel slabs or rods that are segmented, before elution, according to a marker gel (the method is analogous to preparative thin-layer or paper chromatography). More information in this respect can be gained from Refs. 290–292.

6.10.4 Preparative isoelectric focusing

6.10.4.1 Preparative isoelectric focusing in a density gradient

This type of isoelectric focusing is fairly common, apparently because of the commercial availability of the necessary equipment (LKB, Bromma, Sweden). The apparatus consists of a column provided with a finger inside and a jacket outside that ensure efficient cooling. The density gradient is established in the free space between the two cooling systems. The basic and acidic electrode solutions are linked with the ends of the density gradient. The cathode is usually composed of a wire annulus in the upper part of the apparatus, while the anode penetrates through the cooling finger to the bottom of the density gradient.

Density materials used are practically identical with those used on the analytical scale, e.g., sucrose, and less frequently glycerol, sorbitol, ethylene glycol, ficoll and dextran. It should be noted that for the isoelectric fractionation of proteins with p*I* values beyond 8.0 other density material than sucrose should be used (sucrose tends to be dissociated at high pH) (e.g., glycerol) (Fig. 6.29).

The sample may be introduced into the system by preparing the density gradient in an already diluted solution of the sample. In this arrangement, however, a part of the sample protein comes into contact with both electrodes and, moreover, the proteins have to traverse through pH areas that may cause denaturation. Another method is to apply the rather concentrated sample in a position that is close to the expected focusing position. In this case the density of the introduced sample should correspond to the density of the isoelectric focusing system at the position where the sample is loaded. If it is necessary to shorten maximally the running time, then the sample is applied after the pH gradient has been established. At this moment it is necessary to stress that the pH gradient should be chosen carefully as a steep pH gradient may lead to the excessive concentration of separated proteins in narrow bands; then the loading capacity of the system can be easily exceeded and droplet sedimentation of zones may occur. Another problem is tailing of zones that may be caused by protein precipitation, e.g., at their isoelectric points. This is in practice overcome by a suitable choice of polarity of the focusing system, by the method of sample introduction, and by the choice of the method of elution.

Most of the preparative scale isoelectric focusing separations are done at 1600 or 2000 V in 110 and 440 ml columns (commercially available). After the run has ended the column is emptied (before a certain volume of water is pumped on top of the column because hydrostatically forced emptying does not result in constant flow rate due to density differences). Recommended flow rates are 60 ml/h or 240 ml/h for the 110 ml and 440 ml columns, respectively.

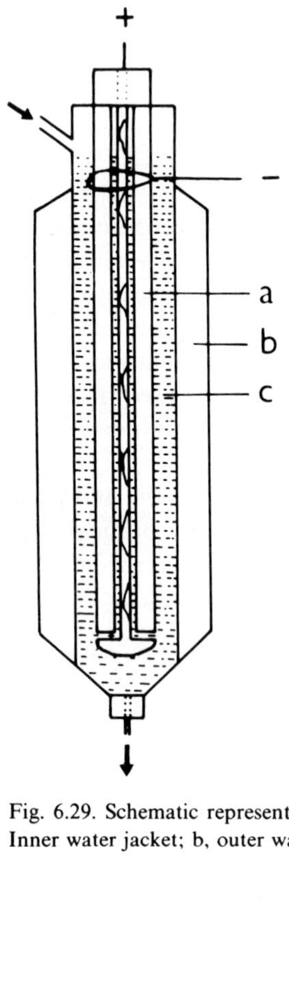

Fig. 6.29. Schematic representation of a column designed for preparative isoelectric focusing (LKB). a, Inner water jacket; b, outer water jacket; c, annulus between water jackets.

6.10.5 Preparative flat bed isoelectric focusing

Equipment for this separation procedure is commercially available, e.g., from Desaga (Heidelberg, FRG) or LKB (Bromma, Sweden). It consists of a horizontal trough with a glass plate and a cooling block. Plate dimensions 20 × 20 × 1 cm and 40 × 20 × 1 cm are available (Fig. 6.30). A homogenous suspension of the swollen gel (Sephadex G-50 or G-200 Superfine, Pharmacia, Uppsala) is introduced into the trough and excess of water is removed in a stream of air. The trough is then placed on the cooling block and connections are made with the electrode jars by means of paper wicks. These wicks are soaked in appropriate electrode buffers, e.g., 1 M H_2SO_4 and 2 M ethylene diamine. The loading capacity of this technique is high, and separations of up to 10 g of protein have been reported. Loading of the sample can be done by admixing it to the gel suspension before it is poured onto the trough, or it may be introduced as a zone in any position of the bed both before and after establishment of the pH gradient. Precipitation of proteins does not interfere so much with separation as with other isoelectric focusing procedures, as the precipitates are trapped within the gel. Elution of separated zones is done by sectioning the gel, transferring the section into small columns and eluting them by the elution buffer. Information about the position of individual zones can be obtained by the printing technique: a sheet of filter paper is pressed onto the gel layer and stained by the conventional procedure. The position of the zones within the gel is obtained by comparison with the print.

6.10.5.1 Continuous flow isoelectric focusing

The apparatus for continuous flow-through electrophoresis with slight modification is used for this procedure (see page 487). It consists of two glass plates (e.g.

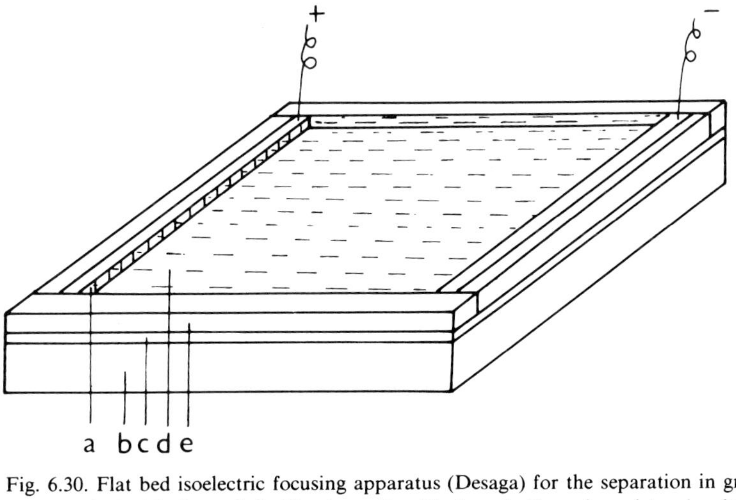

Fig. 6.30. Flat bed isoelectric focusing apparatus (Desaga) for the separation in granulated gel. a, Filter paper soaked with electrode buffer; b, cooling block; c, bottom glass plate; d, gel layer; e, trough.

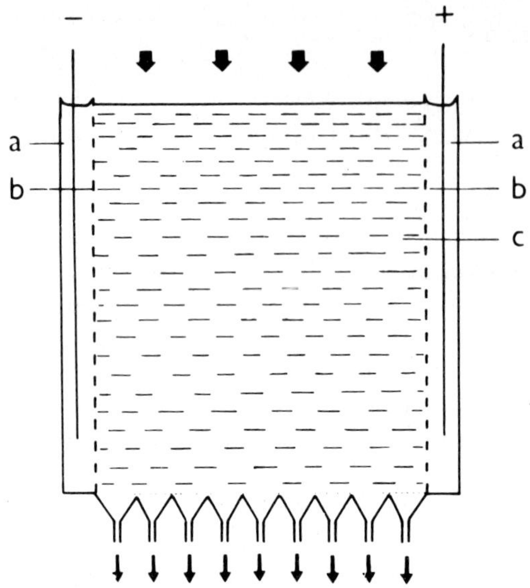

Fig. 6.31. Schematic arrangement of an apparatus for continuous flow isoelectric focusing. a, Electrode compartment; b, semi-permeable membrane; c, running space.

23 × 30 × 0.3 cm) filled with a suitable support preventing convection (e.g., Sephadex, polyacrylamide, sucrose) (Fig. 6.31). It is also possible to work without the anticonvection medium (free-flow), however in this case the space between the two plates must be more narrow (0.3–0.5 mm). The electrode compartments on both sides are separated by a dialysis membrane. The ampholyte-containing buffer is loaded via a multichannel proportional pump to ensure equal feeding of the system. Two other pumps ensure the exchange of electrode buffers. Since in most apparatuses effective cooling is provided, the system may be easily operated at high voltage gradients (50–100 V/cm). The capacity of the system is high: up to 500 mg can be separated per day. Sample loading is effected in a certain position of the plate and the sample can be applied along the whole width of the plate.

6.10.6 Preparative isotachophoresis

The advantage of isotachophoresis from the point of view of preparative scale operations is the increase in resolution with increasing load. A higher load increases the length of individual zones. Separation is carried out in an adapted LKB-preparative electrophoresis column (Uniphor) and the following buffer systems are recommended by Svendsen [293].

(A) *Gel buffers* (leading electrolyte, stock solutions)

(A-1) pH 4.0

Glacial acetic acid	30 ml
Tris	1 g
Water	to 100 ml
TEMED	300 μl

(A-2) pH 4.35

Glacial acetic acid	3 ml
Tris	1.7 g
Water	to 100 ml
TEMED	300 μl

(A-3) pH 4.8

Glacial acetic acid	30 ml
Tris	32 g
Distilled water	to 100 ml
TEMED	300 μl

(A-4) pH 6.0

MES (2-(*N*-morpholino)ethanesulphonic acid)	7.3 g
Tris	1.5 g
Distilled water	to 100 ml
TEMED	300 μl

(A-5) pH 6.2

MES	7.3 g
Tris	20 g
Distilled water	to 100 ml
TEMED	300 μl

(A-6) pH 6.25

1 M H_3PO_4	30 ml
Tris	4.0 g
Distilled water	to 100 ml
TEMED	300 μl

(A-7) pH 7.05

1 M H_3PO_4	30 ml
Tris	6.0 g
Distilled water	to 100 ml
TEMED	300 μl

(A-8) pH 8.1

1 M H_3PO_4	30 ml
Tris	12.0 g
Distilled water	to 100 ml
TEMED	300 μl

(A-9) pH 8.5

1 M H_3PO_4	30 ml

Tris	22.0 g
Distilled water	to 100 ml
TEMED	300 μl

All buffers are diluted with distilled water 1 : 9, stability of pH must be checked after dilution.

(B) *Elution electrolyte* (identical with the anode buffer)

(B-1) pH 7.1

1 M H_2SO_4	121 ml
Tris	32.0 g
Distilled water	to 4000 ml

(C) *Terminating electrolyte and cathode buffer*

(C-1) pH 8.9

EACA (ε-aminocaproic acid)	60.0 g
Tris	3.0 g
Distilled water	to 2000 ml

(D) *Gel solutions*

(D-1)

Acrylamide	38.0 g
DATD (*N*, *N'*-diallytartardiamide)	2.0 g
Distilled water	to 100 ml

(D-2)

Acrylamide	38.8 g
Bis	1.2 g
Distilled water	to 100 ml

It is recommended to treat the gel solutions batchwise with a mixed-bed ion exchange resin and remove the ion exchanger by filtration. The treated gel solutions are stable in a cold room over a period of several weeks.

(E) *Catalyst solution*

(E-1)

Riboflavin 5'-phosphate	6.0 mg
Distilled water	to 100 ml

(E-2)

Ammonium dioxypersulphate	200 mg
Distilled water	to 100 ml

The catalyst should be freshly prepared every week. Storage in dark bottles.
The mixture is prepared according to the following scheme.

Gel solution (either D-1 or D-2)	6.25 ml
Gel buffer stock solution	5.0 ml
E-1	5.0 ml
E-2	5.0 ml
Distilled water	to 50 ml

The gels are overlayered with distilled water and polymerized by UV illumination. Polymerized gels can be stored for several weeks provided that the overlaying water is replaced by gel buffer diluted 1 : 9, and the gel is sealed. Acidic buffers should be polymerized for a prolonged period of time (3 h).

Preparative isotachophoresis can also be done without supporting gel using a capillary system. The procedure was described by Arlinger [293]; the principle is obvious from Fig. 6.32.

6.10.7 Continuous flow through electrophoresis

This procedure represents a preparative version of zone electrophoresis. The apparatus and technique has been previously elaborated [294,295]; separation itself is carried out either in a stream of electrolytic solution or on a sheet of cardboard (curtain electrophoresis). In the free flow version the separation is carried out in a cell formed by two parallel glass plates (50 × 50 cm) situated 0.5–1.0 mm apart. It is necessary to ensure an equal and laminar flow of the electrolyte, which is carried out by feeding the buffer through a multichannel peristaltic pump. The sample is continuously applied in the middle near the upper edge of the cuvette (or paper sheet). The electrophoretic separation occurs transversally between vertical electrodes located on the right and left hand side of the separation cuvette (Fig. 6.33). The separated fractions are collected at the lower end of the cell by a system of small communicating vessels or by a multi-channel pump. In the version using paper the

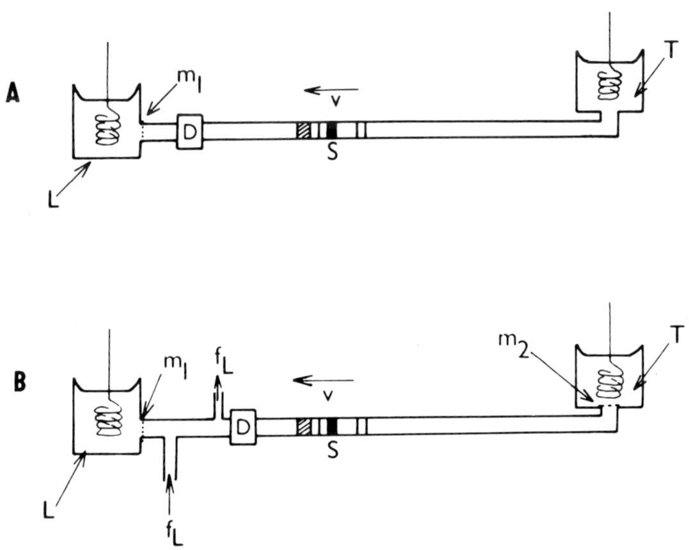

Fig. 6.32. The principle of preparative capillary isotachophoresis. (A) Schematic representation of the analytical equipment. (B) The preparative version with the additional flow of the leading electrolyte. L, Leading electrolyte; T, terminating electrolyte; D, UV detector; S, sample; v, migration velocity; m_1, m_2, semipermeable membranes; f_L, flow of L to elute the sample zones.

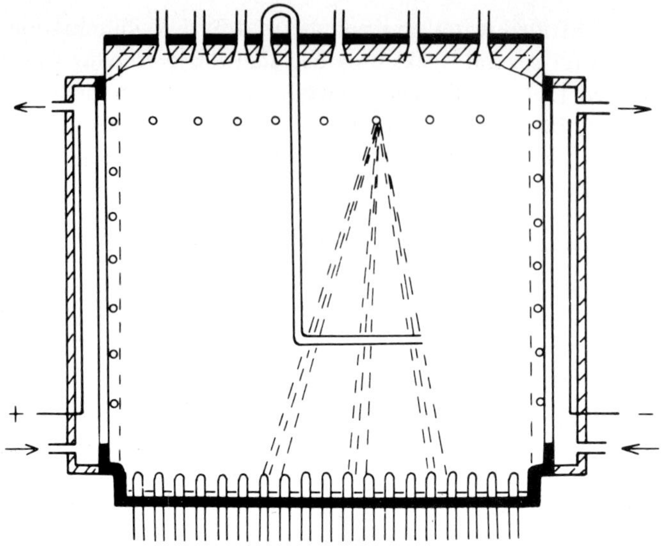

Fig. 6.33. Schematic representation of a cell for liquid film continuous electrophoresis.

lower end of the cardboard sheet is cut to form a multi V-shaped comb and fractions are collected into test tubes placed below the individual V's of the comb.

In the carrier-free version the buffer flowing through the cuvette is linked through a membrane with the electrode chambers. Ion exchange membranes are used for this purpose as long as the electrical resistance of such membranes is small, and therefore there is not a voltage gradient jump at the boundary of the separating cell and the electrode chamber. When cardboard is used as electrophoretic support the paper sheet is extended so that it is directly attached to the electrode compartments. The electrode buffer is steadily replaced within the electrode jars by means of a pump. A highly efficient cooling system must be incorporated both in the support and support free version. In the carrier type of continuous electrophoresis the cardboard is not in contact with a cooling plate and the amount of Joules is removed with a stream of cold air. The losses of the solvent due to evaporation are compensated by another wetted paper sheet, the temperature of which is about 5°C higher than that of the separating cardboard. In the free flow system one of the glass plates forming the separation cell is attached by a thin layer of silicon oil to a copper plate to which Peltier cooling elements are mounted.

The version without support has, especially, the advantage of circumventing the problem of recovering the separated material from the supporting medium; therefore, the procedure appears suitable for the separation of substances sensitive to interaction with the support. Diffusion can be hindered by incorporating 10% sucrose into the buffer solution. The system can also be used for the separation of cells and cellular particles that are too big to be separated by any other electrophoretic method.

The apparatus of Hannig is most well known; however apparatuses of cylindrical or disc shape, or exploiting the effects of the magnetic field, have also been described (for review see Ref. 296).

6.11 Drying of polyacrylamide gels

It has been mentioned in the preceding Section that for some operations it is necessary to use dried polyacrylamide gels. Several methods have been elaborated for removal of water from the gel matrix. Thus, for instance, the longitudinally sliced gel is put on a moistened filter paper and both are placed on a small pore suction filter. The gel surface is covered by a sheet of plastic film and the edges are sealed by means of a rubber gasket and silicone grease. Drying is effected by suction and irradiating the gel from a distance by an infrared lamp.

Another procedure was described by Woodworth and Clark [297]: the gel is put into a solution containing 90% ethanol and 5% 1,2-propanediol, the gel is then covered from both sides with two sheets of cellulose membrane and placed on a bed of dried silica gel. It is also possible to cover the gel with porous hydrophilic polyethylene and dry the polyacrylamide gel in vacuo [298]. According to Reid and Bieleski [299] the gel slab is divided into two halves, placed onto a glass plate and molten 2% agar is poured over the gel. Agar is dried under an infrared lamp and the polyacrylamide layer is dried by standing at room temperature. Microgels can be dried by freeze drying [300]. For other methods see Ref. 301.

References

1 Cremer, D. and Tiselius, A. (1950) Biochem. Z. 320, 273.
2 Durrum, E.L. (1950) J. Am. Chem. Soc. 72, 2943.
3 Kraus, K. and Smith, G. (1950) J. Am. Chem. Soc. 72, 4329.
4 McDonald, H., Urbin, M. and Williamson, M. (1950) J. Am. Chem. Soc. 112, 227.
5 Turba, F. and Enenkel, H. (1950) Naturwissenschaften 37, 93.
6 Grassman, W. and Hannig, K. (1950) Naturwissenschaften 37, 496.
7 Grassman, W. and Hannig, K. (1955) Hoppe Seyler's Z. Physiol. Chem. 290, 1.
8 Williams, F.G., Pickels, E.G. and Durrum, E.L. (1955) Science 121, 829.
9 Michl, H. (1951) Monatsh. Chem. 82, 489.
10 Ryle, A.P., Snager, F., Smith, L.F. and Kitai, R. (1955) Biochem. J. 60, 541.
11 Gross, D. (1961) J. Chromatogr. 5, 1.
12 Blackburn, S. (1965) Methods Biochem. Anal. 13, 1.
13 Werner, G. and Westphal, O. (1955) Angew. Chem. 67, 251.
14 Prusík, Z. and Keil, B. (1960) Collect. Czech. Chem. Commun. 25, 2049.

15 Ostrowski, W. and Keawczyk, A. (1966) Acta Biochim. Polon. 13, 121.
16 Wieland, T. and Pfleiderer, G. (1955) Angew. Chem. 67, 257.
17 Durrum, E.L. (1951) J. Colloid. Sci. 6, 274.
18 Strain, H.H. (1952) Anal. Chem. 24, 356.
19 Mikeš, O. and Holeyšovský, V. (1958) Collect. Czech. Chem. Commun. 23, 524.
20 Hartley, B.S. (1970) Biochem. J. 119, 805.
21 Naughton, M.A. and Hagopian, H. (1962) Anal. Biochem. 3, 276.
22 Ingram, U.M. (1958) Biochim. Biophys. Acta 28, 539.
23 Chen, K.C.S. and Krause, R.M. (1975) Anal. Biochem. 69, 180.
24 McDonald, H.J. and Urbin, M.C. (1953) Fed. Proc., Fed. Am. Soc. Exp. Biol. 12, 243.
25 McDonald, H.J., Lappe, R.J., Marbach, E.P., Spitzer, R.H. and Urbin, M.C. (1955) Ionography: Electrophoresis in Stabilized Media, Year Book Publ. Chicago, IL.
26 McDonald, H.J., Ribeiro, L.P. and Banaszak, L.J. (1959) Anal. Chem. 31, 825.
27 Kolin, A. (1953) Science 117, 134.
28 Kowalczyk, J. and Pompowski, T. (1961) Wiad. Chem. 15, 107.
29 Katsuno, A. (1961) Bull. Yamaguchi Med. Sch. 8, 1.
30 Nakamura, S. (1966) Cross Electrophoresis — Its Principle and Applications, pp. 28–38, Igaku Shoin, Tokyo and Elsevier, Amsterdam.
31 Nakamura, S., Takeo, K., Tanaka, K. and Ueba, T. (1960) Hoppe Seyler's Z. Physiol. Chem. 318, 115.
32 Grassman, W. and Hübner, L. (1953) Naturwissenschaften 40, 272.
33 Nakamura, S., Takeo, K. and Sasaki, I. (1963) Hoppe Seyler's Z. Physiol. Chem. 328, 139; ibid (1963) 334, 95.
34 Bickel, M.H. and Bovet, D. (1962) J. Chromatogr. 8, 466.
35 Kohn, J. (1957) Biochem. J. 65, 9.
36 Přistoupil, T.I. (1966) Nature (London) 212, 121.
37 Přistoupil, T.I. (1966) Biochim. Biophys. Acta 117, 475.
38 Yaron, A. and Sober, H.A. (1965) Anal. Biochem. 12, 173.
39 Huehns, E.R. and Jakubovic, O. (1960) Nature (London) 186, 729.
40 Alberti, G. (1966) Chromatogr. Rev. 8, 246.
41 Quereshi, M., Quereshi, S.Z., Gupta, J.P. and Rathore, H.S. (1972) Sep. Sci. 7, 615.
42 Quereshi, M. and Israeli, A.H. (1968) Anal. Chim. Acta 41, 523.
43 De, A.K. and Pal, B.K. (1980) Sep. Sci. Technol. 15, 155.
44 De, A.K. and Chakraborty, P. (1981) Electrophoresis 2, 330.
45 Sanger, F., Brownlee, G.G. and Barrell, B.G. (1965) J. Mol. Biol. 13, 373.
46 Murray, K. (1970) Biochem. J. 118, 831.
47 Crossley, J., Mansbridge, J. and Williamson, R. (1974) Biochem. J. 131, 299.
48 Dube, S.K., Marcher, K.A., Clark, B.F.C. and Cory, S. (1969) Eur. J. Biochem. 8, 244.
49 Southern, E.M. and Mitchell, A.R. (1971) Biochem. J. 123, 613.
50 Konrad, M.W. (1973) Anal. Biochem. 54, 213.
51 Porter, A.G., Hindley, J. and Milleter, M.A. (1974) Eur. J. Biochem. 41, 413.
52 Contreras, R. and Fiers, W. (1975) Anal. Biochem. 67, 319.
53 Van Ormondt, H. and Hattmann, S. (1976) Anal. Biochem. 74, 207.
54 Lankowitz, A.M. and Luck, D.J.L. (1976) J. Biol. Chem. 251, 3081.
55 Briand, J.P., Jonard, G., Guilley, H., Richards, K. and Hirth, L. (1977) Eur. J. Biochem. 72, 453.
56 Edström, J.E. (1953) Biochim. Biophys. Acta 12, 361.
57 Edström, J.E. (1956) Biochim. Biophys. Acta 22, 378.
58 Marchalonis, J.J. and Nossal, G.J.V. (1968) Proc. Natl. Acad. Sci U.S.A. 61, 860.
59 Tufts, B.J. (1961) Anal. Chim. Acta 25, 322.
60 Van Dijk, Ijsseling, F.P. and Loman, H. (1962) Anal. Chim. Acta 27, 563.
61 Matioli, G.T. and Niewisch, H.B. (1965) Science 150, 1824.
62 Hydén, H., Bjustam, K. and McEven, B. (1966) Anal. Biochem. 17, 1.

63 Grunbaum, B.W. and Kirk, P.L. (1960) Anal. Chem. 32, 564.
64 Grunbaum, B.W., Zec, J. and Durrum, E.L. (1963) Microchem. J. 7, 41.
65 Ostrowski, W. (1979) In Electrophoresis: A Survey of Techniques and Applications, Part A, Techniques (Deyl, Z., Everaerts, F.M., Prusík, Z. and Svendsen, P.J., Eds.), Elsevier, Amsterdam, Oxford, New York.
66 Paul, M.H. and Durrum, E.L. (1952) J. Am. Chem. Soc. 74, 4721.
67 Oehme, F. and Rauschenbach, I. (1956) Chem. Tech. 8, 21.
68 Krawczyk, A. and Dziembor, E. (1971) Bull. Acad. Polon. Sci. 19, 81.
69 Preetz, W. and Pfeifer, H.L. (1969) J. Chromatogr. 41, 500.
70 Gordon, T.H., Thornburg, W. and Werum, L.N. (1962) J. Chromatogr. 9, 44.
71 Brattsten, I., Synge, R.L.M. and Watt, W.B. (1965) Biochem. J. 97, 678.
72 Bagdasarian, M., Matheson, N.A., Synge, R.L.M. and Youngson, M.A. (1964) Biochem. J. 91, 91.
73 Thorum, W. and Mehl, E. (1968) Biochim. Biophys. Acta 160, 132.
74 Demus, H. and Mehl, E. (1970) Biochim. Biophys. Acta 203, 291.
75 Pusztai, A. and Watt, W.B. (1971) Biochim. Biohpys. Acta 251, 158.
76 Pusztai, A. and Watt, W.B. (1973) Anal. Biochem. 54, 58.
77 Pinder, J.C., Staynov, D.Z. and Gratzer, W.B. (1974) Biochemistry 13, 5373; ibid (1974) 13, 5367.
78 Gassen, H.G. (1969) J. Chromatogr. 39, 147.
79 Hanson, L.A., Johansson, B.G. and Rymo, L. (1966) Clin. Chim. Acta 14, 391.
80 Alberti, G., Grassini, G. and Trucco, R. (1962) J. Electroanal. Chem. 3, 283.
81 Honig, E.P. and Ketelaar, J.A.A. (1966) Trans. Faraday Soc. 62, 190.
82 Arnikar, H.J. (1957) C.R. Acad. Sci. 344, 2241.
83 Katelaar, J.A.A. and Kivak, J.C.T. (1967) J. Phys. Chem. 71, 1149.
84 Alberti, G., Conte, A. and Allulli, S. (1965) J. Chromatogr. 18, 564.
85 Dole, V.P. (1945) J. Am. Chem. Soc. 67, 1119.
86 Alberty, R.A. (1950) J. Am. Chem. Soc. 72, 2367.
87 Nichol, J.C. (1950) J. Am. Chem. Soc. 72, 2367.
88 Tondeur, D. and Dodds, J.A. (1972) J. Chem. Phys. 3, 441.
89 Brouwer, G. and Postema, G.A. (1970) J. Electrochem. Soc. 117, 7; ibid (1970) 117, 874.
90 Longsworth, L.G. (1959) In Electrophoresis (Theory, Methods and Applications) (Bier, M., Ed.) p. 137, Academic Press, New York.
91 Everaerts, F.M. and Beckers, J.L. In Electrophoresis, A Survey of Techniques and Applications, Part A, Techniques (Deyl, Z., Everaerts, F.M., Prusík, Z. and Svendsen, P.J., Eds.) p. 156, Elsevier, Amsterdam.
92 Smithies, O. (1955) Biochem. J. 61, 629.
93 Hrkal, Z. (1979) In Electrophoresis, A Survey of Techniques and Applications, Part A, Techniques (Deyl, Z., Everaerts, F.M., Prusík, Z. and Svendsen, P.J., Eds.), p. 113, Elsevier, Amsterdam.
94 Weber, K. and Osborn, M. (1969) J. Biol. Chem. 244, 4406.
95 Ornstein, L. (1964) Ann. N.Y. Acad. Sci. 121, 321.
96 Davis, B. (1964) Ann. N.Y. Acad. Sci. 121, 404.
96a Ferguson, K.A. (1964) Metabolism 13, 985.
97 Besaw, L.C., Dalbey, P.L. and Lillewick, H.A. (1972) Anal. Biochem. 50, 632.
98 Choules, G.L. and Zimm, B.H. (1965) Anal. Biochem. 13, 336.
99 Matson, C.F. (1965) Anal. Biochem. 13, 294.
100 Gelfi, C. and Righetti, P.G. (1981) Electrophoresis 2, 213.
101 Gelfi, C. and Righetti, P.G. (1981) Electrophoresis 2, 220.
102 Maurer, H.R. and Dati, F.A. (1971) In Protides and Biological Fluids (Peeters, H., Ed.) p. 569, Pergamon Press, Oxford.
103 Maurer, H.R. and Dati, F.A. (1972) Anal. Biochem. 46, 19.
104 Margolis, J. and Kenrick, K.G. (1968) Anal. Biochem. 25, 347.
105 Reynolds, J.A. and Tanford, C. (1970) J. Biol. Chem. 245, 5161.
106 Shapiro, A.L., Vinuela, E. and Maizel, J.V. (1967) Biochem. Biophys. Res. Commun. 28, 815.

107 Manwell, C. (1977) Biochem. J. 165, 487.
108 Reynold, J.A. and Tanford, C. (1970) Proc. Natl. Acad. Sci. U.S.A. 66, 1002.
108a Kurian, P., Gersten, D.M., Suhocki, P.V. and Ledley, G. (1981) Electrophoresis 2, 184.
109 Williams, J.G. and Gratzer, W.B. (1971) J. Chromatogr. 57, 121.
110 Deyl, Z. (1979) In Electrophoresis, A Survey of Techniques and Applications, Part A, Techniques (Deyl, Z., Everaerts, F.M., Prusík, Z. and Svendsen, P.J., Eds.), p. 45, Elsevier, Amsterdam.
111 Gordon, A.H. (1975) Electrophoresis of Proteins in Polyacrylamide and Starch Gels, North Holland/American Elsevier, Amsterdam, New York.
112 Hjelmeland, L.M. and Chrambach, A. (1981) Electrophoresis 2, 1.
112a Rothe, G.M. and Purkhambaba, H. (1981) Electrophoresis 2, 33.
113 O'Farell, P.H. (1971) J. Biol. Chem. 250, 4007.
114 Kaltschmidt, E. and Wittmann, H.G. (1970) Anal. Biochem. 36, 401.
115 Hukeshoven, J. and Dernick, R. (1981) Electrophoresis 2, 91.
116 Altland, K., Rauh, S. and Hackler, R. (1981) Electrophoresis 2, 148.
117 Lam, K.S. and Kasper, C.B. (1980) Anal. Biochem. 108, 220.
118 Sammons, D.W., Adams, L.D. and Nishizawa, E.E. (1981) Electrophoresis 2, 135.
119 Hjertén, S. (1961) Biochim. Biophys. Acta 53, 514.
120 Hoch, H. and Lewallen, C.G. (1977) Anal. Biochem. 78, 312.
121 Johansson, B.G. (1972) J. Clin. Lab. Invest. 29, Suppl. 124, 71.
122 Peacock, A.C. and Dingman, C.W. (1968) Biochemistry 7, 668.
123 Grabar, P. and Williams, C.A. (1953) Biochim. Biophys. Acta 10, 193.
124 Laurell, C.P. (1967) In Protides and Biological Fluids (Peeters, H., Ed.) p. 499, Pergamon Press, Oxford.
125 Laurell, C.P. (1965) Anal. Biochem. 10, 358.
126 Clarke, H.G.M. and Freeman, T. (1967) In Protides and Biological Fluids (Peeters, H., Ed.) p. 503, Pergamon Press, Oxford.
127 Krøll, J. (1968) Scand. J. Clin. Lab. Invest. 22, 112.
128 Svendsen, P.J. (1979) In Electrophoresis, A Survey of Techniques and Applications, Part A, Techniques (Deyl, Z., Everaerts, F.M., Prusìk, Z. and Svendsen, P.J., Eds.) p. 133, Elsevier, Amsterdam.
129 Arquembourg, P.C. (1975) Immunoelectrophoresis, Theory, Methods, Identification, Interpretation, S. Karger, Basel.
130 Svensson, H. (1961) Acta Chem. Scand. 15, 325.
131 Rilbe, H. (1973) Ann. N.Y. Acad. Sci. 209, 11.
132 Catsimpoolas, N. (1979) In Electrophoresis, A Survey of Techniques and Applications, Part A, Techniques (Deyl, Z., Everaerts, F.M., Prusík, Z. and Svendsen, P.J., Eds.) p. 167, Elsevier, Amsterdam.
133 Svensson, H. (1962) Arch. Biochem. Biophys. Suppl. 1, 132.
134 Vesterberg, O. and Svensson, H. (1966) Acta Chem. Scand. 20, 820.
135 Svensson, H. (1962) Acta Chem. Scand. 16, 456.
136 Vesterberg, O. (1973) Ann. N.Y. Acad. Sci. 209, 23.
137 Bates, L.S. and Degoe, C.W. (1972) J. Chromatogr. 73, 296.
138 Chrambach, A., Doerr, P., Finlayson, G.R., Miles, L.E.M., Sherins, R. and Rodbard, D. (1973) Ann. N.Y. Acad. Sci. 209, 44.
139 Catimpoolas, N. (1973) Anal. Biochem. 54, 66.
140 Gasparic, V. and Rosengren, A. (1975) In Isoelectric Focusing (Arbuthnott, J.P. and Beeley, J.A., Eds.) p. 178, Butterworths, London.
141 Låås, T., Olsson, I. (1981) Electrophoresis 2, 235.
142 Catsimpoolas, N. (1970) Sep. Sci. 5, 523.
143 Wrigley, C.V. (1971) In New Techniques in Amino Acid, Peptide and Protein Analysis (Niederwieser, A. and Pataki, G., Eds.) p. 291, Ann Arbor Sci. Publ., Ann Arbor, MI.
144 Kinzhofer, A. and Radola, B.J. (1981) Electrophoresis 2, 174.

145 Leaback, D.H. and Rutter, A.C. (1968) Biochem. Biophys. Res. Commun. 32, 447.
146 Awdeh, Z.L., Williamson, A.R. and Askonas, B.A. (1968) Nature (London) 219, 66.
147 Catsimpoolas, N. (1968) Anal. Biochem. 26, 480.
148 Radola, B.J (1969) Biochim. Biophys. Acta 194, 335; ibid (1973) 235, 412.
149 Catsimpoolas, N. (1973) Sep. Sci. 8, 71.
150 Haglund, H. (1971) Methods Biochem. Anal. 19, 64.
151 Catsimpoolas, N. (1979) In Electrophoresis, A Survey of Techniques and Applications, Part A, Techniques (Deyl, Z., Everaerts, F.M., Prusík, Z. and Svendsen, P.J., Eds.) p. 167, Elsevier, Amsterdam.
152 Lundahl, P. and Hjertén, S. (1973) Ann. N.Y. Acad. Sci. 209, 94.
153 Hjertén, S. (1967) Chromatogr. Rev. 9, 122.
154 Bours, J. (1976) In Isoelectric Focusing (Catsimpoolas, N., Ed.) p. 209, Academic Press, New York.
155 Dale, G. and Latner, A.L. (1968) Lancet 847.
156 Catsimpoolas, N. (1969) Biochim. Biophys. Acta 27, 365.
157 Kenrick, K.G. and Margoles, J.J. (1970) Anal. Biochem. 33, 204.
158 Catsimpoolas, N. (1969) Sci. Tools 16, 1.
159 Riley, R.F. and Coleman (1968) J. Lab. Clin. Med. 72, 714.
160 Catsimpoolas, N. (1969) Immunochemistry 6, 501.
161 Rotbøl, C.L. (1970) Clin. Chim. Acta 29, 101.
162 Laurel, B.L. (1965) Anal. Biochem. 10, 358.
163 Carrel, S., Theilkaes, L., Skvaril, S. and Barandun, S. (1969) J. Chromatogr. 45, 483.
164 Grabar, P. and Burtin, P. (1964) Immunoelectrophoretic Analysis, Elsevier, New York.
165 Catsimpoolas, N. (1974) In Electrophoresis and Isoelectric Focusing in Polyacrylamide Gel (Allen, R.C. and Maurer, H.R., Eds.) p. 174, Walter de Gruyter, Berlin.
166 Catsimpoolas, N. (1975) In Isoelectric Focusing (Arbuthnott, J.P. and Beeley, J.A., Eds.) p. 58, Butterworths, London.
167 Catsimpoolas, N. (1976) In New Developments in Separation Methods (Grushka, E., Ed.) p. 79, Marcel Dekker, New York.
168 Martin, A.J.P. and Everaerts, F.M. (1967) Anal. Chim. Acta 38, 223.
169 Arlinger, L. and Routs, R. (1970) Sci. Tools 17, 21.
170 Verheggen, Th.P.E.M. (1972) J. Chromatogr. 64, 185.
171 Everaerts, F.M. (1979) In Electrophoresis, A Survey of Techniques and Applications, Part A, Techniques (Deyl, Z., Everaerts, F.M., Prusík, Z. and Svendsen, P.J., Eds.), p. 193, Elsevier, Amsterdam.
172 Deml, M., Boček, P. and Janák, J. (1975) J. Chromatogr. 106, 283.
173 Vestermark, A. (1973) Acta Chim. Scand. 27, 377.
174 Sjödin, B. and Vestermark, A. (1972) J. Chromatogr. 73, 219.
175 Verstermark, A. and Sjödin, B. (1972) J. Chromatogr. 73, 211; ibid (1972) 72, 588.
176 Routs, R.J. Thesis Electrolyte Systems in Isotachophoresis and Their Application to Some Protein Separations, Universitetsforlaget, University of Technology, Eindhoven.
177 Catsimpoolas, N. (1972) Biochim. Biophys. Acta 285, 287.
178 Davies, H. and Rosenbaum, T. (1972) LKB Application note No. MLB/an-21, p. 1.
179 Hořejší, V. (1981) Anal. Biochem. 112, 1.
180 Bøg-Hansen, T.C., Bjerrum, O.J. and Brogren, C.H. (1977) Anal. Biochem. 81, 78.
181 Bøg-Hansen, T.C. and Takeo, K. (1980) Electrophoresis 1, 67.
182 Hořejší, V. (1979) J. Chromatogr. 178, 1.
183 Hořejší, V. and Tichá, M. (1981) J. Chromatogr. 216, 43.
184 Hořejší, V. and Kocourek, J. (1974) Methods Enzymol. 34, 178.
185 Siepmann, R. and Stegemann, H. (1967) Naturwissenschaften 54, 116.
186 Takeo, K. and Nakamura, S. (1972) Arch. Biochem. Biophys. 153, 1.
187 Hořejší, V., Tichá, M. and Kocourek, J. (1977) Biochim. Biophys. Acta 499, 290.
188 Čeřovský, V., Tichá, M., Hořejší, V. and Kocourek, J. (1980) J. Biochem. Biophys. Methods 3, 163.

189 Hořejší, V., Tichá, M., Tichý, P. and Holý, A., Anal. Biochem., in press.
190 Hořejší, V. and Tichá, M. (1981) Anal. Biochem. 116, 22.
191 Müller, W., Hattesohl, I., Schuetz, H-J. and Meyer, G. (1981) Nucleic Acids Res. 9, 95.
192 Dravid, A.R., Fredén, H. and Larsson (1969) J. Chromatogr. 41, 53.
193 Watkin, J.E. and Miller, R.A. (1970) Anal. Biochem. 34, 424.
194 Tsanev, R.G. and Staynov, D. (1964) Biokhimiya 29, 1126.
195 Loening, U.E. (1967) Biochem. J. 102, 251.
196 Allen, R.C. and Jamieson, G.R. (1966) Anal. Biochem. 16, 450.
197 Hannig, K. and Wirth, H. (1968) Z. Anal. Chem. 243, 522.
198 Hydén, H., Bjurstam, K. and McEwen, B. (1966) Anal. Biochem. 17, 1.
199 Petrakis, P.L. (1969) Anal. Biochem. 28, 416.
200 Weidekamm, E., Wallach, D.F.H., Neurath, P., Flückiger, R. and Hendriks, J. (1974) Anal. Biochem. 58, 217.
201 Eisinger, J. (1971) Biochem. Biophys. Res. Commun. 44, 1135.
202 Grant, D.H., Miller, J.N. and Burns, D.T. (1973) J. Chromatogr. 79, 267.
203 Easton, D.M., Lipner, H., Hines, J. and Leif, R.C. (1971) Anal. Biochem. 39, 478.
204 Isenberg, I., Smerdon, M.J., Cardenas, J., Miller, J., Schaup, H.W. and Bruce, J. (1975) Anal. Biochem. 69, 531.
205 Ragland, W.L., Pace, J.L. and Kemper, D.L. (1974) Anal. Biochem. 59, 24.
206 Le Pecq, J.B. and Paoletti, C. (1966) Anal. Biochem. 17, 100.
207 Karsten, U. and Wollenberger, A. (1972) Anal. Biochem. 46, 135.
208 Shenk, T.E., Rhodes, C., Rigby, P.W.J. and Berg, P. (1975) Proc. Natl. Acad. Sci. USA 79, 989.
209 Jovin, T.M. (1973) In Methods in Enzymology (Grossman, L. and Moldave, K., Eds.) Vol. 21, p. 179, Academic Press, New York.
210 Dahlberg, A.E., Lund, E. and Kjeldgaard, N.O. (1973) J. Mol. Biol. 78, 627.
211 Schwabe, C. (1966) Anal. Biochem. 17, 201.
212 Ahoroth, A. and Mutt, V. (1970) Anal. Biochem. 37, 125.
213 Lushnikov, A.A. and Balandin, I.G. (1971) Lab. Delo 9, 522.
214 Holmberg, H. and Hutternen, P. (1973) Anal. Biochem. 620.
215 Farmer, R., Turano, P. and Turner, W.J. (1966) J. Chromatogr. 24, 204.
216 Maurer, H.R. (1971) Disc Electrophoresis and Related Techniques of Polyacrylamide Gel Electrophoresis, Walter de Gruyter, Berlin, New York.
217 Östner, V. and Hultin, T. (1968) Biochim. Biophys. Acta 154, 376.
218 Peterson, J.I. (1969) Anal. Biochem. 31, 204.
219 Prusík, Z. (1968) J. Chromatogr. 32, 191.
220 Ward, S. (1970) Anal. Biochem. 33, 259.
221 Burger, W.C. and Wardrip, W.O. (1965) Anal. Biochem. 13, 580.
222 Nagy, L.K., Rogerson, B. and Tomkuss, N. (1966) Nature 212, 293.
223 Frey, M.D. and Radola, B.J. (1982) Electrophoresis 3, 27.
224 Adams, D.L. and Sammons, D.W. (1981) Electrophoresis 71, 155.
225 Switzer, C.R., Merril, C.R. and Shifrin, S. (1979) Anal. Biochem. 98, 231.
226 Ochs, D.C., McConkey and Sammons, D.W. (1981) Electrophoresis 2, 304
227 Oakley, B.R., Kirsch, D.R. and Morris, N.R. (1980) Anal. Biochem. 105, 361.
228 Merril, C.R., Goldmann, D. and Ebert, M. (1981) Proc. Natl. Acad. Sci. USA in press.
229 Goldman, D. and Murill, C.R. (1982) Electrophoresis 3, 24.
230 John, D.W. and Miller, L.L. (1965) Lab. Invest. 14, 1402.
231 Hayes, D.H. and Barritault, D.S. (1973) Anal. Biochem. 53, 272.
232 Rogers, A. (1973) Techniques of Autoradiography, 2nd edition, Elsevier, Amsterdam, New York, Oxford.
233 Prescott, K.M. and David, G.S. (1974) Anal. Biochem. 57, 232.
234 Tykva, R. and Votruba, I. (1972) Anal. Biochem. 50, 18.
235 Bonner, W.M. and Laskey, R.A. (1974) Eur. J. Biochem. 46, 83.

236 Ramsden, D.B. (1972) J. Chromatogr. 67, 382.
237 Boyd, J.B. and Mitchell, H.K. (1966) Anal. Biochem. 14, 441.
238 Creyton, M.O. and Trevithick, R.J. (1972) Anal. Biochem. 50, 255.
239 Takahashi, M., Ogino, T. and Baba, K. (1969) Biochim. Biophys. Acta 174, 183.
240 Warner, J.R. (1971) J. Biol. Chem. 246, 447.
241 Sung, M.T., Dixon, G.H. and Smithies, O. (1971) J. Biol. Chem. 246, 1358.
242 McKenzie, R.M. and Gholson, R.M. (1973) Anal. Biochem. 54, 17.
243 Koening, E. and Brattgård, S.O. (1963) Anal. Biochem. 6, 424.
244 Kalberer, F. and Rutschman, J. (1961) Helv. Chim. Acta 44, 1956.
245 Jeffay, H. and Alvarez, J. (1961) Anal. Biochem. 2, 506.
246 Oliverio, V.T., Denham, C. and Davidson, J.D. (1962) Anal. Biochem. 4, 188.
247 Peterson, J.I. (1969) Anal. Biochem. 31, 204.
248 Peterson, J.I., Wagner, F., Siegel, S. and Nixon, W. (1969) Anal. Biochem. 31, 189.
249 McEwen, B.S. (1968) Anal. Biochem. 25, 172.
250 Gupta, G.N. (1966) Anal. Chem. 38, 1356.
251 Young, R.W. and Fulhorst, H.W. (1965) Anal. Biochem. 11, 389.
252 Tischler, P.V. and Epstein, C.J. (1968) Anal. Biochem. 22, 89.
253 Choules, G.L. and Zimm, B.H. (1965) Anal. Biochem. 13, 336.
254 Spear, P.G. and Roizman, B. (1968) Anal. Biochem. 26, 197.
255 Anker, H.S. (1970) FEBS Letters 7, 293.
256 Anderson, L.E. and McClure, W.O. (1973) Anal. Biochem. 51, 173.
257 Bockemüller, W. and Rebling, R. (1952) Chem. Eng. Tech. 24, 617.
258 Carlson, L.A. (1954) Acta Chem. Scand. 8, 510.
259 Flodin, P. and Porath, J. (1954) Biochim. Biophys. Acta 13, 175.
260 Porath, J. (1956) Biochim. Biophys. Acta 22, 151
261 Porath, J., Lindner, E.B. and Justedt, S. (1958) Nature 182, 244.
262 Sorof, S., Young, E.M., Spence, M.M. and Fatterman, P.L. (1960) Biochim. Biophys. Acta 38, 559.
263 Porath, J., Hjertén, S. (1968) In Methods in Biochemical Analysis (Glick, D., Ed.) Vol. 9, p. 193, Interscience, New York.
264 Porath, J. (1968) In Methods in Immunology and Immunochemistry (Williams, C.A. and Chase, M.W., Eds.) Vol. 2, p. 62, Academic Press, New York.
265 Hjertén, S. (1963) J. Chromatogr. 12, 510.
266 Reuter, W. (1959) Biochem. Z. 331, 337.
267 Wober, W. and Alupovič, P. (1971) Eur. J. Biochem. 19, 340.
268 Weil, P.A. and Hampel, A. (1973) Biochemistry 12, 4361.
269 Wieme, R.J. (1965) Agar Gel Electrophoresis, Elsevier, Amsterdam.
270 Ressler, N. (1958) Nature 182, 463.
271 Gordon, A.H. and Louis, L.N. (1967) Anal. Biochem. 21, 190.
272 van Kreel, B.K., Pijnenburg, A.M.C.M., van Eijk, H.G. and Leijnse, B. (1971) Clin. Chim. Acta 32, 103.
273 Bont, W.S., Geels, J. and Rezelman, G. (1969) Anal. Biochem. 27, 99.
274 Dudman, N.P.B. and Zerner, B. (1974) Anal. Biochem. 57, 14.
275 Coy, D.C. and Wuu, T.C. (1971) Anal. Biochem. 44, 174.
276 Jovin, T., Chrambach, A. and Naughton, M.A. (1964) Anal. Biochem. 9, 351.
277 Malicinski, G.M. (1970) Anal. Biochem. 38, 288.
278 Marceau, M., Blais, R. and Balaux, N. (1975) Anal. Biochem. 68, 17.
279 Brownstone, A.D. (1976) Anal. Biochem. 70, 572.
280 Kawata, H., Chase, M.W., Elyjiv, R. and Machek, E. (1971) Anal. Biochem. 39, 93.
281 Reznick, A.Z., Allen, H.J. and Winzler, R.J. (1973) Anal. Biochem. 52, 395.
282 Schenkein, I., Levy, M. and Weiss, P. (1968) Anal. Biochem. 25, 387.
283 Padjak, W. (1973) Clin. Chim. Acta 48, 113.
284 Hjertén, S., Jerstedt, S. and Tiselius, A. (1965) Anal. Biochem. 11, 211; ibid p. 219.

285 Stegeman, H. (1970) Anal. Chem. 252, 165; ibid (1972) 261, 388.
286 Mann, M.B. and Huang, P.C. (1969) Anal. Biochem. 32, 138.
287 Jacobson, A. and Lodish, H.F. (1973) Anal. Biochem. 54, 513.
288 Ray, D.K., Troisi, R.M. and Rappaport, H.P. (1969) Anal. Biochem. 32, 322.
289 Voytek, P. (1972) Anal. Biochem. 47, 629.
290 Schnaitman, C.A. (1973) Arch. Biochem. Biophys. 155, 541.
291 Sulitzeanu, D., Slavin, M. and Yecheskeli, E. (1967) Anal. Biochem. 21, 57.
292 Zeiger, R.S., Salomon, R. and Peacock, A.C. (1971) Biochemistry 10, 4219.
293 Arlinger, L. (1979) In Electrophoresis, A Survey of Techniques and Applications, (Deyl, Z., Everaerts, F.M., Prusík, Z. and Svendsen, P.J., Eds.) Elsevier, Amsterdam.
294 Grassman, W. and Hannig, K. (1950) Naturwissenschaften 37, 397.
295 Grassman, W. and Hannig, K. (1949) Ger. Pat. 805399, May 24th.
296 Prusík, Z. (1979) In Electrophoresis, A Survey of Techniques and Applications, Part A, Techniques, (Deyl, Z., Everaerts, F.M., Prusík, Z. and Svendsen, P.J., Eds.) p. 229, Elsevier, Amsterdam.
297 Woodworth, R.C. and Clark, L.G. (1967) Anal. Biochem. 18, 295.
298 Lim, R., Huang, J.J. and Davis, G.A. (1969) Anal. Biochem. 29, 48.
299 Reid, M.S. and Bieleski, R.L. (1968) Anal. Biochem. 22, 274.
300 Neuhoff, V. (1968) 18, 35.
301 Herrick, H.E. and Lawrence, J.M. (1965) Anal. Biochem. 12, 400.

Deyl (ed.) Separation Methods

CHAPTER 7

Field-flow fractionation

JOSEF JANČA

Institute of Analytical Chemistry, Czechoslovak Academy of Sciences, 611 42 Brno, Czechoslovakia

7.1 Introduction

In the beginning of 1966 Giddings [1] published a communication in which he described a new separation concept based on a coupling of concentration and flow non-uniformities. Separation is caused by differential migration in a narrow channel in which a fluid flows unidirectionally. A physical or chemical field that acts across the channel, e.g., temperature gradient, electrical, magnetic or gravitational forces, chemical potential gradient, etc., induces a concentration gradient in the direction across the channel, i.e., usually in the direction of the axis of the field gradient. The velocity of the flow of the fluid in the channel also varies in the direction across the channel. This velocity gradient is caused by viscosity and other effects, accompanying flow processes. The method was called field-flow fractionation (FFF). Utilization of this concept, particularly for the separations of macromolecules and particles, is of exceptional significance.

Already in his first communication Giddings assumed the possibilities of programming the field strength, programming the flow, forming the shape of the velocity profile by varying geometric characteristics of the channel, using casual channel packing, and thus combining the separation based on the above concept with the chromatographic technique, and applying gases and liquids as mobile carriers, etc. So far only some of these ideas have been verified experimentally in practice, and associated problems have been solved to a rather a limited extent even from the theoretical viewpoint.

FFF is similar to chromatography in many respects, especially as far as both dynamic aspects and experimental equipment and procedures are concerned. However, all of the processes associated with the separation take place in the fluid phase and there is no stationary phase which would play an active part in the separation process. This simpleness is, however, characteristic of classical FFF only and not of its combination with the chromatographic technique, e.g., with the use of a channel packed with a chromatographic bed. This is why FFF is sometimes classified as a one-phase chromatography [2–6]. The absence of the stationary phase that has a large surface area, which is a condition of efficient separation in chromatography, can be of fundamental significance for fractionations of materials of biological origin. These materials often are very sensitive to a type and intensity of interactions with active surfaces of the packings of chromatographic columns, and on contact with large surface areas of these packings they can denaturate in an undesirable way. The surface of the FFF channel plays no active part in the separation process except that it creates hydrodynamic resistance necessary for the establishment of the velocity gradient, and keeps together the separation system of the FFF channel mechanically. The total surface area of the channel is lower, by several orders of magnitude, than active surface area of the chromatographic column with a comparable separation capacity and, moreover, it can be deactivated in a suitable manner so that even eventual weak and undesirable interactions with the substance under separation are suppressed to the minimum. The strength of the physical field applied can be varied within a very wide range continuously, starting from very fine fields up to very strong fields that induce intensive transversal migration of substances under separation.

As mentioned above, FFF is of particular import for fractionations of macromolecules [7] within a wide range of molecular masses, starting from several thousands of daltons up to 10^{12}, and even perhaps to 10^{15} [8], even for particles in submicron and micron range, and for organized structures, such as cells and microorganisms, etc. [4,9–11]. None of the more modern chromatographic techniques possesses such a flexibility.

For the reasons mentioned above, FFF appears to be a very promising methodology for fractionation and separation of polymeric systems and particles, and is noted also for its high resolution [12].

FFF has alternatively been called 'polarization chromatography' [13]. This name does, however, not give a true picture of the fundamental principle of the separation, i.e., of coupling of field-induced concentration and flow non-uniformities, in such a descriptive and elegant manner as does the original name field-flow fractionation.

7.2 Principle of FFF

The principle of FFF can be explained best with the aid of Fig. 7.1. A lateral field acting across a narrow channel, composed usually of two planparallel walls, interacts with molecules or particles of a solute and compresses them to one of the channel walls in the direction of x-axis perpendicular to this wall. Hence a concentration gradient is established in the direction of the x-axis. This concentration gradient induces a diffusion flow in the reverse direction. After a certain time a steady state has been reached and the distribution of the solute across the channel can be characterized by a mean layer thickness l. At a laminar isothermal flow of a Newtonian fluid along a narrow channel, usually a parabolic velocity profile is formed inside the channel. It means that the molecules or the particles of the solute are transported in the direction of the longitudinal axis of the channel at varying

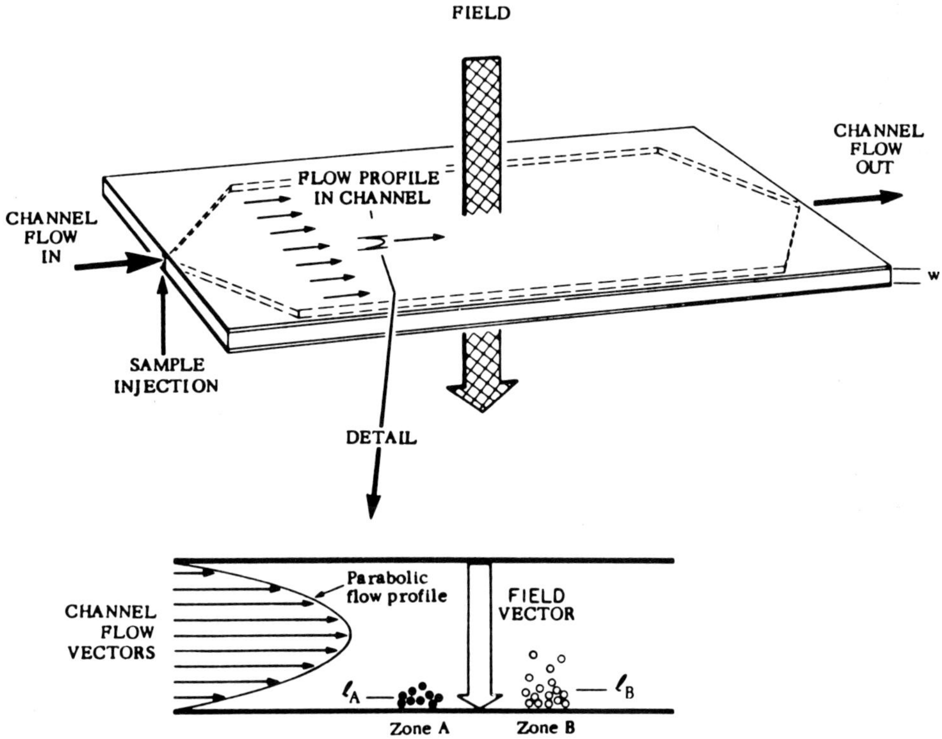

Fig. 7.1. Principle of field-flow fractionation. Reprinted from Ref. 78, by courtesy of Marcel Dekker, Inc.

velocities, depending on the distance from the channel wall, in which they occur. For instance, if there are in the channel two solute species, A and B, which are distributed across the channel owing to the field effects to different extent which can be expressed quantitatively in terms of various magnitudes, l_A and l_B, then the solute species that is compressed to a lesser extent, solute B in the present case, moves along the channel on average more rapidly, is retained less, and zones of solute species A and B are eluted in different times. By this mechanism an effective separation is effectuated.

Larger molecules or particles are usually compressed closer to the wall and thus retained more. Hence the elution sequence proceeds from the smallest molecules or particles to the largest ones. A simple geometry of the fractionation channel, permitting a relatively simple mathematical description of the separation process, is one of the advantages of FFF. The range of retentions is also fairly variable, theoretically from zero retention, determined by a mere passage of the solute through the channel, to a total retention when the solute is completely compressed to the channel wall. A retention, in the range from zero to that determined from a 10–50-fold void volume of the channel, is encountered in practice.

7.3 Theoretical backgrounds of FFF

7.3.1 Retention

A non-equilibrium theory of FFF [14] was elaborated on the principles parallel to the principles of the non-equilibrium theory of chromatography [15]. The solute is displaced in a moving fluid by a combined action of the flow and the fields applied. As long as no flow occurs in the system, a concentration gradient induced by the field will be balanced by diffusion which will lead, after a certain time, to a steady state or to a quasi-equilibrium concentration, c'. As soon as the flow is activated, the quasi-equilibrium will be permanently disturbed. The distribution of concentrations across the channel can be described by the relationship

$$c'(x) = c'_0 \exp(xU/D) \tag{1}$$

In the present coordinate system, x designates the distance from the wall at which the solute accumulates, U is a negative magnitude that characterizes the velocity of the migration induced by the field, D is the diffusion coefficient. Constant c'_0 is the concentration at the coordinate $x = 0$. On defining $l = D/|U|$, Eqn. 1 can be written

$$c'(x) = c'_0 \exp(-x/l) \tag{2}$$

where l is the mean layer thickness defined above. The average velocity of the zone in the axial direction z is given by the following equation

$$V = \langle c'(x) v(x) \rangle / \langle c(x) \rangle \tag{3}$$

where $v(x)$ is the actual velocity of the stream line at the coordinate x, brackets $\langle\,\rangle$ designate average values. Retention ratio R is then defined by the equation

$$R = V/\langle v(x) \rangle = \langle c'(x) \cdot v(x) \rangle / \langle c'(x) \rangle \langle v(x) \rangle \tag{4}$$

For the isothermal, isoviscous flow of a Newtonian liquid, between two parallel infinite planes, that is not affected by any outer field, it applies [16] that

$$R = 6\lambda \left[\coth(2\lambda)^{-1} - 2\lambda \right] \tag{5}$$

where $\lambda = l/w$ (w = distance between channel walls). This equation is a basic theoretical relationship that describes retention in FFF quantitatively. In the limit it holds that

$$\lim_{\lambda \to 0} R = 6\lambda \tag{6}$$

In a number of practical applications of FFF the approximation given by Eqn. 6 is fully justified. Hence it can be seen that the relationship between the values of R and

λ is often very simple. For a better illustration of the physical meaning of λ, it can be written [4]

$$U = F/f \text{ and } D = R^*T/f \tag{7}$$

and hence

$$\lambda = R^*T/Fw \tag{8}$$

where F is the effective force acting by the field on 1 mol of the solute, f is the friction coefficient, R^* is the gas constant, and T absolute temperature. Eqn. 8 demonstrates the fact that λ is the ratio of the thermal energy and energy related to the acting field.

7.3.2 Zone spreading

Zone spreading is characterized quantitatively by a height equivalent to a theoretical plate, H [16]

$$H = \sigma^2/L = 2D/R\langle v(x)\rangle + \mathscr{X} w^2\langle v(x)\rangle/D + \sum H_i \tag{9}$$

where σ is the standard deviation of the concentration zone just before the elution, L is channel length and $\mathscr{X}$ is a dimensionless parameter. The first term in Eqn. 9 describes the longitudinal diffusion, the second non-equilibrium effects, and the third the sum of different contributions following from relaxation processes, finite width of the injected sample zone, etc. Eqn. 9 is analogous to the description of chromatographic processes. However, with respect to the character of FFF a term corresponding to eddy diffusion in the classical chromatographic column, that is packed with the stationary phase, is not included in the above equation. The term that describes non-equilibrium processes [17], derived theoretically [14], is a most significant contribution to the total value of H in Eqn. 9. As follows from the non-equilibrium theory [14,16,17], dimensionless parameter $\mathscr{X}$ is expressed as

$$\mathscr{X} = 2D\langle c'(x)\cdot\varepsilon\cdot v(x)\rangle/\{\langle c'(x)\cdot v(x)\rangle w^2\cdot\langle v(x)\rangle\cdot(\mathrm{d}\ln c'/\mathrm{d}z)\} \tag{10}$$

The complex relationship for $\mathscr{X}$ has been discussed in detail [17]. In the limit when λ tends to zero, it holds simply [16]

$$\lim_{\lambda\to 0} \mathscr{X} = 24\,\lambda^3 \tag{11}$$

if λ tends to infinity, then it holds [17]

$$\lim_{\lambda\to\infty} \mathscr{X} = 1/105 \tag{12}$$

7.3.3 Relaxation

Immediately after the injection into the channel, the solute is distributed across the channel homogeneously. Only due to the action of the field does the concentration gradient start its formation until the steady state is reached. The time period until the establishment of the quasi-equilibrium is called the relaxation time. Relaxation processes contribute to zone-spreading according to the relationship [11,16]

$$H_r = (17\,n/140\,L)\,(\lambda w^2 \langle v(x)\rangle / D)^2 \tag{13}$$

where n is an effective number of relaxation processes along the channel, usually equal to one. Their contribution is often negligible. When the total zone-spreading increases undesirably it can be eliminated by stopping the flow through the channel after the injection for the time period that makes it possible to obtain the quasi-equilibrium. Relaxation processes occurring after the injection of the solute obviously also affect the retention. If the relaxation time of the solute, t_r, is exactly defined as the time necessary for overcoming the distance between the centre of the channel and the centre of gravity of the quasi-equilibrium zone, then the relationship between the apparent retention ratio, R', and theoretical R, i.e., not including the relaxation processes, is given by the following equation [16]

$$1/R' = 1/R - n\langle v(x)\rangle t_r (1 - R)/RL \tag{14}$$

7.3.4 Optimization of FFF

By analysing Eqn. 9 it was found [18] that the minimal obtainable value of H, i.e., the maximal efficiency with regard to the flow-rate, is given by an approximate relationship

$$H_{\min} \doteq R \cdot w \tag{15}$$

and the optimal flow-rate corresponding to this is

$$\langle v(x)\rangle_{\mathrm{opt}} = \sqrt{18} \cdot D/R^2 w \tag{16}$$

The above relationships show that $H_{\min}$ decreases rapidly as the retention increases, i.e., with a decreasing retention ratio R, and hence the best resolution will be obtained for solutes that are retained most. For them the optimal flowrates will be the highest ones. It follows from the above that flow programming, i.e., a gradual increase, would make it possible on the one hand to work in the range of the optimal parameters permanently, and to decrease the time of the analysis on the other. The speed of the FFF analysis can be expressed as the maximal number of theoretical plates, $\dot{N}_{max}$, generated per unit of time [18]

$$\dot{N}_{max} = D/4w^2\lambda^2 \tag{17}$$

It is obviously desirable that the values of w and λ be suppressed to the minimum, i.e., the field strength increased, and D brought to the maximum.

In 1977 Krishnamurthy and Subramanian published an exact theoretical analysis of FFF [19], based on their generalized dispersion theory. Without touching on the details of a complicated mathematical treatment, with the aid of which they solved the problems of both the separation and dispersion processes that occur in the FFF channel during the complete separation from the injection to the elution, let us only say in general that their solution makes it possible to explain some experimental artefacts in detail. These artefacts could not be explained by means of the non-equilibrium theory of FFF mentioned above, which is based on some asymptotic assumptions. Perhaps the most important discrepancy between the theory and the experimental data is that the zone spreading that is observed is considerably larger than the spreading predicted by the theory.

Rigorous convective diffusion theory [19] doubtlessly brings a valuable contribution to the exact mathematical analytical description of FFF. On the other hand, it must be realized that under real experimental conditions a number of non-ideal conditions exist, such as imperfect smoothness of the surface of the FFF channel walls, and others which can cause fundamental deviations from the theory, of the data observed. The above, and a number of other possible conditions have not been considered by any of these theories. It is the simpleness and the easy distinction of the physical conception of the derived relationships that are, in spite of some simplifying asymptotic assumptions, an advantage of the non-equilibrium theory.

Jayaraj and Subramanian [20] further extended the original analysis [19] by a detailed theoretical study of relaxation phenomena in FFF. Using a numerical

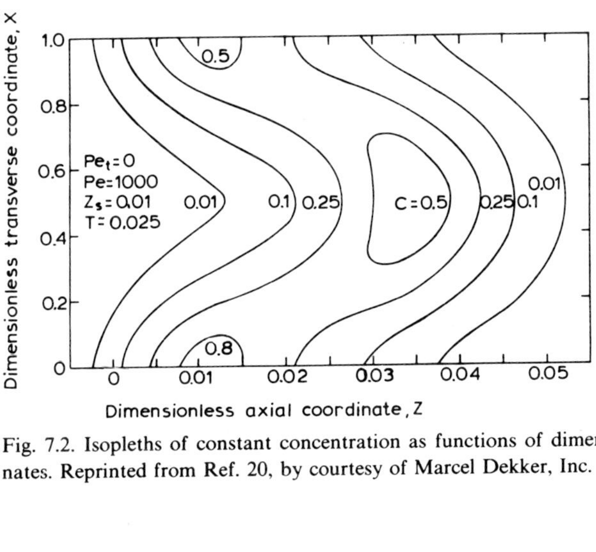

Fig. 7.2. Isopleths of constant concentration as functions of dimensionless transverse and axial coordinates. Reprinted from Ref. 20, by courtesy of Marcel Dekker, Inc.

solution, they were able to model the processes occurring in the FFF channel in various phases of the development. For an illustration, let us discuss some of their results, shown in Figs. 7.2 to 4. Fig. 7.2 shows the distribution of the concentrations in the zone of the solute soon after its injection into the channel with no field effect ($Pe_t = 0$) in dimensionless coordinates $X = x/w$ and $Z = z/wPe$, where Pe is Peclet's number (see further text), demonstrated by isopleths, i.e., by the lines connecting the points at identical concentrations. $Pe_t = u_t w/D$ in Fig. 7.2 is the transversal Peclet's number, u_t is the transversal migration velocity, $Pe = \langle v(x) \rangle w/D$ is axial Peclet's number, $Z_s = z_s/wPe$ is the dimensionless length of the solute cloud, z_s is the length of this cloud and $T = Dt/w^2$ is dimensionless time. Zones of the high concentration are not only visible in the vicinity of the channel centre, but also in the vicinity of its walls. In the vicinity of the channel centre the velocity gradient is low and thus axial dispersion is minimal. Although the velocity gradient is high in the vicinity of the wall, the actual velocity is very low and for this reason the axial dispersion is also relatively low.

Fig. 7.3 illustrates concentration profiles, calculated with the aid of three different models of FFF [16,19,20], and thus, substantially, a comparison of the non-equilibrium model with the dispersion model. The dimensionless mean concentration, C_m, in Fig. 7.3 is defined by the relationship

$$C_m(T, Z) = \int_0^1 C(T, X, Z)\, dX \tag{18}$$

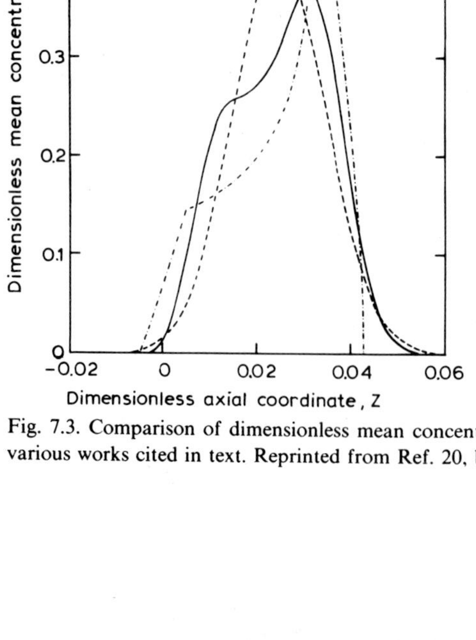

Fig. 7.3. Comparison of dimensionless mean concentration against dimensionless axial coordinate from various works cited in text. Reprinted from Ref. 20, by courtesy of Marcel Dekker, Inc.

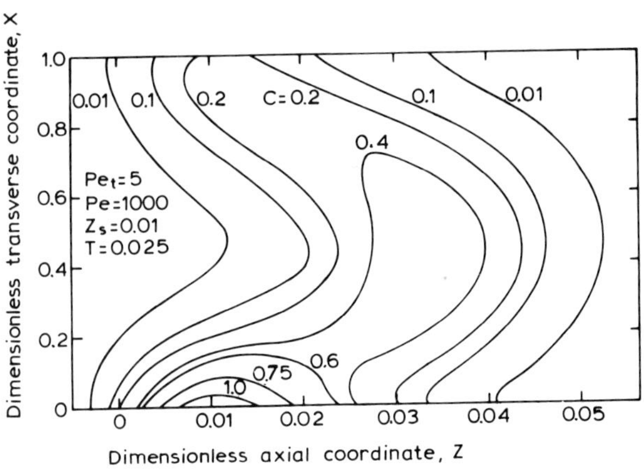

Fig. 7.4. Isopleths of constant concentration as functions of dimensionless transverse and axial coordinates. Reprinted from Ref. 20, by courtesy of Marcel Dekker, Inc.

Hence the average concentration in the channel cross-section is obtained. The curves, calculated with the aid of the models from Refs. 16 and 20, are thus fairly similar in shape. Fig. 7.4 illustrates a situation similar to that in Fig. 7.2. However, with the action of the field ($Pe_t \neq 0$). It is obvious from the figure that the solute accumulates in the vicinity of one of the walls ($X = 0$). It is further obvious that the concentration distributions in various cross-sections of the channel will be rather complex functions. The results of Ref. 20 indicated a complex dependence of the concentration profiles across the channel on the axial coordinate during the relaxation.

7.4 FFF Subtechniques

The principal arrangement of the experimental equipment for FFF is, except for the FFF channel proper, identical with the arrangement for liquid chromatography, i.e., substantially it is composed of a reservoir of the solvent, pump, injection system; a chromatographic column is replaced with a FFF channel followed by a detector and a registration device. Various FFF subtechniques differ from each other by the character of the field applied. Their survey, including applications published, which defines also a certain range of molecular weights is presented in Fig. 7.5.

As can be seen from Fig. 7.5, FFF is applicable in an exceptionally wide range of molecular masses. As particles are usually characterized by their diameters, d_p, these diameters were, because of the uniformity of the scale of molecular masses, M, converted according to the following relationship [8]

$$M = \pi N \rho d_p^3/6 \qquad (19)$$

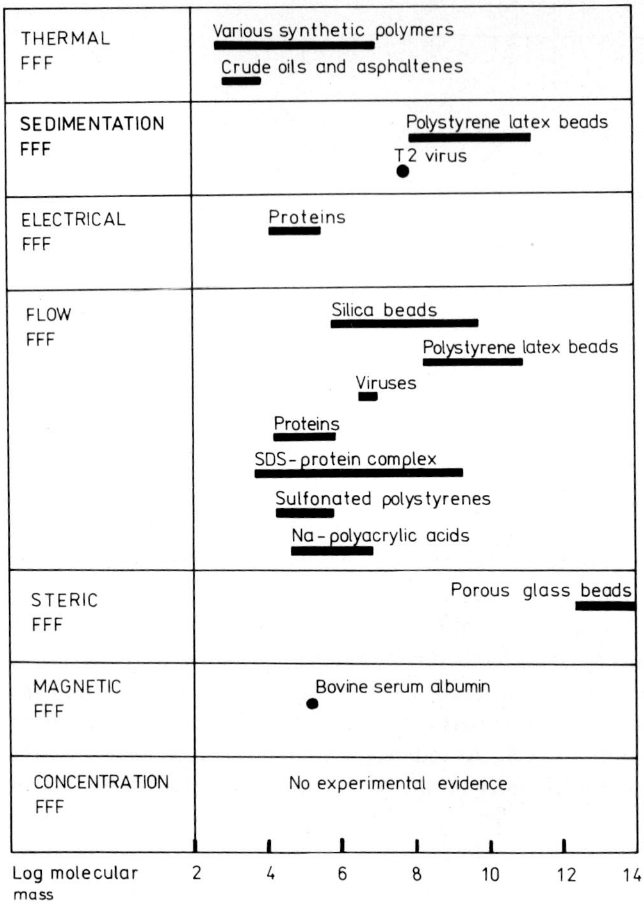

Fig. 7.5. Survey of practical applications of field-flow fractionation subtechniques.

where N is Avogadro's number and ρ the mean density of the particle. Further text will treat various subtechniques of FFF in more detail.

7.4.1 Thermal FFF

Thermal field-flow fractionation (TFFF) belongs to the historically oldest subtechniques of FFF. It is based on the principle of thermal diffusion. In early works [21,22] a basic experimental arrangement was described and a successful fractionation of polystyrene (PS) standards with narrow distribution of molecular masses was demonstrated. A channel for TFFF is relatively simple. It is composed of two metallic blocks with highly polished surfaces, which clamp a spacer. The upper of the two blocks is heated electrically, the lower one is cooled with water. The channel shape, with dimensions of ca. 1–2.54 cm × 50–300 cm, is cut into the spacer, usually 50–250 μm thick. A common temperature gradient between the walls is 20–100°C. Some fundamental theoretical and experimental aspects of TFFF, associated with retention and dispersion, were studied in papers by Giddings and co-workers [16,23]. The value of λ is expressed quantitatively by the following relationship [16]

$$\lambda = [w(\alpha/T) \cdot dT/dx]^{-1} \tag{20}$$

where α is dimensionless thermal diffusion factor, associated with the thermal diffusion coefficient, D_T, according to the relationship

$$\alpha = D_T \cdot T/D \tag{21}$$

From the theoretical viewpoint TFFF is the most complex subtechnique. A complication is brought about by the fact that, owing to the temperature gradient across the channel, the flow is not isoviscous and, as a result of this, the velocity profile is not parabolic [23]. An exact analysis of the shape of the velocity profile in TFFF with regard to non-isothermal flow was presented by Westermann-Clark [24]. The applicability of TFFF to the measurements of thermal diffusion factors [25] was demonstrated. Theoretical comparison of TFFF and gel permeation chromatography for the separation of polymers pointed out a number of advantages of TFFF concerning, in the first place, the fractionating power [26]. TFFF was further used to study thermal diffusion of PS in various solvents differing in their thermodynamic quality [27]. Experimental factors that would improve this method, after being optimized, were studied in a number of other papers. The use of the pressurized system, permitting the extension of the operational range of the temperatures of the TFFF channel, provided an effective fractionation of low-molecular mass PS (with molecular masses of several hundred Da) [28]. On the other hand, potentialities of this method were shown even for the fractionation of polymers with extremely high molecular masses [8] of up to 10^{12} Da. Temperature gradient programming made it possible to improve TFFF in such a way that samples of PS standards under study could be fractionated in a wide range of molecular masses from 4000 up to 7 000 000 Da in a single experiment, with the range of retention [29] acceptable from the viewpoint of both experimental and time requirements. Miniaturization of the channel for TFFF and some other design modifications made it possible to speed up the total analysis substantially, so that the total time might be reduced to from several tens of seconds up to several minutes [30].

The subsequent study was oriented at the theoretical explanation of the factors that cause and affect zone spreading in TFFF [31]. Contributions of non-equilibrium and polydispersity of polymer samples under study to the total peak width were studied, as was the possibility of determining the precise polydispersity of the polymer by measuring peak width at various linear velocities of the solvent, and by extrapolating to zero velocity, i.e., by eliminating the contribution of non-equilibrium processes [32]. An improved separation in TFFF can be obtained by using thermogravitational effect, i.e., by using thermal convection in a non-horizontal channel. The resulting velocity profile formed under such conditions has a more complicated non-parabolic shape [33]. All of the experimental results mentioned above were obtained with the use of PS samples. Only recently has a work been published, showing decisively that TFFF is also applicable to fractionation of other polymers [34]. Martin [35] demonstrated an advantage of a coupling of the TFFF channel with the photogoniodiffusometer for low-angle laser light scattering for the analysis of polymers. Janča and Klepárník [36] published data showing potentialities of TFFF for the determination of molecular mass distribution of polymers and a procedure for an exact interpretation of experimental results of TFFF.

7.4.2 Sedimentation FFF

Sedimentation field-flow fractionation (SFFF) belongs, besides TFFF, to the oldest subtechniques of FFF. It is of great importance for biochemical applications. It has already been predicted conceptually in Giddings' work [1], and Berg and co-workers [37–39] published their results independently. Either natural gravitational or centrifugal forces in the centrifuge serve here as an effective field. As various authors chose different experimental arrangements, these will be described in connection with the discussion of respective works.

For the value of λ of spherical particles it holds [40]

$$\lambda = 6kT/\pi d_{\mathrm{p}}^{3} G w \Delta\rho \tag{22}$$

where G is gravitational or centrifugal acceleration, k is Boltzmann constant, and $\Delta\rho$ is the difference in the densities of the particles and the solvent used. Berg and Purcell [37] presented the first elementary theoretical analysis of fractionation of particles by using gravitational or centrifugal forces in the centrifuge. In their first experimental paper [38] they described the fractionation of PS latex with particle sizes of 0.796 and 1.305 μm. Their experimental arrangement was quite simple. Along the surface of a glass plate situated in a closed vessel a liquid flows in a layer 0.15 cm deep. The separation was observed with the aid of photographing. It was found that the surface layer of the liquid does not move, which was explained by the formation of the surface monolayer of a detergent. The velocity profile formed corresponded to the geometry of the channel with two walls. The time of the experiment was very long, 76–125 h. In their subsequent paper they described the separation of R 17 *E. coli* bacteriophage with a molecular mass of 4×10^6 Da in the centrifuge. The experimental arrangement consisted of a vertical cylinder, 7.5 cm in diameter, 52 cm long, that rotated round its central axis. The liquid was introduced from above on the internal wall of the cylinder where, due to a centrifugal force, formed an even layer with a thickness of 259–295 μm, and was taken down into a detector. Though the internal surface of the liquid was again free, i.e., only on contact with the helium atmosphere in which the cylinder rotated so that aerodynamic noise at relatively high rpm might be suppressed, it was found that the velocity profile again corresponded to the flow in the channel with two fixed walls. The time of the analysis was substantially shorter in this instance, approximately 4–12 h.

Giddings and co-workers [40] described in their experimental work dealing with SFFF, the device shown in Fig. 7.6. The channel 0.0635 cm thick and 2.54 cm wide was coiled along the internal wall of the centrifuge basket to a length of 45.7 cm. In their paper [40] the basic theoretical and experimental aspects of SFFF were discussed and the fractionation of a series of monodisperse spherical PS latex (range of particle diameters of 0.0907–0.4808 μm), was demonstrated. At the same time it was pointed out that it is not feasible to use a channel with a circular cross-section, because under such geometrical conditions parasitic secondary flows, perpendicular to the channel axis, are formed owing to the action of the centrifugal force. These secondary flows impair retention substantially and enlarge zone spreading; they can even lead to an entire loss of the separation. In the following paper [41] a theory of programmed SFFF using the programming of the intensity of centrifugal gravitational field and the programming of the density of the solvent was elaborated. Utilization of programming expands considerably the range of molecular masses which can be fractionated in a single run. The programming of the field strength in SFFF was effectuated by decreasing gradually the number of revolutions of the centrifuge in the course of the analysis, and the programming of the solvent density by increasing gradually concentration of saccharose in water which was used as a solvent. The fractionation was demonstrated by using PS latex with particle diameters of 0.1756–0.3117 μm. An excellent agreement was obtained between the parameters calculated theoretically and those found experimentally [41].

Another type of programming — programming of the flow — was elaborated theoretically and verified experimentally in the succeeding work by Giddings and

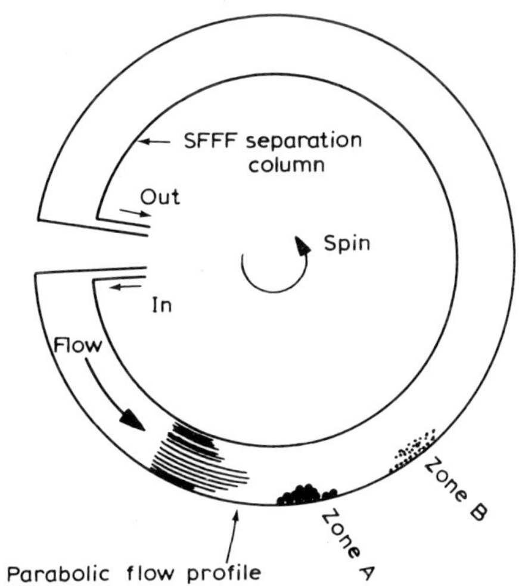

Fig. 7.6. Principle of sedimentation FFF. Reprinted with permission from Ref. 40, Copyright 1974 American Chemical Society.

co-workers [42]. Extension and narrowing of the channel, slowing down the flow and an increase in the field strength applied [13] provided a high resolution of particle separation by the SFFF method. In this way the separation of PS latex was obtained in the range of particle diameters of 0.091–0.982 μm in a single measurement, i.e., in a thousand-fold range of masses [43]. An application of SFFF to separation of biological particles appeared to be very advantageous [44]. In addition to the separation it was possible that absolute dimensions or molecular masses of the particles under analysis might also be calculated from the retention data. In the above example molecular mass of T 2 bacteriophage was determined.

Yau and Kirkland [45] also dealt with the programming of SFFF in the analysis of the particle size distribution. A new design of the equipment for SFFF [46] allowed to work with very high intensities of the centrifugal field, up to 15 000 $\times$ g at 12 000 rpm. The character of SFFF makes this technique very attractive, particularly for biological applications.

7.4.3 Electrical FFF

Electrical field-flow fractionation (EFFF) is one of most sophisticated experimental subtechniques of FFF. This may be the reason why relatively fewer papers have been published about it, in spite of the first work [47] having been published as early as in 1972. With this subtechnique, the field is induced by electrical voltage causing electrical current across the channel. The walls of the channel for EFFF are composed of two semi-permeable membranes permitting the passage of small ions and separating the channel space from the electrode compartment. Consequently, an undisturbed, altogether homogeneous electrical field will be induced inside the channel. A schematic diagram of the EFFF channel is shown in Fig. 7.7. The actual dimensions of the channel are approximately 0.0356 $\times$ 2.54 $\times$ 43 cm. Dimensionless quantity λ is determined by the electrophoretic mobility, μ, electrical field strength,

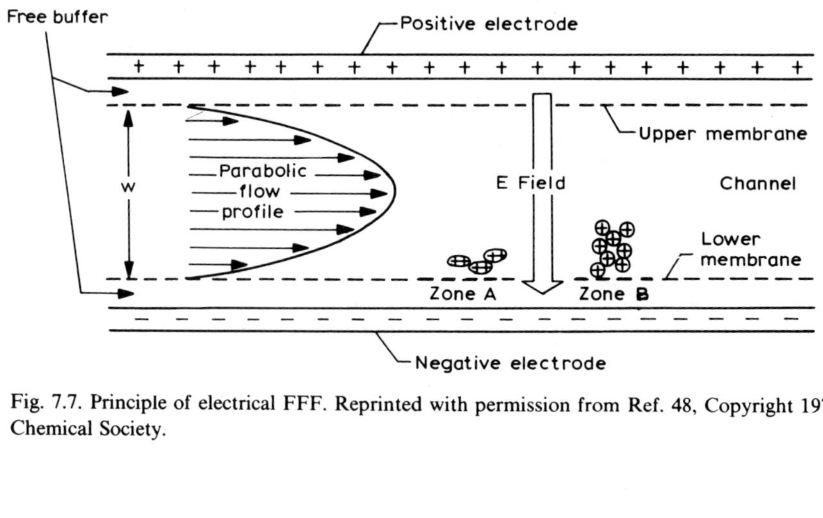

Fig. 7.7. Principle of electrical FFF. Reprinted with permission from Ref. 48, Copyright 1976 American Chemical Society.

E, diffusion coefficient and channel width according to the relationship [48]

$$\lambda = D/\mu E w \tag{23}$$

In the first paper [47] in 1972, the principle of EFFF was described qualitatively and the method was applied to the analysis of some proteins — albumin, lysozyme, haemoglobin and γ-globulin — with the use of 0.02 M sodium acetate solution and acetic acid at pH 4.5, and 0.02 M Tris(hydroxymethyl)aminomethane solution with acetic acid at pH 8.0 as buffering solutions. The time required for the separation was approximately 120 min. Certain deviations in the retentions calculated theoretically and the retentions measured experimentally were observed. In a subsequent paper [48] the theory of EFFF was elaborated in more detail, the experimental arrangement using regenerated cellulose as a semi-permeable flexible membrane that composed the channel walls was described and, again, some proteins were fractionated. The anomalies observed, i.e., discrepancy between the theoretical retention and theoretical dispersion data and the experimental values, were explained on the basis of electro-osmotic effects that are caused, among others, also by an uneven expansion of the semi-permeable walls of the channel.

A further explanation of these deviations, observed during the fractionation of the above proteins at pH 4.5 was presented by Subramanian and co-workers [49] as a result of the electrical field gradient in the vicinity of the membrane interface.

In order to eliminate the disadvantages following from the use of flexible membranes, Giddings et al. [50] designed a new channel in which both of the semi-permeable membranes of cellulose acetate were carried by polyethylene frist, 0.318 cm thick. A substantially better mechanical stiffness of the channel walls was thus obtained. This gave rise to a better reproducibility and a better agreement between the theory and experiments for the separation of native proteins. The attempts to separate denaturated proteins and PS latex were also made, which, however, were not successful.

Reis and Lightfoot [13] also treated the separation of proteins by using a method from the EFFF class in spite of their calling it 'electropolarization chromatography'. Besides a theoretical description of the separation they developed also an experimental apparatus. The channel was composed of a circular semipermeable tube with an inside diameter of 0.43 mm and a length of 50 cm; the electrical field was applied perpendicularly to the central axis of the channel.

EFFF, as well as the separation by direct electrophoretic methods, can be manipulated by varying experimental conditions, such as pH, ionic strength, etc. These variations affect electrophoretic mobilities of charged particles and, to a lesser extent, also diffusion coefficients. The part of the electrophoretic mobility, μ, in the electrophoresis corresponds to the part of the ratio D/μ in EFFF. This is why the solutes that show only minor differences in mobilities (however, differing substantially in D) can be separated by means of EFFF successfully, in spite of the fact that their electrophoretic resolution is in this instance poor. Hence EFFF and direct electrophoretic methods complement each other. The fact that high voltage gradients per unit of length are obtained at low absolute values of the voltage across the channel is an experimental advantage of EFFF. The heat, generated due to high voltage values, is an interfering element that impairs separation characteristics of direct electrophoretic methods.

7.4.4 Flow FFF

Flow field-flow fractionation (FFFF) has been until now the most universally used subtechnique of FFF. With this subtechnique, the flow of the solvent, perpendicular to the flow of the basic medium in the channel, is an external field. The earliest works belonging to this class of FFFF were published by Lee and co-workers [51,52], and were called one-phase chromatography [51] or ultrafiltration-induced polarization chromatography [52]. Their experimental arrangement consisted of a circular tube, 0.02 cm in diameter, made of semi-permeable material permitting penetration of the solvent through the walls, but preventing the penetration of a high-molecular mass solute. They elaborated a basic theoretical model of the separation in this arrangement and demonstrated it in practice for the separation of blue-dextran and human plasma [51], bovine serum albumin and some polydextrans [52].

Giddings and co-workers [53,54] designed the FFFF channel in a classical manner, i.e., using two planparallel semi-permeable membranes, and a thin (0.38 mm) spacer into which the shape of the channel was cut. They developed theoretical bases of FFFF and fractionated successfully a series of monodisperse spherical PS latex and a number of proteins. They reached an excellent agreement between the theoretical assumptions of the retention and the experimental results. As with other FFF subtechniques, the agreement between the theoretical and the experimental characteristics of dispersion was poorer. The problem further requires a more extensive study.

In FFFF the perpendicular flow having the velocity U acts on all of the solutes uniformly. For this reason the separation in FFFF is determined only by the differences in the values of the diffusion coefficient, D, or the friction coefficient, f. The retention parameter, λ, is then determined by the relationship [9]

$$\lambda = R^{*}TV^{\circ}/3\pi N\eta V_{c}w^{2}d \tag{24}$$

where V_c is the volumetric perpendicular flow (inducing the field), η is the viscosity of the medium, V° is the dead volume owing to the channel, and d is the effective Stokes' diameter.

The effect of relaxation on the retention and resolution in FFFF was studied in further detail in the subsequent work [55]. A substantial improvement in the resolution of the fractionation of f2 virus was proved as long as the stop-flow technique was being applied after the injection into the channel within the relaxation time sufficient for the establishment of the quasi-equilibrium.

FFFF can be applied as a dialysation or ultrafiltration cell [56] to a continuous separation. For instance, two components, of which one is capable of permeating through a semi-permeable membrane, whereas the other is not, can be separated effectively by selecting the proper proportions among various experimental parameters, such as geometry and dimensions of the channel, the ratio of the flow along the channel, and across the channel through the membranes creating the walls. The operation of one such unit was demonstrated in practice for the isolation of low molecular mass ethylene blue from bovine serum albumin [56]. High versality in the use of FFFF was shown by examples of effective separations of various solutes having particular characteristics. Various viruses [57] and a number of proteins [58] were separated, purified and characterized from the viewpoint of diffusivity; colloid silica gel samples [59] with particle diameters of 0.01–0.13 μm were fractionated and analyzed in this manner. FFFF complements SFFF as far as size distribution analysis is concerned [10].

FFFF of water-soluble polyelectrolytes, sulphonated polystyrene and sodium salt of polyacrylic acid proved the applicability of this subtechnique even to the separation of macromolecules [60]. A complication which must be solved is represented by a concentration dependence of the effective dimensions of polyelectrolyte solutions. For this reason it may be difficult to interpret the fractograms obtained so that they may result in a distribution curve of molecular masses of the polymer studied.

7.4.5 Steric FFF

Steric field-flow fractionation (steric FFF) occupies, among other subtechniques of FFF, rather an exceptional position. It utilizes the upper limit of the field strength applied, at which rather a modified retention mechanism occurs. The particles are compressed permanently closer to the wall as the field strength applied increases. At the instant when the mean distance of Brownian motion is less than the particle radius, a, steric FFF takes place. The name steric describes the fact that the mean layer thickness is controlled by steric exclusion of the particles from the layer adjacent to the wall. Hence larger particles migrate into the stream lines of higher velocities of the solvent than smaller particles do and are, consequently, transferred more rapidly. To illustrate the mechanism of the separation by steric FFF, its principle is shown schematically in Fig. 7.8. Giddings [61] treated theoretical aspects of steric FFF and its comparison with the mechanism of hydrodynamic chromatography. He derived the limit relationship for $\lambda_a \to 0$ for the value of R, which is the real condition with respect to the character of steric FFF mentioned above

$$R = 6\lambda_a(1 - \lambda_a) \tag{25}$$

where $\lambda_a = a/w$.

Theoretically it is possible that any effective field — electrical, sedimentation, etc. — may be applied to the steric FFF mode. However, the gravitational field represents the most practical means of the utilization of the principle of steric FFF

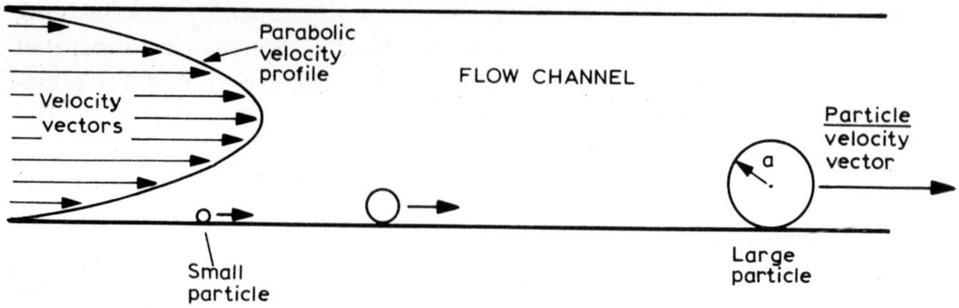

Fig. 7.8. Principle of steric FFF. Reprinted from Ref. 62, by courtesy of Marcel Dekker, Inc.

for fractionations of 1–100 μm particles. Experimental evidence for the applicability of steric FFF was presented by Giddings and Myers [62], who carried out the fractionation of glass beads of 10–32 μm in diameter. The column was composed of a 0.127 mm thick spacer clamped between two glass plates. The actual channel (10×860 mm) was cut into this spacer. Various types of the chromatographic spherical packing were fractionated and characterized from the viewpoint of dimensions in the subsequent work [63]. In this case, a dependence of the retention ratio, R, on the flow-rate, which was not predicted by the theory, was observed.

By inclining the transversal axis of the channel and by injecting the sample into the upper part of the channel, particles under separation were carried and slid towards the lower part of the channel where collection 'pockets' were placed along the channel [64]. The particles that were carried along the channel slid at the same time to lower parts of the channel and were trapped in the 'pockets'. Smaller, slower particles were trapped a longer distance from the injection port, larger particles, moving more quickly were carried and trapped nearer to the injection port. Continuous fractionation of particles could be obtained in this manner by selecting the channel design properly.

Caldwell and co-workers [65] explained the dependence of R on the flow-rate [63] by the existence of lift forces. Steric FFF represents a further principal progress in the methodology of FFF, and permits a simultaneous extension of applications into the range of large-diameter particles (hundreds of microns). Its applications to fractionation of cells, microorganisms, etc., in biochemistry may be expected.

7.4.6 Magnetic FFF

Magnetic field-flow fractionation (MFFF) has been the youngest subtechnique of FFF. So far the only work [66] dealing with MFFF defined elementary theoretical principles of the separation, and demonstrated in practice retentions of bovine serum albumin in the presence of nickel(II) ions in a magnetic field of 400 G. A coiled Teflon capillary with an inside diameter of 0.15 cm and length of 304 cm was used as a channel. In the absence of nickel(II) ions no retention was observed.

For the value of λ of spherical particles the relationship was derived

$$\lambda = (8r/w)(kT/\mu_p \Delta H)^2 \tag{26}$$

where r is particle radius, ΔH is the gradient of the magnetic field strength, and μ_p is the magnetic permeability.

Although the results obtained with the aid of MFFF are of preliminary significance only, the assumption formulated in the work cited above that this subtechnique of FFF is potentially a very promising method for the analysis, particularly of objects of biological character, is justified.

7.4.7 Concentration FFF

So far we have been treating FFF subtechniques making use of physical forces of a certain type to induce migration of molecules or particles of the solute. Concentration field-flow fractionation (CFFF) has been the only subtechnique of FFF that makes use of a concentration gradient in order to induce effective chemical forces or chemical field [67] for the separation. With this subtechnique it is a concentration gradient of a mixed solvent across the channel that is the effective field inducing a gradient of chemical potential with respect to the solute. The chemical potential gradient is $d\mu^\circ/dx$. It follows from the theory [67] that the value of λ is then

$$\lambda = R^*T/\Delta\mu_c^\circ \tag{27}$$

where $\Delta\mu_c^\circ = (d\mu^\circ/dx) \cdot w$ is the total increment of the chemical potential across the channel. If the ratio of the concentrations near both walls is $\alpha' = c_o/c_w$, then it holds for the retention ratio, R,

$$R = 6/\ln \alpha' \left[\frac{\alpha' + 1}{\alpha' - 1} - \frac{2}{\ln \alpha'} \right] \tag{28}$$

It was found by analyzing Eqn. 28 that, for an effective fractionation, α' must be at least 10–100. Moreover, additional conditions necessary, with respect to the concentration gradient in the channel and to the total flux of the solute or the total concentration difference between the reservoirs of the mixture, were derived theoretically.

The design of the channel for CFFF again consists of two semi-permeable membranes forming the walls of the channel. The reservoirs of the mixed solvent at various concentrations of active components are adjoined to these walls from the opposite sides.

The solubility of proteins in alcoholic solutions of various salts was studied. On the basis of this study the conditions for the concentration gradient across the channel that are required for CFFF could be determined. In addition, permeabilities of various membranes were studied in order to define the conditions of CFFF more precisely. The above study led to the finding that, for a successful utilization of CFFF, the design of the FFF channel must be modified substantially in order that all of the assumptions that are discussed in the work and that affect the success of the separation might be satisfied. CFFF has been the most difficult subtechnique to realize,but there exists a prospect that, owing to its unique retention mechanism, the effort required for its practical realization and application will be made.

7.5 Prospects of FFF

In the preceding sections a survey of the results obtained in the development of the FFF methodology in the course of the fiteen years since its establishment is presented. Except in rare instances, there has been no comparison of FFF subtechniques with other separation methods capable of solving similar analytical problems. The methods that are comparable, as to the range of the applications to the solution of actual separation problems, are in most instances older. Consequently, their methodology has been experimentally more developed and elaborated in detail elsewhere. In spite of this it is possible to state in general that a number of efficient separations have already been obtained with the aid of FFF, surpassing other separation methods in many parameters.

Hence FFF can be considered to be a relatively young separation method, the improvement, utilization and expansion of which is to be expected. Several recent works have suggested both theoretically and practically further possibilities that can be provided by FFF.

An increase in the retention and the capacity of the FFF channel, and an increase in the selectivity, can be obtained by modifying the surface of the channel wall on which the solute is accumulated with the aid of transversal barriers as shown by Giddings et al. [68]. These barriers form spaces in which the solvent does not move, and where the solute can permeate both in and out by diffusion only. Consequently, the fractionation characteristics mentioned above are improved. The channels established transversally could be used to trap even the second phase, and to combine thus the action of field strength and the partition between the phases. Preliminary results were obtained in experiments with the fractionation of PS standards by the TFFF method using the channel with transversal slits [68]. The results proved, in principle, the applicability of this system.

Subramanian [69] published an interesting work, describing the separation method, making use of a perpendicular field across a part of the channel that establishes concentration distribution across the channel without the flow, and later on, separation with the aid of the flow without the field action. By its character, this method can be classified as a category of FFF. However, it differs from FFF in some basic aspects. The field acts in a short part of the channel only, and the fluid does not move at that point. As the fluid moves, the separation itself is performed owing to hydrodynamic phenomena only; the field being no longer in action. By selecting properly the experimental conditions, i.e., intensity and the time of the field action, channel length and the solvent flow-rate, a high efficiency of fractionations can be reached within a relatively short time, as shown by a theoretical analysis of the problem [69].

Except TFFF, all of the other subtechniques of FFF involved the establishment of a parabolic velocity profile. If the profile is not parabolic and is asymmetric with regard to the longitudinal central axis of the channel, and can be described by a general function of a polynomial type, both theoretical retention and zone spreading will correspond to this distribution of velocities across the channel. Theoretical analysis of this general problem was performed by Martin and Giddings [70].

Retention could be controlled by intentional variations in the shape of the velocity profile inside the channel. Janča and Giddings [71] showed a prospective possibility of utilizing non-Newtonian behaviour of some liquids. They used the flexible three-parameter equation of Ellis, describing non-Newtonian phenomena, to perform a theoretical analysis and derived the dependence of R on λ for different conditions of the non-Newtonian flow. This phenomenon could be utilized actively to increase the selectivity of various FFF separations, particularly with the solutes, retained more strongly, i.e., with low values of λ. The utilization of variations in the shape of the velocity profile in a single separation run was also suggested [71], i.e., programming of the values that are decisive for the range of non-Newtonian behaviour of the liquid applied.

An extension of FFF to the separation of nonspherical particles and the influences of the wall effect have been studied both theoretically and experimentally by Gajdos and Brenner [72]. High-resolution polymer separations have been achieved with a TFFF channel of a new construction [73]. Retention and zone spreading equations in SFFF have been verified [74]. Recent studies of electroretention of proteins [75,76] have resulted in more detailed understanding of previously observed retention anomalies in EFFF. Several papers have demonstrated many possibilities of application of the FFF subtechniques in various fields of chemistry, biology and technology [77–79].

The papers published so far have illustrated well the fact that FFF methodology has a wide range of possibilities that have not been used as yet. In contrast to the first decade of its existence a greater number of research laboratories have started to investigate this methodology and its practical applications [80–81]. It can therefore be expected that the elementary knowledge of the theoretical, experimental and application fields of FFF will be extended and elaborated in the years to come.

Acknowledgements

I am indebted to Dr. J. Janák for stimulating discussions.

References

1 Giddings, J.C. (1966) Sep. Sci. 1, 123.
2 Giddings, J.C. 157th National Meeting, American Chemical Society, Minneapolis, April 13–18, 1969, Anal. 003, ACS, Washington, 1969.
3 Giddings, J.C., Myers, M.N., Lin, G.C. and Martin, M. (1977) J. Chromatogr. 142, 23.
4 Giddings, J.C., Fisher, S.R. and Myers, M.N. (1978) Am. Lab. 10, 15.
5 Giddings, J.C. Proc. 6th Discuss. Conf. Macromol., IUPAC, Prague, July 1978.
6 Giddings, J.C. (1979) Pure Appl. Chem. 51, 1459.
7 Grushka, E., Caldwell, K.D., Myers, M.N. and Giddings, J.C. (1974) In Separation and Purification Methods, Vol. 2, p. 127 (Perry, E.S., Van Oss, C.J. and Grushka, E., Eds.) Marcel Dekker, New York.
8 Giddings, J.C. (1976) J. Chromatogr. 125, 3.
9 Giddings, J.C., Myers, M.N., Yang, F.J.F. and Smith, L.K. (1976) In Colloid and Interface Science (Kerker, M., Ed.) Vol. 4, Academic Press, New York.
10 Giddings, J.C., Myers, M.N. and Moellmer, J.F. (1978) J. Chromatogr. 149, 501.
11 Giddings, J.C., Myers, M.N., Caldwell, K.D. and Fisher, S.R. (1980) In Methods of Biochemical Analysis p. 79, (Glick, D., Ed.) Vol. 26.
12 Giddings, J.C. (1973) J. Chem. Educ. 50, 667.
13 Reis, J.F.G. and Lightfoot, E.N. (1976) AIChE J. 22, 779.
14 Giddings, J.C. (1968) J. Chem. Phys. 49, 81.
15 Giddings, J.C. (1965) Dynamics of Chromatography, Marcel Dekker, New York.
16 Hovingh, M.E., Thompson, G.H. and Giddings, J.C. (1970) Anal. Chem. 42, 195.
17 Giddings, J.C., Yoon, Y.H., Caldwell, K.D., Myers, M.N. and Hovingh, M.E. (1975) Sep. Sci. 10, 447.
18 Giddings, J.C. (1973) Sep. Sci. 8, 567.
19 Krishnamurthy, S. and Subramanian, R.S. (1977) Sep. Sci. 12, 347.
20 Jayaraj, K. and Subramanian, R.S. (1978) Sep. Sci. Technol. 13, 791.
21 Thompson, G.H., Myers, M.N. and Giddings, J.C. (1967) Sep. Sci. 2, 797.
22 Thompson, G.H., Myers, M.N. and Giddings, J.C. (1969) Anal. Chem. 41, 1219.
23 Myers, M.N., Caldwell, K.D. and Giddings, J.C. (1974) Sep. Sci. 9, 47.
24 Westermann-Clark, G. (1978) Sep. Sci. Technol. 13, 819.
25 Giddings, J.C., Hovingh, M.E. and Thompson, G.H. (1970) J. Phys. Chem. 74, 4291.
26 Giddings, J.C., Yoon, Y.H. and Myers, M.N. (1975) Anal. Chem. 47, 126.
27 Giddings, J.C., Caldwell, K.D. and Myers, M.N. (1976) Macromolecules 9, 106.
28 Giddings, J.C., Smith, L.K. and Myers, M.N. (1975) Anal. Chem. 47, 2389.
29 Giddings, J.C., Smith, L.K. and Myers, M.N. (1976) Anal. Chem. 48, 1587.
30 Giddings, J.C., Martin, M. and Myers, M.N. (1978) J. Chromatogr. 158, 419.
31 Smith, L.K., Myers, M.N. and Giddings, J.C. (1977) Anal. Chem. 49, 1750.
32 Martin, M., Myers, M.N. and Giddings, J.C. (1979) J. Liq. Chromatogr. 2, 147.
33 Giddings, J.C., Martin, M. and Myers, M.N. (1979) Sep. Sci. Technol. 14, 611.
34 Giddings, J.C., Myers, M.N. and Janča, J. (1979) J. Chromatogr. 186, 37.
35 Martin, M. and Hes, J., 13th Int. Symp. on Chromatography, Cannes, June 30–July 4, 1980, p. 1, P5-1.
36 Janča, J. and Klepárník, K (1981) Sep. Sci. Technol. 16, 657.
37 Berg, H.C. and Purcell, E.M. (1967) Proc. Nat. Acad. Sci. U.S.A. 58, 862.
38 Berg, H.C., Purcell, E.M. and Stewart, W.W. (1967) Proc. Nat. Acad. Sci. U.S.A. 58, 1286.
39 Berg, H.C. and Purcell, E.M. (1967) Proc. Nat. Acad. Sci. U.S.A. 58, 1821
40 Giddings, J.C., Yang, F.J.F. and Myers, M.N. (1974) Anal. Chem. 46, 1917.
41 Yang, F.J.F., Myers, M.N. and Giddings, J.C. (1974) Anal. Chem. 46, 1924.
42 Giddings, J.C., Caldwell, K.D., Moellmer, J.F., Dickinson, T.H., Myers, M.N. and Martin, M. (1979) Anal. Chem. 51, 30.
43 Yang, F.J., Myers, M.N. and Giddings, J.C. (1977) J. Colloid Interface Sci. 60, 574.
44 Giddings, J.C., Yang, F.J.F. and Myers, M.N. (1975) Sep. Sci. 10, 133.

45 Yau, W.W. and Kirkland, J.J., 13th Int. Symp. on Chromatography, Cannes, June 30–July 4, 1980, p. 1, P-4.
46 Kirkland, J.J., Yau, W.W., Doerner, W.A. and Grant, J.W. (1980) Anal. Chem. 52, 1944.
47 Caldwell, K.D., Kesner, L.F., Myers, M.N. and Giddings, J.C. (1972) Science 176, 296.
48 Kesner, L.F., Caldwell, K.D., Myers, M.N. and Giddings, J.C. (1976) Anal. Chem. 48, 1834.
49 Subramanian, R.S., Jayaraj, K. and Krishnamurthy, S. (1978) Sep. Sci. Technol. 13, 273.
50 Giddings, J.C., Lin, G.Ch. and Myers, M.N. (1976) Sep. Sci., 11, 553.
51 Lee, H.L., Reis, J.F.G., Dohner, J. and Lightfoot, E.N. (1974) AIChE J. 20, 776.
52 Lee, H.L. and Lightfoot, E.N. (1976) Sep. Sci. 11, 417.
53 Giddings, J.C., Yang, F.J. and Myers, M.N. (1976) Science 193, 1244.
54 Giddings, J.C., Yang, F.J. and Myers, M.N. (1976) Anal. Chem. 48, 1126.
55 Yang, F.J., Myers, M.N. and Giddings, J.C. (1977) Anal. Chem. 49, 659.
56 Giddings, J.C., Yang, F.J. and Myers, M.N. (1977) Sep. Sci. 12, 499.
57 Giddings, J.C., Yang, F.J. and Myers, M.N. (1977) J. Virol. 21, 131.
58 Giddings, J.C., Yang, F.J. and Myers, M.N. (1977) Anal. Biochem. 81, 395.
59 Giddings, J.C., Lin, G.C. and Myers, M.N. (1978) J. Colloid Interface Sci. 65, 67.
60 Giddings, J.C., Lin, G.C. and Myers, M.N. (1978) J. Liq. Chromatogr. 1, 1.
61 Giddings, J.C. (1978) Sep. Sci. Technol. 13, 241.
62 Giddings, J.C. and Myers, M.N. (1978) Sep. Sci. Technol. 13, 637.
63 Giddings, J.C., Myers, M.N., Caldwell, K.D. and Pav, J.W. (1979) J. Chromatogr. 185, 261.
64 Myers, M.N. and Giddings, J.C. (1979) Powder Technol. 23, 15.
65 Caldwell, K.D., Nguyen, T.T., Myers, M.N. and Giddings, J.C. (1979) Sep. Sci. Technol. 14, 935.
66 Vickrey, T.M. and Garcia-Ramirez, J.A. (1980) Sep. Sci. Technol. 15, 1297.
67 Giddings, J.C., Yang, F.J. and Myers, M.N. (1977) Sep. Sci. 12, 381.
68 Giddings, J.C., Smith, L.K. and Myers, M.N. (1978) Sep. Sci. Technol. 13, 367.
69 Subramanian, R.S. (1978) J. Colloid Interface Sci. 63, 49.
70 Martin, M. and Giddings, J.C. (1981) J. Phys. Chem. 85, 727.
71 Janča, J. and Giddings, J.C. (1981) Sep. Sci. Technol. 16, 805.
72 Gajdos, L.J. and Brenner, H. (1978) Sep. Sci. Technol. 13, 215.
73 Giddings, J.C., Martin, M. and Myers, M.N. (1981) J. Polym. Sci. Polym. Phys. Ed. 19, 815.
74 Karaiskakis, G., Myers, M.N., Caldwell, K.D. and Giddings, J.C. (1981) Anal. Chem. 53, 1314.
75 Chiang, A.S., Kmiotek, E.H., Langan, S.M., Noble, P.T., Reis, J.F.G. and Lightfoot, E.N. (1979) Sep. Sci. Technol. 14, 453.
76 Shah, A.B., Reis, J.F.G., Lightfoot, E.N. and Moore, R.E. (1979) Sep. Sci. Technol. 14, 475.
77 Myers, M.N., Graff, K.A. and Giddings, J.C. (1980) Nuclear Technol. 51, 147.
78 Giddings, J.C., Graff, K.A., Myers, M.N. and Caldwell, K.D. (1980) Sep. Sci. Technol. 15, 615.
79 Caldwell, K.D., Nguyen, T.T., Giddings, J.C. and Mazzone, H.M. (1980) J. Virol. Methods 1, 241.
80 Inagaki, H. and Tanaka, T. (1980) Anal. Chem. 52, 201.
81 Martin, M. and Reynaud, R. (1980) Anal. Chem. 52, 2293.

Subject index